COUNCIL OF WAR

A HISTORY OF THE JOINT CHIEFS OF STAFF

1942–1991

By Steven L. Rearden

Opinions, conclusions, and recommendations expressed or implied within are solely those of the contributors and do not necessarily represent the views of the Defense Department or any other agency of the Federal Government. Cleared for public release; distribution unlimited.

Portions of this book may be quoted or reprinted without permission, provided that a standard source credit line is included. NDU Press would appreciate a courtesy copy of reprints or reviews.

First printing, July 2012

Cover image: Meeting of the Joint Chiefs of Staff on November 22, 1949, in their conference room at the Pentagon. From left to right: Admiral Forrest P. Sherman, Chief of Naval Operations; General Omar N. Bradley, Chairman, Joint Chiefs of Staff; General Hoyt S. Vandenberg, U.S. Air Force Chief of Staff; and General J. Lawton Collins, U.S. Army Chief of Staff. Department of the Army photograph collection.

Contents

Foreword ix

Preface xi

Chapter 1. THE WAR IN EUROPE 1
    The Origins of Joint Planning 2
    The North Africa Decision and Its Impact 9
    The Second Front Debate and JCS Reorganization 12
    Preparing for *Overlord* 15
    Wartime Collaboration with the Soviet Union 18

Chapter 2. THE ASIA-PACIFIC WAR AND THE
BEGINNINGS OF POSTWAR PLANNING 29
    Strategy and Command in the Pacific 29
    The China-Burma-India Theater 33
    Postwar Planning Begins 38
    Ending the War with Japan 43
    Dawn of the Atomic Age 46

Chapter 3. PEACETIME CHALLENGES 59
    Defense Policy in Transition 61
    Reorganization and Reform 64
    War Plans, Budgets, and the March Crisis of 1948 69
    The Defense Budget for FY 1950 76
    The Strategic Bombing Controversy 81

Chapter 4. MILITARIZING THE COLD WAR 95
    Pressures for Change 95
    The H-Bomb Decision and NSC 68 98
    Onset of the Korean War 102
    The Inch'on Operation 105
    Policy in Flux 108
    Impact of the Chinese Intervention 111
    MacArthur's Dismissal 113
    Europe—First Again 116

Chapter 5. EISENHOWER AND THE NEW LOOK 133
    The 1953 Reorganization 134
    Ending the Korean War 137
    A New Strategy for the Cold War 140
    Testing the New Look: Indochina 146

	Confrontation in the Taiwan Strait	149
	The "New Approach" in Europe	152
	NATO's Conventional Posture	156
	Curbing the Arms Race	158
Chapter 6.	CHANGE AND CONTINUITY	173
	Evolution of the Missile Program	174
	The Gaither Report	177
	The "Missile Gap" and BMD Controversies	179
	Reorganization and Reform, 1958–1960	183
	Defense of the Middle East	190
	Cuba, Castro, and Communism	196
	Berlin Dangers	199
Chapter 7.	KENNEDY AND THE CRISIS PRESIDENCY	211
	The Bay of Pigs	213
	Berlin under Siege	216
	Laos	221
	Origins of the Cuban Missile Crisis	224
	Showdown over Cuba	228
	Aftermath: The Nuclear Test Ban	233
Chapter 8.	THE MCNAMARA ERA	245
	The McNamara System	245
	Reconfiguring the Strategic Force Posture	247
	NATO and Flexible Response	251
	The Skybolt Affair	253
	Demise of the MLF	255
	A New NATO Strategy: MC 14/3	258
	The Damage Limitation Debate	261
	Sentinel and the Seeds of SALT	267
Chapter 9.	VIETNAM: GOING TO WAR	277
	The Roots of American Involvement	277
	The Road to an American War	281
	The Gulf of Tonkin Incident and Its Aftermath	284
	Into the Quagmire	292
Chapter 10.	VIETNAM: RETREAT AND WITHDRAWAL	305
	Stalemate	305
	Tet and Its Aftermath	310
	Nixon, the JCS, and the Policy Process	313

	Winding Down the War	316
	Back to Airpower	321
	The Christmas Bombing Campaign	324
	The Balance Sheet	326
Chapter 11.	DÉTENTE	335
	SALT I	336
	Shoring Up the Atlantic Alliance	342
	China: The Quasi-Alliance	347
	Deepening Involvement in the Middle East	351
Chapter 12.	THE SEARCH FOR STRATEGIC STABILITY	365
	The Peacetime "Total Force"	365
	Modernizing the Strategic Deterrent	367
	Targeting Doctrine Revised	371
	SALT II Begins	375
	Vladivostok	378
	Marking Time	381
Chapter 13.	THE RETURN TO CONFRONTATION	391
	Carter and the Joint Chiefs	391
	Strategic Forces and PD-59	394
	SALT II	397
	NATO and the INF Controversy	400
	The Arc of Crisis	403
	Rise of the Sandinistas	407
	Creation of the Rapid Deployment Force	408
	The Iran Hostage Rescue Mission	411
Chapter 14.	THE REAGAN BUILDUP	421
	Reagan and the Military	421
	Forces and Budgets	425
	Military Power and Foreign Policy	429
	The Promise of Technology: SDI	432
	Arms Control: A New Agenda	438
Chapter 15.	A NEW RAPPROCHEMENT	449
	Debating JCS Reorganization	449
	The Goldwater-Nichols Act of 1986	454
	NATO Resurgent	457
	Gorbachev's Impact	459
	Terrorism and the Confrontation with Libya	462

  Showdown in Central America..................................464
  Tensions in the Persian Gulf....................................467
  Operation *Earnest Will*..469

Chapter 16. ENDING THE COLD WAR..............................479
  Policy in Transition..479
  Powell's Impact as Chairman....................................481
  The Base Force Plan...485
  Operations in Panama..489
  The CFE Agreement..493
  START I and Its Consequences..................................495

Chapter 17. STORM IN THE DESERT...............................505
  Origins of the Kuwait Crisis....................................505
  Framing the U.S. Response......................................508
  Operational Planning Begins....................................510
  The Road to War..515
  Final Plans and Preparations....................................518
  Liberating Kuwait: The Air War..................................522
  Phase IV: The Ground Campaign..................................525
  The Post-hostilities Phase......................................528

Chapter 18. CONCLUSION...537

Glossary...549

Index..555

About the Author...585

Foreword

Established during World War II to advise the President on the strategic direction of the Armed Forces of the United States, the Joint Chiefs of Staff (JCS) continued in existence after the war and, as military advisers and planners, have played a significant role in the development of national policy. Knowledge of JCS relations with the President, the Secretary of Defense, and the National Security Council is essential to an understanding of the current work of the Chairman and the Joint Staff. A history of their activities, both in war and peacetime, also provides important insights into the military history of the United States. For these reasons, the Joint Chiefs of Staff directed that an official history of their activities be kept for the record. Its value for instructional purposes, for the orientation of officers newly assigned to the JCS organization, and as a source of information for staff studies is self-apparent.

Council of War: A History of the Joint Chiefs of Staff, 1942–1991 follows in the tradition of volumes previously prepared by the Joint History Office dealing with JCS involvement in national policy, the Korean War, and the Vietnam War. Adopting a broader view than earlier volumes, it surveys the JCS role and contributions from the early days of World War II through the end of the Cold War. Written from a combination of primary and secondary sources, it is a fresh work of scholarship, looking at the problems of this era and their military implications. The main prism is that of the Joint Chiefs of Staff, but in laying out the JCS perspective, it deals also with the wider impact of key decisions and the ensuing policies.

Dr. Steven L. Rearden, the author of this volume, holds a bachelor's degree from the University of Nebraska and a Ph.D. in history from Harvard University. His association with the Joint History Office dates from 1996. He has written and published widely on the history of the Joint Chiefs of Staff and the Office of the Secretary of Defense, and was co-collaborator on Ambassador Paul H. Nitze's book *From Hiroshima to Glasnost: At the Center of Decision—A Memoir* (1989).

This publication has been reviewed and approved for publication by the Department of Defense. While the manuscript itself is unclassified, some parts of documents cited in the source notes may remain classified. This is an official publication of the Joint History Office, but the views expressed are those of the author and do not necessarily represent those of the Joint Chiefs of Staff or the Department of Defense.

—John F. Shortal
Brigadier General, USA (Ret.)
Director for Joint History

Preface

Shortly after arriving at Fort McPherson, Georgia, in 1989, to head the U.S. Army Forces Command (FORSCOM), General Colin L. Powell put up a framed poster of the Reverend Dr. Martin Luther King, Jr., a present from Dr. King's widow, in the main conference room. On it were inscribed Dr. King's words: "Freedom has always been an expensive thing." Dr. King had in mind the sacrifices of the civil rights movement, of which he had been a major catalyst, in the 1950s and 1960s. But to Powell, a career Army officer who would soon leave FORSCOM to become the 12th Chairman of the Joint Chiefs of Staff, Dr. King's words had a broader, deeper meaning. Not only did he find them applicable to the civil rights struggle, but also he felt they spoke directly to the entire American experience and the central role played by the Armed Forces in preserving American values—freedom first among them.[1]

For the Joint Chiefs of Staff (JCS), the defense of freedom began with their creation as a corporate body in January 1942 to deal with the growing emergency arising from the recent Japanese attack on Pearl Harbor. Thrust suddenly into the maelstrom of World War II, the United States found itself ill-prepared to coordinate a global war effort with its allies or to develop comprehensive strategic and logistical plans for the deployment of its forces. To fill these voids, President Franklin D. Roosevelt established the JCS, an ad hoc committee of the Nation's senior military officers. Operating without a formal charter or written statement of duties, the Joint Chiefs functioned under the immediate authority and direction of the President in his capacity as Commander in Chief. A committee of coequals, the JCS came as close as anything the country had yet seen to a military high command.

After the war the Joint Chiefs of Staff became a permanent fixture of the country's defense establishment. Under the National Security Act of 1947, Congress accorded them statutory standing, with specific responsibilities. Two years later they acquired a presiding officer, the Chairman, Joint Chiefs of Staff, a statutory position carrying statutory authority that steadily increased over time. While often criticized as ponderous in their deliberations and inefficient in their methods, the JCS performed key advisory and support functions that no other body could duplicate in high-level deliberations. Sometimes, like during the Vietnam War in the 1960s, their views and recommendations carried less weight and had less impact than at other times. But as a rule their advice, representing as it did a distillation of the Nation's top military leaders' thinking, was impossible to ignore. Under legislation enacted in 1986, the Joint Chiefs' assigned duties and responsibilities passed almost in toto to the Chairman, who became principal military advisor to the President, the Secretary of Defense, and the National Security Council. But even though their corporate advisory role was over, the Joint Chiefs retained their statutory standing and continued to meet regularly as military advisors to the Chairman.

The history of the Joint Chief of Staff parallels the emergence of the United States in a great-power role and the growing demands that those responsibilities placed on American policymakers and military planners. During World War II, the major challenge was to wage a global war successfully on two fronts, one in Europe, the other in Asia and the Pacific. Afterwards, with the coming of an uneasy peace, the JCS faced new, less well-defined dangers arising from the turbulent relationship between East and West known as the Cold War. The product of long-festering political, economic, and ideological antagonisms, the Cold War also saw the proliferation of nuclear weapons and soon became an intense and expensive military competition between the United States and the Soviet Union. Though the threat of nuclear war predominated, the continuing existence of large conventional forces on both sides heightened the sense of urgency and further fueled doomsday speculation that the next world war could be the last. A period of recurring crises and tensions, the Cold War finally played out in the late 1980s and early 1990s, not with the cataclysmic confrontation that some people expected, but with the gradual reconciliation of key differences between East and West and eventually the collapse of Communism in Europe and the implosion of the Soviet Union.

The narrative that follows traces the role and influence of the Joint Chiefs of Staff from their creation in 1942 through the end of the Cold War in 1991. It is, first and foremost, a history of events and their impact on national policy. It is also a history of the Joint Chiefs of Staff themselves and their evolving organization, a reflection in many ways of the problems they faced and how they elected to address them. Over the years, the Joint History Office has produced and published numerous detailed monographs on JCS participation in national security policy. There has never been, however, a single-volume narrative summary of the JCS role. This book, written from a combination of primary and secondary sources, seeks to fill that void. An overview, it highlights the involvement of the Joint Chiefs of Staff in the policy process and in key events and decisions. My hope is that students of military history and national security affairs will find it a useful tool and, for those so inclined, a convenient reference point for further research and study.

Like most authors, I have numerous obligations to recognize. For their willingness to read and comment on various aspects of the manuscript, I need to thank Dr. Samuel R. Williamson, Jr., former Vice Chancellor and Professor of History Emeritus of Sewanee University; Dr. Lawrence S. Kaplan, Professor of History Emeritus of Kent State University; Dr. Donald R. Baucom, former Chief Historian of the Ballistic Missile Defense Organization; Dr. Wayne W. Thompson of the Office of Air Force History; and Dr. Graham A. Cosmas of the Joint History Office. I am also extremely grateful to the people at the Information Management Division of the Joint Chiefs of Staff, in particular Ms. Betty M. Goode and Mr. Joseph R. Cook, for their help in the documentation and clearance process. I am especially indebted to Molly Bompane and the Army Heritage and Education Center for their outstanding pictorial support. I would like to thank Richard Stewart of the Center of Military History for the use of the Army's art. The production of this book would not have been possible without the able advice and assistance of NDU Press Executive Editor Dr. Jeffrey D. Smotherman and Senior Copy Editor Mr. Calvin B. Kelley.

I am also deeply indebted to Dr. Edward J. Drea and Dr. Walter S. Poole who contributed in more ways than I can begin to enumerate. Both are long-standing friends and colleagues whose unrivaled knowledge, wisdom, and insights into military history and national security affairs have been sources of inspiration for many years. I want to thank Frank Hoffman of NDU Press for his faith in and support of this project. My heaviest obligations are to the two Directors for Joint History who made this book possible—Brigadier General David A. Armstrong, USA (Ret.), who initiated the project, and his successor, Brigadier General John F. Shortal, USA (Ret.), who saw it to completion. They were unstinting in their encouragement, support, and human kindness.

Lastly, I need to thank my wife, Pamela, whose patience and love were indispensible.

—Steven L. Rearden
Washington, DC
March 2012

NOTE

1 Colin L. Powell, with Joseph E. Persico, *My American Journey* (New York: Random House, 1995), 399–400.

British and American Combined Chiefs of Staff with President Roosevelt and Prime Minister Churchill at the Second Quebec Conference, September 1944. (front row, left to right) General George C. Marshall, Chief of Staff, U.S. Army; Admiral William D. Leahy, Chief of Staff to the Commander in Chief; President Franklin D. Roosevelt; Prime Minister Winston S. Churchill; Field Marshal Sir Alan F. Brooke, Chief of the Imperial General Staff; Field Marshal Sir John Dill, Chief of the British Joint Staff Mission to the United States; (back row, left to right) Major General Leslie C. Hollis, Secretary of the Chiefs of Staff Committee; General Sir Hastings Ismay, Prime Minister Churchill's Military Assistant and Representative to the Chiefs of Staff Committee; Admiral Ernest J. King, Chief of Naval Operations; Air Chief Marshal Sir Charles Portal; General Henry H. Arnold, Commanding General, Army Air Forces; and Admiral Sir Andrew B. Cunningham, First Sea Lord.

Chapter 1

THE WAR IN EUROPE

During the anxious gray winter days immediately following the Japanese attack on Pearl Harbor, Franklin D. Roosevelt confronted the most serious crisis of his Presidency. Now engaged in a rapidly expanding war on two major fronts—one against Nazi Germany in Europe, the other against Imperial Japan in the Pacific—he welcomed British Prime Minister Winston S. Churchill to Washington on December 22, 1941, for 3 weeks of intensive war-related discussions. Code-named ARCADIA, the meeting's purpose, as Churchill envisioned it, was to "review the whole war plan in the light of reality and new facts, as well as the problems of production and distribution."[1] Overcoming recent setbacks, pooling resources, and regaining the initiative against the enemy became the main themes. To turn their decisions into concrete plans, Roosevelt and Churchill looked to their senior military advisors, who held parallel discussions. From these deliberations emerged the broad outlines of a common grand strategy and several new high-level organizations for coordinating the war effort. One of these was a U.S. inter-Service advisory committee called the Joint Chiefs of Staff (JCS).[2]

ARCADIA was the latest in a series of Anglo-American military staff discussions dating from January 1941. Invariably well briefed and meticulously prepared for these meetings, British defense planners operated under a closely knit organization known as the Chiefs of Staff Committee, created in 1923. At the time of the ARCADIA Conference, its membership consisted of the Chief of the Imperial General Staff, General Sir Alan F. Brooke (later Viscount Alanbrooke), the First Sea Lord, Admiral of the Fleet Sir Dudley Pound, and the Chief of the Air Staff, Air Chief Marshal Sir Charles Portal. They reported directly to the Prime Minister and the War Cabinet and served as the government's high command for conveying directives to commanders in the field.[3]

Prior to ARCADIA nothing comparable to Britain's Chiefs of Staff Committee existed in the United States. As Brigadier General (later General) Thomas T. Handy recalled the situation: "We were more or less babes in the wood on the planning and joint business with the British. They'd been doing it for years. They were experts at it and we were just starting."[4] The absence of any standing coordinating mechanisms on the U.S. side forced the ARCADIA participants to improvise if they were to assure future inter-Allied cooperation and collaboration. Just before adjourning on January 14, 1942, they established a consultative body known as the Combined Chiefs of Staff (CCS), composed of the British chiefs and their American "opposite numbers." Since the British chiefs had their headquarters in London, they designated the senior members of the British Joint Staff Mission (JSM) to the United States, a tri-Service organization, as their day-to-day representatives to the CCS in Washington. Thereafter, formal meetings of the Combined Chiefs (i.e., the British chiefs and their American opposite numbers) took place only at summit conferences attended by the President and the Prime Minister. Out of a total of 200 CCS meetings held during the war, 89 were held at these summit meetings.[5]

U.S. membership on the CCS initially consisted of General George C. Marshall, Chief of the War Department General Staff; Admiral Harold R. Stark, Chief of Naval Operations (CNO); Admiral Ernest J. King, Commander in Chief, U.S. Fleet; and Lieutenant General Henry H. Arnold, Chief of the Army Air Forces and Deputy Chief of Staff for Air. Though Arnold's role was comparable to Portal's, he spoke only for the Army Air Forces since the Navy had its own separate air component.[6] Shortly after the ARCADIA Conference adjourned, President Roosevelt reassigned Stark to London as Commander, U.S. Naval Forces, Europe, a liaison job, and made King both Chief of Naval Operations and Commander in Chief, U.S. Fleet. In this dual capacity, King became the Navy's senior officer and its sole representative to the CCS.[7] To avoid confusion, the British and American chiefs designated collaboration between two or more of the nations at war with the Axis powers as "combined" and called inter-Service cooperation by one nation "joint." The U.S. side designated itself as the "Joint United States Chiefs of Staff," soon shortened to "Joint Chiefs of Staff."

THE ORIGINS OF JOINT PLANNING

Though clearly a prudent and necessary move, the creation of the Joint Chiefs of Staff was a long time coming. By no means was it preordained. When the United States declared war on the Axis powers in December 1941, its military establishment consisted of autonomous War and Navy Departments, each with a subordinate air arm. Command and control were unified only at the top, in the person of President Franklin D. Roosevelt in his constitutional role as Commander in Chief. Politically astute and charismatic, Roosevelt dominated foreign and defense affairs and insisted on exercising close personal control of the Armed Forces. The creation of the Joint Chiefs of Staff effectively reinforced his authority. Often bypassing the Service Secretaries, he preferred to work directly with the uniformed heads of the military Services. From 1942 on, he used the JCS as an extension of his powers as

Commander in Chief. The policy he laid down stipulated that "matters which were purely military must be decided by the Joint Chiefs of Staff and himself, and that, when the military conflicted with civilian requirements, the decision would have to rest with him."[8] In keeping with his overall working style, his relations with the chiefs were casual and informal, which allowed him to hold discussions in lieu of debates and to seek consensus on key decisions.[9]

Below the level of the President, inter-Service coordination at the outset of World War II was haphazard. Officers then serving in the Army and the Navy were often deeply suspicious of one another, inclined by temperament, tradition, and culture to remain separate and jealously guard their turf. Not without difficulty, Marshall and King reached a modus vivendi that tempered their differences and allowed them to work in reasonable harmony for most of the war.[10] Their subordinates, however, were generally not so lucky. Issues such as the deployment of forces, command arrangements, strategic plans, and (most important of all) the allocation of resources invariably generated intense debate and friction. As the war progressed, the increasing use of unified theater commands, bringing ground, sea, and air forces under one umbrella organization, occasionally had the untoward side-effect of aggravating these stresses and strains. According to Sir John Slessor, whose career in the British Royal Air Force brought him into frequent contact with American officers during and after World War II, "The violence of inter-Service rivalry in the United States in those days had to be seen to be believed and was an appreciable handicap to their war effort."[11]

Inter-Service collaboration before the war rested either on informal arrangements, painstakingly worked out through goodwill as the need arose, or on the modest achievements of the Joint Army and Navy Board. Established in 1903 by joint order of the Secretaries of War and Navy, the Joint Board was responsible for "conferring upon, discussing, and reaching common conclusions regarding all matters calling for the co-operation of the two Services."[12] By the eve of World War II, the Board's membership consisted of the Army Chief of Staff, Deputy Chief of Staff, Chief of the War Plans Division, Chief of Naval Operations, Assistant Chief of Naval Operations, and Director of the Naval War Plans Division.[13]

The Joint Board's main functions were to coordinate strategic planning between the War and Navy Departments and to assist in clarifying Service roles and missions. Between 1920 and 1938, the board's major achievement was the production of the "color" plans, so called because each plan was designated by a particular color. Plan Orange was for a war with Japan.[14] But after the Munich crisis in the autumn of 1938, with tensions rising in both Europe and the Pacific, the board began to consider a wider range of contingencies involving the possibility of a multifront war simultaneously against Germany, Italy, and Japan. The result was a new series of "Rainbow" plans. The plan in effect at the time of Pearl Harbor was Rainbow 5, which envisioned large-scale offensive operations against Germany and Italy and a strategic defensive in the Pacific until success against the European Axis powers allowed transfer of sufficient assets to defeat the Japanese.[15]

To help assure effective execution of these plans, the Joint Board also sought a clearer delineation of Service roles and missions. A contentious issue in the best of times, roles and missions became all the more divisive during the interwar period owing to the limited funding available and the emergence of competing land- and sea-based military aviation systems. The board addressed these issues in a manual, *Joint Action of the Army and Navy (JAAN)*, first published in 1927 and revised in 1935, with minor changes from year to year thereafter. The doctrine incorporated into the *JAAN* called for voluntary cooperation between Army and Navy commanders whenever practicable. Unity of command was permitted only when ordered by the President, when specifically provided for in joint agreements between the Secretaries of War and Navy, or by mutual agreement of the Army and Navy commanders on the scene. For want of a better formula, the *JAAN* simply accepted the status quo and left controversial issues like the control of airpower divided between the Services, to be exploited as their respective needs dictated and resources allowed.[16]

After 1938, with the international situation deteriorating, the Joint Board became increasingly active in conducting exploratory studies and drafting joint strategic plans (the Rainbow series) where the Army and the Navy had a common interest. For support, the board relied on part-time inter-Service advisory and planning committees. The most prominent and active were the senior Joint Planning Committee, consisting of the chiefs of the Army and Navy War Plans Divisions, which oversaw the permanent Joint Strategic Committee and various ad hoc committees assigned to specialized technical problems, and the Joint Intelligence Committee, consisting of the intelligence chiefs of the two Services, which coordinated intelligence activities. Despite its efforts, however, the Joint Board never acquired the status or authority of a military command post and remained a purely advisory organization to the military Services and, through them, to the President.[17]

While the limitations of the Joint Board system were abundantly apparent, there was little incentive prior to Pearl Harbor to make significant changes. The most ambitious reform proposal originated in the Navy General Board and called for the creation of a joint general staff headed by a single chief of staff to develop general plans for major military campaigns and to issue directives for detailed supporting plans to the War and Navy Departments. First broached in June 1941, this proposal was referred to the Army and Navy Plans Divisions where it remained until after the Japanese attack. Public reaction to the Pearl Harbor catastrophe, allegedly the result of faulty inter-Service communication, flawed intelligence, and divided command, led Admiral Stark in late January 1942, to rescue the joint general staff paper from the oblivion of the Plans Divisions and to place it on the Joint Board's agenda. Here it encountered strong opposition from Navy representatives, its erstwhile sponsors. Upon further reflection, they declared it essentially unworkable. Their main objection was that such a scheme would require a corps of staff officers, which did not exist, who were thoroughly cognizant of all aspects of both Services. Army representatives favored the plan but did not push it in light of the Navy's strong opposition. Discussion of the matter culminated at a Joint Board meeting on March 16, where the members, unable to agree, left it "open for further study."[18]

By the time the Joint Board dropped the joint general staff proposal, the Joint Chiefs of Staff were beginning to emerge as the country's de facto high command. This process resulted not from any directive issued by the President or emergency legislation enacted by Congress, but from the paramount importance of forming common cause with the British Chiefs of Staff on matters of mutual interest and the strategic conduct of the war. As useful as the Joint Board may have been as a peacetime planning mechanism, it had limited utility in wartime and was not set up to function in a command capacity or to provide liaison with Allied planners. Though still in its infancy, the Combined Chiefs of Staff system was already exercising a pervasive influence on American military planning, thanks in large part to the easy and close collaboration that quickly developed between General Marshall and the senior British representative, Sir John Dill.[19] As the CCS system became more entrenched, it demanded a more focused American response, which only the organizational structure of the Joint Chiefs of Staff could provide.

The Joint Chiefs held their first formal meeting on February 9, 1942, and over the next several months gradually absorbed the Joint Board's role and functions.[20] To support their work, the Joint Chiefs established a joint staff that comprised a network of inter-Service committees corresponding to the committees making up the Combined Chiefs of Staff. Initially, only two JCS panels—the Joint Staff Planners (JPS) and the Joint Intelligence Committee (JIC)—had full-time support staff, provided by remnants of the Joint Board. Most of those on the other joint committees served in a part-time capacity and appeared on the duty roster as "associate members," splitting their time between their Service responsibilities and the JCS. A few officers, designated "primary duty associate members," were considered to be full-time. Owing to incomplete records, no one knows for sure how many officers served on the Joint Staff at any one time during the war. Committees varied in size,

5

from the Joint Strategic Survey Committee, which had only three members, on up to the Joint Logistics Committee, which once had as many as two hundred associate members.[21] Money to support the Joint Chiefs' operations, including the salaries for about 50 civilian clerical helpers, came from the War and Navy Departments and an allocation from the President's contingent fund.[22]

Figure 1–1.

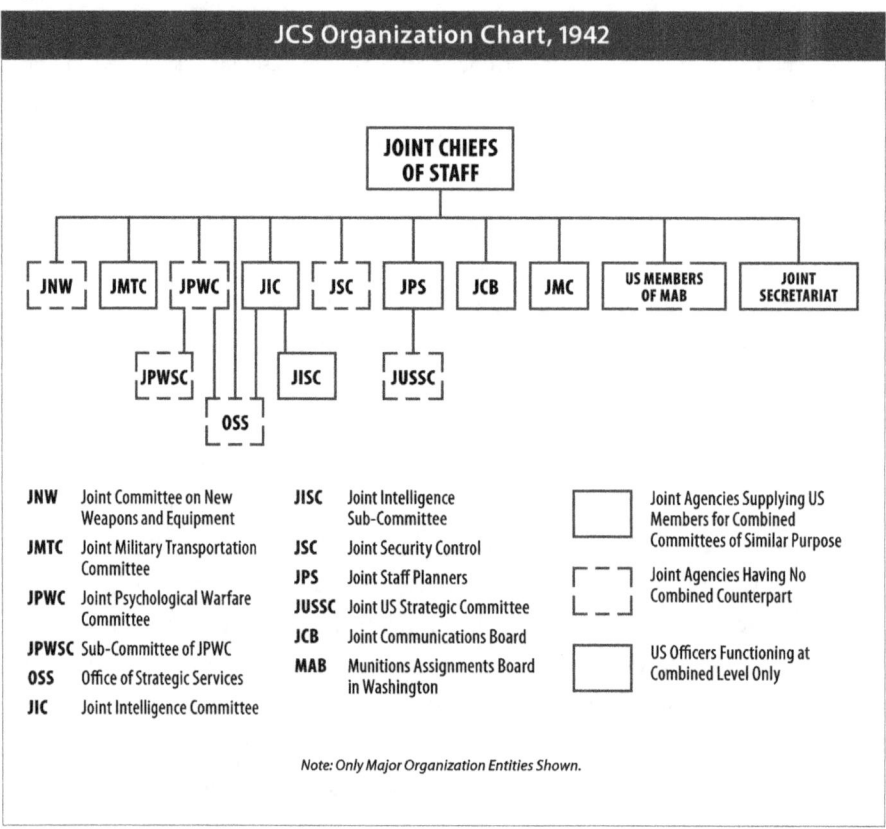

Initially modeled on the CCS system, the JCS organization gradually departed from the CCS structure to meet the Joint Chiefs' unique requirements. During 1942 the Joint Chiefs added three subordinate components without CCS counterparts—the Joint New Weapons Committee, the Joint Psychological Warfare Committee, and the Office of Strategic Services (OSS). The first two were part-time bodies providing advisory support to the Joint Chiefs in the areas of weapons research and wartime propaganda and subversion. The third was an operational and research agency that specialized in espionage and clandestine missions behind enemy lines. Though the OSS fell under the jurisdiction of the Joint Chiefs of Staff, it had its own director, William J. Donovan, who reported directly to the President.[23] Between 1943 and March 1945, the JCS organization expanded further to include the Army-Navy Petroleum Board and separate committees dealing with production and supply matters, postwar political-military planning, and the coordination of civil affairs in liberated and occupied areas.

Wartime membership of the Joint Chiefs was completed on July 18, 1942, when President Roosevelt appointed Admiral William D. Leahy as Chief of Staff to the Commander in Chief. The inspiration for Leahy's appointment came from General Marshall, who suggested to the President in February 1942 that there should be a direct link between the White House and the JCS, an officer to brief the President on military matters, keep track of papers sent to the White House for approval, and transmit the President's decisions to the JCS. As the President's designated representative, he could also preside at JCS meetings in an impartial capacity.[24]

President Roosevelt initially saw no need for a Chief of Staff to the Commander in Chief. Likewise, Admiral King, fearing adverse impact on Navy interests if another officer were interposed between himself and the President, opposed the idea. It was not until General Marshall suggested appointing Admiral Leahy, an old friend of the President's and a trusted advisor, that Roosevelt came around.[25] The Admiral, who had retired as Chief of Naval Operations in 1939, was just completing an assignment as Ambassador to Vichy, France. The appointment of another senior naval officer was perhaps the only way of gaining Admiral King's endorsement, since it balanced the JCS with two members from the War Department and two from the Navy.

A scrupulously impartial presiding officer, Leahy never became the strong representative of JCS interests that Marshall hoped he would be. In Marshall's view, Leahy limited himself too much to acting as a liaison between the JCS and the White House. Still, he played an important role in conveying JCS recommendations and in briefing the President every morning.[26] In no way was his position comparable to that later accorded to the Chairman, Joint Chiefs of Staff. In meetings with the President or with the Combined Chiefs of Staff, Leahy was rarely the JCS spokesman. That role usually fell to either General Marshall, who served as the leading voice on strategy in the European Theater, or Admiral King, who held sway over matters affecting the Pacific.

Though considerable, the Joint Chiefs' influence over wartime strategy and policy was never as great as some observers have argued. According to historian Kent Roberts Greenfield, there are more than 20 documented instances in which Roosevelt overruled the chiefs' judgment on military situations.[27] While the chiefs liked to present the President with unanimous recommendations, they were not averse to offering a "split" position when their views differed and then thrashing out a solution at their meetings with the President. During the first year or so of the war, the President's special assistant, Harry Hopkins, also regularly attended these meetings. Rarely invited to participate were the Service Secretaries (Secretary of War Henry L. Stimson and Secretary of the Navy Frank Knox) and Secretary of State Cordell Hull, all of whom found themselves marginalized for much of the war. But despite their close association, the President and the Joint Chiefs never developed the intimate, personal rapport Churchill had with his military chiefs. Between Roosevelt and the JCS, there was little socializing. Comfortable and productive, their relationship was above all professional and businesslike.[28]

Even though the Joint Chiefs of Staff functioned as the equivalent of a national military high command, their status as such, throughout World War II, was never established in law or by Executive order. Preoccupied with waging a global war, they paid scant attention to the question of their status until mid-1943 when they briefly considered a charter defining their duties and responsibilities. The only JCS member to evince strong interest in a charter was Admiral King, who professed to be "shocked" that there was no basic definition of JCS duties and responsibilities. In the existing circumstances, he doubted whether the JCS could continue to function effectively. Admiral Leahy took exception. "The absence of any fixed charter of responsibility," he insisted, "allowed greater flexibility in the JCS organization and enabled us to extend its activities to meet the changing requirements of the war." He pointed out that, since the JCS served at the President's pleasure, they performed whatever duties he saw fit; under a charter, they would be limited to performing assigned functions. Initially, General Marshall sided with Admiral Leahy but finally became persuaded, in the interests of preserving JCS harmony, to support issuance of a charter in the form of an Executive order.[29]

The Joint Chiefs approved the text of such an order on June 15, 1943, and submitted it to the President the next day. The proposed assignment of duties was fairly routine and related to ongoing activities of advising the President, formulating military plans and strategy, and representing the United States on the Combined Chiefs of Staff.[30] Still, the overall impact would have been to place the JCS within a confined frame of reference, and arguably restrict their deliberations to a specific range of issues. Satisfied with the status quo, the President rejected putting the chiefs under written instructions. "It seems to me," he told them, "that such an order would provide no benefits and might in some way impair flexibility of operations."[31] As a result, the Joint Chiefs continued to manage their affairs throughout the war without a written definition of their functions or authority, but with the tacit assurance that President Roosevelt fully supported their activities.

THE NORTH AFRICA DECISION AND ITS IMPACT

While the ARCADIA Conference of December 1941–January 1942 confirmed that Britain and the United States would integrate their efforts to defeat the Axis, it left many details of their collaboration unsettled. The agreed strategic concept that emerged from ARCADIA was to defeat Germany first, while remaining on the strategic defensive against Japan. Recognizing that limited resources would constrain their ability to mount offensive operations against either enemy for a year or so, the Allied leaders endorsed the idea of "tightening the ring" around Germany during this time by increasing lend-lease support to the Soviet Union, reinforcing the Middle East, and securing control of the French North African coast.[32]

To augment this broad strategy, the CCS in March 1942 adopted a working understanding of the global strategic control of military operations that divided the world into three major theaters of operations, each comparable to the relative interests of the United States and Great Britain. As a direct concern to both parties, the development and execution of strategy in the Atlantic-European area became a combined responsibility and, as such, the region most immediately relevant to the CCS. Elsewhere, the British Chiefs of Staff, working from London, would oversee strategy and operations for the Middle East and South Asia, while the Joint Chiefs of Staff in Washington would do the same for the Pacific and provide military coordination with the government of Generalissimo Chiang Kai-shek in China.[33]

British and American planners agreed that the key to victory was the Soviet Union, which engaged the bulk of Germany's air and ground forces. "In the last analysis," predicted Admiral King, "Russia will do nine-tenths of the job of defeating Germany."[34] Keeping the Soviets actively and continuously engaged against Germany thus became one of the Western Allies' primary objectives, even before the United States formally entered the war.[35] Within the JCS-CCS organization that emerged following the ARCADIA Conference, developing a "second front" in Western Europe quickly emerged as a priority concern, both to relieve pressure on the Soviets and to demonstrate the Western Allies' sincerity and support. Unlike their American counterparts, however, British defense planners were in no hurry to return to the Continent. Averse to repeating the trench warfare of World War I, and with the Soviet Union under a Communist regime that Churchill despised, British planners proved far more cautious and realistic in entertaining plans for a second front.

The Joint Chiefs assumed that initially their main job would be to coordinate the mobilization and deployment of a large army to Europe to confront the Germans directly, as the United States had done in World War I. As General Marshall put it, "We should never lose sight of the eventual necessity of fighting the Germans in Germany."[36] By mid-March 1942, the consensus among the Joint Chiefs was that they should press their British allies for a buildup of forces in the United Kingdom for the earliest practicable landing on the Continent and restrict deployments in the Pacific to current commitments. But they adopted no timetable for carrying out these operations and deferred to the War Department General Staff to come up with a concrete plan for invading Europe. At this stage, the Joint Chiefs of Staff were a new and novel organization, composed of officers from rival Services who were still unfamiliar with one another and uneasy about working together. As a result, the most effective and efficient strategic planning initially was that done by the Service staffs, with the Army taking the lead in shaping plans for Europe and the Navy doing the same for the Pacific.[37]

The impetus for shifting strategic planning from the Services to the corporate oversight of the JCS was President Roosevelt's decision in July 1942 to postpone a Continental invasion and, at Churchill's urging, to concentrate instead on the liberation of North Africa. Personally, Roosevelt would have preferred a second front in France, and in the spring of 1942 he had sent Marshall and Harry Hopkins to London to explore the possibility of a landing either later in the year or in 1943. Though the British initially seemed receptive to the idea and endorsed it in principle, they raised one objection after another and insisted that the time was not ripe for a landing on the Continent. Pushing an alternate strategy, they favored a combined operation in the Mediterranean.[38] Based on the production and supply data he received, Roosevelt ruefully acknowledged that the United States would not be in a position to have a "major impact" on the war much before the autumn of 1943.[39] Eager that U.S. forces should see "useful action" against the Germans before then, he became persuaded that North Africa would be more feasible than a landing in France. The upshot in November 1942 was Operation *Torch*, the first major offensive of the war involving sizable numbers of U.S. forces.[40]

While not wholly unexpected, the *Torch* decision had extensive ripple effects. The most immediate was to nullify a promise Roosevelt made to the Soviets in May 1942 to open a second front in France before the end of the year.[41] A bitter disappointment in Moscow, it was also a major rebuff for Marshall and War Department planners who had drawn up preliminary Continental invasion plans. One set, called SLEDGEHAMMER, was for a limited "beachhead" landing in 1942; another, called BOLERO-ROUNDUP, was for a full-scale assault on the northern coast of France in mid-1943.[42] Unable to contain his disappointment, Marshall told the President that he was "particularly opposed to 'dabbling' in the Mediterranean in a wasteful logistical way."[43] In Churchill's view, however, an invasion of France was too risky and premature until the Allies brought the U-boat menace in the Atlantic under control, had greater mastery of the air, and American forces were battle-tested. In the interests of unity, Churchill continued to assure his Soviet and American allies that he supported a cross-Channel invasion of Europe in 1943. But as a practical matter, he seemed intent on using the invasion of North Africa to protect British interests east of Suez and as a stepping stone toward further Anglo-American operations in the Mediterranean that would "knock Italy out of the war."[44]

Churchill's preoccupation with North Africa and the Mediterranean reflected a time-honored British tradition that historians sometimes refer to as "war on the periphery," in contrast to the more direct American approach involving the massing of forces, large-scale assaults, and decisive battles. Limited in manpower and industrial capability, the British had historically preferred to avoid direct confrontations and had pursued strategies that exploited their enemies' weak spots, wearing them down through naval action, attrition, and dispersion of forces. In World War I, the British had departed from this strategy with disastrous results that gave them the sense of having achieved a pyrrhic victory. Committed to avoiding a repetition of the World War I experience, Churchill and his military advisors preferred to let the Soviets do most of the fighting (and dying) against Germany, while Britain and the United States concentrated on eviscerating Germany's "soft underbelly" in the Mediterranean. Although Churchill fully intended to undertake an Anglo-American invasion of Europe, he expected it to follow in due course, once Germany was worn down and on the verge of defeat.[45]

Following the planning setbacks they experienced in the summer of 1942, the Joint Chiefs sought to regroup and regain the initiative, starting with a clarification of overall strategy. Their initial response was the creation in late November 1942 of the Joint Strategic Survey Committee (JSSC), an elite advisory body dedicated to long-range planning. Composed of only three senior officers, the JSSC resembled a panel of "elder statesmen," representing the ground, naval, and air forces, whose job was to develop broad assessments on "the soundness of our basic strategic policy in the light of the developing situation, and on the strategy which should be adopted with respect to future operations." In theory, Service affiliations were not to interfere with or prejudice their work. The three chosen to sit on the committee—retired Lieutenant General Stanley D. Embick of the Army, Major General Muir S. Fairchild of the Army Air Corps, and Vice Admiral Russell Willson—served without other duties and stayed at their posts throughout the war.[46]

Early in December 1942, the JSSC submitted its first set of recommendations, a three-and-a-half-page overview of Allied strategy for the year ahead. In surveying future options, the committee sought to keep the war focused on agreed objectives. Assuming that the first order of business remained the defeat of Germany, the JSSC recommended freezing offensive operations in the Mediterranean and transferring excess forces from North Africa to the United Kingdom as part of the buildup for an invasion of Europe in 1943. The committee also urged continuing assistance to the Soviet Union, a gradual shift from defensive to offensive operations in the Pacific and Burma, and an integrated air bombardment campaign launched from bases in England, North Africa, and the Middle East against German "production and resources."[47]

Here in a nutshell was the first joint concept for a global wartime strategy, marshaling the efforts of land, sea, and air forces toward common goals. All the same, it was a highly generalized treatment and, as such, it glossed over the impact of conflicting Service interests. At no point did it attempt to sort out the allocation of resources, by far the most controversial issue of all, other than on the basis of broad priorities. Challenging one of the paper's core assumptions, Admiral King doubted whether a landing in Europe continued to merit top priority. King maintained that, with adoption of the *Torch* decision and the diversions that operation entailed, the Anglo-American focus of the war had shifted from Europe to the Mediterranean and Pacific. King wanted U.S. plans and preparations adjusted accordingly, with more effort devoted to the Pacific and defeating the Japanese.[48] Meeting with the President on January 7, 1943, the Joint Chiefs acknowledged that they were divided along Service lines. As Marshall delicately put it, they "regarded an operation in the north [of Europe] more favorably than one in the Mediterranean but the question was still an open one."[49] Despite nearly a year of intensified planning, the JCS had yet to achieve a working consensus on overall strategic objectives.

THE SECOND FRONT DEBATE AND JCS REORGANIZATION

Faced with indecision among his military advisors, Roosevelt gravitated to the British, who had worked out definite plans and knew precisely what they wanted to accomplish. At the Casablanca Conference in January 1943, he gave in to Churchill's insistence that the Mediterranean be accorded "prime place" and that a move against Sicily (Operation *Husky*) should follow promptly upon the successful completion of Operation *Torch* in North Africa.[50] To placate the Americans, the British agreed to establish a military planning cell in London to begin preliminary preparations for a cross-Channel attack. But with attention and resources centered on the Mediterranean, a Continental invasion was now unlikely to materialize before 1944. Knowing that a further postponement would not go down well in Moscow, Roosevelt proposed—and Churchill grudgingly agreed—that the United States and Britain issue a combined public declaration of their intent to settle for nothing less than "unconditional surrender" of the Axis powers.[51]

A further result of the Casablanca Conference—one with significant but unintended consequences for the future of the Joint Chiefs—was the endorsement of an intensive combined bombing campaign against Germany. This decision fell in line with the recent recommendations of the Joint Strategic Survey Committee and was widely regarded as an indispensable preliminary to a successful invasion of France. Under the agreed directive, however, first priority was not the destruction of the enemy's military-industrial complex, as some air power enthusiasts had advocated, but the suppression of the German submarine threat, which was taking a horrific toll on Allied shipping.[52] Still, American and British air strategists had long sought the opportunity to demonstrate the potential of airpower and greeted the decision as a step forward, even as they disagreed among themselves over the relative merits of daylight precision bombing (the American approach) versus nighttime area bombing (the British strategy). The impact on the JCS was more long term and subtle. Previously, as the senior Service chiefs, Marshall and King had dominated JCS deliberations. Now, with strategic bombing an accepted and integral part of wartime strategy, Arnold assumed a more prominent role of his own, becoming a true coequal to the other JCS members in both rank and stature by the war's end.[53]

For the Joint Chiefs and the aides accompanying them, the Casablanca Conference was, above all, an educational experience that none wanted to repeat. Traveling light, the JCS had kept their party small and had arrived with limited backup materials. In contrast, the British chiefs had brought a very complete staff and reams of plans and position papers. Admiral King found that whenever the CCS met and he or one of his JCS colleagues brought up a subject, the British invariably had a paper ready.[54] Brigadier General Albert C. Wedemeyer, the Army's chief planner, had a similar experience. At each and every turn he found the British better prepared and able to outmaneuver the Americans with superior staff work. "We came, we listened and we were conquered," Wedemeyer told a colleague. "They had us on the defensive practically all the time."[55]

The Joint Chiefs of Staff returned from the Casablanca Conference with less to show for their efforts than they hoped and determined to apply the lessons they learned there. In practice, that meant never again entering an international conference so ill-prepared or understaffed. To strengthen the JCS position, General

Marshall arranged for Lieutenant General Joseph T. McNarney, Deputy Army Chief of Staff, to oversee a reorganization of the joint committee system, with special attention to developing more effective joint-planning mechanisms. The main bottleneck was in the Joint Staff Planners, a five-member committee that had fallen behind in its assigned task of providing timely, detailed studies on deployment and future operations. The new system, introduced gradually during the spring of 1943, reduced the range and number of issues coming before the Joint Staff Planners and transferred logistical matters to the Joint Administrative Committee, later renamed the Joint Logistics Committee.[56]

Under McNarney's reorganization, nearly all the detailed planning functions previously assigned to the Joint Staff Planners became the responsibility of a new body, the Joint War Plans Committee (JWPC), which functioned as a JPS working subcommittee. Thenceforth, the JPS operated in more of an oversight capacity, reviewing, amending, and passing along the recommendations they received from the Joint War Plans Committee. The JWPC drew its membership from the staffs of the chiefs of planning for the Army, the Navy, and the Air Staff. Under them was an inter-Service "planning team" of approximately 15 officers who served full time without other assigned duties. The directive setting up the JWPC reminded those assigned to it that they were now part of a joint organization and to conduct themselves accordingly by going about their work and presenting their views "regardless of rank or service."[57]

The first test of these new arrangements came at the TRIDENT Conference, held in Washington in May 1943 to develop plans and strategy for operations after the invasion of Sicily during the coming summer. By then, King had grudgingly resigned himself to the inevitability of a cross-Channel invasion and agreed with Marshall that further operations in the Mediterranean should be curbed. King viewed the British preoccupation there as a growing liability that had the potential of preventing the Navy from stepping up the war against Japan. Based on naval production figures, King estimated that by the end of 1943, the Navy would begin to enjoy a significant numerical superiority over the Japanese in aircraft carriers and other key combatants. To take advantage of that situation, the CNO proposed a major offensive in the Central Pacific and secured JCS endorsement just before the TRIDENT Conference began. But with the British dithering in the Mediterranean and a firm decision on the second front issue still pending, King could easily find his strategic initiative jeopardized.[58]

At TRIDENT, for the first time in the war, the Joint Chiefs obtained the use of procedures that worked to their advantage. Namely, they insisted on an agenda and some of the papers developed by the Joint War Plans Committee in lieu of those offered by the British, who had controlled the "paper trail" at Casablanca.[59] As often as possible during TRIDENT, King tried to shift the discussion to the Pacific. But the dominating topic was the choice between continuing operations in the Mediterranean or opening a second front in northern France. With President Roosevelt's concurrence and with Marshall doing most of the talking, the Joint Chiefs pressed the British for a commitment to a cross-Channel attack no later than the spring of 1944. The deliberations were brisk and occasionally involved what historian Mark A. Stoler describes as "some private and very direct exchanges." Six months earlier British views would probably have prevailed. But with improved staff support behind them, the JCS were now more than able to hold their own.[60]

A crucial factor in the Joint Chiefs' effectiveness was a carefully researched feasibility study by the JWPC showing that there would be enough landing craft to lift five divisions simultaneously (three in assault and two in backup), making the cross-Channel operation feasible.[61] Forced to concede the point, the British agreed to begin moving troops (seven divisions initially) from the Mediterranean to the United Kingdom. While accepting a tentative target date of May 1, 1944, for the invasion, the British sidestepped a full commitment by insisting on further study. The JCS also wanted to limit additional operations in the Mediterranean to air and sea attacks. But out of the ensuing give-and-take, the British prevailed in obtaining an extension of currently planned operations against Sicily onto the Italian mainland, in Churchill's words, "to get Italy out of the war by whatever means might be best."[62]

A significant improvement over the Joint Chiefs' previous performance, TRIDENT demonstrated the utility and effectiveness of Joint Staff work over reliance on separate and often uncoordinated Service inputs. From then on, preparations for inter-Allied conferences became increasingly centralized around the Joint Staff, with the Joint War Plans Committee the focal point for the development of the necessary planning papers and inter-Service coordination.[63] The emerging dominance of the JCS system was largely the product of necessity and rested on a growing recognition as the war progressed that at the high command level as well as in the field, joint collaboration was more successful than each Service operating on its own.

PREPARING FOR *OVERLORD*

Even though the Joint Chiefs secured provisional agreement at the TRIDENT Conference to begin preparations for an invasion of France, it remained to be seen whether the British would live up to their promise. Reports from London indicated that Churchill was "rather apathetic and somewhat apprehensive" about a firm commitment to invade Europe and that he would press next for an invasion of Italy, followed by operations against the Balkans.[64] Even though a campaign on the

Italian mainland would delay moving troops and materiel to England for the invasion, Churchill had made a convincing argument that Italy would fall quickly and not pose much of a diversion. With U.S. and British forces currently concentrated in Sicily and North Africa, the JCS acknowledged that it made sense to take advantage of the opportunity before moving forces en masse to England. Still, they were adamant that the operation be limited and not go beyond Rome, lest it jeopardize plans for the invasion of northern France.[65]

At the first Quebec Conference (QUADRANT) in August 1943, Churchill, Roosevelt, and the Combined Chiefs of Staff confirmed their intention to attack Italy and attempted to reconcile continuing differences over a landing on the northern French coast, now code-named Operation *Overlord*. Despite pledges made at the TRIDENT Conference, Churchill and the British chiefs procrastinated, prompting several heated exchanges and some "very undiplomatic language" by Admiral King, who considered the British to be acting in bad faith.[66] At one point the CCS cleared the room of all subordinates and continued the discussion off the record. The sense of trust and partnership appeared to be eroding on both sides. While professing their commitment to *Overlord*, the British objected to an American proposal to give the invasion of France "overriding priority" and wanted to delay the repositioning of troops as agreed at TRIDENT so campaigns in the Mediterranean could proceed without serious disruption. Working a compromise, the Combined Chiefs agreed to make *Overlord* the "primary" Anglo-American objective in 1944, but couched the decision in ambiguous language that left open the possibility of further operations in the Mediterranean.[67] Once back in London, Churchill assured the War Cabinet that the QUADRANT agreement on *Overlord* notwithstanding, he would continue to insist on "nourishing the battle" in Italy as long as he remained in office.[68]

At that stage in the war, Churchill and the British Chiefs of Staff still viewed themselves as the "predominant partner" in the Western alliance. Yet it was a role they were less equipped to play with each passing day. By mid-1943, with the mobilization and stepped-up industrial production initiated since 1940 beginning to bear fruit, the United States was steadily overtaking Britain in manpower and materiel to become the preeminent military power within the Western alliance. One consequence was to give the U.S. chiefs a larger voice and stronger leverage within the CCS system, much to the consternation of the British.[69] Meetings of the Combined Chiefs of Staff, as evidenced by the discussions at TRIDENT and QUADRANT, were becoming more and more confrontational. Clearly frustrated, Sir Alan Brooke, Chief of the Imperial General Staff, lamented that he and his British colleagues were no longer able "to swing those American Chiefs of Staff and make them see daylight."[70]

With tensions mounting between the American and British military chiefs over *Overlord*, a showdown was only a matter of time. It finally came at the Tehran Conference in late November 1943, the first "Big Three" summit of the war. During the trip over aboard the battleship *Iowa*, the Joint Chiefs had the opportunity to discuss among themselves and with the President the issues they should raise and the approach they should take, so when the conference got down to business, the American position was unambiguous. Stopping in Cairo to meet with Generalissimo Chiang Kai-shek, the Chinese leader, Roosevelt, Churchill, and the Combined Chiefs of Staff took time out to review the status of planning for the invasion of France. Though Churchill again paid lip service to *Overlord*, calling it "top of the bill," he also outlined his vision for expanding military operations into northern Italy, Rhodes, and the Balkans. Roosevelt and the Joint Chiefs, feeling that now was not the time to debate these issues, simply turned a collective deaf ear.[71]

At Tehran, with the Soviets present, the Joint Chiefs left no doubt that launching *Overlord* was their first concern, then sat back while the senior Soviet military representative, Marshal Klementy Voroshiloff, interrogated Brooke and his British colleagues on why they wanted to devote precious time and resources on "auxiliary operations" in the Mediterranean.[72] In the plenary sessions with Roosevelt and Soviet leader Marshal Josef Stalin, Churchill fell under intense pressure to shelve his plans for the Mediterranean and to throw unequivocal support behind the invasion. To improve the prospects of success, Stalin offered to launch a major offensive on the Eastern Front in conjunction with the landings in France. Outnumbered and outmaneuvered, Churchill grudgingly acknowledged that it was "the stern duty" of his country to proceed with the invasion. At long last, the British commitment to *Overlord* had become irrevocable. Though the JCS were elated at the outcome, the British chiefs were visibly distraught and immediately began picking away at the invasion plan's details as if they could make it disappear or change the decision.[73]

Confirmation that *Overlord* would go forward signaled a major turning point in the war. The beginning of the end in the West for Hitler's Germany, it also affirmed the emergence of the United States as leader of the Western coalition, with the Joint Chiefs of Staff firmly ensconced as the senior military partners. Even the supreme commander of the operation was to be an American. Though General Marshall had wanted the job, it went instead to a former subordinate and protégé, General Dwight D. Eisenhower, who presided over what became one of the most truly integrated and successful international command structures in history. All the same, with the United States contributing the larger share of the manpower and much, if not most, of the materiel to the operation, British involvement took on a diminished appearance. Except for a brief gathering in London in early June 1944 timed roughly to coincide with the D-Day invasion, the JCS had little need for further full-dress meetings of the Combined Chiefs of Staff. In fact, they did not see their British counterparts again until, at Churchill's insistence, they reassembled at a second Quebec Conference in September 1944. A year later, with the war over, the CCS quietly became for the most part inactive. Though it met occasionally over the next few years, its postwar contributions were never enough to make much difference, and on October 14, 1949, by mutual agreement, it was finally dissolved.[74]

The decision to proceed with *Overlord*, giving it priority over all other Anglo-American operations against Germany, marked the culmination of grand strategic planning in the European theater. Once the troops landed in Normandy on June 6, 1944, it was up to Eisenhower and his British deputy, General Bernard Law Montgomery, and their generals to wage the battles that would bring victory in the West. Had it not been for the JCS and their determination to see the matter through, the invasion might have been postponed indefinitely, and the results of the war could have been quite different. In a very real sense, the Tehran Conference and the *Overlord* decision marked the Joint Chiefs' coming of age as a mature and reliable organization. Out of that experience emerged a decidedly improved and more effective planning system within the JCS organization and a better appreciation among the chiefs themselves of what they could accomplish by working together. A turning point in the history of World War II, the *Overlord* decision was thus also a major milestone in the progress and maturity of the Joint Chiefs of Staff.

WARTIME COLLABORATION WITH THE SOVIET UNION

In contrast to the many contacts and close collaboration the Joint Chiefs enjoyed with their British counterparts through the Combined Chiefs of Staff system, their access to the Soviet high command remained limited throughout World War II. The "Grand Alliance," as Churchill called it, brought together countries—the United States and Great Britain, on the one hand, the Soviet Union, on the other—which, until recently, had viewed one another practically as enemies. Divided prior to the war by politics and ideology, they found it expedient in wartime to concert their efforts toward a common objective—the defeat of Nazi Germany—and little else. While idealists like Roosevelt hoped a new postwar relationship would emerge from the experience, promoting peaceful coexistence between capitalist and Communist systems, realists like Churchill remained skeptical. All agreed that it was a unique and uneasy partnership that was difficult to manage.

The bond holding the Grand Alliance together was, from its inception, the unique relationship among its "Big Three" leaders—Roosevelt, Churchill, and

Stalin—who remained in regular direct contact throughout the war. As a rule, Stalin managed high-level contacts himself and discouraged his generals from becoming overly friendly with their Western counterparts. Churchill followed a similar practice. While professing friendship and cooperation, he showed little inclination to share military information with the Soviets or to take them into his confidence. Although Roosevelt was more forthcoming, he too recognized that, at bottom, the Grand Alliance was a marriage of convenience and declined to bring Stalin in on the biggest secret of the war—that the United States was building an atomic bomb—perhaps because he knew that Soviet espionage agents had passed that information along to Moscow sometime in 1943.[75]

Given the ground rules that tacitly governed the Grand Alliance, East-West military collaboration followed a loose and haphazard course. Though they tried from time to time, JCS planners could find little common ground for creating anything comparable to the Combined Chiefs of Staff to help coordinate East-West military operations.[76] Occasionally, they floated proposals to exchange observers at the field command headquarters level. But there was not much interest from the British and even less from the Soviets.[77] The collaboration that developed derived either from ad hoc arrangements or initiatives mounted through the military missions assigned to the American Embassy in Moscow and tended to be more concerned with logistical matters and lend-lease aid than with coordinating the conduct of the war.

Despite the difficulties inherent in dealing with the Soviets, Roosevelt was determined to demonstrate American goodwill and solidarity of purpose. Brushing aside Churchill's penchant for caution, he exhorted the Joint Chiefs to explore ways of helping the Soviets, even if it meant diverting scarce war resources from other urgent tasks. Yet whatever the JCS could do was limited. As a practical matter, the Eastern Front was too distant and remote for most of the war for them to contemplate stationing substantial military forces there. Nor was it clear whether U.S. forces would have been welcome, given Stalin's aversion to foreign influences.[78] Small deployments of aircraft were another matter, however, and from mid-1942 on, the JCS found themselves peppered with proposals from various sources, including the White House, to provide the Soviets with supply planes and to establish an Anglo-American combat air force in the Caucasus. At the time, German forces had resumed the offensive and for a while there was a glimmer of interest from Stalin. But as the Soviet military position improved, Stalin's enthusiasm waned and the project died.[79]

While the Western powers poured large quantities of material assistance into the Soviet Union, Stalin insisted that the best help they could provide was opening a second front in Western Europe to draw off some of the pressure on the Red army in the East. Churchill maintained that, by concentrating on North Africa, Italy, and the Mediterranean, the Western Allies were already accomplishing much the same thing. Unconvinced, the JCS regarded these operations as sideshows that were perhaps annoying to the Germans but a drain on Allied resources and indecisive by nature. Moreover, the longer the Allies delayed a landing in France, the more opportunity it gave the advancing Russian forces to expand and consolidate Moscow's political influence across Europe.[80]

After the QUADRANT Conference of August 1943, with the prospects for *Overlord* on the rise, the JCS redoubled their efforts to improve contacts and collaboration with the Soviet high command, initially to enlist their promised assistance in diverting German units away from the Normandy invasion area and eventually to prod them into the war against Japan. With these objectives in mind, they sought to upgrade their liaison capabilities with the Soviets and in the fall of 1943 named Major General John R. Deane to head a new joint American military mission in Moscow, reporting directly to the JCS.[81] At the same time, President Roosevelt named W. Averell Harriman, who had been instrumental in setting up the lend-lease program, to replace the ineffectual Admiral William H. Standley as Ambassador to the Soviet Union. Until recently the U.S. secretary of the Combined Chiefs of Staff, Deane was familiar with the current state of thinking in Washington and the status of Allied war plans. At the time he arrived, he recalled, collaboration with the Soviets was "a virgin field" and military coordination "almost nonexistent."[82] Though he found the Soviets to be guarded in their dealings with Westerners, he saw no reason to doubt their commitment to the war and "felt certain" they would enter the conflict against Japan once Germany was defeated.[83]

During the year and a half he spent in Moscow, Deane experienced one frustration after another and kept the Joint Chiefs up to date on every agonizing detail. Though there were a few modest successes, a shuttle bombing agreement of questionable military value foremost among them, he never detected any serious interest on the Soviets' part in establishing a full military dialogue or partnership. Indeed, as the war progressed and as victory over the Germans became more certain, Deane noticed a progressive falling off of Soviet cooperation—so much so that by December 1944 he was expressing serious apprehension over the future of U.S.-Soviet relations. "Everyone will agree on the importance of collaboration with Russia," Deane told Marshall. "It won't be worth a hoot, however, unless it is based on mutual respect and made to work both ways."[84] Impressed by Deane's sobering assessments, Marshall passed them along to the White House without any discernible effect.[85]

THE WAR IN EUROPE

Deane's sentiments reflected a growing sense of unease about the Soviets that permeated JCS deliberations from late 1943 on. The Joint Chiefs got their first close-up look at Stalin and his generals at the Big Three Tehran Conference in November 1943 and came away with mixed impressions. Though judged to be tough-minded and determined, the Soviet generals also appalled members of the JCS with their superficial appreciation of modern military science, most notably their lack of understanding of the difficulties of amphibious operations. As far as Stalin and his generals were concerned, a cross-Channel attack was like fording a river.[86] But with a war yet to be won and the Joint Chiefs eager to nail down a Soviet commitment to join the fight against Japan, they were not inclined to judge the Soviets too harshly.[87]

This view began to change during the early part of 1944, as rumors spread that the Soviets, now on the verge of expelling German troops from their territory, might seek a separate peace. Also around the same time, the JCS received a barrage of reports from Harriman and Deane in Moscow and OSS sources, warning of waning Soviet interest in military collaboration with the West owing to diplomatic friction over the political makeup of Eastern Europe after the war.[88] With *Overlord* only a few months away, the chiefs' concern was considerable, to say the least. About the only immediate source of leverage was to curb shipments under the lend-lease program, which General Marshall described as "our trump card . . . to keep the Soviets on the offensive in connection with the second front."[89] President Roosevelt, however, strongly opposed any avoidable disruptions in assistance, lest they adversely affect U.S.-Soviet relations or the conduct of the war. In September 1944, with *Overlord* a fait accompli, he vetoed any immediate changes in the program.[90]

The Joint Chiefs adopted a wait-and-see attitude toward their Soviet allies for the duration of the war in Europe. By the time of the Yalta Summit Conference in February 1945, they had come to the conclusion, as General Marshall put it, that closer liaison with the Soviet general staff would be "highly desirable" but not absolutely essential.[91] Where the JCS still wanted the Soviets engaged was in Manchuria to keep the Japanese Kwantung army there from reinforcing the home islands against a U.S.-led invasion.[92] Accordingly, they urged President Roosevelt to use his influence with Stalin to overcome what they characterized as Soviet "administrative delays" that were thwarting the implementation of "broad decisions" about U.S.-Soviet collaboration.[93] But with U.S. forces now moving relentlessly across the Pacific, JCS planners were increasingly skeptical whether access to Soviet air and naval bases in Siberia—a requirement once thought to be crucial to an invasion of the Japanese home islands—would make any difference.

The diminished need for Soviet bases and other support was soon reflected in President Harry S. Truman's "get tough" approach toward the Soviets following Roosevelt's death in April 1945. With a U.S. victory in the Pacific now more probable than ever, Truman was less forbearing than Roosevelt in putting pressure on Moscow to live up to its wartime political agreements facilitating free elections in Eastern Europe.[94] Worried that the new President might go too far, Leahy and Marshall reminded him that the wartime agreements Roosevelt had reached were subject to interpretation and that JCS planning still assumed Soviet participation in the war against Japan. With these caveats before him, Truman soon moderated his criticism of the Soviets. Yet owing to the sharp tone and substance of some of his complaints about Soviet behavior, the wartime alliance showed clear signs of breaking down.[95]

That closer wartime cooperation and collaboration between the Joint Chiefs and the Soviet high command could have helped to avoid this outcome is highly unlikely. Stalin's main concerns throughout the war in Europe were to eradicate the threat posed by Nazi Germany and to solidify as much of his control as possible over Eastern Europe, making it in effect a cordon sanitaire between the Soviet Union and the West. With these objectives in mind, the level of cooperation that Stalin sought (and was prepared to accept) was always more specific than general and invariably revolved around the issues of additional aid and the opening of a second front in France. While the JCS did what they could to promote better Soviet-American relations, their options were limited and became even more so as the war progressed. Eventually, the JCS came to see cooperation and collaboration with Moscow as a one-way street. As a rule, General Marshall recalled, the Soviets were "delicate . . . jealous, and . . . very, very hard to preserve a coordinated association with."[96] Regarded by Churchill and others as a marriage of convenience to begin with, the Grand Alliance was probably lucky that it lasted as long as it did and certainly was not destined to survive much beyond the end of the war.

NOTES

1. Churchill to Roosevelt, December 9, 1941, in Francis L. Loewenheim, Harold D. Langley, and Manfred Jonas, eds., *Roosevelt and Churchill: Their Secret Wartime Correspondence* (New York: Saturday Review Press, 1975), 169.
2. Winston S. Churchill, *The Grand Alliance* (Boston: Houghton Mifflin, 1950), 686–687.
3. William G.F. Jackson and Lord Bramall, *The Chiefs: The Story of the United Kingdom Chiefs of Staff* (London: Brassey's, 1992), traces the origins and evolution of the British Chiefs of Staff. Field Marshal Sir John Dill attended the ARCADIA Conference in Brooke's place. Dill then stayed behind in Washington to head the British Joint Staff

Mission. He died in November 1944 and was succeeded by Field Marshal Sir Henry Maitland Wilson.

4   Quoted in Andrew Roberts, *Masters and Commanders: How Four Titans Won the War in the West, 1941–1945* (New York: Harper, 2009), 71.

5   Memo by Combined Chiefs of Staff, January 14, 1942, "Post-Arcadia Collaboration," ABC-4/CS4, *World War II Inter-Allied Conferences* (Washington, DC: Joint History Office, 2003, on CD-ROM). (Hereafter cited as *World War II Conference Papers*); Jackson and Bramall, 224; "Combined Chiefs of Staff (CCS) committee," in I.C.B. Dear and M.R.D. Foot, eds., *The Oxford Companion to World War II* (Oxford: Oxford University Press, 1995), 254.

6   Admiral King turned down a suggestion to have the Navy's senior aviation officer, Rear Admiral John H. Towers, Chief of the Bureau of Aeronautics, join the group. King believed Towers' presence was unnecessary since naval air units were fully integrated into the operating fleets King commanded.

7   Jeffrey G. Barlow, *From Hot War to Cold: The U.S. Navy and National Security Affairs, 1945–1955* (Stanford, CA: Stanford University Press, 2009), 4, 13–17.

8   Notes Taken at Meeting in Executive Offices of the President, November 25, 1942, box 29, Map Room File, Roosevelt Library.

9   See William Emerson, "Franklin Roosevelt as Commander-in-Chief in World War II," *Military Affairs* 22 (Winter 1958–1959), 201.

10 See Eric Larrabee, *Commander in Chief: Franklin Delano Roosevelt, His Lieutenants, and Their War* (Annapolis: Naval Institute Press, 2004), 105–106.

11 John Slessor, *The Central Blue: The Autobiography of Sir John Slessor, Marshal of the RAF* (New York: Praeger, 1957), 494.

12 Quoted in Ray S. Cline, *Washington Command Post: The Operations Division* (Washington, DC: Center of Military History, 1990 reprint), 44.

13 Vernon E. Davis, *History of the Joint Chiefs of Staff in World War II: Organizational Development* (Washington, DC: Historical Division, Joint Secretariat, Joint Chiefs of Staff, 1972), I, 28.

14 See Edward S. Miller, *War Plan Orange: The U.S. Strategy to Defeat Japan, 1897–1945* (Annapolis: Naval Institute Press, 1991); and Henry G. Gole, *The Road to Rainbow: Army Planning for Global War, 1934–1940* (Annapolis: Naval Institute Press, 2003).

15 Louis Morton, "Germany First: The Basic Concept of Allied Strategy in World War II," in Kent Roberts Greenfield, ed., *Command Decisions* (Washington, DC: Center of Military History, 1984), 11–47.

16 Joint Board, *Joint Action of the Army and the Navy* (Washington, DC: U.S. Government Printing Office [GPO], 1927, rev. ed. 1935). Provisions relating to unity of command date from changes incorporated in 1938, as directed in J.B. Serial No. 350, November 30, 1938.

17 Cline, 44–47; Davis, I, 27–59.

18 Davis, I, 239–252.

19 Roberts, 76–78.

20 JCS Minutes, 1st Meeting, February 9, 1942, RG 218, CCS 334 (2-9-42); Davis, I, 229.

21 Davis, II, 506–08.

22 Memo, McFarland to King, January 4, 1945, "H.R. 5604, To Provide for the Permanent Establishment of the JCS and Joint Secretariat," CCS 334 (12-80-44).

23 Thomas F. Troy, *Donovan and the CIA: A History of the Establishment of the Central Intelligence Agency* (Washington, DC: Center for the Study of Intelligence, 1981), 117–153, 427–428.

24 Forrest C. Pogue, *George C. Marshall: Ordeal and Hope, 1939–1942* (New York: Viking Press, 1965), 298–301.

25 Henry L. Stimson and McGeorge Bundy, *On Active Service in Peace and War* (New York: Harper & Bros., 1947), 414.

26 Forrest C. Pogue, "The Wartime Chiefs of Staff and the President," in Monte D. Wright and Lawrence J. Paszek, eds., *Soldiers and Statesmen* (Washington, DC: Office of Air Force History, 1973), 71.

27 Kent Roberts Greenfield, *American Strategy in World War II: A Reconsideration* (Malabar, FL: Krieger Publishing, 1982 reprint), 51–52.

28 Pogue, "Wartime Chiefs," 72–73. Marshall recalled that the first time he visited Hyde Park, NY, Roosevelt's home, was for the President's funeral in April 1945.

29 William D. Leahy, *I Was There* (New York: Whittlesey House, 1950), 102; Davis, II, 439–445.

30 "Charter: Joint Chiefs of Staff," June 15, 1943, JCS 202/24.

31 Letter, FDR to Leahy, July 16, 1943, JCS 415.

32 Maurice Matloff and Edwin M. Snell, *Strategic Planning for Coalition Warfare, 1941–1942* (Washington, DC: Office of the Chief of Military History, 1953), 97–119.

33 Cline, 101–102.

34 Quoted in Larrabee, 187.

35 See Mark A. Stoler, *The Politics of the Second Front: American Military Planning and Diplomacy in Coalition Warfare, 1941–1943* (Westport, CT: Greenwood Press, 1977), 14–22.

36 JCS Minutes, 4th Meeting, March 7, 1942, RG 218, CCS 334 (2-9-42).

37 Matloff and Snell, 161; Grace Person Hayes, *The History of the Joint Chiefs of Staff in World War II: The War Against Japan* (Annapolis: Naval Institute Press, 1982), 113–114.

38 Roberts, 137–166; George F. Howe, *Northwest Africa: Seizing the Initiative in the West* (Washington, DC: Office of the Chief of Military History, 1957), 10–14.

39 Memo, Roosevelt to Marshall, August 24, 1942, box 4, President's Secretary's File, Roosevelt Library.

40 Stimson and Bundy, 425–126; Memo, Roosevelt to Hopkins, Marshall, and King, July 24, 1942, box 4, President's Secretary's File, Roosevelt Library.

41 Sherwood, 563.

42 Gordon A. Harrison, *Cross-Channel Attack* (Washington, DC: Office of the Chief of Military History, 1951), 15–16; Matloff and Snell, 177.

43 Minutes, Meeting Held at the White House, December 10, 1942, box 29, Map Room File, Roosevelt Library.

44 Llewellyn Woodward, *British Foreign Policy in the Second World War* (London: Her Majesty's Stationery Office [HMSO], 1971), II, 462.

45 See Maurice Matloff, "Allied Strategy in Europe, 1939–1945," in Peter Paret, ed., *Makers of Modern Strategy: From Machiavelli to the Nuclear Age* (Princeton, NJ: Princeton University Press, 1986), 677–702; and Richard Overy, *Why the Allies Won* (New York: W.W. Norton, 1995), 141.

46 Davis, II, 373–375. At the request of Admiral King, the JSSC's charter made provision for a fourth member to represent naval aviation. King failed to follow up, however, and the position was never filled.

47 Report by JSSC to JCS, "Basic Strategic Concept for 1943," December 11, 1942, JCS 167.

48 Mark A. Stoler, *Allies and Adversaries: The Joint Chiefs of Staff, the Grand Alliance, and U.S. Strategy in World War II* (Chapel Hill: University of North Carolina Press, 2000), 90–91.

49 Minutes of a Meeting at the White House, January 7, 1943, President's Secretary's File, Roosevelt Library; U.S. Department of State, *Foreign Relations of the United States: Conferences at Washington, 1941–1942, and Casablanca, 1943*, 505–514.

50 Winston S. Churchill, *The Hinge of Fate* (Boston: Houghton, Mifflin, 1950), 676.

51 Maurice Matloff, *Strategic Planning for Coalition Warfare, 1943–44* (Washington, DC: Office of the Chief of Military History, 1959), 37; Churchill, *Hinge of Fate*, 686–687.

52 "Bomber Offensive from the UK," January 21, 1943, CCS 166/1/D, *World War II Conference Papers*, 88–89.

53 Wesley Frank Craven and James Lea Cate, eds., *The Army Air Forces in World War II*, vol. II, *Europe: Torch to Pointblank, August 1942 to December 1943* (Washington, DC: Office of Air Force History, 1983), 274–307.

54 Hayes, 363.

55 Letter, Wedemeyer to Handy, January 20, 1943, quoted in Davis, II, 436.

56 Cline, 234–242.

57 "Charter: Joint War Plans Committee," May 11, 1943, JCS 202/14.

58 Ernest J. King and Alfred Muir Whitehill, *Fleet Admiral King: A Naval Record* (New York: W.W. Norton, 1952), 491–495. See also JPS Report, May 7, 1943, and decision on May 8, 1943, "Strategic Plan for the Defeat of Japan," JCS 287 and JCS 287/1.

59 Cline, 219–220.

60 Stoler, *Allies and Adversaries*, 120–121.

61 Matloff, *Strategic Planning, 1943–44*, 132.

62 Matloff, *Strategic Planning, 1943–44*, 126–145; Forrest C. Pogue, *George C. Marshall: Organizer of Victory* (New York: Viking Press, 1973), 193–213; CCS Minutes, TRIDENT, 1st meeting, May 12, 1943, *World War II Conferences*, 253; "Final Report to the President and Prime Minister," May 25, 1943, CCS 242/6.

63 Cline, 222.

64 Minutes of Meeting Held at the White House between the President and the Chiefs of Staff, August 10, 1943, box 29, Map Room File, Roosevelt Library.

65 Memo, JPS to JCS, August 5, 1943, "Strategic Concept for the Defeat of the Axis in Europe," JCS 444; Memo, JSSC to JCS, August 5, 1943, revised per JCS discussion August 6, 1943, "QUADRANT and European Strategy," JCS 443.

66 Leahy Diary entry, August 14, 1943, Leahy Papers, Library of Congress (microfilm); Leahy, *I Was There*, 175.

67 Final Report to the President and Prime Minister, August 24, 1943, CCS 319/5, *World War II Conference Papers*; Cline, 224–225.

68 David Dilks, ed., *The Diaries of Sir Alexander Cadogan, 1938–1945* (New York: G.P. Putnam's Sons, 1972), 570–571.

69 Roberts, 431–435.

70 Arthur Bryant, *Triumph in the West, 1943–1945* (London: Collins, 1959), 59, 71.

71 Sextant Conference, Minutes of 2d Plenary Meeting, held at Villa Kirk, Cairo, November 24, 1943, *World War II Conference Papers*.

72 Minutes, Military Conference Between U.S.A., Great Britain, and U.S.S.R., November 29, 1943, EUREKA Conference, ibid.

73 Overy, 143; Bryant, 88–101; Pogue, *Organizer of Victory*, 316–318.

74 JCS Info Memo 687, October 14, 1949, "Dissolution of the Combined Chiefs of Staff Organization," U, CCS 334 (2-9-46) sec. 1, RG 218.

75 See Martin J. Sherwin, *A World Destroyed: The Atomic Bomb and the Grand Alliance* (New York: Vintage Books, 1977), 102–104.

76 The most ambitious proposal was a recommendation by the Joint Strategic Survey Committee in the late summer of 1944 to replace the CCS with a tripartite "United Chiefs of Staff." The British gave it a chilly reception and the project quickly died. See JSSC Report, "Machinery for Coordination of U.S.-Soviet-British Military Effort," August 31, 1944, JCS 1005/1.

77 See Memo by U.S. CoS to CCS, August 17, 1944, "Machinery for Coordination of U.S.-Soviet-British Military Effort," JCS 1005; and Report by JPS, November 17, 1944, "Liaison Between Theater Commanders and the Russian Armies," JCS 1005/3.

78 Adam B. Ulam, *Expansion and Coexistence: The History of Soviet Foreign Policy, 1917–67* (New York: Praeger, 1967), 319.

79 Robert E. Sherwood, *Roosevelt and Hopkins* (New York: Enigma Books, 2001), 616; Matloff and Snell, 329–336.

80 Stoler, *Allies and Adversaries*, 124–132. See also Annex A, "Relationship of Russia to U.S. Global Military Situation," to Memo, JSSC to JCS, September 16, 1943, JCS 506; Sherwood, 748–749; and James F. Schnabel, *The Joint Chiefs of Staff and National Policy, 1945–1947* (Washington, DC: Joint History Office, 1996), 7–8.

81 "Instructions to Members of U.S. Military Mission to U.S.S.R.," October 5, 1943, JCS 506/1; Matloff, *Strategic Planning, 1943–44*, 290–291.

82 John R. Deane, *The Strange Alliance* (Bloomington: Indiana University Press, 1973), 47.

83 Stoler, *Allies and Adversaries*, 166.

84 Letter, Deane to Marshall, December 2, 1944, in Deane, 84.

85 Pogue, *Organizer of Victory*, 530–531.

86 Comments by Soviet Marshal Voroshiloff, Minutes of Military Conference between U.S.A., Britain, and U.S.S.R., November 29, 1943, *World War II Conferences*, 539; King and Whitehill, 518.

87 Pogue, *Organizer of Victory*, 311–313.
88 Stoler, *Allies and Adversaries*, 182–183.
89 Memo, Marshall to Roosevelt, March 31, 1944, quoted in Matloff, *Strategic Planning, 1943–44*, 497.
90 Letter, Roosevelt to Marshall, September 9, 1944, JCS 771/8.
91 Argonaut Conference, Minutes of Meeting Held in the President's Sun Room, Livadia Palace, February 4, 1945, box 29, Map Room File, Roosevelt Library.
92 Matloff, *Strategic Planning, 1943–44*, 500–501.
93 Memo, JCS to President Roosevelt, January 23, 1945, U.S. Department of State, *Foreign Relations of the United States: Diplomatic Papers—the Conferences at Malta and Yalta, 1945* (Washington, DC: GPO, 1955), 396–400.
94 John Lewis Gaddis, *The United States and the Origins of the Cold War, 1941–1947* (New York: Columbia University Press, 1972), 198–243.
95 Pogue, 578–581; Stoler, *Allies and Adversaries*, 237–238.
96 Marshall interview, quoted in Pogue, 574.

Burma Road, 1944–1945

Chapter 2

THE ASIA-PACIFIC WAR AND THE BEGINNINGS OF POSTWAR PLANNING

The Joint Chiefs' greatest accomplishment in World War II was planning and executing a two-front war, one in the European-Atlantic theater and the other in the Asia-Pacific region. Even though the agreed Anglo-American strategy gave primary importance to defeating Germany, the attack on Pearl Harbor and Japan's rapid advances during the early stages of the war created a political and military environment that focused heavy attention on the Pacific and Far East. For the first year or so of the war, bolstering the American posture there consumed as much, if not more, of the Joint Chiefs' energy as Europe. At the same time, the absence of an agreed long-range wartime strategy made it practically impossible for JCS planners to draw a clear distinction between primary and secondary theaters. As a result, by the end of 1943, deployments of personnel were practically the same (1.8 million) against Japan as against Germany.[1] Thereafter, as the United States stepped up its preparations for Operation *Overlord* and as the Allies brought the German submarine threat in the Atlantic under control, the buildup in the United Kingdom accelerated quickly, overshadowing the allocation of resources elsewhere. But with such a substantial concentration of personnel and other assets in Asia and the Pacific from the outset, it was practically impossible for the Joint Chiefs to draw and maintain a clear distinction in priorities.

STRATEGY AND COMMAND IN THE PACIFIC

To wage the Pacific war, the Joint Chiefs adopted somewhat different command procedures than they used in the European and Mediterranean theaters. In Europe, the lines of command and control followed in accordance with the decision taken by President Roosevelt and Prime Minister Churchill immediately after Pearl Harbor to pool their resources and to pursue a common strategy. For the North Africa–Mediterranean campaigns and for the invasion of France, the Allies established combined unified commands, which operated under directives issued by the

Combined Chiefs of Staff. The Supreme Commander for the invasion of Europe, General Dwight D. Eisenhower, took his orders from the CCS (which were relayed to him via the War Department) and presided over an integrated staff that was both multinational and multi-Service in its composition.[2]

Command arrangements in the Pacific evolved differently, owing to the predominant role played by the Navy in that theater, the Combined Chiefs' limited participation, and decisions taken during the initial stages of the war to split the theater into two parts. Shortly after Pearl Harbor General Marshall persuaded Admiral King to endorse the creation of a combined Australian-British-Dutch-American Command (ABDACOM) for the Southwestern Pacific in hopes of mobilizing greater resistance.[3] The Japanese surge continued and ABDACOM soon fell apart, leaving command relationships in the South Pacific in a shambles. From this unpleasant experience (and a later one involving difficulties with the British over protection of Anglo-American convoys crossing the Atlantic), King resolved never again to be drawn into a combined or unified command arrangement if he could possibly avoid it. Unity of command, King insisted, was highly overrated and definitely "not a panacea for all military difficulties" as some "amateur strategists"—a veiled reference to Secretary of War Henry L. Stimson—seemed to believe.[4]

King's solution to command problems in the Pacific lay in a division of responsibility, approved by the Joint Chiefs with little debate on March 16, 1942, that created two parallel organizations: a Southwest Pacific Area command under General Douglas MacArthur, bringing together a patchwork of U.S. ground, sea, and air forces with the remnants of the ABDACOM, and a Pacific Ocean Area command under Admiral Chester W. Nimitz, composed predominantly of Navy and Marine Corps units.[5] In 1944, a third Pacific command emerged, organized around the Twentieth Air Force, which operated under the authority of the JCS, with General Arnold as its executive agent. King would have preferred a single joint command for the Pacific, but he knew that if he pushed for one, it would probably go to MacArthur rather than to a Navy officer. MacArthur was practically anathema to the Navy, and Nimitz, the leading Navy candidate for the post, was junior to MacArthur and still relatively unknown.[6] Unlike the ABDACOM, which had fallen under the Combined Chiefs of Staff, these new commands were the exclusive responsibility of the United States and reported directly to the Joint Chiefs, the presence of Australian and other foreign forces under MacArthur notwithstanding. Though joint organizations, composed of ground, air, and naval forces, they were not, strictly speaking, "unified" or integrated commands: MacArthur's staff was almost entirely Army; Nimitz's predominantly Navy. One byproduct of the new command structure was the establishment of the JCS "executive agent" system, using the Service chiefs as go-betweens. Thus, in relaying orders and other communications, Marshall dealt directly with MacArthur and King with Nimitz.[7]

From the outset, the two original commands conducted separate and different types of wars. MacArthur's principal aim was to redeem his reputation and liberate the Philippines, where he had suffered an ignominious defeat early in 1942. Promising "I shall return," he launched an ambitious campaign, first to contain, then to roll back the Japanese in the Southwest Pacific. With aircraft carriers in short supply, he turned to Lieutenant General George C. Kenney, commander of the Fifth Air Force, to supply the bulk of his combat air support from a motley force of land-based fighters and bombers, many of them cast-offs from other theaters.[8] For naval support he relied on Vice Admiral Thomas C. Kinkaid, commander of the Seventh Fleet. Working with limited resources and in a hostile climate where tropical diseases could be as lethal as the Japanese, MacArthur developed a leap-frog strategy that took him up the northeastern coast of New Guinea and eventually back to the Philippines. Nimitz's concept of the war centered on the interdiction of Japanese shipping and the destruction of the Japanese fleet as the keys to victory. Cautious and reserved by nature, he was initially skeptical of the idea—pressed upon him by King after the Casablanca Conference—that the Navy should, in effect, revive the old War Plan Orange and concentrate its efforts on strategic objectives in the Central Pacific. Seeking a war-winning strategy, King proposed a thrust through the Marshalls and Marianas, spearheaded by fast carrier task forces and Marine Corps amphibious assault units. Though Nimitz went along with the idea, he and his planning staff at Pearl Harbor insisted on refinements that included recapturing and holding the Aleutian Islands and neutralizing the Gilberts to give U.S. warships the benefit of land-based air protection.[9] As it turned out, Nimitz moved more slowly than King originally envisioned, chiefly because he synchronized his advance to progress more or less in unison with MacArthur's march up through New Guinea and Admiral William F. Halsey's campaign in the Solomon Islands, thereby optimizing his assets and assuring the protection of his western flank.[10]

King assured Nimitz as he embarked upon the Central Pacific strategy that he would enjoy substantial numerical superiority over the Japanese fleet. Indeed, a critical factor in King's advocacy of the plan was his knowledge that the Navy would soon have a "new" fleet in the Pacific, the product of a naval construction program inaugurated in 1940 and hurried along after Pearl Harbor.[11] Among the first of these ships to take up station in the Pacific during the second half of 1943 were a half-dozen of the new 27,000-ton *Essex*-class attack carriers. Built to accommodate nearly a hundred planes each, these ships gave Nimitz the capability of launching carrier bombing strikes comparable to land-based aviation. By the end of the year, he had a force of over 700 carrier-based aircraft, many of them improved models, and a growing fleet of ships, half of them built since the beginning of the war.[12]

Additional support for Nimitz's push into the Central Pacific came from the Army Air Forces (AAF), who saw an opportunity to use island bases in the Marianas to launch B–29 attacks against Japan. Until mid-1943, the Air Staff had concentrated on China as the primary staging area for its B–29s, which were new high-altitude, long-distance, very heavy bombers that the AAF expected to deploy in large numbers against Japan during the second half of 1944. Owing to problems of supplying bases in China and protecting them against expected Japanese counterattacks, however, Air Staff planners began to look elsewhere. With the emergence of Nimitz's Central Pacific strategy, they refocused their efforts there.[13] Although the Joint Chiefs tried from time to time to develop an overall war plan for the Pacific, the divided command in the theater made it virtually impossible. Invariably, the decisions that emerged from Washington represented compromises, resulting in "an *ad hoc* approach to Pacific strategy."[14] Friction between MacArthur and Nimitz was endemic to the Pacific theater and required frequent intervention from Marshall and King. At the same time, in CCS meetings with the British, King often pursued what amounted to a separate agenda. Technically, the CCS exercised no responsibility for the Pacific, but because the demands of the various theaters regularly impinged on each other, the Combined Chiefs took it upon themselves to review plans for Asia and the Pacific while developing strategy for Europe and the Mediterranean. At the wartime summit conferences and in routine contacts in Washington, King's blatant Anglophobia and persistence in promoting the Navy's interests in the Pacific became practically legendary. Of the Americans they dealt with, King was by far the most unpopular with the British. Yet he also proved remarkably effective at getting what he wanted. In Grace Person Hayes's estimation, he was clearly "the JCS member whose influence upon the course of events in the Pacific was greatest."[15]

In contrast to other aspects of the war, there were relatively few sharp disagreements among the JCS over the merits of one course of strategy in the Pacific over another. Marshall had no objection to the Navy's Central Pacific strategy as long as it was logistically feasible and did not crowd MacArthur out of the picture.[16] Moreover, none of the chiefs wanted to see a stalemate develop that could prolong the Pacific conflict into 1947 or 1948 and lead to war-weariness at home. By 1943, the JCS agreed that a predominantly defensive posture in the Pacific was incompatible with American interests and that the tide had turned sufficiently to allow for the transition to an "offensive-defensive" philosophy. As the arrival of the many new ships and planes in the Pacific suggested, increased industrial production at home was finally making a difference by offering a broader range of options on the battle front.[17] These matters came to a head at the first Quebec Conference

(QUADRANT) in mid-August 1943. Though the chiefs' number-one goal at First Quebec was to firm up the British commitment to *Overlord*, stepping up the war in the Pacific was a close second. Applying a mathematical formula approach (a technique he enjoyed using), King proposed a worldwide boost in the allocation of resources from 15 to 20 percent in the Pacific, a 5 percent increase that would translate into one-third more available resources and only a 6 percent drop in supplies to Europe.[18] The British knew that, as a rule, the Joint Chiefs used exceedingly conservative production and supply estimates, so that in all likelihood an increase in the allocation to the Pacific would mean little or no change elsewhere. Though the CCS never officially approved King's formula, the British members were well aware that there was not much they could do if the Americans elected to abide by it. Turning to an alternative approach, the conference wound up approving an American plan increasing the tempo of operations in the Pacific at such a rate as to assure the defeat of Japan within 12 months of Germany's surrender or collapse.[19] Thus, by mid to late 1943, though not exactly on a par with the war in Europe, the war in the Pacific was steadily gathering momentum and recognition that the outcome there was no less important than victory in Europe. The chiefs knew that long, drawn-out wars tended to sap morale at home and have unforeseen political side-effects. Consequently, they hoped to lay the groundwork for the defeat of Japan well in advance and make it happen as quickly as possible once Germany surrendered. The chiefs assumed that, to carry out this strategy, they would need to move troops from Europe to the Pacific as fast as possible and mass forces on an unprecedented scale. Little did they realize that, when that moment arrived, they would have in their hands a new weapon—the atomic bomb—that would not only facilitate Japan's surrender more abruptly than anyone realized, but usher in a new era in warfare at the same time.

THE CHINA-BURMA-INDIA THEATER

With Europe and the Pacific commanding most of the attention and resources, problems in the China-Burma-India Theater (CBI) took a distinctly secondary place in the Joint Chiefs' strategic calculations. Under the division of responsibility adopted by the Combined Chiefs of Staff in March 1942, the United States provided military coordination with the government of China, while Britain saw to the defense of Burma and India. The only American combat formations assigned to the CBI during the war were the Galahad commando unit (Merrill's Marauders) formed near the end of 1943, and the XX Bomber Command, consisting of four B–29 groups that operated mainly from Chengtu in southwest China in 1944–1945. Otherwise, the U.S. presence consisted of noncombat personnel involved in construction projects, training and advisory functions, and logistical support for China under the lend-lease aid program.

China's need for assistance had grown steadily since the outbreak of its undeclared war with Japan in 1937. Forced by the invading Japanese to abandon its capital at Nanking, the Chinese Nationalist government of Generalissimo Chiang Kai-shek had relocated to the interior. Operating out of Chongqing, Chiang had used his well-established political connections in Washington to mobilize American public opinion and congressional support for his cause. Prohibited under the 1937 Neutrality Act from providing direct military assistance, the Roosevelt administration arranged several large loans that allowed Chiang to buy arms and equipment to bolster his military capabilities. But with graft and corruption permeating Chiang's government, much of the financial help from Washington was wasted. By the time the United States entered the war in December 1941, Chiang's regime was near collapse. At the ARCADIA Conference, with Japanese forces moving practically at will across East Asia and the Pacific, Roosevelt and Churchill sought to boost Chiang's morale and shore up his resistance by inviting him to become supreme commander of a new China Theater. Inclusion of nonwhite, non-Christian China in the Grand Alliance helped the Western Allies undercut Japanese propaganda about "Asia for the Asiatics" and reduced the chances of World War II being seen as a racial conflict.[20] The offer carried with it no promise of additional assistance or immediate support, but it struck Roosevelt as a logical first step toward realizing his vision that China should emerge from the war as "a great power." Chiang promptly accepted and, to seal the deal, asked the United States to appoint an American officer to be his chief of staff, in effect his military second in command.[21]

To assist Chiang as his chief of staff, Marshall and Secretary of War Stimson eventually settled on Lieutenant General Joseph W. Stilwell, an "Old China Hand" whom Marshall had once described as "qualified for any command in peace or war."[22] Gifted in learning languages, Stilwell was fluent in Mandarin Chinese, which he mastered during his numerous tours of duty in the Far East, dating from 1911, and intensive language training in the 1920s. But he had a prickly personality and soon grew contemptuous of Chiang, whom he regarded as an ineffectual political leader and inept as a general. As the military attaché to the U.S. Embassy in China from 1935 to 1939, Stilwell had deplored Chiang's lack of preparedness for dealing with the Japanese and had developed a tempered respect for Chiang's Communist rivals, led by Mao Zedong, who seemed determined to mount resistance to the Japanese with whatever limited resources they could from their power base in the countryside.[23]

Stilwell embarked on his mission with virtually no strategic or operational guidance. His only instructions were a generalized set of orders issued by the War Department early in February 1942. While the Army General Staff and the JCS routinely affirmed the importance of the CBI, they consistently treated it as a low priority. Preoccupied with Europe and the Pacific, the JCS had little inclination and even fewer resources for waging a war on the China mainland. Only Marshall and Arnold took a personal interest in Chinese affairs—Marshall because he had spent 3 years in China during the interwar period and was a personal friend of Stilwell's, and Arnold because of the AAF's heavy commitment of men and equipment for supply operations and planned B–29 deployments. The most important military uses the JCS could see for China were as a base for future air operations against Japan and as a source of manpower for confronting and holding down large segments of the Japanese army. But it was unlikely that the AAF would make much use of China as a base of operations until the Navy completed its advance across the Pacific and could provide secure lines of supply and communications. Until then, as the senior American officer in the CBI, Stilwell was to oversee the distribution of American lend-lease assistance, train the Chinese army, and wage war against the Japanese with whatever U.S. and Chinese forces might be assigned to him.[24]

Stilwell arrived in Asia in April 1942, just as the military situation was going from bad to worse. The success of four Japanese divisions in attacking Burma, routing the British-led defenders and forcing them back into India, effectively cut the last remaining overland access route—the Burma Road—to China. For nearly the remainder of the war, from June 1942 until January 1945, China was virtually isolated from the rest of the world except via air. Though Stilwell had a replacement route known as the Ledo Road (renamed the Stilwell Road in 1945) under construction by the end of the year, it took over 2 years of arduous work in a torturous climate and terrain to complete. Of the 15,000 U.S. Servicemen who helped to build the Ledo Road, about 60 percent were African-Americans.[25] Meantime, supplies and equipment had to be flown into China from bases in India over the Himalayas (the "Hump") at considerable risk and cost. Eventually, the effort diverted so many American transport aircraft that, in General Marshall's opinion, it significantly prolonged the Allied campaigns in Italy and France.[26]

Logistics were only one of Stilwell's problems. Most difficult of all was establishing a working relationship with the Generalissimo, whose autocratic ways, intricate political connections, and lofty expectations clashed with Stilwell's coarse manner and business-like determination. Stilwell may have been the wrong choice for the job, but whether anyone else could have done better is open to question. Never a great admirer of Chiang to begin with, Stilwell became even less so as the war progressed. Rarely did he acknowledge the extraordinary political pressures under which Chiang operated or what some Chinese scholars now see as Chiang's accomplishments in the strategic management of his forces.[27] In Stilwell's private diary, published after the war, the full depth of his contempt for Chiang became apparent in his numerous references to the Generalissimo by the nickname "Peanut." In fact, Stilwell and Chiang rarely saw one another. Stilwell spent most of his time in India training Chinese troops, while Chiang stayed in Chongqing.

The number one task that Stilwell and the JCS faced in China was to develop a capability to fight the Japanese; for Chiang the situation was more complex. Though he held the titles of president and generalissimo, he exercised limited authority over a group of independently minded generals, politicians, and war lords. Apart from the threat posed by the invaders, he also faced the likelihood of a showdown after the war with his archrival, Mao Zedong, leader of the Chinese Communists, who styled themselves as being in the forefront of the resistance to Japanese aggression. In fact, Nationalist forces put up as much if not more resistance to the Japanese than the Communists and suffered significantly heavier casualties. But on balance, it was Mao who emerged as most committed to the war. Saving his best troops for the postwar period, Chiang often ignored Stilwell's military advice and listened instead to an American expatriate and former captain in the Army Air Corps, Claire L. Chennault, who convinced Chiang that airpower could defeat the Japanese. An innovator in tactical aviation during the interwar years, Chennault led a flamboyant group of American volunteer aviators known as the "Flying Tigers." Recalled to active duty in April 1942, Chennault was eventually promoted to major general. Meanwhile, the Flying Tigers were absorbed into the Army Air Forces, becoming part of the Fourteenth Air Force in 1943. Though technically subordinate to Stilwell, Chennault often used his close connections with Chiang and his personal friendship with President Roosevelt to bypass Stilwell's authority.[28]

Despite the frustration and setbacks, Stilwell achieved some remarkable results. His most notable accomplishment was establishing the Ramgarh Training Center in India's Bihar Province, which served as the hub of his efforts to train and modernize the Chinese army. At Ramgarh, Stilwell initiated practices and policies that the JCS adopted as standard procedure for U.S. military advisory and assistance programs in the postwar period. By placing American commanders and staff officers with Chinese units, creating Service training schools, and indoctrinating Chinese forces in the use of U.S. arms and tactics, Stilwell helped to bring a new degree of professionalism to the Chinese Nationalist army. In the process, he created a system that saw extensive use in Korea, Vietnam, and other countries in later years. By the time Stilwell was recalled in 1944, he had trained five Chinese divisions that he considered to be on a par with those in the Japanese army, and was in the process of producing more, both at Ramgarh and in China.[29]

At the first Quebec Conference in August 1943, the Combined Chiefs of Staff agreed the time had come to make plans for liberating Burma (thereby reopening the Burma Road to China) and the other parts of Southeast Asia the Japanese had conquered the year before. To organize the campaign, the CCS established a Southeast Asia Command (SEAC), with Lord Louis Mountbatten as supreme commander. Earlier, when asked to contribute forces to the operation, Chiang had indicated that he would never allow a British officer to command Chinese troops. To get around this problem, the CCS named Stilwell as Mountbatten's deputy, thus adding yet another layer of responsibility to his difficult mission.[30] With the decision to launch the Burma offensive, the Joint Chiefs, through the CCS, became more actively and directly engaged in CBI affairs than at any time to that point in the war. Even so, the British chiefs left no doubt that they were determined to have their way in Southeast Asia, just as the JCS insisted on running the war in the Pacific.[31] Propping up Chiang, whose importance and role in the war Churchill dismissed as "minor," did not fit the British agenda. At the Cairo Conference (SEXTANT) in November 1943, Mountbatten and the British chiefs apprised Chiang of a change of plans for the Burma operation that would lessen the role of Chinese forces and thus reduce his projected allocation of shipments over the Hump.[32] To assuage Chiang's disappointment, Roosevelt promised to equip and train 90 Chinese divisions, but avoided setting specific dates for initiating and completing the project.[33] Around this same time, the Air Staff became convinced that bomber bases in China would be too vulnerable and difficult to maintain, and began eyeing Formosa or the Marianas as alternate staging sites for their B–29s. While the deployment of B–29s to China (Operation *Matterhorn*) went ahead in April 1944 as planned, the JCS cut the force in half, from eight bombardment groups to four, due to supply limitations.[34]

Coupled with the actions approved earlier at SEXTANT, the chiefs' decision curbing B–29 deployments confirmed China's fate as a secondary theater of the war. Bitter and indignant, Chiang became ever more critical of Stilwell and insisted—to Stilwell's and the Joint Chiefs' dismay—on micromanaging Chinese military operations in East China and Burma. Reverses followed on practically every front. At the same time, Chiang remained intent on preserving his authority and refused to listen when Stilwell proposed opening contacts with Mao and diverting lend-lease aid to Chinese Communist forces fighting the Japanese north of the Yellow River.[35] By then, Roosevelt was also having second thoughts about Chiang's leadership. At Marshall's instigation, the President urged Chiang in September 1944 to give Stilwell "unrestricted command" of all Chinese forces.[36] Though Chiang acknowledged that he might be willing to make concessions, he refused to have anything more to do with Stilwell and demanded his recall. Seeing no alternative, Roosevelt reluctantly acquiesced and in October 1944, Stilwell's mission ended.[37]

Following Stilwell's departure, the Joint Chiefs made no attempt to find a successor and decided to abolish the CBI. In its place they created two new commands: the China Theater, which they placed under Lieutenant General Albert C. Wedemeyer, Mountbatten's deputy chief of staff; and the India-Burma Theater, which went to

Lieutenant General David I. Sultan, formerly Stilwell's second in command. The decision to break up the CBI was supposed to make Wedemeyer's task easier, but in reality it did no such thing. Though Wedemeyer served as the Generalissimo's chief of staff, the cooperation he received from Chiang was only marginally better than Stilwell had gotten. Revised instructions issued by the Joint Chiefs on October 24, 1944, were largely the product of Marshall's hand and implicitly urged Wedemeyer to exercise utmost caution. Barred from exercising direct command over Chinese forces, he could only "advise and assist" the Generalissimo in the conduct of military operations.[38]

Marshall correctly surmised that the wartime problems Stilwell and the Joint Chiefs experienced with Chiang Kai-shek were only a foretaste of the future. Roosevelt's desire to make China a great power and Chiang's eagerness to assume the leadership role fueled expectations that could never be fulfilled. Chiang's regime was too weak politically and too corrupt to play such a part. Preoccupied with preparing for the expected postwar showdown with his Communist rivals, Chiang hoarded his resources rather than trying to defeat the Japanese. The JCS were as interested as anyone in seeing a stable and unified China emerge from the war, but they were averse to making commitments and expending resources that might jeopardize operations elsewhere. China, meanwhile, remained a strategic backwater. While some American planners, Marshall foremost among them, hoped for better to come after Japan surrendered, they were not overly optimistic as a group.

POSTWAR PLANNING BEGINS

Despite setbacks in Asia and the steady but slow progress in pushing the Japanese back across the Pacific, the Joint Chiefs detected definite signs by mid-1943 that the global tide of battle was turning in the Allies' favor and that victory over the Axis would soon be in sight. Assuming a successful landing on the northern French coast in the spring of 1944, the Combined Chiefs of Staff at the QUADRANT Conference had estimated for planning purposes that the war in Europe would be over by October 1944.[39] While this proved to be an overly optimistic prediction, it did help draw attention to issues that the Joint Chiefs thus far had largely ignored: the need for policies and plans on the postwar size, composition, and organization of the country's Armed Forces, and similar actions on postwar security and other political-military arrangements.

Preoccupied with the war, the Joint Chiefs were averse to firm postwar commitments until they had a clearer idea of the outcome. A case in point was their reticence concerning the postwar organization and composition of the Armed Forces, an issue they knew was bound to provoke inter-Service friction and sharp debate. In the aftermath of Pearl Harbor and the other setbacks early in the war, there was a growing sense within the military and the public at large that a return to the prewar separateness of the Services was out of the question and that the postwar defense establishment should be both bigger and better prepared for emergencies. In assessing postwar requirements, the Joint Chiefs agreed that the country needed a larger, more flexible, and more effective standing force. Where differences arose was over its size, the assignment of roles and missions to its various components, and its overall structure—in short the fundamental issues that differentiated each Service.[40]

The Joint Chiefs of Staff discussed these issues from time to time during the war but made little headway in the absence of a consensus on postwar defense organization and the possibility that the Armed Forces might adopt a system of universal military training (UMT).[41] In consequence, JCS planning to determine the optimum size, composition, and capabilities of the postwar force amounted to a compilation of requirements generated by the Services themselves, based on their own perceived needs and assessments. These uncoordinated estimates projected a permanent peacetime military establishment of 1.6 million officers and enlisted personnel organized into an Army of 25 active and Reserve divisions, a 70-group Air Force emphasizing long-range strategic bombardment, a Navy of 321 combatant vessels in the active fleet, including 15 attack carriers and 3,600 aircraft, and a Marine Corps of 100,000 officers and enlisted personnel.[42]

Whether the Services would achieve these goals depended, among other things, on the kind of defense establishment that would emerge after the war. The most outspoken on the need for postwar organizational reform—and the first to propose a course of action—was General Marshall, whose strong views grew out of his experiences with the hasty and chaotic demobilization that followed World War I and the Army's chronic underfunding during the interwar years. Expecting money to be tight again after the war, Marshall foresaw the return to a relatively small standing army and endorsed UMT as a means of expanding it rapidly in an emergency. To make better use of available funds, he also urged improved management of the Armed Forces, and in November 1943 he tendered a plan for JCS consideration to create a single unified department of war. Arguing that the current JCS-CCS committee structure was cumbersome and inefficient, Marshall proposed more streamlined arrangements stressing centralized administration, "amalgamation" of the Services, and unity of command.[43] Arnold and King were lukewarm toward the idea and favored tabling the matter until after the war. While Arnold agreed with Marshall on the need for postwar reorganization, his first priority was to turn the Army Air Forces into a separate coequal service. At King's suggestion, the chiefs sidestepped the issues Marshall had raised by referring them to the Joint Strategic Survey Committee for study "as soon as practicable."[44]

By the spring of 1944, emerging congressional interest in postwar military organization compelled the JCS to revisit the issue sooner than they wanted to. At the suggestion of the Joint Strategic Survey Committee, the chiefs appointed an inter-Service fact-finding panel chaired by Admiral James O. Richardson to carry out an in-depth appraisal.[45] In April 1945, after a 10-month investigation conducted largely through interviews, the committee overwhelmingly endorsed unifying the Armed Forces under a single department of national defense. Though composed of separate military branches for land, sea, and air warfare, the unified Department would have a single civilian secretary. A uniformed chief of staff would oversee military affairs and act as the Department's liaison with the President, performing a role similar to Admiral Leahy's. The committee's lone dissenter was its chairman, Admiral Richardson. As a harbinger of the bitter debates to come, he proclaimed the plan "unacceptable" on the grounds that a single department was likely to be dominated by the Army and the Air Force and could end up short-changing the Navy and stripping it of its air component. Arguing essentially for the status quo, Richardson urged restraint until the "lessons" of the recent war had been "thoroughly digested." Until then, he favored preserving the Joint Chiefs of Staff and their committee structure in their current form and using that as the basis for expanding inter-Service coordination after the war.[46]

Even though the Joint Chiefs had authorized the Richardson Committee study, they could reach no consensus on its findings. Rather than resolving differences, the study had exacerbated them, revealing a sharp cleavage between the War Department members (Marshall and Arnold), who favored the single department approach, and the Navy members (Leahy and King), who preferred the current system. Unable to come up with a unanimous recommendation, the JCS agreed to disagree and on October 16, 1945, sent their "split" opinions to the White House. While the debate over Service unification was far from over, the JCS took no further part in it as a corporate body.[47]

A similar sense of trepidation characterized the Joint Chiefs' approach to political-military affairs. Initially, Admiral Leahy, the President's military Chief of Staff, believed it inappropriate for officers in the armed Services to offer opinions on matters outside their realm of professional expertise. Convinced that the JCS should tread carefully, he objected as a rule to military involvement in "political" matters.[48] Actually, Leahy's position at the White House drew him into daily contact with military issues having political and diplomatic impact, as had his recent assignment as Ambassador to Vichy, France. Nonetheless, Leahy's outlook was fairly typical of military officers of his generation, whose mindsets were rooted in a professional ethos and concept of civil-military relations dating from the late 19th century. Once in place, this attitude was hard to dislodge.[49]

The Joint Chiefs became caught up in political-military affairs not because they wanted to, but because they had no choice. Like his military advisors, President Roosevelt put the needs of the war first and preferred to relegate postwar issues relating to a peace settlement and other political matters to the back burner. This approach worked for a while, but by the Tehran Conference of November 1943, the pressure was beginning to build for the administration to clarify its position on a growing number of subjects. As an overall solution, Roosevelt put his faith in the creation of a new international security organization—the United Nations (UN)—to sort out postwar problems. But there were many issues that would need attention before the UN was up and running. At the same time, Roosevelt's deteriorating health—carefully shielded from the public—left him with less and less stamina, so that by the spring of 1944, his workdays were down to 4 hours or less.[50] In those circumstances, it was often up to the Joint Chiefs of Staff to help fill the void by contributing to the postwar planning process. For most of the war, the Joint Chiefs had neither their own organization for political-military affairs nor ready access to interagency machinery for handling such matters. At the outset of the war, the only formal mechanism for interdepartmental coordination was the Standing Liaison Committee, composed of the Army Chief of Staff, the Chief of Naval Operations, and the Undersecretary of State. Established in 1938, the Standing Liaison Committee operated under a vague charter that gave it broad authority to bring foreign policy and military plans into harmony. Its main contribution was to give the military chiefs an opportunity to learn trends in State Department thinking, and vice versa. Rarely did it deal with anything other than political and military relationships in the Western Hemisphere. After Pearl Harbor, it met infrequently, finally going out of business in mid-1943.[51]

In the absence of formal channels, coordination between the Joint Chiefs and the foreign policy community became haphazard. To help bridge the gap, the JCS accepted an invitation from the State Department to establish and maintain liaison through the Joint Strategic Survey Committee, initially to further the work of State's Postwar Foreign Policy Advisory Committee.[52] Seeking to expand these contacts, the Joint War Plans Committee recommended in late May 1943 that the State Department designate a part-time representative to advise the joint staff, arguing that it was "impossible entirely to divorce political considerations from strategic planning." Going a step further, Brigadier General Wedemeyer, a key figure in the Army's planning staff, thought State should have an associate member on the Joint Staff Planners who could also participate in JCS meetings "when papers concerned with national and foreign policies are on the agenda."[53]

Nothing immediately came of these proposals. But by spring 1944, the chiefs found themselves taking a closer look at the question of political-military consultation.

Their first concerns were to provide guidance to the European Advisory Commission (EAC), an ambassadorial-level inter-Allied committee operating from London, with a mandate to make recommendations on the termination of hostilities, and to help settle a growing list of disputes between the Western powers and the Soviet Union over the future political status of Eastern Europe. In assessing the prospects for a durable peace, the Joint Chiefs cautioned the State Department in May 1944 that the "phenomenal" wartime surge in Soviet military and economic power could make for trouble in devising effective security policies in the postwar period. In particular, the chiefs saw a high probability of friction between London and Moscow that could require U.S. intervention and mediation. While the chiefs downplayed the likelihood of a conflict between the Soviet Union and the West, they acknowledged that should one erupt, "we would find ourselves engaged in a war which we could not win even though the United States would be in no danger of defeat and occupation." Far more preferable, in the chiefs' view, would be the maintenance of "the solidarity of the three great powers" and the creation of postwar conditions "to assure a long period of peace."[54]

With growing awareness that postwar problems would require a greater measure of attention, the Joint Chiefs in June 1944 created the Joint Post-War Committee (JPWC) under the Joint Strategic Survey Committee, to work with State and the EAC on surrender terms for Germany and to prepare studies and recommendations on postwar plans, policies, and other problems as the need arose.[55] The JPWC proved a disappointment, however, due to its inability to process recommendations in a timely manner.[56] The problem was especially acute with respect to the development of a coherent policy on the postwar treatment of Germany, an issue brought to the fore by rumors of Germany's impending collapse in the early fall of 1944 and the intervention in the policy process of the President's close personal friend, Secretary of the Treasury Henry Morgenthau, Jr. Lest Germany rise again to threaten the peace of Europe, Morgenthau proposed severely restricting its postwar industrial base, and at the second Quebec Conference (OCTAGON), in September 1944, he persuaded Roosevelt and Churchill to embrace a plan calling for Germany to be converted into a country "primarily agricultural and pastoral in its character."[57] There followed a lengthy debate, with Secretary of War Stimson leading the opposition to the Morgenthau plan, that left the policy toward Germany in limbo for the next 6 months. Eventually, a watered-down version of the Morgenthau plan prevailed, in part because its hands-off approach toward the postwar German economy appealed to the JCS and civil affairs officers in the War Department as the easiest and most expeditious policy to administer in light of requirements for redeploying U.S. forces from Europe to the Pacific.[58]

To help break the impasse over the treatment of Germany and to avoid similar bottlenecks in the future, the Secretaries of State, War, and Navy created a committee of key subordinates to oversee political-military affairs. Activated in December 1944, the State-War-Navy Coordinating Committee (SWNCC) operated at the assistant secretary level and resembled an interagency clearinghouse. By January 1945, it had functioning subcommittees on Europe, the Far East, Latin America, and the Near and Middle East. An Informal Policy Committee on Germany (IPCOG), organized separately to accommodate the Treasury's participation, handled German affairs. To simplify administration, SWNCC and IPCOG shared the same secretariat.[59] Though the JCS played little part in the policy debate over Germany, their command and control responsibilities gave them authority over the U.S. military occupation, which was run under a JCS directive (JCS 1067).[60]

The postwar treatment of Germany was only one of a growing list of political-military issues involving the JCS as the war wound down. By the time of the second Big Three conference at Yalta, in February 1945, the only military-strategic issue of consequence on the chiefs' agenda was the timing of the Soviet entry into the war against Japan. Otherwise, as the chiefs' pre-conference briefing papers suggest, JCS attention focused either on immediate operational matters growing out of strategic decisions taken earlier, or pending administrative, political, and diplomatic issues that were expected to arise from Germany's surrender, the allocation of postwar zones of occupation in Germany and Austria, shipping requirements for the redeployment of Allied forces, and disarming the Axis. Less than 6 months later, when the Big Three resumed their deliberations at Potsdam, their third and final wartime summit conference, political and diplomatic issues clearly dwarfed military and strategic matters. JCS planners, in preparing for the conference, were hard pressed to find enough topics to fill the Combined Chiefs of Staff expected agenda, not to mention a meeting with the Soviet military chiefs.[61]

Throughout most of World War II, the Joint Chiefs viewed themselves as, first and foremost, a military planning and advisory body to the President. But as they prepared to enter the postwar era, they found their mandate changing to encompass not only military plans and strategy, but also related issues with definite political and diplomatic implications. To be sure, as the postwar era beckoned, the Joint Chiefs still had an abundance of military and related security matters before them. Never again, however, would military policy and foreign policy be the separate and distinct entities they had seemed to be when the war began.

ENDING THE WAR WITH JAPAN

While addressing problems of the coming peace, the Joint Chiefs of Staff still faced difficult wartime decisions, none more momentous than those affecting the final stages of the war in the Pacific. Since the early days of the war, the Joint Chiefs had pursued a double-barreled strategy against Japan that allowed MacArthur to conduct operations in New Guinea and the Bismarck Islands, while Nimitz rolled back the Japanese in the Central Pacific. Under the agreed worldwide allocation of shipping and landing craft set by the CCS, Nimitz's operations had a prior claim over MacArthur's whenever there were conflicts over timing of operations and the allocation of resources. But by early 1944, as the two campaigns began to converge, a debate developed on how and where to conduct future operations. At issue was whether to follow MacArthur's advice and make the liberation of the Philippines the primary objective in the year ahead, or to follow a plan favored by Nimitz of bypassing the Philippines for the most part and concentrating on the Marianas as a stepping stone toward seizing Formosa, from which U.S. forces could link up with the Chinese for the final assault on Japan.[62]

Of the options on the table, the Joint Chiefs considered the Formosa strategy the most likely to succeed in bringing U.S. forces closer to Japan and shortening the war.[63] To carry it out effectively, however, they would have to reconsider the dual command arrangements that had prevailed since the start of the war and to adopt a single, comprehensive Pacific strategy, something that neither MacArthur nor Nimitz was yet ready to accept. Most intransigent of all was MacArthur. Treating the Formosa operation as a diversion, MacArthur insisted that the liberation of the Philippines was a "national obligation." With a strong personal interest in the outcome, he was determined to see the expulsion of the Japanese from the entire Philippine archipelago through to the end.[64]

In July 1944, President Roosevelt paid a personal visit to Pearl Harbor for face-to-face meetings with MacArthur and Nimitz "to determine the next phase of action against Japan." The only JCS member to accompany him was Admiral Leahy, whose part in the deliberations was minor. In fact, the discussions were inconclusive; by the time they ended, President Roosevelt seemed inclined to support MacArthur's position. Nimitz took the hint and, shortly after the conference adjourned, he directed his staff to take a closer look at attacking Okinawa as a substitute for invading Formosa.[65] While King and Leahy continued to hold out for Formosa, a shortage of support troops and the prospects of a lengthy campaign there persuaded the Joint Staff Planners by late summer 1944 that the prudent course was to postpone a final decision on Formosa pending the outcome of initial operations in the southern Philippines.[66] This became, in the absence of the Joint Chiefs' ability to settle on a better solution, the accepted course of action and more or less assured MacArthur that he could move on to liberate the rest of the Philippines in due course. The coup de grace was Nimitz's decision, which he conveyed to King at a face-to-face meeting in San Francisco in September 1944, to shelve plans for a Formosa invasion and to focus on taking Okinawa. With this, the die was cast and on October 3, 1944, the

JCS approved a directive to MacArthur setting December 20 as the target date for invading Luzon and marching on to Manila.[67]

Clearly, in this instance, the views of the theater commanders had prevailed over those of the Joint Chiefs, an increasingly common phenomenon in the latter stages of the war and a preview of the influential role that combatant commanders would play in the postwar era. Left unresolved and somewhat obscured by the Philippines-versus-Formosa imbroglio was the final strategy for the defeat of Japan and whether to plan a full-scale invasion of the Japanese home islands. Initial discussion of these issues dated from the summer of 1944 when, in response to a preliminary review of options by the Joint War Plans Committee, Admiral Leahy mentioned the possibility of bringing about Japan's surrender through intensive naval and air action rather than through a landing of troops.[68] Over the following months, as MacArthur moved up the Philippines and Nimitz prepared his attack against Okinawa, Japan's situation steadily deteriorated. By late 1944–early 1945, with the home islands now within reach of Twentieth Air Force's B–29s operating from the Marianas and with the Navy conducting an unrelenting war at sea and a naval blockade, the outcome of the conflict was no longer in doubt. Though Japan's armed forces could still mount tenacious resistance, they were clearly engaged in a losing cause.

As the pressure on Japan mounted, so did conjecture within the joint staff about the means of achieving victory. Prodded by their superiors, Navy planners were especially reluctant to consider an invasion inevitable until air and naval attacks and the blockade had run their course. To Leahy, King, and Nimitz, it seemed "that the defeat of Japan could be accomplished by sea and air power alone, without the necessity of actual invasion of the Japanese home islands by ground troops."[69] Weighing the pros and cons, the Joint Staff Planners acknowledged in late April 1945 that while a case could indeed be made for a strategy of blockade and saturation bombardment, prudence dictated moving ahead with preparations for an invasion as the most likely course of action to assure Japan's unconditional surrender.[70]

On May 10, 1945, the Joint Chiefs gave the go-ahead for planning to continue for the invasion, while noting several objections and reservations raised by Admiral King.[71] The overall concept (code-named DOWNFALL) was a collaborative effort between the joint staff and the major Pacific commands. It called for the attack to take place in two stages: an initial invasion of southern Kyushu (Operation *Olympic*) toward the end of 1945, followed by a landing in the spring of 1946 on Honshu (Operation *Coronet*) in the vicinity of the Tokyo (Kanto) Plain, once reinforcements arrived from Europe. Still to be decided were final command arrangements, which the JCS had neatly sidestepped during the Philippines-versus-Formosa debate. Avoiding the issue once again, the chiefs in early April 1945 approved an interim assignment of responsibilities, under which MacArthur would serve as commander in chief of all Army land forces while Nimitz commanded all theater naval forces. Strategic air assets would remain essentially as they were since the creation of the Twentieth Air Force a year earlier, under the strategic direction of the Joint Chiefs of Staff, with General Arnold as executive agent, but available to General MacArthur as needed.[72] The Joint Chiefs expected the looming invasion of Japan to be their biggest operation of the war, dwarfing the D-Day invasion of Europe. Anticipating strong resistance, Operation *Olympic* proposed a 12-division assault force, with 8 divisions in reserve. *Coronet* would be even bigger, with 14 divisions in the initial invasion and 11 more in following echelons. By comparison, the D-Day landings at Normandy had involved an initial assault force of eight divisions—five American, two British, and one Canadian. Altogether, *Olympic* and *Coronet* would require more than a million ground troops, 3,300 aircraft, and over 1,000 Navy combatant vessels.[73]

Missing from these plans were hard estimates of U.S. casualties. Those under consideration at the time were extrapolated from earlier Pacific campaigns by the Joint War Plans Committee, which predicted U.S. losses ranging from 25,000 killed and 105,000 wounded for an invasion of Kyushu alone, to 46,000 dead and 170,000 wounded for attacks on Kyushu and the Tokyo Plain combined.[74] To draw off defenders, the joint staff in May–June 1945 put together a deception plan (*Broadaxe*) to convince the Japanese that there would be no invasion prior to 1946, or until U.S. forces had consolidated control of Formosa, the China coast, and Indochina, and the British had liberated Sumatra.[75] Yet even if the deception worked, Admiral King believed that an invasion of the home islands would still meet stronger resistance than any previously encountered and that the joint staff should calculate its casualty figures accordingly.[76] In view of the methodological problem Admiral King raised, the Joint Staff Planners decided to withhold an estimate of casualties, stating only that losses were "not subject to accurate estimate" but would be at least on a par with those elsewhere in the Pacific Theater, which tended to be higher than in Europe.[77]

DAWN OF THE ATOMIC AGE

Also absent from U.S. invasion plans was an assessment of the impact of the atomic bomb, still a super-secret project outside the purview of the joint staff. Launched in October 1939, the atomic bomb program had come about as insurance against research being done in Nazi Germany, where scientists a year earlier had demonstrated a process known as "nuclear fission." While the Germans were apparently slow to grasp the full importance of what they had achieved, their colleagues elsewhere in Europe and the United States speculated that, under properly controlled conditions, nuclear fission could produce enormous explosive power. Among those alarmed by the German breakthrough were Leo Szilard, a Hungarian expatriate, and Enrico Fermi, a refugee from Mussolini's Fascist Italy, both living in the United States. Unable to interest the Navy Department in a program of stepped-up nuclear research, they persuaded Albert Einstein, the celebrated physicist, to send a letter (written by Szilard) to President Roosevelt, drawing attention to the German experiment and suggesting the possibility of "extremely powerful bombs of a new type." Roosevelt agreed that the United States needed to act, and from that point forward the program grew steadily to become the Manhattan Engineer District (MED), with the War Department covertly funding and overseeing the effort.[78]

The Joint Chiefs of Staff learned of the atomic bomb project individually, at different times during the course of the war. The first to be brought in on the secret was General Marshall, who became involved in 1941 as a member of the President's Top Advisory Group, which was nominally responsible for overseeing the program.[79] Marshall told Admiral King about the project late in 1943, but according to King, the subject was still too sensitive to be placed on the chiefs' agenda or discussed at meetings.[80] General Arnold had suspected for some time that something was afoot, and received confirmation from the MED director, Brigadier General Leslie R. Groves, in July 1943. Toward the end of March 1944, Groves gave Arnold a more in-depth description of the project and a list of tentative requirements.[81] The last to learn about the bomb was Admiral Leahy, who was not apprised until September 1944 when he attended the second Quebec Conference. Afterwards, he received a full briefing at the President's home in Hyde Park, New York, by Vannevar Bush, Director of the Office of Scientific Research and Development and scientific coordinator of the project.[82]

Whether the Manhattan Project would yield a workable weapon was an open question for much of the war. Convinced that the project had merit, Bush assured President Roosevelt as early as July 1941 that the explosive potential of an atomic bomb would be "thousands of times more powerful" than any conventional weapon and that its use "might be determining."[83] Leahy, on the other hand, scoffed at Bush's claims and thought the effort would never amount to much. "The bomb will never go off," he insisted, "and I speak as an expert in munitions."[84] Even though the other members of the JCS appeared not to share Leahy's skepticism, they were still cautious and knew better than to incorporate a nonexistent weapon into their strategic calculations. Nor was it clear, even if the bomb worked, exactly when it would be available and in what quantities. According to Groves, the earliest date for a prototype was around August 1, 1945, with a second bomb to follow 5 months later.[85] As it turned out, the first atomic test took place July 16, 1945, 2 months after Germany's capitulation and well into the planning cycle for the invasion of Japan. Until then, lacking confirmation of the bomb's capability, the JCS could count on nothing more than an expensive program wrapped in secrecy that might or might not change the course of history.

Despite JCS uncertainty over whether the bomb would work, preparations for its possible use received top priority from March 1944 onward, when Groves briefed Arnold on the project. Expecting the bomb to be of considerable size and weight, Groves speculated that, for delivery purposes, it might be necessary to use a British Lancaster heavy bomber, the largest plane of its kind in the Allied inventory, which could carry a payload of up to 22,000 pounds. Arnold strenuously objected to using a British plane and insisted that the AAF could provide a suitable delivery platform from a modified B–29. From this discussion emerged Project SILVERPLATE, which produced the 14 specially configured B–29s that made up 313th Bombardment Wing of 509th Composite Group, the unit that carried out the attacks on Hiroshima and Nagasaki in August 1945.[86]

Composed of carefully selected top-rated pilots and crews, 509th was the most elite unit in the Army Air Forces. Eventually it became part of the Twentieth Air Force, though for all practical purposes it operated independently and was responsible to Groves and the MED. As a composite group, 509th carried with it most of its own logistical support and was by design a stand-alone organization. Training began in early September 1944 in utmost secrecy at Wendover Field, an isolated air base in western Utah within easy reach of the MED's weapons research laboratory at Los Alamos, New Mexico. Crews concentrated on learning to drop two different weapons—a cylindrical uranium bomb called "Little Boy" and a rotund plutonium bomb called "Fat Man." The initial plan was to use nuclear bombs against Germany. But as it became apparent that the war in Europe might end before they were ready, 509th turned its attention to the Pacific in December 1944 and spent the next 2 months conducting test flights over Cuba to familiarize crews with terrain similar to Japan's. In May 1945, advance elements of the 509th began arriving at their staging base on Tinian, one of the Marianas, to dig the pits from which the bombs would be hoisted into the planes. Pilots and crews arrived soon thereafter and by late July were executing combat test strikes over Japan with high-explosive projectiles of the Fat Man design.[87]

Roosevelt's death on April 12, 1945, left the fate of the Manhattan Project in the hands of his successor, Harry S. Truman. Though a bomb had yet to be manufactured and tested, the project was far enough along that Truman was reasonably certain it would succeed. What remained to be seen was how powerful the explosive device would be. In early May, on Secretary of War Stimson's initiative, Truman authorized the War Department to create an interdepartmental Interim Committee to recommend policies and plans for using the bomb and related issues.[88] Separately, a committee of technical experts chaired by Groves began to assemble a list of targets.

Omitted from both groups was any formal JCS representation, though Marshall received regular updates from Stimson on the Interim Committee's progress and even attended one of its meetings on May 31, 1945. How much, if any, of this information Marshall conveyed to the other chiefs is unknown. According to Groves, the omission of the Joint Chiefs was intentional, to preserve security and, no less important, to avoid having to deal with Leahy's negative views.[89]

With the atomic bomb still in gestation and blanketed in secrecy, the Joint Chiefs continued to ignore it in their plans for ending the war with Japan. Meeting with the new President and the Service Secretaries on June 18, 1945, they described in some detail the preparations for the invasion, discussed the probability of heavy casualties, and agreed that Soviet intervention would be desirable but not essential for winning the war. Characterizing Japan's situation as "hopeless," the JCS estimated that it would only worsen under the continuing onslaught of the blockade and accompanying air and naval bombardment. In Marshall's opinion, however, air and sea attacks would not suffice to bring about a Japanese surrender, a view in which Admiral King now grudgingly concurred. What caused King to come around is not apparent from the official record, but it may have been recent ULTRA radio intercepts, to which all at the meeting had access. These indicated an accelerated buildup of Japanese forces on Kyushu and a feverish determination by the Japanese high command to mount a last-ditch stand using heavily dug-in forces and suicide air attacks.[90] Despite sending out peace feelers, the Japanese showed no sign of giving up. Instead, the military leaders appeared intent on inflicting such heavy damage and casualties on the United States that it would see the futility of further fighting and seek a negotiated peace. Even skeptics like King seemed to agree that an invasion was the only viable option for obtaining Japan's surrender. Truman was visibly distraught over the prospects of a bloodbath, but by the time the meeting broke up he saw no other choice and ordered planning for the Kyushu operation to proceed.[91]

Whether the use of nuclear weapons as a possible alternative to an invasion was discussed at this meeting is unclear. While the formal minutes make no mention of the atomic bomb, they indicate an interest on Stimson's part in finding a political solution for ending the war and an off-the-record discussion of "certain other matters."[92] Assistant Secretary of War John J. McCloy, who accompanied Stimson, recalled raising the issue of sending the Japanese an ultimatum, urging them to surrender or be subjected to a "terrifyingly destructive weapon." McCloy remembered that the JCS were "somewhat annoyed" by his interference and veiled reference to the bomb, but that President Truman "welcomed it" and directed that such a political initiative be set in motion. However, Secretary of the Navy James V. Forrestal, who was also present, had no recollection of McCloy's remarks and reckoned that the discussion McCloy had in mind took place at another time.[93]

Planning for military action against Japan now followed a two-track course, one along the lines laid out by the Joint Chiefs in preparation for an invasion, the other driven by the gathering momentum of the Manhattan Project. Both came together at the Potsdam Conference (TERMINAL) in July–August 1945, where Truman and the JCS received word of the successful test shot held near Alamogordo, New Mexico. By then, Truman had also received the recommendations of the Interim Committee, which favored using the bomb if the experiment succeeded. The expense of having developed the bomb in the first place, the potential diplomatic leverage it offered in dealing with the Russians, and last but not least the elimination of the need for a bloody invasion, all doubtless weighed heavily on Truman's mind. Once he had confirmation that the bomb would work, the decision to use it became almost automatic.[94] Looking back, Leahy and King strongly disagreed with the President's choice. Insisting that the enemy's collapse was only a matter of time, they considered attacks with atomic weapons excessive and unnecessary. Still, there is no evidence that either stepped forward to propose a different course. If Leahy and King objected at the time, they kept their reservations to themselves.[95]

The only JCS member who seriously considered an alternative course of action was Marshall. Like King and Leahy, Marshall hoped the Japanese would see the light and surrender, making use of the atomic bomb unnecessary. The difficulty arose in finding a way of bringing the Japanese around. During the Interim Committee's deliberations prior to Potsdam, Marshall and Stimson discussed the possibility of issuing an explicit warning before dropping the bomb or of confining its use to a demonstration over uninhabited terrain. But they could see no practical way of assuring that the Japanese would be sufficiently awed by either a warning or a demonstration shot to draw the logical conclusion and concede defeat.[96] According to his biographer, Forrest C. Pogue, Marshall's main concern was to wind up the war quickly with as few casualties as possible to either side; on this basis he came to the conclusion that if the test at Alamogordo turned out to be a success, the bomb should be used against targets in Japan.[97]

The attacks that followed, destroying Hiroshima on August 6, 1945, with the Little Boy gun-type uranium bomb and Nagasaki, 3 days later, with the Fat Man plutonium implosion bomb, forced Japanese military leaders to acknowledge that they had no countermeasures to the Americans' new weapons. In between these attacks, on August 8, 1945, the Soviet Union declared war on Japan and invaded Manchuria. For years, historians debated whether the atomic bombs were decisive in bringing the war to an end. Recently, however, a Japanese scholar has conjectured that while it was the atom bomb that convinced the Japanese high command that the war was lost, it was not until the Soviets invaded Manchuria that Japan's civilian leadership came to the same conclusion, since without the USSR there was no one left to mediate an end of the war. In other words, a convergence of events—the atomic bombing of Japan and the Soviet Union's entry into the war at the same time—provided the catalyst for Japan's surrender.[98] Yet of these two sets of events, it was the use of the atomic bomb that produced the most lasting impressions—tens of thousands killed and injured, two cities destroyed, and an entire nation lying at the mercy of another. Without question, the attacks on Hiroshima and Nagasaki confirmed the predictions of Stimson, Groves, and others associated with the Manhattan Project that atomic weapons were indeed more awesome in their destructive power than any existing weapon. Whether they would revolutionize warfare and produce, as Stimson predicted, "a new relationship of man to the universe," was another matter.[99]

Shortly after the attacks, at the chiefs' request, the Joint Strategic Survey Committee presented its assessment of the atomic bomb's military and strategic impact. At issue was whether, as some military analysts were beginning to speculate, atomic weapons would preclude the need for sizable conventional forces after the war. Though duly impressed with the atomic bomb's destructive power, the committee pointed out that these weapons were as yet too few in number, too expensive and difficult to produce, and too hard to deliver to be used in anything other than special circumstances. In view of these unique characteristics, the committee doubted whether atomic weapons would render conventional land, sea, and air forces obsolete, though they might change the "relative importance and strength of various military components." Any immediate changes were apt to be minor, however, as long as the United States enjoyed a monopoly on the bomb. This situation could change if other industrialized countries—the United Kingdom and the Soviet Union—wanted to devote the time and resources to developing nuclear weapons. The most dangerous and destabilizing situation that the Joint Strategic Survey Committee could foresee was if the Soviet Union acquired the bomb. Even so, the committee downplayed the likelihood of a dramatic transformation in modern warfare resulting from the proliferation of nuclear technology. It pointed out that the development of "new weapons" had been continuous throughout history and that the advent of one new weapon invariably produced something equally effective to counter it.[100]

Thus, as the war drew to a close, the Joint Chiefs found themselves entering the uncharted realm of atomic war, somewhat reassured that the apocalypse predicted by Stimson and likeminded others had been postponed, yet cautious and uneasy at the same time. No less unsettling was the Joint Chiefs' own uncertain future as an organization. At the outset of World War II, the Joint Chiefs of Staff had not existed. By 1945, they were an established fixture atop the largest, most powerful military machine in history. Despite inter-Service friction and competition, the JCS had found that working together produced better results than working separately. A corporate advisory and planning body, they reported directly to the President and were at the center of decision throughout the conflict. Operating without a formal charter, the Joint Chiefs were at liberty to conduct business as needed to meet the requirements of the war. With the onset of peace, this free-wheeling style was sure to change. Still, few seriously contemplated a postwar defense establishment in which the Joint Chiefs of Staff, or some comparable organization, did not loom large.

NOTES

1 Maurice Matloff, *Strategic Planning for Coalition Warfare, 1943–44* (Washington, DC: Office of the Chief of Military History, 1959), 398.

2 See Forrest C. Pogue, *The Supreme Command* (Washington, DC: Center of Military History, 1954), 41–42.

3 "Directive to the Supreme Commander in the ABDA Area as Approved by the President and PM," January 10, 1942, ABC-4/5; and "Procedure for Assumption of Command by General Wavell," January 16, 1942, ABC-4 C/S 3, both in *World War II Conference Papers*.

4 King quoted in Robert W. Love, Jr., *History of the U.S. Navy, 1942–1991* (Harrisburg, PA: Stackpole Books, 1992), 14. See also Ernest J. King and Walter Muir Whitehill, *Fleet Admiral King: A Naval Record* (New York: W.W. Norton, 1952), 368–372.

5 Minutes, JCS 6th Meeting, March 16, 1942, RG 218, CCS 334 (3-16-42).

6 Louis Morton, "Pacific Command: A Study in Interservice Relations," in Harry R. Borowksi, ed., *The Harmon Memorial Lectures in Military History, 1959–1987* (Washington, DC: Office of Air Force History, 1988), 134.

7 Louis Morton, *Strategy and Command: The First Two Years* (Washington, DC: Office of the Chief of Military History, 1962), 244–250.

8 Williamson Murray and Allan R. Millett, *A War to Be Won* (Cambridge: Harvard University Press, 2000), 207.

9 Report by JPS, August 6, 1943, "Specific Operations in the Pacific and Far East, 1943–44," JCS 446. See also Steven T. Ross, *American War Plans, 1941–45* (London: Frank Cass, 1997), 68–69.

10 Henry H. Adams, "Fleet Admiral Chester W. Nimitz," in Michael Carver, ed., *The War Lords* (Boston: Little, Brown and Co., 1976), 411; E.B. Potter, *Nimitz* (Annapolis: Naval Institute Press, 1976), 235–56, 279–297.

11 King and Whitehill, 491–493.

12 Love, II, 199–200; Morton, *Strategy and Command*, 447–453.

13 Wesley Frank Craven and James Lea Cate, eds., *The Army Air Forces in World War II*, vol. V, *The Pacific: Matterhorn to Nagasaki* (Washington, DC: Office of Air Force History, 1983; reprint), 3–32; Morton, *Strategy and Command: The First Two Years*, 602–603.

14 Ross, *American War Plans, 1941–45*, 50.

15 Grace Person Hayes, *The History of the Joint Chiefs of Staff in World War II: The War Against Japan* (Annapolis: Naval Institute Press, 1982), 725.

16 Forrest C. Pogue, *George C. Marshall: Organizer of Victory, 1943–1945* (New York: Viking Press, 1973), 206.

17 Matloff, 231.

18 Minutes, CCS 107th Meeting, August 14, 1943, *World War II Conference Papers*; King and Whitehill, 483–484.

19 "Progress Report to the President and Prime Minister," August 27, 1943, CCS 319/2 (revised), *World War II Conference Papers*.

20 See Tohmatsu Haruo, "The Strategic Correlation between the Sino-Japanese and Pacific Wars," in Mark Peattie, Edward J. Drea, and Hans van de Ven, eds., *The Battle for China: Essays on the Military History of the Sino-Japanese War of 1937–1945* (Stanford, CA: Stanford University Press, 2011), 424.

21 Herbert Feis, *The China Tangle* (Princeton, NJ: Princeton University Press, 1953), 11–13; Charles F. Romanus and Riley Sunderland, *Stilwell's Mission to China* (Washington, DC: Center of Military History, 2002, reprint), 61–63; Joint Planning Committee Report to Chiefs of Staff, January 10, 1942, "Immediate Assistance to China," U.S. ABC-4/6, ARCADIA Conference, *World War II Conference Papers*.

22 Quoted in Barbara W. Tuchman, *Stilwell and the American Experience in China, 1911–45* (New York: Macmillan, 1970), 125.

23 Eric Larrabee, *Commander in Chief: Franklin Delano Roosevelt, His Lieutenants, and Their War* (Annapolis: Naval Institute Press, 1987), 513–515.

24 Hayes, 80; Feis, *China Tangle*, 15–16.

25 Ulysses Lee, *The Employment of Negro Troops* (Washington, DC: Center of Military History, 2000), 610.

26 Charles F. Romanus and Riley Sunderland, *Stilwell's Command Problems* (Washington, DC: Center of Military History, 1987; reprint), 454.

27 See Zang Yunhu, "Chinese Operations in Yunnan and Central Burma," in Peattie, Drea, and van de Ven, eds., *Battle for China*, 386–391.

28 Craven and Cate, I, 504–505; IV, 436–443.

29 Theodore H. White, ed., *The Stilwell Papers* (New York: William Sloane Associates, 1948), 136–138; Romanus and Sunderland, *Stilwell's Mission to China*, 212–221; Romanus and Sunderland, *Stilwell's Command Problems*, 471.

30 Matloff, 238.

31 Hayes, 518.

32 Winston S. Churchill, *Closing the Ring* (Boston: Houghton Mifflin, 1951), 328; Minutes, 1st Plenary Meeting, Villa Kirk, Cairo, November 23, 1943, *World War II Conference Papers*.

33 Matloff, 350–351.

34 Report by JPS, April 6, 1944, and decision on April 10, 1944, "VLR Bombers in the War Against Japan," JCS 742/6. Like all B-29 units in World War II, XX Bomber Command was part of the Twentieth Air Force, which reported to the JCS. Bombing missions

from China began in July 1944 and concentrated initially on Japanese industrial targets in Manchuria, gradually expanding to targets in Japan as XX Bomber Command gained experience. Chiang's government objected, however, to what it regarded as a diversion of resources; it wanted the fuel and munitions used by XX Bomber Command to go to Chennault's air force. By the end of January 1945, B–29 bombing operations from China ceased owing to the deteriorating security situation. See Craven and Cate, V, 3–32, 92–131.

35 Feis, *China Tangle*, 192. In fact, no lend-lease aid ever reached the Communists. See Zhang Baijia, "China's Quest for Foreign Military Aid," in Peattie, Drea, and van de Ven, eds., *Battle for China*, 299.

36 Roosevelt to Chiang, September 16, 1944, in U.S. Department of State, *Foreign Relations of the United States: The Conference at Quebec, 1944* (Washington, DC: GPO, 1972), 465.

37 Romanus and Sunderland, *Stilwell's Command Problems*, 468–471.

38 Feis, *China Tangle*, 201–202; Pogue, *Organizer of Victory*, 478–479; Romanus and Sunderland, *Time Runs Out in CBI* (Washington, DC: Center of Military History, 1999; reprint), 15.

39 Matloff, 240.

40 JCS Statement, "Basis for the Formulation of a Military Policy," September 20, 1945, JCS 1496/3.

41 Army and AAF planners were emphatic that the lack of guidance on UMT and postwar organization was a major impediment; Navy planners were less convinced. See Memo, Arnold to JCS, September 7, 1945, "Reorganization of National Defense, JCS 749/17; and Memo, King to JCS, September 10, 1945, "Reorganization of National Defense," JCS 749/18.

42 Memo, Marshall to JCS, September 19, 1945, "Interim Plan for the Permanent Establishment of the Army of the United States," with enclosures, JCS 1520; Memo, Arnold to JCS, October 2, 1935, "Interim Plan for the Permanent Military Establishment of the United States," JCS 1478/4; *Annual Report of the Secretary of the Navy, 1945* (Washington, DC: Department of the Navy, January 10, 1946), 3.

43 Memo, Marshall to JCS, November 2, 1943, "A Single Dept of War," JCS 560.

44 Memo, King to JCS, November 7, 1943, "A Single Department of War in the Post-War Period," JCS 560/1; Herman S. Wolk, *The Struggle for Air Force Independence, 1943–1947* (Washington, DC: Air Force History and Museums Program, 1997), 40 and passim.

45 Report by JSSC, March 8, 1944, "Reorganization of National Defense," JCS 749; Enclosure to Letter, Leahy (for JCS) to Secretaries of War and Navy, May 9, 1944, JCS 749/6. See also the Memo by Richardson Committee, October 19, 1944, "Tentative: Origin and Activities of the JCS Special Committee for the Reorganization of National Defense," JCS 749/14.

46 The Special Committee's recommendations and Richardson's dissenting opinions are filed together in "Report of the Joint Chiefs of Staff Special Committee for Reorganization of National Defense," April 11, 1945, JCS 749/12.

47 Memo, Leahy to Truman, October 16, 1945, "Reorganization of National Defense," JCS 749/29.

48 See Mark A. Stoler, *Allies and Adversaries: The Joint Chiefs of Staff, the Grand Alliance, and U.S. Strategy in World War II* (Chapel Hill: University of North Carolina Press, 2000), 130, 138.

49 For a fuller discussion, see Samuel P. Huntington, *The Soldier and the State: The Theory and Politics of Civil-Military Relations* (New York: Random House, 1957), still the classic work of the subject. Leahy was not very good at practicing what he preached. Toward the end of the war, for example, he successfully argued for modification of the Japanese surrender terms to allow retention of the emperor, an issue heavy in political implications.

50 See Robert H. Ferrell, *The Dying President: Franklin D. Roosevelt, 1944–1945* (Columbia: University of Missouri Press, 1998), 72–73 and passim.

51 Ernest R. May, "The Development of Political-Military Consultation in the United States," *Political Science Quarterly* 70 (June 1955), 172–173; Ray S. Cline, *Washington Command Post: The Operations Division* (Washington, DC: Center of Military History, 1990; reprint), 41–42.

52 Harley Notter, *Postwar Foreign Policy Preparation* (Washington, DC: U.S. Department of State, 1949), 76, 124–125.

53 Cline, *Washington Command Post*, 317.

54 Letter, Leahy to Hull, May 16, 1944, in U.S. Department of State, *Foreign Relations of the United States: The Conference of Berlin (The Potsdam Conference), 1945*, 2 vols. (Washington, DC: GPO, 1960), I, 264–266. This series hereafter cited as *FRUS*.

55 "Charter: Joint Post-War Committee," enclosure to Note by the Secretaries, June 7, 1944, JCS 786/2.

56 Cline, *Washington Command Post*, 325.

57 Memo Initialed by Roosevelt and Churchill, March 15, 1944, U.S. Department of State, *Foreign Relations of the United States: The Conference of Quebec, 1944* (Washington, DC: GPO, 1972), 466–467. The words were actually Churchill's, the sentiments Morgenthau's.

58 John Lewis Gaddis, *The United States and the Origins of the Cold War, 1941–1947* (New York: Columbia University Press, 1972), 114–132.

59 Steven L. Rearden, "American Policy Toward Germany, 1944–1946" (Ph.D. Thesis, Harvard University, 1974), 176–177.

60 Paul Y. Hammond, "Directives for the Occupation of Germany: The Washington Controversy," in Harold Stein, ed., *American Civil-Military Decisions: A Book of Case Studies* (Birmingham: University of Alabama Press, 1963), 311–460; Earl F. Ziemke, *The U.S. Army in the Occupation of Germany, 1944–1946* (Washington, DC: Center of Military History, 1975), 98–108, 208–224.

61 Hayes, 713–721.

62 Matloff, 453–459; Hayes, 603–604.

63 Robert Ross Smith, *Triumph in the Philippines* (Washington, DC: Office of the Chief of Military History, 1963), 4–8.

64 Message, MacArthur to Marshall, June 18, 1944, quoted in Hayes, 606.

65 Douglas MacArthur, *Reminiscences* (New York: McGraw-Hill, 1964), 196–198; D. Clayton James, *The Years of MacArthur*, vol. II, *1941–1945* (Boston: Houghton Mifflin, 1975), 529–533; Smith, 9–11.

66 Matloff, 484–485.

67 Smith, 16; Hayes, 623–624.

68 Matloff, 487.

69 King and Whitehill, 598; Hayes, 702; William D. Leahy, *I Was There* (New York: Whittlesey House, 1950), 245, 384–385; and Michael S. Sherry, *The Rise of American Air Power: The Creation of Armageddon* (New Haven: Yale University Press, 1987), 172–173.

70 JPS Report, "Pacific Strategy," April 25, 1945, JCS 924/15.

71 Decision on JCS 924/15, "Pacific Strategy," May 10, 1946.

72 JCS Directive, April 3, 1945, "Command and Operational Directives for the Pacific," JCS 1259/4; Craven and Cate, V, 676–684.

73 Directive to CINC, U.S. Army Forces, Pacific, CINC U.S. Pacific Fleet, CG Twentieth AF, May 25, 1945, "Directive for Operation 'Olympic,'" JCS 1331/3; Report by JPS, June 16, 1945, "Details of the Campaign Against Japan," JCS 1388. John Ray Skates, *The Invasion of Japan: Alternative to the Bomb* (Columbia: University of South Carolina Press, 1994), looks at invasion planning in depth.

74 See Report by the JWPC, June 15, 1945, "Details of the Campaign Against Japan," JWPC 369/1, in Douglas J. MacEachin, *The Final Months of the War With Japan: Signals Intelligence, U.S. Invasion Planning, and the A-Bomb Decision* (Washington, DC: Center for the Study of Intelligence, 1998), 11–12 and Document No. 5. See also Barton J. Bernstein, "A Postwar Myth: 500,000 U.S. Lives Saved," *Bulletin of the Atomic Scientists* 42 (June–July 1986), 38–40.

75 Thaddeus Holt, *The Deceivers: Allied Military Deception in the Second World War* (New York: Scribner, 2004), 746–749.

76 Memo, CNO to JCS, May 2, 1945, "Pacific Strategy," JCS 924/16.

77 Report by JPS, June 16, 1945, "Details of the Campaign Against Japan," JCS 1388.

78 Letter, Einstein to Roosevelt, August 2, 1939; Letter, Roosevelt to Einstein, October 19, 1939, Safe File, PSF, Roosevelt Library. See also Leo Szilard, "Reminiscences," in *Perspectives in American History* 2 (1968), 94–116.

79 Richard G. Hewlett and Oscar E. Anderson, Jr., *A History of the United States Atomic Energy Commission*, vol. I, *The New World* (Washington, DC: U.S. Atomic Energy Commission, 1962), 46, 77. The other members of the Top Advisory Committee were President Roosevelt, who served as chairman, Vice President Henry A. Wallace, Secretary of War Stimson, Vannevar Bush, and James B. Conant. Never once did the full committee meet. See McGeorge Bundy, *Danger and Survival* (New York: Random House, 1988), 45–46.

80 King and Whitehill, 620–621.

81 Walton S. Moody, *Building a Strategic Air Force* (Washington, DC: Air Force History and Museums Program, 1996), 7; H.H. Arnold, *Global Mission* (New York: Harper and Bros., 1949), 491; Leslie R. Groves, *Now It Can Be Told* (New York: Harper and Bros., 1962), 253; and Vincent C. Jones, *Manhattan: The Army and the Atomic Bomb* (Washington, DC: Center of Military History, 1985), 519.

82 Leahy, *I Was There*, 265, 269.

83 Letter, Bush to Roosevelt, July 16, 1941, quoted in Marchtin J. Sherwin, *A World Destroyed: The Atomic Bomb and the Grand Alliance* (New York: Vintage Books, 1977), 36–37.

84 Leahy quoted in Harry S. Truman, *Year of Decisions* (Garden City, NY: Doubleday, 1955), 11. Leahy, *I Was There*, 429, 440, acknowledges that he misjudged the bomb.

85 Memo, Groves to Marshall, December 30, 1944, "Atomic Fission Bombs," U.S. Department of State, *Foreign Relations of the United States: Diplomatic Papers—The Conferences at Malta and Yalta, 1945* (Washington, DC: GPO, 1955), 383–384.

86 Groves, *Now It Can Be Told*, 253–254.

87 Jones, 519–528; Groves, *Now It Can Be Told*, 253–262; "History of 509th Composite Group, 313th Bombardment Wing, Twentieth Air Force, Activation to 15 August 1945" (MS, August 31, 1945, Maxwell AFB, Alabama), 45–50.

88 Entry, May 2, 1945, Meeting with Truman, Stimson Diary, Library of Congress (microfilm).

89 Hewlett and Anderson, 344–345, 356–357; Groves, *Now It Can Be Told*, 271.

90 MacEachin, 6–9; Edward J. Drea, *MacArthur's ULTRA: Codebreaking and the War against Japan, 1942–1945* (Lawrence: University Press of Kansas, 1992), 204–210; Richard B. Frank, *Downfall: The End of the Imperial Japanese Empire* (New York: Random House, 1999), 197–213.

91 Minutes of Meeting Held at the White House, June 18, 1945, filed with JCS 1388.

92 Ibid.

93 Walter Millis, ed., *The Forrestal Diaries* (New York: Viking Press, 1951), 70–71; John J. McCloy, *The Challenge to American Foreign Policy* (Cambridge: Harvard University Press, 1953), 40–43.

94 See Robert H. Ferrell, ed., *Off the Record: The Private Papers of Harry S. Truman* (New York: Harper and Row, 1980), 55–56; and entry, July 18, 1945, Stimson Diary, also cited in U.S. Department of State, *Foreign Relations of the United States: The Conference of Berlin, 1945*, 2 vols. (Washington, DC: GPO), II, 1361.

95 Leahy, *I Was There*, 441; King and Whitehill, 621.

96 Bundy, *Danger and Survival*, 75; Herbert Feis, *The Atomic Bomb and the End of World War II* (Princeton: Princeton University Press, 1966), 45–46.

97 Forrest C. Pogue, *George C. Marshall: Statesman, 1945–1959* (New York: Viking Penguin, 1987), 24–25.

98 Tsuyoshi Hasegawa, *Racing the Enemy: Stalin, Truman, and the Surrender of Japan* (Cambridge: Harvard University Press, 2005), 177–214.

99 Stimson's comments in Notes of Interim Committee Meeting, May 31, 1945, Misc. Historical Documents Collection, Truman Papers, Truman Library.

100 Report by JSSC to JCS, October 30, 1945, "Over-all Effect of Atomic Bomb on Warfare and Military Organization," JCS 1477/1.

A conference of Secretary of Defense James Forrestal with the Joint Chiefs of Staff was held at the U.S. Naval War College, Newport, Rhode Island, on August 21–22, 1948. Shown at the conference table are, left to right, Major General Alfred M. Gruenther, USA, Director, Joint Staff; General Hoyt S. Vandenberg, Chief of Staff, U.S. Air Force; Admiral Louis E. Denfeld, Chief of Naval Operations; General Omar N. Bradley, Chief of Staff, U.S. Army; Secretary of Defense James Forrestal (at head of table); Lieutenant General Albert C. Wedemeyer, Director of Plans and Operations, U.S. Army; Vice Admiral Arthur W. Radford, Vice Chief of Naval Operations; and Lieutenant General Lauris Norstad, Deputy Chief of Operations, U.S. Air Force.

Chapter 3

PEACETIME CHALLENGES

World War II confirmed that high-level strategic advice and direction of the Armed Forces were indispensable to success in modern warfare. These accomplishments, however, did not assure the Joint Chiefs of Staff a permanent place in the country's defense establishment. Indeed, as the war ended, the demobilization of the Armed Forces and the country's return to peacetime pursuits pointed to a shift in priorities that diminished the chiefs' role and importance. Yet even though the JCS may have been shorn of some of the power and prestige they enjoyed during the conflict, they remained a formidable organization, served by some of the best talent in the Armed Forces, and thus a key element in the immediate postwar development of national security policy.

The postwar fate of the Joint Chiefs of Staff initially rested in the hands of one individual: President Harry S. Truman. A sharp contrast in style and work habits to his patrician predecessor, Truman was the epitome of down-to-earth Middle America. Born and raised in northwest Missouri, he had served as the captain of a National Guard artillery unit in World War I. After the war, he returned to Missouri, tried his hand in the haberdashery business, failed, and turned to politics, becoming a fringe part of the notorious Pendergast "machine" of Kansas City. Elected to the U.S. Senate in 1934, he worked hard and developed a reputation as a fiscal conservative, ever protective of the taxpayers' money. When Roosevelt decided to drop Vice President Henry A. Wallace from the ticket in 1944, he turned to Truman to be his running mate, even though the two barely knew one another. After the election, they rarely met or conversed by phone.[1]

As Commander in Chief, Truman was almost the antithesis to Roosevelt. Preferring a structured working environment, he conducted business with the Joint Chiefs on a more formal basis and usually met with them in the presence of the Service Secretaries or, later, the Secretary of Defense. As a rule, he got along better with Army and Air Force officers than Navy officers. His bête noire was the Marine Corps, which he once accused as having "a propaganda machine that is almost the equal of Stalin's."[2] Once the wartime emergency was over, Truman found his time and attention increasingly taken up with domestic chores, which reduced his contacts with the chiefs. Still, he had the utmost respect for members of the Armed Forces and often named retired or former military officers to what were normally considered civilian positions.[3] Highest of all in Truman's estimation was General George C. Marshall, to whom he turned repeatedly for help as his special representative to China from 1945 to 1946, as Secretary of State from 1947 to 1949, and as Secretary of Defense from 1950 to 1951. But he tempered the military's influence with close control of the defense budget and a strong emphasis on civilian authority in key areas such as atomic energy.

Truman had no intention of keeping the Joint Chiefs of Staff in existence any longer than it took Congress to enact legislation unifying the armed Services. Throwing his support behind a War Department proposal drawn up to Marshall's specifications toward the end of the war, Truman favored replacing the JCS with a uniformed chief of staff presiding over an "advisory body" of senior military officers who would be part of a single military department.[4] The idea had mixed appeal in Congress, however, where several leading members complained that it could lead to a "Prussian-style general staff" and dilute civilian control of the military. Increasingly popular on Capitol Hill was a competing proposal sponsored by Secretary of the Navy James Forrestal. Under the Navy plan, the JCS would remain intact and form part of a network of interlocking committees promoting cooperation and coordination for national security on a government-wide scale.[5] Pending resolution of the unification debate, Truman opted for the status quo.

Thus, the Joint Chiefs continued to operate much as they had during the war, though at a reduced level of activity, with fewer personnel in the organization and with new membership. Having accomplished their job, most of the wartime members elected to retire soon after the war. Their successors were officers who had held significant U.S. or Allied commands. The first to leave was General of the Army Marshall, who stepped down as Chief of Staff in November 1945 to make way for General of the Army Dwight D. Eisenhower, leader of the D-Day invasion of Normandy and Supreme Allied Commander in Europe. A month later, Fleet Admiral Chester W. Nimitz succeeded Fleet Admiral Ernest J. King as Chief of Naval Operations. And in March 1946, General Carl Spaatz, Commander of the Eighth and Twentieth Air Forces and a key architect of the strategic bombing campaigns against Germany and Japan, succeeded General of the Army Henry H. Arnold as Commanding General, Army Air Forces. The only hold-over was Fleet Admiral William D. Leahy, who continued to serve until illness forced his retirement in March 1949, at which time the position he occupied as Chief of Staff to the Commander in Chief lapsed.

PEACETIME CHALLENGES

DEFENSE POLICY IN TRANSITION

At the outset of the postwar era in 1945, the Joint Chiefs of Staff viewed the prospects for an enduring peace with growing apprehension. Even though Germany and Japan were no longer a threat, a new danger arose from the Soviet Union, now the leading power on the Eurasian landmass, whose "phenomenal" increase in military and economic strength gave the JCS cause for concern.[6] Never an overly close partnership, the Grand Alliance began dissolving even before the war was over. Factors that made the future uncertain in the Joint Chiefs' eyes included an uneasy modus vivendi over the postwar treatment of Germany and Soviet insistence on German reparations, the spread of Communist control in Eastern Europe, disputes over Venezia Giulia at the northern end of the Adriatic, political instability in Greece, Soviet demands for political and territorial concessions from Turkey and Iran, and the impasse over the control of atomic energy. None of these issues alone need have caused undue alarm. Taken together, however, they formed an ominous pattern that suggested to the chiefs a fundamental divergence of interests that could result in an adversarial relationship.[7]

Unsettled relations with the Soviet Union reinforced what the Joint Chiefs of Staff had been saying for some time about the need for a strong postwar defense posture. But in the immediate aftermath of the war, the trend was in the opposite direction, as the country embarked on one of the most rapid and thorough demobilizations in history. Bowing to strong public and congressional pressure to "bring the boys home," the War and Navy Departments discharged veterans pell-mell, shrinking the Armed Forces from 12 million in June 1945 to 1.5 million 2 years later. Operating on a conservative economic philosophy that gave priority to balancing the budget and reducing debt, President Truman ordered sharp reductions in Federal spending that included the wholesale cancellation of war-related contracts, curbs on military outlays, and strict ceilings on future military expenditures.[8]

While cutting deeply into the effective combat capabilities of the Armed Forces, the posthaste demobilization and limitations on military spending left the JCS uneasy over the country's defense posture. To be sure, the chiefs recognized that funding for defense would be tight after the war. Convinced, however, that the United States had been woefully unprepared prior to Pearl Harbor, the JCS believed that Congress and the American public should be willing to support a level of military readiness well above that of the interwar period. Under a broad blueprint of postwar requirements, the JCS argued that U.S. forces should have the resources to carry out their increased peacetime responsibilities and to respond effectively during the initial stages of a future war.[9] Some, like General Marshall, saw universal military training as the solution to the country's long-term defense needs. But after

Hiroshima and Nagasaki, UMT steadily lost ground to more technologically-oriented solutions, with reliance on airpower and "new weapons" like the atomic bomb foremost among them. Whether that reliance should be on land-based airpower or carrier-based aviation or both became one of the most contentious defense issues of the immediate postwar period.

At the center of the emerging postwar debate over military policy was the atomic bomb, a weapon of awesome proven destructive power but uncertain prospects. Despite the enormous wartime effort to develop the bomb, production of fissionable materials (uranium-235 and plutonium) dropped quickly once the war was over, as most of the scientists and technicians recruited for the Manhattan Project returned to their civilian pursuits. Refinements in weapon design virtually ceased and bomb production slowed to a snail's pace. Sketchy and incomplete records suggest that by the latter part of 1946 there were between six and nine nuclear cores in the atomic stockpile—an exceedingly small arsenal by later standards but still a sufficient number, President Truman believed, "to win a war."[10]

In the immediate aftermath of World War II, the Truman administration had no incentive to keep the atomic bomb program at its wartime level of production and efficiency. As the war ended, the prevailing belief in many quarters was that atomic energy would be taken out of the hands of the military and that nuclear weapons would be banned, just as poison gas was after World War I. The notion of civilian control had an appealing ring and gave rise to legislation in 1946 establishing the Atomic Energy Commission (AEC). A civilian body appointed by the President with the advice and consent of the Senate, the AEC acquired complete authority over the Nation's nuclear program, from the production of fissionable material and the manufacture of bombs to the custody and control of finished weapons. In support of the commission's activities, Congress also established a nine-member General Advisory Committee to provide scientific and technical guidance, and a Military Liaison Committee (MLC), to assure coordination between the commission and the Armed Forces.[11]

In contrast, the movement to ban the bomb, or at least to place it under some form of international supervision, produced far less definitive results. Intense policy debates, starting in the autumn of 1945, extended into the following spring. The outcome was the Baruch Plan, placed before the United Nations in June 1946, under which the United States offered to give up its nuclear monopoly in exchange for a stringent regime of international controls and inspections. A magnanimous gesture, the Baruch Plan was too intrusive to suit the Soviets, who declared it unacceptable "either as a whole or in [its] separate parts." As an alternative, Moscow proposed a flat prohibition on nuclear weapons with a vague promise of inspections sometime in the future. A UN special committee voted overwhelmingly to accept the Baruch Plan, but the Soviet Union and Communist-controlled Poland abstained, leaving the plan's fate up in the air.[12]

Throughout the deliberations leading to announcement of the Baruch Plan, the Joint Chiefs maintained a guarded attitude that endorsed international controls in principle as a desirable long-term goal, but with strong reservations attached to giving up any atomic secrets until outstanding international issues had been fully vetted and resolved.[13] This line of reasoning remained the JCS core position on arms control and disarmament for the duration of the Cold War. But in 1945, the chances of overcoming the chiefs' objections and of enlisting their support for a stringent regime of international control were probably better than they ever were again. Regarded by the JCS as a special weapon with limited applications, the atomic bomb had yet to acquire a permanent niche in their military planning and was in many ways a disruptive presence that the chiefs could have done without. Later, as the Services launched expensive acquisition and training programs to integrate nuclear weapons into their equipment inventories, and as national policy came to rely heavily on a strategy of nuclear deterrence, the chances of making sweeping changes in the JCS position faded. But until then, the chiefs were actually more flexible and open-minded than most critics gave them credit.

While awaiting the outcome of the international control debate, the Joint Chiefs sought a clearer picture of the atomic bomb's military potential. Having seen from the results of Hiroshima and Nagasaki what nuclear weapons could do to targets on land, they obtained President Truman's approval in January 1946 to explore the atomic bomb's effect on targets at sea.[14] Planning and preparations for Operation *Crossroads* took place under the auspices of the Joint Staff Planners, who named a six-member ad hoc inter-Service subcommittee headed by Lieutenant General Curtis E. LeMay to coordinate the effort. Almost immediately, quarrels erupted between AAF and Navy representatives over the placement of the target ships and other details, turning *Crossroads* into yet another arena of inter-Service strife. A joint task force led by Vice Admiral William H.P. Blandy eventually carried out the operation, but like the LeMay committee, it had to contend with a good deal of inter-Service bickering and competition.[15]

The *Crossroads* tests were unique in several respects. First, they were the only nuclear experiments organized and conducted under the authority of the Joint Chiefs of Staff; and second, they received an extraordinarily high level of publicity, in sharp contrast to the restricted nature of subsequent nuclear experiments carried out by the AEC. Despite strong political pressure to cancel the tests lest they interfere with the debate in the UN, President Truman refused, citing the waste of

$100 million if they failed to proceed. The ensuing experiments, involving 42,000 Servicemen, took place in July 1946 at Bikini Atoll in the Pacific and rendered mixed results. The first weapon, an air-dropped, Nagasaki-type bomb, missed the aim point by 1,500 yards. Sinking only a few of the ships in the target area, it did relatively minor damage to the rest. But a second bomb, detonated under water, was more impressive and left the members of a JCS evaluation board convinced that atomic weapons had the potential for achieving decisive results in future wars. "If used in numbers," the board found, "atomic bombs not only can nullify any nation's military effort, but can demolish its social and economic structure and prevent their reestablishment for long periods of time."[16]

Still, the *Crossroads* tests had little immediate impact on JCS plans or military policy. Although the Joint Chiefs recognized that atomic bombs, like other new weapons (e.g., jet aircraft and long-range guided missiles), could have a significant bearing on the conduct of future wars, the ongoing deliberations in the UN over international controls, coupled with the limited availability of fissionable materials, effectively ruled out a defense posture resting to any great extent, if at all, on nuclear weapons. This did not stop the Army Air Forces, acting on their own, from making informal arrangements in the summer of 1946 with the British to modify bases in England for air-atomic missions (the Spaatz-Tedder Agreement).[17] Nor did it deter the Navy from commissioning design studies for a new generation of flush-deck "super carriers" dedicated to nuclear warfare.[18] But in looking ahead, the Joint Chiefs and their Joint Staff Planners clung to the view that wars of the future would be much like the one they had just finished, engaging large conventional armies, navies, and air forces. The only major difference the JCS could see was that the next time, the enemy would probably be the Soviet Union.[19]

REORGANIZATION AND REFORM

Foremost among the issues needing to be addressed in framing a postwar defense policy was the reorganization of the Armed Forces, including a settlement of the controversial unification issue, a clarification of command arrangements, and a re-articulation of Service roles and missions. Unable to arrive at an agreed position on unification, the Joint Chiefs told President Truman in October 1945 that they had no corporate wisdom to offer and would defer to Congress and the administration to make the necessary adjustments.[20] As the senior officers of their respective Services, however, all JCS members remained actively engaged in the debate. Even Admiral Leahy, who had no Service responsibilities and who viewed himself as above the fray, took a position from time to time, invariably in support of the Navy.

In consequence, it was almost impossible for tensions generated by the unification quarrel not to spill over into JCS deliberations on other matters.

Though the Joint Chiefs sidestepped involvement in the unification controversy, they could not avoid two related matters—the establishment of a unified command plan, and the redefinition of Service functions in light of the experience of World War II, new technologies, and the changing nature of modern warfare. In addressing the first, the chiefs overcame their differences to establish a flexible command structure which, while far from perfect, proved remarkably adaptable to the tests of time. But in dealing with the roles and missions issue, they made little headway and eventually ceded this pivotal responsibility to others.

The unified command plan was the outgrowth of the extensive and generally successful use of joint and combined "supreme commands" in World War II, and the realization that, with the occupation of Germany and Japan and other responsibilities, the United States would have joint military obligations abroad for the indefinite future. Even before the war ended, the Joint Chiefs envisioned retention of the unified command system in peacetime, and by June 1945 they were taking steps to transform General Eisenhower's combined headquarters in Europe into a unified U.S. command, a relatively easy task since most of the forces involved were ground and air units under the War Department.[21]

The picture was more complex in the Pacific. There, the impetus for change came early in 1946 from the Navy, which sought to consolidate what were at the time far-flung command arrangements. Adopted by the JCS the previous April as an interim measure, the existing setup adhered to MacArthur's dictum that "neither service fights willingly on a major scale under the command of the other."[22] Hence, in allocating command functions, the JCS divided responsibilities between an Army command for all land forces in the theater, and a Navy command for forces at sea. Characterizing these divided command arrangements as "ambiguous" and "unsatisfactory," Admiral Nimitz wanted the JCS to establish a single command for the Pacific encompassing all forces in the area, excluding China, Korea, and Japan.[23] What prompted Nimitz to raise the issue is unclear, though it may have been intended to complement draft legislation submitted by Secretary of the Navy Forrestal asking for an increase in the peacetime authorized strength of the Navy and the Marine Corps. A merger of the two commands would have given the Service in charge a strong claim to a larger budget share. Since the Navy had the predominant interest in the Pacific, Nimitz thought it only logical that the new command should be in Navy hands. Seeing the proposed merger as a blatant power grab, MacArthur, from his headquarters in Tokyo, warned the War Department that it would render Army or AAF units in the area "merely adjuncts" of the Navy.[24]

Hoping to avoid a fractious debate, the Joint Chiefs referred the CNO's proposal to the Joint Staff Planners, whose efforts soon ran aground. The Army and Army Air Forces members insisted on unity of command by the forces involved, while the Navy member urged unity of command by area.[25] Eventually, it took pressure from Congress, which wanted to avoid anything resembling the divided command that existed at Pearl Harbor in 1941, and the direct intervention of Admiral Nimitz and General Eisenhower to settle the matter. All the same, the compromise thus achieved did little more than paper over inter-Service differences that later reappeared. Accepting the Navy's basic premise that unity of command should be by area, Eisenhower proposed extending the system worldwide, to include not only the Pacific but other regions where the United States had significant military assets or military interests. With further fine-tuning by Nimitz, this became the Unified Command Plan (UCP), approved by President Truman in December 1946.[26]

Initially, the UCP called for seven geographic commands and one functional command (known after 1951 as a "specified" command).[27] Implicit in this arrangement was that a senior officer representing the Service with the predominant interest in a particular region or functional activity should head the command. Thus, in Europe the accepted practice (until 2003) came to be that an Army or Air Force officer should exercise command of the theater, while in the Pacific a Navy officer was invariably in charge. The sole functional command recognized in the UCP was the Strategic Air Command (SAC), created by order of General Spaatz in March 1946. SAC comprised the strategic assets of Eighth and Fifteenth Air Forces, 509th Composite Group with its air-atomic capability, and air bombardment units not otherwise assigned. Like Twentieth Air Force in World War II, SAC reported directly to the Joint Chiefs of Staff through the Commanding General, Army Air Forces (later the Chief of Staff of the Air Force), who acted as their executive agent.[28]

The Services could compromise on the UCP because each gave up very little in exchange for official confirmation of their existing geographical equities. Unfortunately, this approach was infeasible when defining overlapping Service functions and sorting out the impact of new technologies on traditional roles and missions. An integral part of the unification debate, the assignment of functions was also highly instrumental in determining the allocation of budget shares among the Services. It seemed only logical, as the successor organization to the Joint Board, which had overseen the assignment of Service functions prior to World War II, that the Joint Chiefs should carry on this task. But with the changes in warfare that had taken place during the war, the traditional formula used by the Joint Board for determining and assigning functions, more or less by the medium in which a Service operated, no longer applied. Quite simply, neat distinctions between land, sea, and air warfare had ceased to exist. But even though the JCS agreed that the old assignments were frayed and outmoded, they were hard-pressed to come up with something better.

The event that brought the roles and missions controversy to a boil was a report by the Joint Strategic Survey Committee to the JCS in February 1946. Intended as a new statement of Service functions, the JSSC report became instead the catalyst for a prolonged and inconclusive debate among the chiefs. Like the Joint Board, the JSSC proposed an assignment of functions organized primarily around the major element in which each Service operated. Where Service functions intersected, however, the committee was often unable to provide unanimous advice. The most contentious points were the Army Air Force's insistence on full control of air transport; the Navy's claim on access to land-based aviation for antisubmarine warfare, as it had in World War II; and the Marine Corps's objections to the Army's efforts to bring amphibious operations under its aegis.[29] The quarreling became so acrimonious and divisive that the Joint Chiefs in June 1946 felt it advisable to suspend their deliberations on roles and missions until such time as "Presidential or legislative action requires that consideration be revived."[30]

Despite the impasse, the Joint Chiefs remained under heavy pressure to compose their differences in order to expedite consideration of a unification bill. Accordingly, in July 1946 they asked the Operations Deputies—Major General Otto P. Weyland of the Army, Major General Lauris Norstad of the Army Air Forces, and Vice Admiral Forrest P. Sherman—to explore a solution.[31] Initially slow work, the pace quickened following a breakthrough meeting at Secretary of the Navy Forrestal's home on November 12, 1946, where Assistant Secretary of War for Air W. Stuart Symington and Vice Admiral Arthur W. Radford, Deputy Chief of Naval Operations (Air), reached a tentative modus vivendi. Based on the discussion that afternoon, Norstad and Sherman agreed to develop a fresh formulation of Service functions and a statement of agreed principles to help jump-start approval of a unification bill that had stalled in Congress. In January 1947, Norstad and Sherman submitted their recommendations to Forrestal and Secretary of War Robert P. Patterson, who then conveyed them to President Truman. Passage of the National Security Act of 1947 followed in July, at which time the President issued an accompanying Executive order delineating Service roles and missions.[32]

The National Security Act was a legislative compromise that combined major elements of the centralized organization the War Department favored, and the decentralized coordinating system the Navy recommended. To unify the armed Services, Congress created a hybrid organization known as the National Military

Establishment (NME) composed of three coequal Service departments (Army, Navy, and Air Force) and a presiding civilian Secretary of Defense, who had a support staff limited to three special assistants. Under the Secretary's authority fell various coordinating bodies: the Research and Development Board (RDB) to advise and assist the Services with policies on scientific research and technology; the Munitions Board (MB) to coordinate production and supply; and the Joint Chiefs of Staff. Now endowed with statutory standing, the Joint Chiefs also acquired a list of assigned functions similar to those in the unused charter of 1943. The law effectively eliminated the role the JCS played in World War II as the country's de facto high command and redefined their mission as a strategic and logistical planning and advisory organization to the President and the Secretary of Defense. Recognizing the chiefs' need for permanent support, Congress authorized a full-time Joint Staff of one hundred officers, drawn in approximately equal number from each Service. President Truman had wanted to replace the JCS with a single military head, but opposition in Congress forced him to drop the idea. The law also created a Cabinet-level National Security Council (NSC) to advise the President on foreign and defense policy, a Central Intelligence Agency (CIA) for the collection, analysis, and distribution of intelligence, and a National Security Resources Board (NSRB) to oversee national mobilization in emergencies.[33]

Figure 3–1.

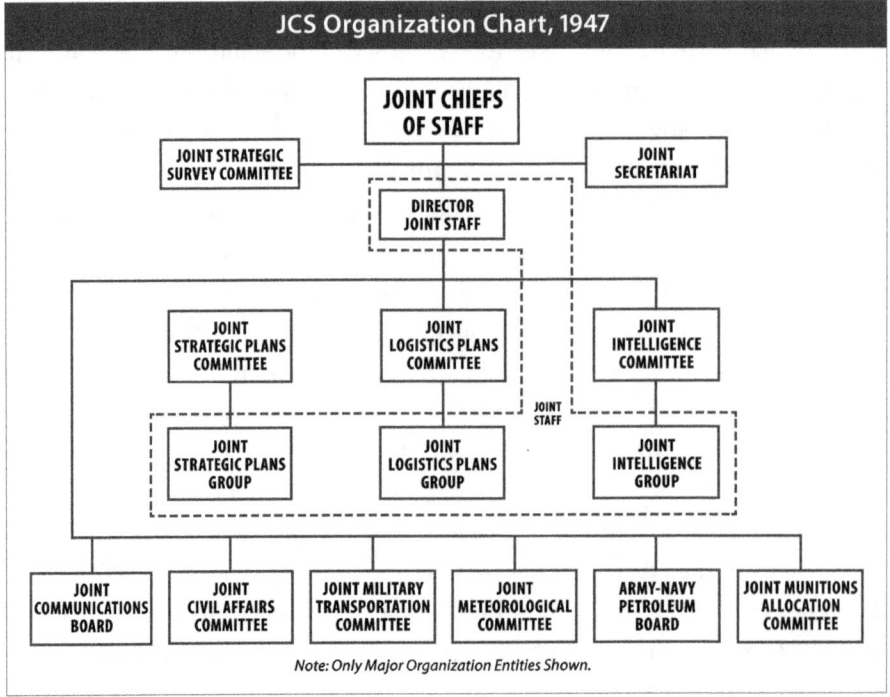

The Executive order (EO 9877) that accompanied the National Security Act was virtually the same statement of Service functions recommended in January by Norstad and Sherman. Where roles and missions overlapped, EO 9877 called on the Services to coordinate their efforts with one another to the greatest extent possible.[34] Between the drafting of the Executive order in January and the passage of the National Security Act in July, however, Congress inserted language into the law that guaranteed the Navy access to "land-based naval aviation" and the Marine Corps a role in amphibious warfare. The net effect was to render key parts of EO 9877 obsolete, opening the door to renewed inter-Service bickering. Secretary of the Navy Forrestal, who became the first Secretary of Defense in September 1947, recognized the problem immediately but needed two contentious conferences with the Joint Chiefs—one at Key West, Florida, in March 1948, and a second at Newport, Rhode Island, the following August—to resolve the problem. These conferences also reaffirmed the practice dating from World War II of allowing the Joint Chiefs to designate one of their members as executive agent for a unified command, a function that effectively preserved the JCS in the chain of command. Drawing on Forrestal's frustrating experience, future Secretaries of Defense relied less on JCS guidance in sorting out roles and missions, and more on the Services to take the necessary steps to reconcile and adjust their differences.[35]

WAR PLANS, BUDGETS, AND THE MARCH CRISIS OF 1948

The National Security Act came into effect on September 18, 1947, a time of escalating tensions with the Soviet Union and dramatic change in American foreign policy. The previous March, in response to the Communist-led insurgency in Greece and Soviet pressure on Turkey, the Truman administration had launched the Greek-Turkish aid program, in the President's words, to prevent "the extension of the iron curtain across the eastern Mediterranean."[36] The following June, Secretary of State Marshall proposed the European Recovery Program (ERP), a large-scale assistance effort aimed at the broader problem of arresting the deteriorating economic and social conditions in Western and Central Europe that were playing into the hands of Communist agitators and Soviet sympathizers. Commenting publicly on these initiatives and the escalation of tensions between Washington and Moscow, journalist Walter Lippmann proclaimed the onset of a "Cold War" between East and West.[37]

As he sought to stem the spread of Communism abroad, President Truman also ordered major changes in the U.S. atomic energy program. Frustrated by the impasse in the United Nations over the Baruch Plan, the President directed the new Atomic Energy Commission in early April 1947 to restore production facilities and to resume the manufacture of nuclear weapons. The President's decision had the strong endorsement of the Joint Chiefs of Staff, who agreed that the time had passed for international control and that the only choice was to resume the production of atomic bombs. Procedures in effect at the time called for the JCS to conduct an annual review of nuclear stockpile requirements and to convey their recommendations, through the Military Liaison Committee, to the AEC. The chiefs tailored their military requirements, stated in numbers of bomb cores, to be roughly commensurate with the AEC's estimate of its annual production capabilities, the standard practice for fixing the size of the nuclear stockpile for the next several years.[38]

With the emerging "strategy of containment" toward the Soviet Union came a sense of unease among the Joint Chiefs over the deterioration of the Nation's military capabilities. Other than resuming the production of nuclear weapons, little had been done since World War II to modernize U.S. forces or improve their effectiveness. The American Military Establishment had shrunk dramatically since the war, and the forces that remained by 1947 were generally understrength, indifferently equipped and trained, and scattered around the globe. Soviet military power, in contrast, was concentrated on the Eurasian landmass and appeared to be largely intact and organized around an estimated ground force of 175 divisions, a figure derived from the order of battle pieced together by German intelligence in World War II.[39] Long-range threat projections developed by the Joint Intelligence Committee between late-1946 and mid-1947 credited the Soviet Union with possessing an overwhelming numerical superiority in conventional forces and the capacity for acquiring nuclear weapons by the early 1950s, if not before. Some in the scientific community thought it would take longer for the Soviets to duplicate the American achievement in atomic energy, but by and large the emerging consensus was that the Soviets were determined to become a nuclear power and that sooner or later they would realize their goal.[40]

Despite the danger signs, the Truman administration initially downplayed the possibility that growing East-West antagonisms and steps taken by Washington to curb Communist expansion might escalate into a military confrontation. The reigning expert on the Soviet threat immediately following World War II was George F. Kennan, a Foreign Service Officer with long experience in the Soviet Union and Director of the State Department's elite Policy Planning Staff. It was Kennan whose 1947 article in *Foreign Affairs*, "The Sources of Soviet Conduct," had given rise to the term "containment" to describe what the administration was trying to achieve vis-à-vis the Soviet Union. Kennan believed that if the United States exerted sufficient economic, political, and diplomatic pressure, it would elicit significant improvements in Soviet behavior. Though Kennan acknowledged that military forces were a vital diplomatic tool, he doubted whether the United States and the Soviet Union would ever go to war. Warning against excessive reliance on armed strength, he preferred small, mobile strike forces that could intervene quickly in crisis situations. For sizing purposes, he favored a defense establishment that could operate effectively in two separate theaters simultaneously, a rule of thumb that would influence U.S. force requirements for decades to come.[41]

Given the Truman administration's preference for nonmilitary solutions and the limited military assets available at the time, the Joint Chiefs saw no urgent need for approving a strategic plan of action against the Soviet Union. During the latter part of World War II, in considering the hypothetical possibility of a future East-West conflict, the Joint Chiefs had concluded that while there was little chance the United States would lose such a war, the likelihood of winning it was exceedingly remote.[42] Acting on its own initiative, the Joint War Plans Committee (JWPC) launched a series of studies code-named PINCHER late in 1945 to explore the problems of waging a war against the Soviet Union. The first fruit of this exercise appeared on March 2, 1946, when the JWPC forwarded a broad concept of operations to the Joint Staff Planners. With refinement, this became the basic concept of operations around which strategic planning revolved for the next several years. Dealing only with the opening stages of a conflict, PINCHER envisioned war breaking out in the eastern Mediterranean or Near East and spreading rapidly across Europe.[43] Arguing that it would be futile for the United States and its allies to try to match Soviet strength on the ground, the JWPC favored a strategic response "more in consonance with our military capabilities and in which we can exploit our superiority in modern scientific warfare methods." Even if such a response failed to defeat the Soviet Union, it would buy time for the United States to mobilize forces, check the Soviet advance, and mount counterattacks.[44]

The first Secretary of Defense, James Forrestal, was an ardent proponent of the new get-tough policy toward the Soviet Union and wanted to give it as much military support as possible. But he was under orders from President Truman to hold the line on defense spending.[45] Hoping to satisfy both requirements, Forrestal looked to the Joint Chiefs to provide an integrated statement of Service requirements for meeting essential national security objectives and an agreed strategic concept, tailored to fit within approved spending limits, to justify those forces.[46] In Forrestal's view, the JCS were the key to the successful implementation of the new unification law, for it was primarily through them that he intended to extend his authority as Secretary of Defense down into the Services.[47]

While it looked good on paper, Forrestal's reliance on the Joint Chiefs proved flawed in practice. Even though the JCS organization had a reputation for highly proficient planning, it had lost much of its edge and efficiency by 1947 through the attrition of veteran personnel and a dwindling pool of suitable replacements. Though the JCS were less affected than other joint agencies (i.e., the MB and the RDB), many able officers were averse to joint duty in Washington lest it cost them command experience in their Services and derail their careers.[48] Limited by law to one hundred officers, the once-mighty Joint Staff now operated at a reduced pace through three groups—the Joint Intelligence Group, Joint Logistics Group, and Joint Strategic Plans Group (formerly the Joint War Plans Committee). With an enormous backlog of business and new requests coming in almost daily from Forrestal's office, the Joint Staff soon found itself with more taskings than it could handle. To augment the Joint Staff, the JCS continued to rely on part-time inter-Service committees of senior officers—the Joint Strategic Plans Committee (which replaced the Joint Staff Planners), the Joint Logistics Committee, the Joint Intelligence Committee, and the Joint Strategic Survey Committee. At Forrestal's urging, Congress increased the size of the Joint Staff from 100 to 210 officers when it amended the National Security Act in 1949. But despite the increase, there always seemed to be more work than the Joint Staff could handle.

The most serious flaw in Forrestal's system lay in the chiefs themselves, whose internal disagreements sapped their cohesion and effectiveness. Some of their quarrels were carryovers from the unification debate or earlier disagreements, like the ongoing battle between the Army and the Marine Corps over amphibious operations. But by far the most visible and contentious issues were those between the Air Force and the Navy over whether long-range, land-based bombers or carrier-based aviation should serve as the country's first line of defense. Now that the production of nuclear weapons had resumed, it seemed clear that the atomic bomb would play a growing role in strategic planning and that the Service with the nuclear mission would get the lion's share of the defense budget. Some, including key figures in Congress and the members of the Finletter Commission, a fact-finding body set up by the White House in 1947 to report on the future of military aviation, assumed that the Air Force had the job sewn up.[49] In fact, the issue was far from settled. While the Air Force had a nuclear-delivery system derived from the SILVERPLATE B-29s of World War II, its capabilities were limited to a handful of planes; thus, its position was not immune to challenge by the Navy.[50]

These disputes were precisely the kinds of quarrels Forrestal had hoped to stifle with an integrated budget process keyed to the development of joint strategic plans. Yet they were practically unavoidable, given the strict spending limits Truman had imposed and Forrestal's reluctance to test his powers as Secretary of Defense against the Joint Chiefs. As Secretary of the Navy, Forrestal had been in the vanguard of those who opposed a closely unified defense establishment. As Secretary of Defense, he found himself in the awkward position of implementing a compromise law he helped to craft but only half-heartedly believed in. Initially, he described himself as a "coordinator" and, in the interests of promoting harmony among the Services, promised to make changes through "evolution, not revolution." He probably never should have taken the job of Secretary of Defense, but when Truman offered it (after Secretary of War Patterson turned it down for personal reasons), he felt duty-bound to accept.[51]

Based on his discussions with the Joint Chiefs and his personal assessments of the international situation, Forrestal became convinced that the President's budget ceilings were too low to fund essential military requirements and to provide a credible defense posture. During his 18 months as Secretary of Defense, he asked Truman twice for more money—in the spring of 1948 and, again, toward the end of the year. On the first occasion, with the help of a crisis atmosphere abroad, he was successful in persuading Truman to lift the ceiling; on the second, despite continuing tensions in Europe, he failed, thereby inadvertently undermining his own authority and credibility.

The immediate occasion that prompted Forrestal's first request for more money was the "March Crisis" of 1948 that followed the Soviet-directed coup against the government of Czechoslovakia the month before. The only country liberated by the Red army that had thus far remained democratic and independent of Soviet domination, Czechoslovakia had tried to steer a course of nonalignment but faced growing pressure from Moscow to curb its contacts with the West. Not only did Czechoslovakia share a common border with the Ukraine; it was also the principal source of high-grade uranium ore for the Soviet atomic bomb project.[52] Beset with growing political turmoil and a general strike organized by Communist-controlled unions, the Czech president, Eduard Beneš, had dismissed his cabinet and turned over all important government posts to Communists, except the foreign ministry, which remained under Jan Masaryk, a popular figure in the West. Within a fortnight, on March 10, Masaryk's body was found on the cement courtyard of the foreign ministry beneath his office window. Czech authorities promptly labeled his death a suicide, but the speculation in the West was that Soviet agents murdered him.[53]

Shortly after the Czech coup, rumors circulated that the Soviets would turn their sights on occupied Germany and try to force the Allied powers out of their enclaves in Berlin. Lending substance to these reports were ominous signs of Soviet troop movements in eastern Germany suggesting a buildup for an invasion of the West. Later, U.S. analysts concluded that these bellicose gestures were a ruse and that there was no "reliable evidence" the Soviets intended military action. All the same, the Intelligence Community refused to rule out the possibility of "miscalculation" by one side or the other leading to an incident that could spark a war.[54]

Toward the end of February 1948, the Director of Army Intelligence, Lieutenant General Stephen J. Chamberlin, paid an unexpected call on General Lucius D. Clay, U.S. Military Governor of Germany, at his Berlin headquarters. Concerned over recent events in Czechoslovakia and Soviet behavior in general, Chamberlin urged Clay to use his considerable influence with the Joint Chiefs and others in Washington to send a "strong message" to stimulate support in Congress for reinstituting the draft and for bolstering other military programs. Clay replied that he had no concrete evidence the Soviets were planning a move. But after sleeping on the matter, he decided to act. On March 5, 1948, he cabled Chamberlin confirming that, while the signs were far from conclusive, he had detected "a subtle change in Soviet attitudes which I cannot define but which now gives me a feeling that [war] may come with dramatic suddenness." Clay's "war warning" message soon leaked to the press, setting off a war scare that had Washington on edge for several weeks.[55]

Based on the intelligence crossing his desk, Truman had known for some time that the Soviets were up to something.[56] Still, Clay's war-warning message caught the President off guard and gave Forrestal and the Joint Chiefs the opportunity to seek an increase in the military appropriations bills for Fiscal Year 1949 then pending in Congress. By then, General Omar N. Bradley had replaced Eisenhower, General Hoyt S. Vandenberg had been named to succeed Spaatz, and Admiral Louis Denfeld had replaced Nimitz. But even with a fresh set of faces the quarreling continued, with the size of the increase and the allocation of funds among the Services the main points in dispute. Some in Congress wanted any additional money to be devoted exclusively to strengthening the Air Force's strategic bombing capability. But it was Forrestal's and Truman's view that the country should have a "balanced" force posture in which all three Services participated on roughly equal terms.

The Joint Chiefs agreed that balanced forces were a laudable objective, but having yet to agree on an integrated strategic concept, they had no basis for identifying deficiencies or recommending an overall plan on how additional money should be allocated. By default, they wound up recommending what each Service unilaterally calculated it needed, a sum well in excess of anything the White House or the Bureau of the Budget (BOB) found acceptable on economic grounds. With an election looming in the fall, Truman was more afraid of inflation at home, fueled by increased military spending, than he was of the Soviets. Nevertheless, the additions he eventually approved in May 1948 increased the military budget by nearly a third and showed Forrestal and the JCS that the President's budget ceilings were not so firm after all.[57]

In addition to boosting the military budget, the March Crisis produced several other outcomes. First, it heightened awareness both in Europe and the United States that the Soviet Union was a potential military threat and needed to be addressed accordingly. Until then, except for a limited military aid program to Greece and Turkey, the Truman administration and Congress had relied on political, economic, and diplomatic initiatives to contain communism and Soviet expansionism; but with the March Crisis came the realization on both sides of the Atlantic that closer military collaboration was a necessary accompaniment to the European Recovery Program.[58] Passed by Congress in May 1948, the Vandenberg Resolution urged the administration to explore a collective security agreement with willing partners in Europe, a process that culminated in April 1949 with the creation of the North Atlantic Treaty Organization (NATO). A major departure from the nonentanglement policy of the past, NATO would be a key element in the Joint Chiefs' military assessments and strategic planning throughout the Cold War and beyond.

The March Crisis also led the JCS to expedite completion of an integrated strategic concept, a major step toward a unified defense budget. The agreed plan, called HALFMOON (later renamed FLEETWOOD), was an outgrowth of the PINCHER series and called for the Strategic Air Command to launch "a powerful air offensive designed to exploit the destructive and psychological power of atomic weapons against the vital elements of the Soviet war-making capacity." Navy carriers would conduct a secondary air offensive from the eastern Mediterranean. But with atomic bombs in short supply, there was no assurance that the Navy would participate in the nuclear phase of the air offensive. Arguing that HALFMOON was overly dependent on SAC's ability to mount nuclear operations, the Navy accepted it only on condition that the JCS treat it as an "emergency" war plan (EWP) and not for long-term force planning beyond the next budget cycle.[59]

A key feature of HALFMOON was the need for overseas bases in Newfoundland, the United Kingdom, and the Cairo-Khartoum area of Northeast Africa from which to mount strikes against the Soviet Union. Keeping alive the "special relationship" developed in World War II, the Joint Chiefs hosted a meeting in Washington for senior British and Canadian planners from April 12 to 21, 1948, to discuss U.S. access to British and Canadian staging points.[60] An inevitable byproduct of U.S. planning, these tripartite discussions were to some extent premature, since President Truman had yet to consent to the HALFMOON plan, transfer the custody of any nuclear weapons from the AEC to the military, or authorize their use. After receiving a JCS briefing on the plan on May 5, 1948, the President asked the Joint Chiefs to prepare a nonnuclear alternative, code-named ERASER. But because of budgetary limitations, Forrestal viewed ERASER as a low priority and later ordered work on it suspended.⁶¹ Confirming the course of action previously discussed, a U.S. Air Force mission of senior officers and planners visiting London later in May assured their RAF colleagues that "all planning was to be based on the use of atomic bombs from the outset including the use of the UK as a base for USAF carrying such bombs."⁶²

As the March Crisis wound down, the Joint Chiefs were gradually making progress toward integrating their requirements and developing a strategic concept to serve as the basis for a postwar defense policy. The emerging centerpiece of this process was the atomic bomb, with the threat of strategic bombardment serving as the country's principal deterrent. While differences persisted among the Services over how this strategy should be interpreted and applied, the overall thrust of what would constitute the American response to Soviet aggression was no longer in doubt. Given the limitations on weapons and equipment under which the Services operated, the JCS were still a very long way from the "massive retaliation" doctrine of the 1950s. Slowly but surely, however, they were moving in that direction.

THE DEFENSE BUDGET FOR FY 1950

Following President Truman's approval of the supplemental defense increase in the spring of 1948, Forrestal and the Joint Chiefs turned their attention to the military budget for Fiscal Year 1950 (July 1, 1949, through June 30, 1950). As the first full set of estimates to be developed since the passage of the National Security Act, the FY 50 budget would be a clear test of the chiefs' ability to perform their assigned strategic planning functions of producing an integrated defense plan within approved spending limits.⁶³ At a meeting with Forrestal and the JCS in May 1948, Truman stated that he wanted new obligational authority (i.e., cash and new contract authority) held under $15 billion. Acknowledging that defense requirements could fluctuate, the President told the chiefs that he would review the situation in September and again in December and make adjustments as needed.⁶⁴ At Forrestal's request, Truman also authorized the new National Security Council to develop a broad statement of national objectives to assist the JCS in developing their estimate of military requirements.⁶⁵ But he cautioned Forrestal against using NSC guidance to override spending limits. "It seems to me," Truman told him, "that the proper thing for you to do is to get the Army, Navy and Air people together and establish a program within the budget limits which have been allowed. It seems to me that is your responsibility."⁶⁶

Whether the international situation would cooperate to hold down military spending remained to be seen. Not only were the Soviets continuing to put pressure on Berlin, but there were also problems in the Middle East that threatened to embroil the United States in a conflict over Palestine, currently a British mandate.

Zionists had long sought to create a Jewish homeland there, and survivors of the Holocaust poured into the area by the thousands in the aftermath of World War II. The partitioning of Palestine into separate Arab and Jewish states had strong popular appeal in the United States and quickly became a crucial part of President Truman's campaign strategy for the 1948 election.[67] The Arab states of the Middle East, however, vowed to resist the Jewish influx with force. Fearing an anti-American backlash across the Arab world, the Joint Chiefs warned against U.S. support of partition on the grounds that it could "gravely prejudice" future access to Middle Eastern oil and compel the United States to wage "an oil-starved war."[68]

For the Joint Chiefs, the issue of most immediate concern was the declared intention of the British to end their mandate in Palestine prematurely and withdraw their forces, which had been serving as a buffer between the Arabs and the Jews. If the British withdrew, the JCS expected the United States to come under intense pressure to intervene as part of a UN peacekeeping operation to prevent Arab armies from slaughtering Jewish refugees and settlers. As it turned out, Jewish defense forces proved more than able to hold their own in defending the new state of Israel. But in the spring of 1948, the threat of another Holocaust appeared imminent.

In what would become a recurring theme for the next several decades, the Joint Chiefs strenuously opposed practically any deepening of U.S. involvement in the Middle East, especially if the United States appeared to be siding with Israel against the Arab states. Based on the size of the British presence in Palestine, the Joint Chiefs estimated that the UN would need to deploy a minimum peacekeeping force of over 100,000 troops (about half from the United States), supported by appropriate air and naval units. To raise the U.S. contribution to such a force, the chiefs notified the President that he would need to seek supplemental appropriations, reintroduce the draft, and order partial mobilization of the Reserves.[69] Suspecting that the chiefs were overdramatizing the situation and inflating their estimates, President Truman refused to rule out the possibility of U.S. intervention. But he took a cautious approach which more or less validated the chiefs' preference for avoiding involvement in the increasingly sensitive Arab-Israeli conflict.[70]

While the situation in Palestine argued for a flexible defense posture resting on a sound conventional base, persistent tensions in Central Europe played into the hands of those who favored reliance on strategic airpower and atomic weapons. Unsuccessful in exacting concessions from the Western powers or forcing their withdrawal from Berlin during the March Crisis, the Soviets turned to more direct measures. On June 19, 1948, they blockaded all access other than by air into the city. General Clay immediately organized an airlift to keep the western sectors of the city in essential supplies, but the longer the standoff went on, the more ominous it became.

By the end of June, the consensus in Washington was that the Western occupying powers—Britain, France, and the United States—should concert their efforts around a show of force and buy time for negotiations backed by a military buildup. Clay wanted to mount an armed convoy to test Soviet resolve, but the Joint Chiefs assessed the risk as too high and Allied forces as too weak to prevail should the Soviets resist.[71] On the other hand, the JCS had no objection to British Foreign Secretary Ernest Bevin's suggestion of a visible reinforcement of American airpower in Europe with B–29s.[72] Approved by President Truman in July, the B–29 augmentation would, in Forrestal's view, give the Air Force much-needed experience and make the presence of these planes "an accepted fixture" to the British public.[73] Encouraged by the success of the operation, Lieutenant General Lauris Norstad, the Air Force Deputy Chief of Staff for Operations, visited Britain in September and arranged to make the deployment permanent, with one B–29 group and one fighter group to be stationed in England at all times. Out of these discussions emerged a tentative agreement by the Air Force to "loan" Britain's Bomber Command an unspecified number of B–29s, and Bomber Command's pledge to place its assets "immediately" under SAC's coordination in the event of war with the Soviet Union.[74]

None of the SAC aircraft deployed to Europe during the Berlin blockade crisis was equipped for atomic operations, a fact the Soviets could easily have deduced from the appearance of the planes, which lacked the enlarged underbelly to accommodate atomic bombs. Even so, it was well known that B–29s carried out the attacks on Hiroshima and Nagasaki. The implied threat these planes represented elevated nuclear weapons to a new level of importance in national policy. Here in embryonic form was the doctrine of nuclear deterrence in practice for the first time. Though the threat may have been hollow, it was sufficient to give the Soviets pause before increasing the pressure and, as one senior Soviet officer later put it, risking "suicide" over Berlin.[75]

Still, without direct access to or control over nuclear weapons, the Joint Chiefs were apprehensive about what could happen if the Soviets called the American bluff. As a result of the stepped-up production program the AEC had initiated the year before, the atomic stockpile stood at around fifty nuclear cores by the summer of 1948.[76] Preliminary results of the recent SANDSTONE experiments, a series of test explosions held at Eniwetok in the Pacific the previous April–May, suggested the feasibility of new design techniques that could increase the size of the stockpile faster than expected and vary the yield of weapons. By demonstrating the feasibility of the "levitated" core, the SANDSTONE experiments confirmed the possibility of yields up to two and a half times larger than the Nagasaki bomb, using less fissionable material. The days of atomic scarcity and handmade bombs were drawing to a close. Thenceforth, the Joint Chiefs would have at their disposal a stockpile of assembly line–produced weapons, more plentiful in number than previously estimated and more varied in type and design.[77]

With U.S. war plans increasingly dependent on the early use of nuclear weapons, the SANDSTONE tests provided the reassurance of a larger and more versatile atomic arsenal than previously imagined. To make the most of the opportunity, Forrestal and the Joint Chiefs became convinced that the time had come to change the custody and control arrangements of nuclear weapons. But after a lengthy White House meeting to examine the matter on July 21, 1948, Truman ruled that custody of nuclear weapons would remain in the hands of the Atomic Energy Commission. A few days later, he told Forrestal that "political considerations" relating to the upcoming Presidential election barred a change of policy at that time.[78] All the same, Truman accepted Forrestal's basic premise that eventually the Services would need more direct access to weapons, and in September he raised no objection when the National Security Council confirmed (NSC 30) that the Armed Forces should expand their training for atomic warfare and integrate nuclear weapons into their regular military planning.[79]

NSC 30 removed the final obstacle to making the air-atomic strategy the centerpiece of postwar American defense policy. Now assured of increased access to weapons and training for their personnel, the Air Force and the Navy moved quickly to expand and refine their capabilities for atomic warfare. For the Navy, this meant pressing ahead with plans for laying the keel of the first in a new generation of super carriers; for the Air Force, it meant bolstering the Strategic Air Command, which continued to have a monopoly on the nuclear mission. A critical factor in preserving the Air Force's dominant position was the appointment of a new SAC commander, Lieutenant General Curtis E. LeMay, who took charge in October 1948, bringing with him a reputation for solving problems and getting results. The architect of the devastating conventional "fire bomb raids" against Japan in World War II, LeMay also had helped to coordinate the 1946 *Crossroads* tests in the Pacific and had thus acquired a working familiarity with nuclear weapons. When he assumed command, SAC had only about 20 atomic-modified B–29s fit for duty. Concentrating on expanding SAC's nuclear capability, LeMay set about eliminating equipment deficiencies and training personnel one group at a time, starting with restoring the 509th to its wartime level of efficiency.[80]

Meanwhile, the budget process for FY 1950 plodded along, with the Berlin situation and the presumed intimidating power of the atomic bomb overshadowing Palestine and other trouble spots where the need for conventional forces predominated. Forrestal continued to favor balanced capabilities, but a detailed analysis of

Service estimates by the Budget Advisory Committee, a tri-Service panel of senior officers chaired by General Joseph T. McNarney, USAF, revealed an enormous gap between the requirements for a balanced force posture and the resources available under the President's budget ceiling.[81] To narrow the difference, the Joint Chiefs reduced the scale and scope of planned operations under the FLEETWOOD (formerly HALFMOON) strategy by eliminating certain Army and Air Force units and deleting the naval air offensive in the eastern Mediterranean. No matter how they priced it, however, the savings from these cuts failed to produce a military budget within the President's spending limit. Convinced that the chiefs had done their best and realizing that they were deadlocked, Forrestal told them on October 15 that he would entertain the proposal of an "intermediate" budget somewhat larger than the President had said he would allow.[82]

To justify the increase, the Joint Chiefs hastily compiled a catalog of commitments that the military budget would have to support. This list was the first in a long line of such statements that the Joint Chiefs would routinely produce during the Cold War to support Service requirements. While the chiefs amply documented the wide range of military obligations the country faced, they fell short of providing a useful framework for assessing military spending. At no point did they put a price tag on U.S. commitments, attempt to link them directly to force requirements, or establish an order of priority for military programs. Given these shortcomings, the chiefs' catalog, while informative, was not very useful as budgetary guidance. Later iterations of these joint planning documents would be similarly defective and would come under sharp criticism from the Office of the Secretary of Defense (OSD) and the White House for failing to sort out and prioritize military requirements. But in view of the consensus-oriented rules under which the JCS operated and the difficulties these procedures posed in allocating resources, a better product was probably unattainable.[83]

A more practical tool for assessing Service requirements was the NSC's evaluation of national security policy (NSC 20/4), which appeared toward the end of November 1948. Prepared mainly by Kennan and State's Policy Planning Staff in response to Forrestal's request for guidance, NSC 20/4 predicted an indefinite period of tension between the United States and the Soviet Union. Cautioning against "excessive" U.S. armaments, the report urged "a level of military readiness which can be maintained as long as necessary as a deterrent to Soviet aggression." These recommendations were not much help to Forrestal in evaluating the relative merits of competing weapons systems or strategic concepts. But they left no doubt that a defense establishment tailored for the long haul and a posture of deterrence would be more in keeping with security needs than one with large, immediate increases for fighting a war that might not materialize.[84]

On December 1, 1948, Forrestal submitted his defense budget for FY50. Actually, he submitted two budgets—one for $14.4 billion that fell within the President's spending ceiling; and a second for nearly $17 billion. (Forrestal dismissed as excessive and unrealistic a third set of estimates, prepared by the JCS, totaling nearly $24 billion.) The first budget, Forrestal explained, would allow for a defense establishment of 10 Regular Army divisions, 287 combatant ships in the Navy, and a 48-group Air Force. The second, which the Secretary of Defense personally endorsed as preferable for national security purposes, would support a defense establishment of 12 divisions, 319 combatant vessels, and 59 air groups. Forrestal added that he had shown these figures to Secretary of State George C. Marshall, who concurred that the larger budget would provide better support for the country's foreign policy.[85] All the same, Truman was unimpressed. Buoyed by his recent come-from-behind victory at the polls, he told the Bureau of the Budget to ignore Forrestal's larger submission. "I don't know why he sent two. The $14.4 billion budget is the one we will adopt."[86]

Refusing to accept the President's decision as final, Forrestal tendered an amended request on December 20 that proposed adding $580 million to fund six additional air bombardment groups in the Air Force. In line with the emerging reliance on air-atomic power as the country's first line of defense, Forrestal argued for the money as the most practical way of addressing the threat posed by "our most probable enemy." Whether he agreed or not with Forrestal's reasoning, Truman continued to give fiscal considerations priority and turned down the Secretary's request without giving it a second thought.[87] Early the following year, in testifying to Congress on the President's 1950 budget, the Joint Chiefs expressed skepticism that it would assure proper readiness in an emergency, but declined to criticize the President for his decision to hold down military spending for fiscal and economic reasons. According to Admiral Denfeld, the budget was "the best division of funds that we could agree on at the time."[88]

THE STRATEGIC BOMBING CONTROVERSY

The strategy and budget debates of 1948 left no doubt that the United States was moving toward a defense posture centered on strategic bombardment with nuclear weapons. While Truman, Forrestal, and other senior administration figures continued to pay lip service to the need for balanced forces, the reality was quite different. Not everyone agreed that reliance on strategic bombing was a sound course to follow, certainly not the Navy, which had its own competing view of strategy and weapons. But in practical terms, the air-atomic strategy had considerable appeal. An intimidating threat, it seemed feasible within the limits of existing technology, had strong bipartisan support in Congress, and could be priced to fit virtually any reasonable spending limit the White House might set. Assuming he had a mandate to proceed, LeMay set about transforming the Strategic Air Command into an all-atomic strike force that grew from a handful of atomic-capable aircraft when he took over in October 1948 to more than 250 a year and a half later. Most of the bombers in SAC's inventory were medium-range B–29s or B–50s (an upgraded version of the B–29), which required overseas bases to reach Soviet targets. A growing number, however, were B–36s that could reach targets in the Soviet Union from bases in the United States.[89]

Affirmation of the air-atomic strategy put major stresses on the JCS, revealing vital shortcomings in their ability to function as a deliberative corporate body. In assessing the chiefs' performance, Forrestal believed a key weakness was the absence of a presiding officer, or chairman, to steer the deliberations. As the only member without Service responsibilities, Admiral Leahy had performed something approximating this function in World War II, but after the war his role and influence had diminished as his health declined. To fill the void, Forrestal persuaded General of the Army Dwight D. Eisenhower, the president of Columbia University in New York, to return to Washington on a part-time basis as his "military consultant." Eisenhower met off and on with the chiefs between mid-December 1948 and late June 1949 and devoted most of his time to war plans and budget matters.[90]

Eisenhower's appointment was a stop-gap measure until Congress could create a permanent position, one of a list of reforms that Forrestal deemed essential for unification to succeed. In December 1948, declaring that his views had changed, Forrestal came out strongly for giving the Secretary of Defense enhanced powers and assistance. Among the measures he proposed was legislative authority to appoint a "responsible head" of the JCS and to increase the size of the Joint Staff.[91] The resulting amendments to the National Security Act took effect in August 1949 and converted the NME into the Department of Defense. In the legislation, Congress added a Chairman, Joint Chiefs of Staff, and gave him "precedence" over all other officers in the Armed Forces. His statutory responsibilities were to preside at JCS meetings, set the agenda, and notify the Secretary of Defense of any disagreements. The Chairman could not vote in JCS deliberations nor could he command any military forces. Clarifying the JCS role in the policy process, Congress designated the Chairman and the Joint Chiefs of Staff collectively as the "principal military advisers" to the President, the Secretary of Defense, and the National Security Council.[92]

By the time the 1949 amendments became law, Forrestal was dead, the victim of an apparent suicide. Frustrated, overworked, and mentally exhausted, he had reluctantly stepped down as Secretary of Defense in March 1949 to make way for his successor, Louis Johnson. A prominent West Virginia attorney, Johnson had been Assistant Secretary of War in the Roosevelt administration and Truman's principal fund-raiser for the 1948 campaign. Johnson's mandate from the President was to bring order and discipline to the Pentagon and make the Services and the JCS toe the line on military spending. Even without the 1949 amendments, Johnson felt he had the power and authority to accomplish his mission. Using the Joint Chiefs less and less, Johnson embraced budgetary procedures that relied more on his own staff to make the tough decisions on military spending and the allocation of resources.[93]

Johnson's first major action as Secretary of Defense came in April 1949, when he cancelled the Navy's new super carrier, the USS *United States*. Incorporating design features derived from the *Crossroads* tests, the *United States* was to be a 65,000-ton, flush-deck carrier capable of accommodating aircraft carrying a 10,000-pound payload, roughly the same as an atomic bomb. Though Johnson strongly endorsed the air-atomic strategy, he acted on economic grounds and believed the Navy's super carrier needlessly duplicated the Air Force's strategic bombing function. His first and foremost aim was to hold down military spending, a goal that became all the more imperative in the summer of 1949, when President Truman disclosed that the defense budget for FY 51 would have to come down to $13 billion to help stave off a recession. An escalation of the quarrel between the Air Force and the Navy soon followed, producing charges and counter charges about the relative merits of long-range bombers versus super carriers, and culminating in a highly publicized congressional investigation. By the autumn of 1949, the senior echelons of the Navy were in open revolt against Johnson's policies and authority.[94]

While these controversies swirled in the public arena, the Joint Chiefs were trying to develop a more rational framework for analyzing the strategic environment and the competing Service claims for rival weapons systems. The impetus behind this effort came from a request by Forrestal in October 1948 for an analysis of two issues: the chances of success of delivering the strategic air offensive contemplated in current war plans, and an evaluation of the effects of SAC's planned air offensive on the Soviet Union's war effort.[95] Forrestal hoped to use the results to help defend his FY 50 budget submission to the President. But owing to the complexity and sensitivity of the issues raised, the Joint Chiefs wanted more time to assure thorough examinations. Initially, the JCS assigned the weapons effects study to an ad hoc body that reported to the Joint Strategic Plans Committee, and the other study, on the chances of success for the air offensive, to the Air Force. When the Air Force replied in December 1948 with a highly generalized boilerplate response, the JCS asked the Weapons Systems Evaluation Group (WSEG), a new technical support organization, to step in.[96]

Lieutenant General Hubert R. Harmon, USAF, a member of the U.S. military staff to the United Nations, chaired the ad hoc weapons-effects study group. A classmate of Eisenhower's and Bradley's at West Point, Harmon had served briefly as commander of Thirteenth Air Force in the South Pacific in World War II. Exactly how or why Harmon came to chair the effort is unclear; however, he had a reputation for being tactful and fair-minded that enhanced the study's objectivity and credibility. To assist him, Harmon assembled an inter-Service team of one Air Force officer, two Navy officers, and two Army officers.[97]

The Harmon committee looked only at SAC's role and the atomic phase of the air offensive, which would take place at the outset of a war. It made no attempt to evaluate the impact of a planned follow-on offensive with conventional bombs, nor did it look at possible Navy contributions under the plan since there was no assurance that the Navy would be allocated nuclear weapons or have the requisite capabilities for delivering them.[98] The committee confirmed that SAC's attacks under the current JCS-approved emergency war plan (now code-named TROJAN) would exact a heavy toll on the Soviet Union. SAC's targets were 70 urban-industrial complexes, with the destruction of Moscow and Leningrad the top priorities. Should all planes and bombs reach their targets (an assumption the WSEG study had yet to test), casualties from the initial attack would be in the vicinity of 2.7 million killed and another 4 million injured. Life for the 28 million survivors in the target areas would be "vastly complicated." The Air Force estimated that the destruction inflicted by the bombing would reduce Soviet industrial production for war-related purposes by 50 percent, with the heaviest impact falling on the petroleum industry. Based on its own separate assessments, the Harmon committee pared this estimate to a drop in production of 30 to 40 percent.

The committee doubted whether the atomic offensive would "seriously impair" ongoing Soviet operations in Western Europe, the Middle East, or the Far East. Large stockpiles of war reserves would allow Soviet forces to operate for some time before the effects of the disruptions to industry caused by the bombing reached the battlefield. Nor was the committee convinced that the planned air attacks would undermine the will and capacity of the Soviet population to resist, a key objective of the EWP. Nevertheless, the committee concluded that the atomic bomb remained "a major element of Allied military strength" and would constitute "the only means of rapidly inflicting shock and serious damage to vital elements of the Soviet war-making capacity." Even if not initially decisive, the crippling effects of nuclear weapons would tilt the balance sooner or later in favor of the West.[99]

Though the Joint Chiefs received the Harmon report in May 1949, they waited until late July to give it to the Secretary of Defense. The reason for the delay was a disagreement over how to handle Air Force objections to the committee's analysis of collateral damage, which failed to consider the impact of fires started by the bombing. General Vandenberg, the Air Force Chief of Staff, wanted the report amended to address this and several other issues the Air Force had raised, whereas Admiral Denfeld thought it should go up the chain of authority as written. Eventually, the Secretary of Defense received the report unchanged, but with a covering note explaining the Air Force's dissenting views.[100]

Only the Secretary of Defense, the Joint Chiefs, and their immediate aides saw the Harmon report. President Truman never received a copy, though he knew of its existence and expressed an interest in seeing it and the WSEG study as well.[101] While Truman wanted economy in defense spending, he also remained a firm believer in a balanced force posture. At this juncture, the President was uneasy over a proposed reapportionment in the Air Force budget to free funds for the procurement of additional B–36s, despite reports that the planes were experiencing significant engine problems. Prodded by the Bureau of the Budget and by his White House naval aide, Rear Admiral Robert L. Dennison, Truman inquired in April 1949 about the status of these studies, telling his staff that he wanted to avoid "putting all of our eggs into one basket." Secretary of Defense Johnson assured the President that when the time was right he would receive a full briefing, but that it could take up to a year for the Pentagon to complete its evaluations.[102]

As Johnson's response suggests, the WSEG study had fallen behind schedule owing to WSEG's start-up problems and disagreements between the Air Force and the Navy over the intelligence data the study should use. WSEG was the brainchild of Vannevar Bush, President Roosevelt's chief scientific advisor on the atomic bomb in World War II and first Chairman of the Research and Development Board (RDB) when that agency acquired statutory status in 1947. According to his biographer, Bush regarded WSEG "as the epitome of the professional partnership between soldiers and scientists that he had tried to foster since 1940."[103] Having worked closely with the JCS in World War II, Bush seriously doubted that they could detach themselves from Service interests and responsibilities, act as a unitary body of strategic advisors, or deal intelligently and effectively with scientific and technical matters. Advocating a greater role for science and scientists in defense affairs, he called for "dispassionate, cold-blooded analysis of facts and trends," and persuaded Secretary Forrestal that there should be "a centrally located, impartial and highly qualified group" to provide the JCS with "objective and competent advice" on current and future weapons systems.[104]

Initially, the Joint Chiefs were concerned that the new organization Bush proposed might infringe on their functions. The Joint Strategic Survey Committee was especially uneasy and warned lest "technical evaluations" become "operational evaluations" that could encroach on JCS responsibilities.[105] But after lengthy discussions with Forrestal and Bush, the JCS finally accepted the WSEG proposal at the Newport Conference in August 1948. Even so, it took until December for the JCS, Forrestal's office, and the RDB to agree on a directive laying out the terms of reference for the group's work, and 6 months more for WSEG to recruit a mixed military-civilian staff. WSEG took up offices in the Pentagon, within the secure restricted area set aside for the Joint Staff and other JCS components on the second level. Many of those who worked for WSEG were alumni of the Manhattan Project in World War II, an indication of how the new organization viewed its mission and where it expected to concentrate its efforts.[106]

Even though the strategic delivery study rated top priority on WSEG's agenda, it did not receive authorization to go forward until late August 1949, when the JCS finally approved intelligence data for the study.[107] At issue was the Air Force contention that Soviet air defenses were technologically substandard and spread too thin to pose a significant obstacle to attacking U.S. bombers.[108] Citing a "dearth of reliable intelligence," Admiral Denfeld challenged this notion and insisted that the Joint Intelligence Committee (JIC) conduct a review.[109] The JIC's preliminary analysis concurred with Denfeld that the Air Force had oversimplified the situation. But in a detailed follow-up report, the committee agreed with the Air Force that by and large Soviet air defenses were second rate. Still, it also pointed to recent improvements in air defense radars that suggested a more complex and effective Soviet air defense environment than the Air Force was anticipating.[110]

In view of the uncertainties surrounding Soviet air defenses, WSEG leaned toward the side of caution and produced a less than favorable report (WSEG R-1) on the chances of success for the planned air offensive. Knowing President Truman's interest in the subject, Secretary Johnson arranged for the WSEG director, Lieutenant General John E. Hull, USA, to hold a briefing at the White House on January 23, 1950, immediately prior to submitting R-1 to the JCS. While calculating that 70 to 85 percent of the attacking aircraft would reach their targets, Hull cited gaps in intelligence and logistical deficiencies that would reduce the effectiveness of the operation. Among SAC's vulnerabilities were a limited aerial refueling capability, competing demands for transport aircraft, and heavy dependence on overseas operating and staging bases. Overall, WSEG estimated that SAC could carry out its mission, but not to the full extent envisioned in current war plans without correcting identifiable deficiencies.[111]

Even though the Hull report presented a conservative view of the chances of complete success for the air offensive, there was no immediate rush to overhaul U.S. war plans or devise a new strategy. Developments on other fronts—the creation of NATO linking the security of Europe to the United States, the recent Communist victory in China, and the discovery that the Soviets had acquired an atomic capability—were shifting the debate on defense and military policy to broader global issues. In many respects, the war plans the Joint Chiefs had so painstakingly developed and refined were becoming irrelevant and obsolete. On the other hand, the preparation of these plans gave the Joint Chiefs a better appreciation for the problems of waging war against the Soviet Union and underscored yet again the critical importance of inter-Service cooperation.

NOTES

1. Robert H. Ferrell, *Harry S. Truman: A Life* (Columbia: University of Missouri Press, 1994), 162–176.

2. Letter, Truman to Gordon McDonough, August 29, 1950, *Public Papers of the Presidents of the United States: Harry S. Truman,* 1950 (Washington, DC: GPO, 1965), 617–618, hereafter cited as *Truman Public Papers.*

3. See Dale R. Herspring, *The Pentagon and the Presidency: Civil-Military Relations from FDR to George W. Bush* (Lawrence: University Press of Kansas, 2005), 54.

4. Message to Congress, December 19, 1945, in Alice Cole et at., eds., *The Department of Defense: Documents on Establishment and Organization, 1944–78* (Washington, DC: Historical Office, Office of the Secretary of Defense, 1978), 7–17.

5. U.S. Senate, Committee on Naval Affairs, *Unification of the War and Navy Departments and Postwar Organization for National Security,* 79:1 (Washington, DC: GPO, 1945).

6. See letter, Leahy to Hull, May, 16, 1944, U.S. Department of State, *Foreign Relations of the United States: The Conference of Berlin (The Potsdam Conference), 1945* (Washington, DC: GPO, 1960), I, 265.

7. Walter S. Poole, "From Conciliation to Containment: The Joint Chiefs of Staff and the Coming of the Cold War," *Military Affairs* 42 (February 1978), 12–16; James F. Schnabel, *The Joint Chiefs of Staff and National Policy, 1945–1947* (Washington, DC: Joint History Office, 1996), 7–32.

8. Robert A. Pollard, *Economic Security and the Origins of the Cold War, 1945–1950* (New York: Columbia University Press, 1985), 20–23.

9. JCS Military Policy Statement, "Basis for the Formulation of Military Policy," September 20, 1945, JCS 1496/3.

10. Christian Brahmstedt, ed., *Defense's Nuclear Agency, 1947–97* (Washington, DC: Defense Threat Reduction Agency, 2002), 1–13; Eban A. Ayers Diary entry, October 14, 1946, in Robert H. Ferrell, ed., *Truman in the White House: The Diary of Eban A. Ayers* (Columbia: University of Missouri Press, 1991), 161; David Alan Rosenberg, "U.S. Nuclear Stockpile, 1945 to 1950," *Bulletin of the Atomic Scientists* 38 (May 19, 1982), 25–30.

11. Richard G. Hewlett and Oscar E. Anderson, Jr., *A History of the United States Atomic Energy Commission,* vol. I, *The New World* (Washington, DC: U.S. Atomic Energy Commission, 1962), 408–427, 482–530.

12. McGeorge Bundy, *Danger and Survival* (New York: Random House, 1988), 166; Hewlett and Anderson, *The New World,* 582–597.

13 Letter, JCS to President, October 23, 1945, "Military Policy as to Secrecy Regarding the Atomic Bomb," JCS 1471/4.

14 Hewlett and Anderson, 581–582.

15 Lloyd J. Graybar, "The 1946 Atomic Bomb Tests: Atomic Diplomacy or Bureaucratic Infighting?" *Journal of American History* 72 (March 1986), 888–907.

16 Brahmstedt, 13–14; JCS Evaluation Board for Operation *Crossroads*, "The Evaluation of the Atomic Bomb as a Military Weapon: Final Report," June 30, 1947, President's Secretary's File (PSF), Truman Library.

17 Walton S. Moody, *Building A Strategic Air Force* (Washington, DC: Air Force History and Museums Program, 1996), 141–142; Ian Clark and Nicholas J. Wheeler, *The British Origins of Nuclear Strategy, 1945–1955* (Oxford: Clarendon Press, 1989), 116.

18 Jeffrey G. Barlow, *Revolt of the Admirals: The Fight for Naval Aviation, 1945–1950* (Washington, DC: Naval Historical Center, 1994), 106–107; David Alan Rosenberg, "American Postwar Air Doctrine and Organization: The Navy Experience," in Alfred F. Hurley and Robert C. Ehrhart, eds., *Airpower and Warfare* (Washington, DC: Office of Air Force History, 1979), 245–278.

19 Schnabel, 70–72.

20 Memo, Leahy to Truman, October 16, 1945, "Reorganization of National Defense," JCS 749/29.

21 Ronald H. Cole et at., *The History of the Unified Command Plan, 1946–1999* (Washington, DC: Joint History Office, 2003), 11.

22 Quoted in Walter S. Poole, "Joint Operations," in Jacob Neufeld, William T. Y'Blood, and Mary Lee Jefferson, eds., *Pearl to V-J Day: World War II in the Pacific* (Washington, DC: Air Force History and Museums Program, 2000), 19.

23 Memo, CNO to JCS, February 1, 1946, "Command Structure in the Pacific Theater," JCS 1259/6.

24 Quoted in Poole, "Joint Operations," 18.

25 Schnabel, 81–82.

26 Memo, CSA to JCS, September 17, 1946, "Unified Command Structure," JCS 1259/12; Memo, CNO to JCS, December 5, 1946, "Unified Command Plan," JCS 1259/26; Memo, JCS to President, December 12, 1946, enclosing Unified Command Plan, JCS 1259/27. Truman approved the plan on December 14, 1946.

27 *History of the UCP*, 11–13. The seven original geographic commands were Far East Command, Pacific Command, Alaska Command, Northeast Command, Atlantic Fleet, Caribbean Command, and European Command.

28 J.C. Hopkins and Sheldon A. Goldberg, *The Development of Strategic Air Command, 1946–1986* (Offutt Air Force Base, NE: Office of the Historian, Strategic Air Command, 1986), 2–3.

29 JSSC Report to the JCS, February 20, 1946, "Missions of the Land, Sea and Air Forces," JCS 1478/8.

30 Schnabel, 113.

31 Herman S. Wolk, *Planning and Organizing the Postwar Air Force, 1943–1947* (Washington, DC: Office of Air Force History, 1984), 157.

32 Walter Millis, ed., *The Forrestal Diaries* (New York: Viking Press, 1951), 221–223; letter, Patterson and Forrestal to Truman, January 16, 1947, in Cole et al., *Documents on Defense*

Organization, 31–33. See also Townsend Hoopes and Douglas Brinkley, *Driven Patriot: The Life and Times of James Forrestal* (New York: Knopf, 1992), 341–343.

33 National Security Act of 1947, July 26, 1947, PL 253, in Cole et al., *Documents on Defense Organization*, 35–50.

34 EO 9877, "Functions of the Armed Forces," in Cole et at., *Documents on Defense Organization*, 267–270.

35 Steven L. Rearden, *History of the Office of the Secretary of Defense: The Formative Years, 1947–1950* (Washington, DC: Historical Office, Office of the Secretary of Defense, 1984), 385–402.

36 Harry S. Truman, *Memoirs*, vol. II, *Years of Trial and Hope* (Garden City, NY: Doubleday, 1956), 100.

37 Walter Lippmann, *The Cold War: A Study in U.S. Foreign Policy* (New York: Harper and Row, 1947).

38 Richard G. Hewlett and Francis Duncan, *A History of the United States Atomic Energy Commission*, vol. 2, *Atomic Shield, 1947–1952* (Washington, DC: Atomic Energy Commission, 1972), 53–55, 196–148; Schnabel, 133–135.

39 Matthew A. Evangelista, "Stalin's Postwar Army Reappraised," *International Security* 7 (Winter 1982/1983), 110–138. Evangelista argues that the JCS consistently overestimated Soviet strength. He attributes this error to a lack of reliable information rather than to an intentional effort to deceive.

40 Samuel R. Williamson, Jr., and Steven L. Rearden, *The Origins of U.S. Nuclear Strategy, 1945–1953* (New York: St. Martin's Press, 1993), 193; Steven T. Ross, *American War Plans, 1945–1950* (London: Frank Cass, 1996), 54; Donald P. Steury, "How the CIA Missed Stalin's Bomb," *Studies in Intelligence*, 199:1 (2005), 19–26.

41 John Lewis Gaddis, *Strategies of Containment* (New York: Oxford University Press, 1982), 39–40; see also David Mayers, *George Kennan and the Dilemmas of U.S. Foreign Policy* (New York: Oxford University Press, 1988), 122–123; and John Lewis Gaddis, *George F. Kennan: An American Life* (New York: Penguin, 2011).

42 Letter, Leahy to Hull, May 16, 1944, derived from JCS 838/1.

43 For the details, see Ross, *American War Plans*, 25–52.

44 Schnabel, 70–75.

45 Millis, 351–352; Melvyn P. Leffler, *A Preponderance of Power: National Security, the Truman Administration, and the Cold War* (Stanford, CA: Stanford University Press, 1992), 221–226. Leffler sees Truman as an essentially ambivalent figure who provided less than dynamic or credible leadership. While committed to preserving the preponderant U.S. position in world affairs, he ignored the gap between commitments and capabilities.

46 During the unification debate, both Forrestal and Secretary of War Patterson had recommended that Congress give the JCS the responsibility under the law to "make recommendations for integration of the military budget." For unexplained reasons, Congress failed to include this provision in the statement of JCS functions when it passed the National Security Act in July 1947.

47 Rearden, *Formative Years*, 311.

48 Ibid., 89–115.

49 U.S. President's Air Policy Commission, *Survival in the Air Age* (Washington, DC: U.S. President's Air Policy Commission, January 1, 1948).

50 Moody, 125–127; Barlow, 105–121.

51 Rearden, *Formative Years*, 35–36; Hoopes and Brinkley, 351–153.

52 Jonathan Haslam, *Russia's Cold War: From the October Revolution to the Fall of the Wall* (New Haven: Yale University Press, 2011), 99.

53 See William R. Harris, "March Crisis 1948, Act I," *Studies in Intelligence* 10, no. 4 (Fall 1966), 1–22; and William R. Harris, "March Crisis 1948, Act II," *Studies in Intelligence* 11, no. 2 (Spring 1967), 9–36 (RG 263, Records of the Central Intelligence Agency, National Archives). Written largely from interviews, these articles, now declassified, provide fascinating insights into the origins of the Czech coup, the March Crisis, and the inner workings of Soviet and U.S. intelligence leading up to the 1948 Berlin blockade.

54 Memo, Hillenkoetter to Truman, March 16, 1948, PSF, Truman Library. See also Donald P. Steury, "Origins of CIA's Analysis of the Soviet Union," in Gerald K. Haines and Robert E. Leggett, eds., *Watching the Bear: Essays on CIA's Analysis of the Soviet Union* (Langley, VA: Center for the Study of Intelligence, 2001), 1–14.

55 Jean Edward Smith, *Lucius D. Clay: An American Life* (New York: Henry Holt and Co., 1990), 166–167; Cable, Clay to Chamberlin, March 5, 1948, in Jean Edward Smith, ed., *The Papers of General Lucius D. Clay: Germany, 1945–1949*, 2 vols. (Bloomington: Indiana University Press, 1974), II, 568.

56 Memo, Hillenkoetter to Truman, December 22, 1947, PSF, Truman Library.

57 Rearden, *Formative Years*, 316–330.

58 Lawrence S. Kaplan, *NATO 1948: The Birth of the Transatlantic Alliance* (Lanham, MD: Rowman and Littlefield, 2007), 192–144, 199–159.

59 "Brief of Short-Range Emergency War Plan 'HALFMOON,'" May 6, 1948, JCS 1844/4; Memo, CNO to JCS, April 5, 1948, "Planning Guidance for Medium-Range Emergency Plan," JCS 1844/2.

60 Kenneth W. Condit, *The Joint Chiefs of Staff and National Policy, 1947–1949* (Washington, DC: Office of Joint History, 1996), 156.

61 Diary Entry, May 6, 1948, Leahy Papers, Library of Congress; Memo, Leahy to JCS, May 13, 1948, "Brief of Short-Range Emergency Plan "HALFMOON," JCS 1844/6; Millis, 158, 161–162; David Alan Rosenberg, "Toward Armageddon: The Foundations of United States Nuclear Strategy, 1945–1961" (Ph.D. diss., University of Chicago, 1983), 110.

62 Notes of an informal meeting between Vice chief of the Air Staff and Maj. Gen. Richard C. Lindsay, May 10, 1948, quoted in Stephen Twigge and Len Scott, *Planning Armageddon: Britain, the United States and the Command of Western Nuclear Forces, 1945–1964* (Amsterdam: Harwood, 2000), 30–31.

63 Warner R. Schilling, "The Politics of National Defense: Fiscal 1950," in Warner R. Schilling, Paul Y. Hammond, and Glenn H. Snyder, *Strategy, Politics, and Defense Budgets* (New York: Columbia University Press, 1962), 155–164.

64 Millis, 135–138.

65 Memo, Forrestal to NSC, July 10, 1948, "Appraisal of the Degree and Character of Military Preparedness Required by the World Situation," *FRUS, 1948*, I, 589–592; letter, Forrestal to Truman, July 10, 1948, ibid., 592–593.

66 Memo, Truman to Forrestal, July 13, 1948, PSF, Truman Library.

67 Alonzo L. Hamby, *Beyond the New Deal: Harry S. Truman and American Liberalism* (New York: Columbia University Press, 1973), 209–211.

68 Memo, JCS to SECDEF, October 10, 1947, "Problem of Palestine," JCS 1684/3.

69 Memo, JCS to Truman, April 19, 1948, "Provision of U.S. Armed Forces in Palestine," JCS 1685/13.

70 Rearden, *Formative Years*, 187–194; Condit, 51–57.

71 Memo, JCS to SECDEF, July 22, 1948, "U.S. Military Course of Action with Respect to the Situation in Berlin," JCS 1907/3.

72 Douglas to SecState, June 26, 1948, *FRUS, 1948*, II, 923–924.

73 Millis, 457.

74 Clark and Wheeler, *British Origins of Nuclear Strategy*, 128–129.

75 John Lewis Gaddis, *We Now Know: Rethinking Cold War History* (Oxford: Clarendon Press, 1997), 198.

76 Rearden, *Formative Years*, 139.

77 Hewlett and Duncan, *Atomic Shield*, 161–165, 174–176; Robert S. Norris, Thomas B. Cochran, and William M. Arkin, "History of the Nuclear Stockpile," *Bulletin of the Atomic Scientists* (August 1985), 106–109. Barlow, 127, discusses the impact of the levitation method.

78 Millis, 460–461; Rearden, *Formative Years*, 425–432.

79 Minutes, 21st Meeting of the National Security Council, September 16, 1948, PSF, NSC Series, Truman Library; NSC 30, "U.S. Policy on Atomic Warfare," September 10, 1948, *FRUS, 1948*, I, 624–628.

80 Moody, 229–233.

81 Condit, 124; Rearden, *Formative Years*, 342.

82 Ibid., 124–131.

83 Memo, JCS to SECDEF, November 2, 1948, "Existing International Commitments Involving the Possible Use of Armed Forces," circulated at NSC 35, November 17, 1948, *FRUS, 1948*, I, 656–662.

84 NSC 20/4, November 23, 1948, "U.S. Objectives with Respect to the USSR to Counter Soviet Threats to U.S. Security," *FRUS 1948*, I, 662–669.

85 Letter, Forrestal to Truman, December 1, 1948, *FRUS 1948*, I, 669–672.

86 Memo, Truman to Webb, December 2, 1948, PSF, Truman Library.

87 USAF CoS Memo, ca. December 16 or 20, 1948, quoted in Williamson and Rearden, *Origins of U.S. Nuclear Strategy*, 95.

88 Denfeld testimony, February 16, 1949, U.S. Congress, Senate, *Hearings: National Military Establishment Appropriations Bill for 1950*, 81:1, Pt. 1, 17.

89 Moody, 265. Other sources give larger numbers. Using JCS historical materials, Ross, *American War Plans, 1945–50*, 139, says that SAC had 521 atomic-capable aircraft by the end of 1949. Actually, as the JCS source makes clear, this was the total number of bomb-

ers in SAC's inventory at the time. Ross erroneously assumes that all SAC bombers had been converted.

90 Diary entry, December 13, 1948, in Robert H. Ferrell, ed., *The Eisenhower Diaries* (New York: W.W. Norton, 1981), 150–151; Condit, 139–152.

91 U.S. National Military Establishment, *First Report of the Secretary of Defense* (Washington, DC: GPO, 1948), 3–4.

92 National Security Act of 1947 as Amended and Approved, August 10, 1949, PL 216, in Cole et al., eds., *Documents on Defense Organization*, 84–106.

93 Rearden, *Formative Years*, 369–376.

94 Paul Y. Hammond, "Super Carriers and B-36 Bombers: Appropriations, Strategy and Politics," in Harold Stein, ed., *American Civil-Military Decisions: A Book of Case Studies* (Birmingham, AL: University of Alabama Press, 1963), 465–564, though dated, is still the best overview of the controversy; Moody, 286–313, is a measured account from the Air Force perspective; Barlow, 182f, and Paolo E. Coletta, *The United States Navy and Defense Unification, 1947–1953* (Newark: University of Delaware Press, 1981), give the Navy perspective.

95 Memo, SECDEF to JCS, October 23, 1948, "Evaluation of Current Strategic Air Offensive Plans," JCS 1952; Memo, SECDEF to JCS, October 25, 1948, "Evaluation of Effect on Soviet War Effort Resulting from Strategic Air Offensive," JCS 1953.

96 John Ponturo, *Analytical Support for the Joint Chiefs of Staff: The WSEG Experience, 1948–1976* (Arlington, VA: Institute for Defense Analyses, 1979), 52–53. For the Air Force response, see the Appendix to Memo, CoS USAF to JCS, December 21, 1948, "Evaluation of Current Strategic Air Offensive Plans," JCS 1952/1.

97 Note by the Secretaries to the JCS, October 25, 1948, JCS 1953; Moody, 295.

98 Details of the planned offensive are in Report, JSPC to JCS, December 13, 1948, "Atomic Weapons Supplement to TROJAN," JCS, 1974.

99 Report by the Ad Hoc Committee (Harmon Committee), May 12, 1949, "Evaluation of Effect on Soviet War Effort Resulting from the Strategic Air Offensive," JCS 1953/1.

100 Memo, CoS USAF to JCS, July 8, 1949, JCS 1953/4; Memo, CNO to JCS, July 19, 1949, JCS 1953/5; Moody, 295.

101 Truman, *Years of Trial and Hope*, 305.

102 Rearden, *Formative Years*, 406–407.

103 G. Pascal Zachary, *Endless Frontier: Vannevar Bush, Engineer of the American Century* (New York: Free Press, 1997), 340.

104 Ponturo, 23–25.

105 Report, JSSC to JCS, February 27, 1948, "Proposed Directive to the RDB," JCS 1812/5.

106 Ponturo, 27–48.

107 Note by the Secretaries to JCS, August 25, 1949, "Joint Intelligence Estimate," JCS 1952/8.

108 See the appendix to Memo, CoS/USAF to JCS, December 21, 1949, "Evaluation of Current Strategic Air Offensive Plans," enclosure to JCS 1952/1.

109 Memo, CNO to JCS, January 11, 1949, "Evaluation of Current Strategic Air Offensive Plans," JCS 1952/2.

110 Rpt, JIC to JCS, March 3, 1949, "Intelligence Aspects of Evaluation of Current Strategic Air Offensive Plans," JCS 1952/4; Report, JIC to JCS, August 25, 1949, "Joint Intelligence Estimate for Basing Operational Evaluation Success of the Strategic Air Offensive," JCS 1952/8.

111 WSEG R-1, "Report on Evaluation of Effectiveness of Strategic Air Operations," February 8, 1950, JCS 1952/11; Ponturo, 74–75, summarizes the White House briefing.

President Harry S. Truman meeting General Douglas MacArthur, USA, Wake Island, October 1950

Chapter 4

MILITARIZING THE COLD WAR

Between 1945 and 1950, relations between the United States and the Soviet Union underwent a 180-degree transformation. Erstwhile allies in the war against Germany and Japan, they became antagonists in a new global rivalry marked by the ominous expansion of Communist power and influence. While the Joint Chiefs of Staff repeatedly urged stronger military power to deal with this situation, their warnings had had limited effect on the Truman administration's fiscal or defense policies. Exercising tight control over military spending, Truman preferred to address the Communist challenge with political, economic, and diplomatic initiatives. Bowing to these realities, the JCS fashioned a defense posture and war plans oriented toward a single contingency—an all-out global conflict. Maintenance of balanced conventional forces with flexible capabilities gave way to reliance on strategic bombardment with nuclear weapons as the country's principal deterrent and first line of defense. Not everyone agreed that this was a sound course or that it adequately addressed the country's increasingly diverse security needs. But at the time, reliance on strategic bombing with nuclear weapons was the country's most practical, effective, and affordable form of defense.

PRESSURES FOR CHANGE

While nonmilitary responses to Soviet expansion had generally met with success, the growing intensity of the Cold War by 1950 was steadily pushing the Truman administration toward an expansion of U.S. military power. Despite its best efforts to avoid it, the "militarization" of the Cold War loomed larger than ever as pressures converged from three directions at roughly the same time: from Europe, where the signing of the North Atlantic Treaty in April 1949 created a new transatlantic community of security interests; from China, where the collapse of Chiang Kai-shek's Nationalist regime ushered in a Communist People's Republic headed by Mao Zedong with apparent designs on extending its power and influence across

Asia; and from the Soviet Union, where the detonation of a nuclear device in late August 1949 ended the American monopoly on the atomic bomb years ahead of predictions. Any one of those events could have triggered substantial alterations in American foreign and defense policy. Taken together, they were the catalysts for a wholesale transformation that would, with the sudden outbreak of the Korea conflict in June 1950, interject military power into the forefront of American responses to the escalating Cold War.

Prior to the Korean War, the administration's only clear-cut commitment embracing the possible use of military force to thwart Communist expansion was the North Atlantic Treaty. During preliminary consideration of the Alliance in the spring of 1948, the Joint Chiefs had endorsed the broad concept of a mutual security pact between Europe and the United States, but had warned against "major military involvement" without adequate preparations.[1] The White House and State Department noted the chiefs' concerns, but as Undersecretary of State Robert A. Lovett explained it, the Alliance's primary function was consultation in support of possible collective action. Like an insurance policy, its immediate role was to bolster Europe's confidence, expedite completion of the Economic Recovery Program, and deter the Soviets.[2]

The principal military component associated with the North Atlantic Treaty Organization (NATO) was the Mutual Defense Assistance Program (MDAP), a companion measure enacted in October 1949 to help rearm the European allies.[3] When the State Department unveiled the program, the Joint Chiefs balked out of concern that the Services might have to pay for MDAP out of their own budgets.[4] Though assured that assistance to NATO through MDAP would be a separate appropriation, the JCS remained uneasy lest it quickly deplete the dwindling war reserves left over from World War II and divert funding for routine military appropriations. In part to guard against NATO becoming a drain on American resources, the Joint Chiefs proposed an elaborate structure of councils, committees, boards, and regional planning groups to give the JCS detailed oversight powers of NATO's activities.[5] Secretary of State Dean Acheson acknowledged that as NATO became more established, pressures were bound to arise for a larger U.S. military role and a more complex organization. But for the time being he saw no pressing need and vetoed the chiefs' plan in preference for a simpler alliance structure that played down direct American military involvement and responsibility.[6]

Meanwhile, the disintegration of Nationalist rule on the China mainland was reshaping the security situation in the Far East. Given the leadership problems and poor performance of Nationalist Chinese forces during World War II, Chiang Kai-shek's collapse came as no surprise to the Joint Chiefs, who never had much confidence in the Generalissimo's ability to lead China out of the war as a great power. But because of China's strategic location, large population, and latent military potential, the JCS were also averse to a Communist takeover of the country and a loss of U.S. influence. As a result, throughout the postwar period, they consistently supported infusions of military aid to prop up the Generalissimo's regime, even as Chiang's rule began to crumble.

Of the President's various advisors, the most reluctant to come to Chiang's rescue was former Army Chief of Staff General George C. Marshall. In November 1945, President Truman had persuaded Marshall to go to China as his special representative. Marshall had served in China in the 1920s as a junior officer, and during World War II he had suffered through Stilwell's ordeal with Chiang. Like Stilwell, he had little confidence in the Generalissimo's leadership, reliability as an ally, or capacity to make effective use of U.S. assistance. But as a loyal soldier he felt duty-bound to accept the mission. Through Marshall's good offices, Truman hoped to broker a power-sharing agreement between Chiang and his Communist rival, Mao Zedong, a nominal ally of the Soviet Union, that would buy time for Chiang to strengthen his position and, with U.S. assistance and logistical support, move his troops into positions where they could effectively confront Mao's forces.[7] Chiang ignored Marshall's advice to seek a political compromise and sought to use his three-to-one advantage in troop strength to achieve a military solution. Exuding confidence, he overextended his forces into North China and Manchuria where they suffered one setback after another.[8]

By 1949, Chiang's military fortunes had declined to such an extent that he was taking steps to relocate his regime from the mainland to the island of Taiwan (Formosa) for what appeared to be a last stand. Short of massive U.S. intervention, the Joint Chiefs saw nothing that might turn the tide. Though they hoped to keep Taiwan (with or without Chiang there) from falling into Communist hands, they did not consider it sufficiently important to merit large-scale military action. The most they would recommend was the deployment of a few ships for deterrence purposes and the use of diplomatic leverage.[9] Since the Nationalist regime had strong political support in Washington, however, the JCS cautioned against abandoning Chiang altogether "at the eleventh hour" and urged the continuation of military assistance as long as Nationalist armies offered organized resistance.[10] Above all, they wanted to keep an American military presence on the China mainland and fought a losing battle with the State Department and the White House to keep the U.S. naval base at Qingdao (Tsingtao) open. Secretary of State Acheson thought the United States should disengage from Chiang as soon as possible and direct its efforts toward a rapprochement with Mao and the Communists. Counseled by the State

Department's "China Hands," Acheson believed it feasible "to detach [China] from subservience to Moscow and over a period of time encourage those vigorous influences which might modify it."[11] But he faced an uphill battle convincing Congress and overcoming the "China Lobby," which wanted stronger measures to resist the spread of communism in the Far East and additional support to save what remained of Chiang's regime.

THE H-BOMB DECISION AND NSC 68

The third and most fateful development that went into reshaping U.S. security perceptions was the discovery, reported to President Truman on September 9, 1949, that the Soviet Union had detonated a nuclear device similar in design to the implosion bomb the United States dropped on Nagasaki 4 years earlier. Without warning, the American nuclear monopoly had ended. The Intelligence Community later determined that the test—"Joe 1"—had taken place on August 29, 1949.[12] While analysts at the Central Intelligence Agency had known for some time that the Soviet Union had an atomic energy program, they miscalculated the Soviet Union's capacity to produce fissionable materials and failed to appreciate either the high priority Stalin attached to acquiring nuclear weapons or the crucial role Soviet espionage played in expediting the project.[13] As a result, they consistently underestimated both the extent of the Soviet effort and when it would come to fruition. Prior to Joe 1, the most recent interagency assessment of the Soviet program, dated July 1, 1949, placed the "probable" date for a Soviet atomic capability in the mid-1953 range, with the "possibility" of a nuclear test as early as mid-1950. Weighing the evidence, the consensus of the Intelligence Community was that the Soviet Union's "first atomic bomb cannot be completed before mid-1951."[14]

While the White House downplayed the achievement, the danger posed by growing Soviet military power was impossible to ignore. Up to that time, the Truman administration had relied implicitly, if not explicitly, on its nuclear monopoly to underwrite its policies. "As long as we can outproduce the world, can control the sea and can strike inland with the atomic bomb," Secretary of Defense Forrestal had once observed, "we can assume certain risks otherwise unacceptable."[15] With that formula now rendered suspect, it was no longer clear whether the United States could continue to mount effective deterrence and containment of the Soviet Union with the military capabilities it had on hand.

The most urgent need was to reassert the American lead in atomic energy. At issue was whether the United States should embark on a "quantum jump" into the unexplored realm of nuclear fusion and the development of "super" bombs based on hydrogen or thermonuclear design. Such weapons in theory could produce yields a thousand times greater than fission bombs. In November 1949, seeking advice on how to proceed, President Truman turned to the "Z Committee" of the National Security Council (NSC), composed of Secretary of State Acheson, Secretary of Defense Johnson, and the Chairman of the Atomic Energy Commission, David E. Lilienthal.[16] As the committee's military advisors, the Joint Chiefs acknowledged that high-yield super bombs would be hard to deliver and therefore would have limited military applications. All the same, the chiefs believed that for political and psychological reasons, it was absolutely imperative to proceed with a determination test. "Possession of a thermonuclear weapon by the USSR," the JCS insisted, "without such possession by the United States would be intolerable."[17] Lilienthal, however, harbored misgivings. Believing the H-bomb morally repugnant, he found the military's growing dependence on nuclear weapons deeply troubling and became convinced that the United States needed increased conventional capabilities and a renewed commitment to obtaining international control of atomic energy more than it needed thermonuclear weapons.[18]

On January 31, 1950, President Truman approved a compromise crafted by Acheson. As the first step, the President directed the AEC to explore the feasibility of the H-bomb, thus setting in motion a research and development program that would culminate on November 1, 1952, with the world's first thermonuclear explosion—a 10 megaton device that completely vaporized the Pacific atoll where the test was held. Meanwhile, he instructed the State and Defense Departments to review the country's basic national security policy.[19] Acheson shared Lilienthal's concern over the military's growing dependence on nuclear weapons, not least of all because he felt it limited diplomatic flexibility. But he also thought the United States had to have the H-bomb because "we do not have any other military program which seems to offer over the short run promise of military effectiveness."[20] In recommending a review of basic policy, Acheson later explained, he hoped to find some middle ground that would restore greater balance to the country's military posture and expand its ability to meet unforeseen contingencies.[21]

The Joint Chiefs embarked on the review with no such preconceptions or expectations. The previous November, Secretary of Defense Johnson had removed Admiral Louis Denfeld as Chief of Naval Operations on grounds of insubordination for his role in the "Revolt of the Admirals," which had challenged Johnson's authority through highly publicized attacks on his economy measures and the Air Force's strategic bombing capabilities.[22] Since then, Johnson had further tightened his control of the Defense Department and military spending. Confirming rumors and press reports, Johnson notified the Joint Chiefs in late February 1950 that the military budget for FY 52 would remain at approximately the same level as that projected for FY 51. Since the Secretary's estimates made no allowance for inflation, except for the Air Force, Johnson's hold-the-line spending policy amounted to a net decrease in programs for the Army and Navy. Using the Secretary's budget guidance as their frame of reference, the Joint Chiefs initially had to assume that any changes the State-Defense review might recommend would be modest at best.[23]

State's participants in the review had other ideas. Though ostensibly a collaborative effort, the dominant influence throughout was the new director of the Policy Planning Staff, Paul H. Nitze. A Wall Street bond trader before World War II, Nitze was well versed in statistics, which, as Vice Chairman of the U.S. Strategic Bombing Survey, he used to great effect in analyzing the results of Hiroshima and Nagasaki in 1945. Joining the State Department after the war, he had emerged as one of State's senior economic analysts and was instrumental in developing the Marshall Plan. A pragmatist and problem-solver by nature, Nitze gave a higher priority to the role of military power in foreign policy than his academically-minded predecessor, George F. Kennan, who had fallen out of favor with Acheson.[24]

JCS contributions to the review group's work came via the Joint Strategic Survey Committee (JSSC), represented by its Air Force member, Major General Truman H. Landon. Nitze recalled that initially Landon presented modest proposals to correct minor deficiencies in the existing force posture. He soon realized, however, "that we were serious about doing a basic strategic review and not just writing some papers which would help people promote special projects of one kind or another." From the quick change in Landon's outlook, Nitze detected that "there was, in fact, a revolt from within" brewing at the Pentagon against Johnson's fiscal policies and strategic priorities.[25]

The review process stretched from mid-February to early April 1950, when the State-Defense review group presented its findings (NSC 68) to the National Security Council. About a third of the report was a close analysis of the Soviet threat, drawn from intelligence estimates that indicated an inordinately large investment by the Soviet Union (up to 40 percent of its gross national product) in military power and war-supporting industries. By mid-1954—the "year of maximum danger" in the report's estimation—the Soviets would have a nuclear stockpile that could threaten serious damage to the United States. Extrapolating motives from capabilities, NSC 68 concluded that "the Soviet Union has one purpose and that is world domination." To frustrate the "Kremlin design," the paper urged the adoption of "a comprehensive and decisive program" resulting in "a rapid and sustained buildup of the political, economic, and military strength of the free world." While NSC 68 strongly endorsed the maintenance of effective nuclear capabilities for deterrence purposes, it also called for significant expansion of conventional air, ground, and sea forces "to the point where we are militarily not so heavily dependent on atomic weapons."[26]

Missing from NSC 68 were any cost estimates for the buildup or a projected allocation of resources among the armed Services. Both omissions were intentional—the first, in order not to frighten off President Truman from accepting the report, the second, to avoid provoking competition and friction within the Pentagon. According to one of his biographers, Acheson wanted to avoid overwhelming Truman with "programmatic details" by offering him instead "a general analysis oriented toward action."[27] Privately, Nitze and others who worked on NSC 68 estimated that it would require expenditures of $35 billion to $50 billion annually over the next 4 years. While Nitze made these calculations known to Acheson, there is no evidence that the Secretary of State conveyed them to Truman. The report conceded that the program would be "costly" and probably would require higher taxes to avoid deficit budgets. But it did not dwell on these points.[28]

Truman, for his part, continued to treat costs as his uppermost concern. Immediately after receiving NSC 68, he directed the creation of an ad hoc committee of economic experts to go over its findings and recommendations.[29] The consensus of this group was that, while the report's proposed course of action would be expensive, it would not place undue burdens on the economy as long as adequate safeguards were in place. The lone dissenting view was from the Bureau of the Budget, which saw adverse consequences for the economy should military spending rise sharply.[30] Truman agreed and said as much during a meeting with his budget director, Frederick J. Lawton, on May 23, 1950. "The President indicated," Lawton noted in his minutes of the meeting, "that we were to continue to raise any questions that we had on this program and that it definitely was not as large in scope as some of the people seemed to think." Translating the President's guidance into hard numbers, the BOB projected NSC 68 increases of $1 billion to $3 billion annually over the next 2 to 3 years.[31]

At the Pentagon, the Joint Chiefs were under similar pressure from Louis Johnson to curb expectations that NSC 68 would result in dramatic increases in military spending. Though Johnson paid lip service to the report, he resented its implied conclusion that the country's defense posture had become enfeebled under his trusteeship and took offense at what he saw as Acheson's unwarranted interference in Defense Department business. Going through all the proper motions, he directed the JCS and the Services to assemble estimates of the "general tasks and responsibilities" mandated under NSC 68, but to bear in mind that until the President indicated otherwise, guidelines and ceilings previously established for the FY 52 budget remained firmly in place.³² Confident that he had the matter in hand, Johnson left Washington on June 12, 1950, accompanied by Chairman of the Joint Chiefs (CJCS) General Omar N. Bradley, USA, for a tour of the Far East to discuss security arrangements for a Japanese peace treaty with General Douglas MacArthur, the theater commander.

On the eve of the Korean War, the fate of NSC 68 remained uncertain. President Truman had yet to approve the report and there were unmistakable signs that if and when he did, it would produce a considerably smaller buildup than its authors intended. The American defense establishment was already far larger and more costly than any country had ever known in peacetime, and to propose significant increases could have provoked a divisive national debate. Although NSC 68 offered ample evidence that the Soviet Union posed a growing threat to Western security, nothing in the report confirmed that spending three, four, or even ten times more on defense would afford better insurance against a Soviet attack than the existing investment of resources. Only after the outbreak of the Korean War would it become clear that the existing defense posture had failed to deter Communist aggression.

ONSET OF THE KOREAN WAR

Like the Soviet nuclear test the previous August, the North Korean invasion of South Korea on June 25, 1950 (Korea time), caught official Washington off guard. Even though NSC 68 had warned policymakers and military planners to be on the alert, no one expected a blatant act of aggression so soon. With most of its limited assets concentrated on Europe, the Intelligence Community had paid relatively little attention to the Far East prior to the North Korean attack. As one Army intelligence officer described the situation, "North Korea got lost in the shuffle and nobody told us they were interested in what was going on north of the 38th parallel." If war broke out or if a Communist takeover occurred, intelligence analysts expected Indochina rather than Korea to be the target.³³

Gathering information on Korea posed special difficulties. Wary of outsiders, MacArthur had banned the OSS from his theater in World War II and was suspicious of allowing its successor, the Central Intelligence Agency (CIA), into his midst after the war. Operating under severe restrictions, the Agency came up with generalized estimates that credited North Korea with limited capabilities for military aggression. As late as June 19, 1950, the CIA predicted that the Communists would confine their actions against the south to propaganda, infiltration, sabotage, and subversion.³⁴ An Army (G-2) intelligence report generated around this same time was more precise in identifying signs of enemy troop movements and the like, but by the time this information reached Washington, the war was in full swing.³⁵

Carefully planned and executed, the North Korean invasion had Stalin's blessing and support and involved approximately 90,000 North Korean troops, armed and trained by the Soviet Union. Early reports were vague, but as the fighting intensified it was apparent that this was no mere border skirmish, as initial reports suggested, but an all-out assault with the ultimate aim of destroying the American-supported Republic of Korea (ROK) in the south and absorbing the Korean Peninsula into the Communist orbit.[36]

Despite the seriousness of the situation, the Joint Chiefs initially saw no grounds for American military intervention, since at the time the United States had no formal defense commitments with South Korea. Divided in 1945 as an expediency at the 38th parallel to facilitate the disarming of Japanese troops by U.S. and Soviet forces, Korea had evolved into two distinct political entities—a Communist regime in the north headed by the Moscow-trained and Soviet-supported Kim Il-song, and a more democratic, U.S.-backed government in the south led by Syngman Rhee.[37] While aware of South Korea's vulnerability, the Joint Chiefs needed the occupation forces stationed there for duty elsewhere and wanted to limit further U.S. involvement. In September 1947, they declared the country to be of "little strategic interest" to the United States, the first step toward withdrawing U.S. troops. Completed in the spring of 1949, the withdrawal left behind large stockpiles of war materiel and a 500-member U.S. Korean Military Advisory Group (KMAG) to train and equip ROK forces against any threat from the north.[38] A few days prior to the invasion, during his trip to Tokyo in June 1950 with Secretary of Defense Johnson, General Bradley discussed the situation with Brigadier General William L. Roberts, USA, who had recently stepped down as KMAG's chief. "The ROK Army," Roberts assured the Chairman, "could meet any test the North Koreans imposed on it."[39]

The Communist success in routing the ROK forces shattered these comfortable assumptions and forced a hasty rethinking of U.S. policy. Like his predecessor during the early stages of World War II, President Truman met regularly with his top advisors and took a hands-on approach to the crisis; but unlike Roosevelt, he turned for advice more to civilians (in this case Secretary of State Dean Acheson) than to the Joint Chiefs of Staff. Owing to earlier decisions leading to the withdrawal of U.S. forces and the downgrading of South Korea's strategic importance, the JCS had not given much thought to the possibility of military action on the Korean Peninsula. When the crisis erupted, they lacked contingency plans for dealing with the emergency and had to improvise with impromptu assessments, personal opinions, and hastily drawn orders for mobilizing and moving forces.[40] Exactly why the Joint Chiefs were so unprepared and slow to respond remains unclear, but it doubtless reflected to some extent their continuing indifference toward Korea's strategic importance and the personnel ceiling under which the Joint Staff operated at the time. Even though the 1949 amendments had doubled the size of the Joint Staff, it remained a relatively small organization with limited capabilities.

Acheson, in contrast, appeared at these meetings with the President fully briefed and prepared, invariably bearing detailed memorandums and lists of recommendations that reflected dedicated staff work. Within hours of the news of the attack, he placed before the President proposals to expedite additional assistance to the South Koreans, to establish a "protective zone" around South Korea with U.S. air and naval forces, and to mobilize international opinion against the attack through the United Nations. Over the next several days, Acheson offered more recommendations, all moving inexorably toward large-scale U.S. military intervention under UN auspices. Six months earlier, Acheson, like the JCS, had more or less written off Korea and the rest of the East Asian mainland. But under the pressure of new events and still smarting from Republican attacks that his policies had "lost" China to the Communists, he had had a change of heart and saw the North Korean attack as a test of American will. "To back away from this challenge, in view of our capacity for meeting it," he wrote in his memoirs, "would be highly destructive of the power and prestige of the United States."[41]

The Joint Chiefs agreed that the North Korean attack challenged American resolve. But they accepted the need for military intervention with the utmost reluctance and initially hoped that air and naval power would suffice. The most readily available ground forces in the region were those of the Eighth Army, whose four divisions were all below authorized strength and short of critical weapons and equipment.[42] More aware than anyone of the constraints imposed by years of frugal defense budgets, the JCS made no attempt to disguise their belief that all-out intervention would be a highly risky business, requiring the mobilization of Reserve and National Guard units and emergency appropriations at a minimum. Should the war spread, warned the Air Force Chief of Staff, General Hoyt S. Vandenberg, the use of nuclear weapons would be the next step, a view shared by other senior commanders.[43] Above all, the JCS hoped to avoid committing U.S. ground troops but stopped short of recommending against such a move. Later, in explaining to Congress how the decision to send troops into Korea had come about, Louis Johnson observed that he and the Joint Chiefs had "neither recommended it nor opposed it."[44]

On Truman's shoulders rested all final decisions. While accepting Acheson's advice that the United States needed to make a forceful stand in Korea, he moved cautiously and intervened in incremental steps. Starting with the authorization of air and sea operations below the 38th parallel on June 26 (Washington time), he progressed to the commitment of U.S. ground forces 4 days later. Showing renewed interest in the fate of Taiwan, he ordered elements of the Seventh Fleet to take up station in the Formosa Strait to deter a resumption of the conflict between Chiang and the Chinese Communists.[45]

While accepting the need for action, Truman resisted the notion that the current emergency might compel a military buildup on the scale proposed in NSC 68. Sidestepping the problem, he inadvertently trivialized the dangers of intervention by publicly describing the North Korean attack as the work of "a bunch of bandits" that a "police action" could handle.[46] His description made it appear the United States could turn back the North Koreans and comfortably meet defense obligations elsewhere. But with the situation continuing to deteriorate, the President notified Congress on July 19, 1950, that at the urging of his military advisors, he was calling up units of the National Guard and would need additional military appropriations and authority to remove the ceiling on the size of the Armed Forces. Even so, he continued to defer action on adopting NSC 68 as administration policy and asked the National Security Council to reassess the report's requirements, with a view to providing recommendations by the beginning of September. Despite the ongoing conflict, he told the Bureau of the Budget that he did not want to place "any more money than necessary at this time in the hands of the Military."[47]

THE INCH'ON OPERATION

Truman believed that if the war in Korea could be contained and won quickly, he might get by with relatively modest increases in defense spending and other security programs. What he did not take into account was General Douglas MacArthur's penchant for independent and unpredictable behavior. American military policy had traditionally given commanders in the field wide latitude to deal with situations as they deemed appropriate. In MacArthur's case, however, there were inherent liabilities in extending this practice too far. During World War II, when the JCS had functioned as a high command, they had been able to exercise a degree of control over MacArthur through the allocation of resources and through the powers they derived from their unique relationship with the President. But from 1947 on, the JCS no longer had such sweeping authority. Meantime, MacArthur operated from his headquarters in Tokyo with a lengthening list of titles, including all-encompassing powers as head of the American occupation and Commander in Chief, Far East (CINCFE), which gave him authority over U.S. land, sea, and air forces throughout the theater. As of July 8, 1950, he also served as the United Nations commander (CINCUNC) in accordance with a UN Security Council resolution.[48]

In Korea, MacArthur found himself waging a war heavy in political overtones which, despite his vast authority, imposed limits on his military flexibility. He responded by treating the policy pronouncements and directives he received from both Washington and the UN as advisory and thus subject to interpretation. Seeking to stem the enemy advance, he ordered the destruction of North Korean airfields a day before President Truman authorized it. By early August 1950, he had antagonized the White House and the State Department with a trip to Taiwan and public statements afterwards (including a proposed message to the Veterans of Foreign Wars, later withdrawn at Truman's insistence) suggesting the restoration of military collaboration and a de facto alliance between Chiang Kai-shek's regime and the United States. His repeated requests for more U.S. combat troops to shore up the South Koreans reflected not simply the gravity of the situation, but also his longstanding contention that policymakers in Washington misunderstood the Far East and underestimated its strategic significance. By and large, the Joint Chiefs were in accord with MacArthur's assessments. But they could sense a showdown coming between MacArthur and the Commander in Chief and had no desire to be caught in the middle.[49]

Despite their differences, Truman and MacArthur both saw the war in Korea as a diversion from larger issues and wanted it brought to a swift conclusion. With this end in mind, MacArthur proposed a counterattack involving a risky large-scale amphibious landing in the enemy's rear. After the contretemps over Taiwan, Truman was so irritated with MacArthur that he gave "serious thought" to replacing him with Bradley. But he dropped the idea because he thought the Chairman would consider it a demotion.[50] Even though he disliked MacArthur personally, Truman needed the general's expertise to execute the counterattack. During World War II, MacArthur had developed and perfected amphibious operations to a fine art, and he proposed to apply his skills again to rout the North Korean People's Army.

The most questionable part of the operation was MacArthur's choice of Inch'on, a port west of Seoul, as the landing site. While a successful invasion there would put UN forces astride enemy supply lines and block a North Korean retreat, extensive mud flats and tidal variations made landing conditions treacherous. "I realize," MacArthur observed at one point while planning the operation, "that Inchon is a 5,000 to 1 gamble, but I am used to taking such odds. We shall land at Inchon and I shall crush them."[51] In fact, the odds were better than MacArthur let on. Thanks to a hastily arranged signals intelligence (SIGINT) intercept program, U.S. code breakers in Washington had succeeded in penetrating North Korean communications in late July 1950. From that point on, MacArthur and the JCS had a fairly full picture of the North Korean order of battle and knew that after weeks of heavy fighting, the North Koreans were running low on replacements and supplies. Most important of all, the intercepted messages disclosed that there were no large enemy units in the Inch'on area to oppose a landing.[52]

Coordination between MacArthur and the JCS for the Inch'on operation was haphazard. In early July 1950, the Joint Chiefs began hearing rumors that MacArthur was planning a counterattack. Despite repeated requests for details, it was not until July 23 that he apprised the JCS of his intentions.[53] MacArthur planned the attack, code-named *Chromite*, for mid-September and needed additional reinforcements which, if granted, would leave only the 82d Airborne Division in the strategic reserve. There followed a succession of high-level conferences at the Pentagon and the White House culminating in the decision to send a JCS delegation headed by General J. Lawton Collins, Army Chief of Staff, and Admiral Forrest P. Sherman, Chief of Naval Operations, to Tokyo to discuss the matter with MacArthur and his staff. Reassured that the Inch'on landing was feasible, albeit risky, they returned to Washington and persuaded their colleagues to agree to allocate the additional units MacArthur wanted. On September 7, the JCS notified MacArthur that he had the authority to proceed.[54]

From this point on, citing operational security needs, MacArthur rarely communicated with the JCS until after the Inch'on operation on September 15, 1950. With access to the same SIGINT that MacArthur and the JCS had, President Truman later insisted that he was not in the least bothered by MacArthur's behavior and had the "greatest confidence" the landing would succeed.[55] As a precaution, however, should the operation fail and a change of commanders become necessary, he gave Bradley a fifth star, reaffirming his authority. At the same time, in a move that many observers considered long overdue, he replaced Louis Johnson as Secretary of Defense and named General George C. Marshall as his successor. An admirer and personal friend of MacArthur's, Johnson was too closely identified with the general for President Truman's comfort, while his economy measures and disagreements with Acheson had become a distinct liability. With the Inch'on operation looming, the President used the occasion to put his house in order for the larger tasks that lay ahead.[56]

As MacArthur predicted, *Chromite* was a stunning success that quickly turned the tide of battle against the North Korean invaders. By the time the operation took place, MacArthur had at his disposal a UN force of nearly 200,000 ground combat troops, including 113,500 Americans, 81,500 South Koreans, and 3,000 British and Filipinos. Within a week, his forces had driven to the outskirts of Seoul, the South Korean capital. On September 27, they linked up with Lieutenant General Walton H. Walker's Eighth Army, which had pushed north from where it had taken up defensive positions near Pusan on the southeastern coast. Seoul fell to the United Nations Command (UNC) on September 28, and the next day MacArthur restored the government of President Syngman Rhee to its capital. By the end of the month, the North Korean army had ceased to exist as an organized fighting force. Still, as much as a third of the 90,000 North Koreans who had participated in the attack and most of the North Korean high command made their way north across the border and began to regroup. At great cost and effort, the UN coalition had thrown the aggressors back, but it was in no position yet to declare total victory.[57]

POLICY IN FLUX

The greatest military triumphs of MacArthur's long career, the Inch'on landing and the ensuing rout of the North Koreans were also a huge relief to Truman and the Joint Chiefs, who had thrown practically everything into the attack the United States could muster on such short notice. The victory, however, left the cupboard bare. Realizing that forces would need to be replenished and rebuilt, both to finish the job in Korea and for general rearmament, President Truman on September 29 took the step he had long postponed—approving NSC 68 and referring it to the Executive departments and agencies "as a statement of policy to be followed over the next four or five years."[58]

Whether President Truman would actually implement NSC 68 to the full extent its authors envisioned remained to be seen. Prior to Inch'on, the Joint Chiefs had assumed that there would probably be an extended conflict in Asia and an open-ended emergency requiring large-scale augmentation elsewhere of the Armed forces. To meet estimated requirements, they projected an active duty defense establishment by the end of FY 54 of 3.2 million uniformed personnel (double the current strength) organized into an Army of 18 divisions, a Navy of nearly 400 combatant vessels (including 12 attack carriers), and an Air Force of 95 wings, with a third of them dedicated to strategic bombardment.[59] But given the Inch'on success, Truman began to doubt whether a defense establishment of such size was needed. When he approved NSC 68, he told the National Security Council, with General Bradley present representing the JCS, that "costs were not final" and that "there were certain things that could be done right now, while others should be studied further."[60]

Truman's ambivalence reflected the continuing uncertainty surrounding the situation in Korea and its impact on American defense obligations elsewhere, Europe especially. Even though MacArthur had the North Koreans on the run, his failure to deliver the coup de grace meant that the conflict could go on indefinitely. The Joint Chiefs had no desire to keep large numbers of U.S. forces tied down in

Korea, but they did not want U.S. troops to leave until the campaign had run its course. At issue was whether to seek modest objectives, such as restoration of the status quo ante, or the complete destruction of the North Korean armed forces and the reunification of Korea under UN authority. Anticipating that UN forces would eventually regain the initiative, State and the JCS had debated this matter at length during July and August 1950, but had been unable to come up with a definitive answer. The best they could recommend was a wait-and-see policy. All agreed, however, that the longer the fighting lasted, the greater the chances of Soviet or Chinese intervention, that the risk would increase significantly if or when UN forces approached the Chinese and Soviet borders, and that MacArthur should be cautioned against launching major military operations north of the 38th parallel without consulting the President.[61]

Inch'on and the ensuing rout of the North Korean army created opportunities that seemed too good to pass up. Toward the end of September 1950, Secretary Marshall advised MacArthur to feel free to continue operations north of the 38th parallel, with the implied objective of liquidating the remnants of the North Korean army. A week later, on October 7, the UN General Assembly passed a resolution reaffirming its desire to unify Korea. Nonetheless, Truman remained uneasy over the possibility of Soviet or Chinese intervention. Unable to persuade MacArthur to return to Washington for consultations, Truman agreed to fly to Wake Island in the Pacific—a 15,000 mile trip—for a hastily arranged review of plans and strategy on October 15. General Bradley was the only JCS member to accompany the President.

Though it lasted barely 2 hours, the Wake Island conference was perhaps the most fateful meeting of the war. Despite SIGINT intercepts indicating a massing of Chinese troops in Manchuria just north of the Yalu River, MacArthur dismissed the possibility that the Chinese might intervene. Should they do so, he was confident that he could defeat them with airpower. "If the Chinese tried to get down to Pyongyang," he said, "there would be the greatest slaughter." Bradley was skeptical, but since the SIGINT intercepts were inconclusive on Chinese intentions, he had no basis for challenging MacArthur's analysis. Convinced that the North Koreans were beaten, MacArthur predicted the end of organized resistance by Thanksgiving, the withdrawal of the Eighth Army to Japan by Christmas, and the redeployment of one of its divisions to Europe in January 1951, leaving two U.S. divisions in Korea for security.[62]

Proclaiming the Wake Island meeting "successful," Truman returned to Washington "highly pleased" with the outcome.[63] Despite its brevity and superficiality, the meeting produced two important results. First, it gave MacArthur a green light to proceed with military operations above the 38th parallel and, implicitly, to use his forces to reunify Korea. And second, it reassured Truman that he had made the right decision to hold back on military spending in anticipation that the war would soon be over. NSC 68 notwithstanding, Truman believed that the buildup had peaked and that the time had come to level off. By early November 1950, the Office of the Secretary of Defense was pressing the Joint Chiefs to reconsider their force-level projections for FY 52 and to reduce manpower requirements to fit within "a realistic military budget."[64]

Meanwhile, MacArthur's spectacular earlier successes were about to prove short-lived. The first hint that he had underestimated the enemy threat came in late October 1950 as UN armies approached the Manchurian border. In a surprising new development, ROK units encountered Chinese forces that expertly concealed their real strength. Based on prisoner interrogations, the Central Intelligence Agency distributed findings in early November 1950 confirming that the Chinese had begun infiltrating around mid-October and now had one and a half or two divisions operating in Korea.[65] (The correct figure was 18 divisions.) MacArthur initially assumed that these troops were part of a limited covert intervention, but within a few days came fresh evidence, as MacArthur characterized it, that the Chinese were "pouring across" the border from Manchuria into North Korea.[66]

MacArthur wanted to isolate the invading Chinese by using U.S. B–29s to bomb the bridges spanning the Yalu River, Korea's frontier with China. In the view of some critics, MacArthur's intention was to expand the war and turn it into a crusade against communism in the Far East. The Joint Chiefs never subscribed to this thesis, but they did worry that an aggressive air campaign extending into Manchuria might give the Soviets an excuse to intervene alongside the Chinese. Consequently, even though the JCS gave MacArthur a free hand to bomb below the Yalu River, they cautioned him to exercise "extreme care" to avoid hitting targets in Manchuria or violating Chinese air space.[67]

While MacArthur and the JCS debated how to handle the Chinese, the UNC advance continued, with some Allied units reaching the Yalu by November 21. Disaster struck 4 days later as the People's Liberation Army unleashed a full-scale offensive, inflicting heavy casualties. As General Bradley described the situation to the President, the Chinese had "come in with both feet."[68] Seeing no other choice, MacArthur ordered an immediate withdrawal back down the peninsula. On November 28, he notified the JCS that he now confronted as many as 200,000 Chinese and 50,000 North Koreans and "an entirely new war."[69] An easy march north to destroy the remnants of the North Korean army and to reunify Korea now became a headlong retreat south.

MILITARIZING THE COLD WAR

IMPACT OF THE CHINESE INTERVENTION

The Chinese intervention changed everything. Almost overnight, JCS planners found themselves scrapping plans to curtail the buildup and developing new ones to accelerate the rearmament program and to expand its base. Instead of using mid-1954 (NSC 68's "year of maximum danger") as their culmination point, the Joint Chiefs, working with OSD and the National Security Council, moved the date up to mid-1952 and reprogrammed manpower and force targets accordingly. Truman, fearing that the costs would bankrupt the country and send the economy into recession, hesitated to commit to a stepped-up effort. But by the end of November 1950, with the Communist onslaught in high gear, he acknowledged that the situation required sweeping action. What was needed, he said, was a more rapid expansion of military power, to "prevent all-out world war and [to] be prepared for it if we can't prevent it."[70]

The ensuing buildup became the largest "peacetime" rearmament in American history up to that time, later surpassed only by the Reagan buildup of the 1980s. From a FY50 base of around $12 billion, defense outlays rose to $20 billion the following year, to $39 billion in FY52, and to $43 billion in FY53, the last budget enacted under the Truman administration. During this same period, Active-duty military personnel increased from 1.4 million to 3.5 million, the Army expanded from 10 to 20 divisions, the Navy grew from 238 major combatant vessels to 401, and the Air Force more than doubled in size from 48 to 98 wings. While the emphasis on nuclear retaliation remained, significant improvements in conventional capabilities signaled the return to a more robust, balanced force posture. In addition, the military assistance program, atomic energy, foreign intelligence, the Voice of America, and Radio Free Europe all received substantial funding increases. Overall, the allocations for defense and related national security programs climbed from 5.1 percent of the country's gross national product (GNP) in FY50 to 14.5 percent in FY53.[71]

With greater resources becoming available, the JCS directed the Joint Staff to step up the preparation of strategic plans that looked beyond the immediate budget cycle in the annual Joint Outline Emergency War Plan (JOEWP). These longer range plans attempted to anticipate the scale of effort for a global war with the Soviet Union and its allies years in advance. The most fully developed long-range plan, known as DROPSHOT, was under consideration when the Korean War began and projected a large-scale conventional mobilization for a war fought along World War II lines in 1957. Never approved, DROPSHOT was withdrawn in February 1951 and superseded by REAPER, a mid-range plan that anticipated a war in 1954. Among its innovations, REAPER attempted to incorporate an active defense of

Europe and to take into account the impact of a nuclear exchange between the United States and the Soviet Union. Inter-Service differences over the allocation of assets, however, left REAPER's approval in limbo. Increased defense spending could ease—but not eliminate—the inter-Service competition for funds and resources.[72]

Given the difficulties of reaching inter-Service agreement and the complexities of trying to develop individual plans to cover all contingencies, the Joint Chiefs decided in July 1952 to phase in new procedures to meet their strategic planning obligations. Under the new system, the JCS embraced a "family" of plans, each updated annually: the Joint Strategic Capabilities Plan (JSCP), which replaced the JOEWP, indicating the disposition, employment, and support of existing forces available to the unified and specified commanders to carry out their missions; the Joint Strategic Objectives Plan (JSOP), estimating Service requirements for the next 3 years; and the Joint Long-Range Strategic Estimate (JLRSE), a 5-year projection of force requirements emphasizing research and development needs.[73] Though subjected to frequent refinements and adjustments, these formats remained the joint strategic planning system until the Goldwater-Nichols Defense Reorganization Act of 1986 compelled a reassessment of planning procedures resulting in the adoption in 1989 of new arrangements vesting sole responsibility for discharging JCS strategic planning functions in the CJCS.[74]

A further consequence of the Korean War buildup was to restore the Joint Chiefs of Staff to a close approximation of the prestige and influence they had enjoyed during and immediately following World War II. With a war in progress, the President needed reliable military advice, and in the aftermath of the Chinese intervention, as MacArthur's views and recommendations became increasingly suspect, Truman turned more and more to the JCS. In fact, the President had been moving in this direction ever since approving a series of reforms in the summer of 1950 to enhance the role of the National Security Council and to improve its coordination with the JCS. Prior to these reforms, the Joint Chiefs had operated on the Council's periphery, with their role confined mainly to commenting on NSC papers referred to them by the Secretary of Defense. Nor had Truman, who had never wanted the NSC in the first place, made more than limited use of it.[75] But with the advent of NSC 68 and the expectation that it would generate additional expenditures, the President decided to upgrade the NSC's capabilities to assess and coordinate programs.[76] In June–July 1950, he approved a reorganization of the NSC staff that included naming former ambassador to Moscow W. Averell Harriman as his special assistant for national security affairs and creating two new interdepartmental advisory bodies—the NSC Senior Staff and a mid-level support group, the Staff Assistants—both with JCS representation. As a result, the Joint Chiefs gained direct access to the NSC's inner workings and a regular voice in the development of NSC products.⁷⁷

Among the reforms that President Truman ordered were curbs on the number of participants at NSC meetings. Convinced that the presence of too many subordinates inhibited discussion, Truman confined attendance to the Council's statutory members and a handful of senior advisors. Rather than having all the chiefs (including the Commandant of the Marine Corps who acquired limited participation in JCS deliberations in 1952) present, Truman asked that only the CJCS, General Bradley, attend on a regular basis.⁷⁸ This practice did not bar the Service chiefs from attending as needed, but it did underscore the Chairman's emerging role as their spokesman and his importance as a key high-level advisor in his own right. Bradley was initially uncomfortable addressing problems from anything other than "a military point of view." But according to Acheson, he gradually came to realize that political, diplomatic, and military issues at the NSC level were often indistinguishable and needed to be dealt with accordingly.⁷⁹

MACARTHUR'S DISMISSAL

Korea was the last war in which the Joint Chiefs were in the chain of command. Under a practice initiated in World War II and reaffirmed by the 1948 Key West agreement, the Service chiefs functioned as executive agents for the JCS. During the Korean War, the Army Chief of Staff, General J. Lawton Collins, served as their executive agent to the Far East Command. It was through him that MacArthur received his orders. But after the Chinese intervention, communications between MacArthur and the JCS became somewhat erratic, and the general's reports were less reliable, requiring Collins to play a more direct and personal role. Collins, soft-spoken with a boyish appearance, was as serious as they came in discharging his duties. A veteran combat commander who had fought in Europe and the Pacific in World War II, Collins was not easily misled or swayed. He visited the theater frequently, toured the battle front, and brought back sound and impartial analyses that the other chiefs and senior policymakers usually found eminently more useful and reliable than MacArthur's often sketchy and slanted reports.

Based on Collins's reports and other information reaching them, the JCS became increasingly skeptical of MarArthur's capacity to discharge his responsibilities. Overly confident after the stunning success of the Inch'on landing, MacArthur was psychologically and militarily unprepared for the setbacks of November–December 1950 brought on by the Chinese intervention. Seeking a freer hand to retaliate, he proposed to bomb targets in Manchuria and to impose a naval blockade against

Communist China. The alternative, he argued, was evacuation of UN forces from Korea. MacArthur never directly requested authority to use atomic weapons, but he implicitly raised the possibility with the JCS on several occasions. He presumably knew of President Truman's decision in the summer of 1950 to stockpile nonnuclear components (bombs minus their nuclear cores) on Guam. Under the current JO-EWP, the JCS intended the Guam stockpile for attacks by the Strategic Air Command against Vladivostok and Irkutsk in the event of general war. But at the first signs of Chinese intervention, the Army General Staff started exploring the tactical use of these weapons in or around Korea and sounding out the State Department on the diplomatic ramifications.[80]

The Joint Chiefs sympathized with MacArthur's predicament and did what they could to protect his freedom of action. But after the Chinese intervention, they were under heavy pressure from the White House and the State Department to localize the war and avoid escalating the conflict. Though they had studied the use of nuclear weapons since the war began, they generally agreed that there were too few targets and too few bombs to make a difference unless faced with a looming "major disaster."[81] Furthermore, administration policy stressed international cooperation and collaboration through the UN, where opinion favored the reunification of Korea, but not if it involved taking risks that could widen the war. The British were especially uneasy, as evidenced by Prime Minister Clement R. Attlee's hasty visit to Washington in early December 1950 in response to rumors that the United States was contemplating the use of nuclear weapons in Korea. Having only begun to develop a nuclear capability, the British saw themselves as yet in no position to take on the Soviets, even as part of an American-led effort.[82] Denied permission to launch operations outside the Korean Peninsula, MacArthur became progressively more frustrated and outspoken, and told the press at one point that his orders from the President and the Joint Chiefs were "an enormous handicap, without precedent in military history."[83]

By late January 1951, Lieutenant General Matthew B. Ridgway, USA, the new commander of the Eighth Army, had reenergized UNC forces with a limited offensive that was driving the enemy north. As of mid-March, UN armies were again in possession of Seoul and had established a relatively stable line across Korea in the vicinity of the 38th parallel. In view of the success of Ridgway's campaign, MacArthur became convinced that, despite their superior numbers, the Chinese were far from invincible and could still be driven out of Korea. Acheson, however, saw the situation differently and persuaded Truman that the time was ripe for negotiations, with the aim of restoring the status quo ante.[84] Around the end of March, MacArthur effectively scuttled Acheson's initiative by publicly issuing a virtual ultimatum that gave the Chinese the choice of an immediate ceasefire or a rapid expansion of the conflict aimed at toppling their regime. MacArthur's statement violated administration policy across the board and set the stage for a showdown with the President. But before the full impact could settle in, another incident occurred—the release on April 5 by House Republican Leader Joseph W. Martin, Jr., of a letter he had recently received from MacArthur urging "maximum counterforce" in Korea and a second front against the Communist Chinese launched from Taiwan. The letter closed with MacArthur's celebrated exhortation: "There is no substitute for victory."[85]

Characterizing MacArthur's letter as the "last straw," Truman moved to relieve him of command on grounds of insubordination.[86] On April 6, 1951, the President met with Acheson, Marshall, Harriman, and Bradley to explore a course of action. Harriman wanted MacArthur's immediate dismissal. But Bradley, deeply distressed, was skeptical whether MacArthur's behavior constituted insubordination, as defined in Army regulations. Buying time, he persuaded Truman to let him discuss the matter with his JCS colleagues as soon as the Army Chief of Staff, General Collins, returned to town.[87]

MacArthur's conduct put the Joint Chiefs in a difficult position. All signs indicated that Truman was going to sack MacArthur. If the chiefs recommended against his relief, they would only be fueling the controversy. In fact, the JCS had lost confidence in MacArthur's leadership and judgment, and wherever feasible were taking steps to work around him. Toward the end of March 1951, they received intelligence that the Soviets had transferred three divisions to Manchuria and were massing aircraft and submarines for a possible attack on Japan or Okinawa. Fearing a major escalation of the war, the Joint Chiefs asked the President to transfer custody of nine nuclear cores from the Atomic Energy Commission to the military for deployment to the western Pacific and to approve an order authorizing CINCFE to carry out retaliatory strikes against enemy air bases in Manchuria and China should the Soviets attack. On April 6 (the same day he met with his senior advisors to discuss MacArthur's future), President Truman approved the draft order and the custody transfer. But instead of placing the bombs under MacArthur's control, he turned them over to the Air Force Chief of Staff, General Hoyt S. Vandenberg. Ordinarily, the JCS would have dispatched the retaliation order immediately to CINCFE. This time, they elected to withhold it and to keep it secret out of concern, as Bradley put it, that MacArthur might "make a premature decision in carrying it out."[88]

The chiefs assembled on Sunday afternoon, April 8, in Bradley's Pentagon office rather than the "Tank" where they conducted official business. Though informal, the proceedings resembled those of a court of inquiry. Weighing the evidence, they talked for 2 hours. In the end, they concluded that, while MacArthur may have been guilty of poor judgment, the case against him for insubordination did not stand up. Even so, they believed the President would be fully within his rights as Commander in Chief to remove MacArthur in the interest of upholding the principle of civilian control of the military. If the President wanted to fire MacArthur, the JCS would not stand in the way. The next morning Bradley and Secretary Marshall conveyed the chiefs' views to the President. Two days later, on April 11, the White House press office revealed that MacArthur was being recalled and that Ridgway would replace him as CINCFE and commander of UN forces.[89]

MacArthur at this time was still a popular and widely respected figure in the United States—a national hero in some circles—and his firing provoked a good deal of outrage. A congressional investigation ensued and for the second time in as many years the Joint Chiefs found themselves explaining and defending their actions on Capitol Hill. This time, however, the hearings were closed to the public. As the inquiry progressed and the substance of its proceedings became known through leaks and edited transcripts, popular support for MacArthur began to sag. The Korean War was dragging on longer than anyone expected and, with casualties and costs continuing to mount, MacArthur's repeated calls for "victory" envisioned sacrifices that fewer and fewer Americans deemed worthwhile. More in line with majority opinion was the administration's determination to seek a negotiated settlement. Attempting to put the matter in perspective, General Bradley told Congress that MacArthur's prescription for victory would have invited an open-ended conflict on the Asian mainland. Had MacArthur's advice prevailed, Bradley added, the United States would have found itself in "the wrong war, at the wrong place, at the wrong time, and with the wrong enemy."[90]

EUROPE—FIRST AGAIN

Following MacArthur's dismissal, the Korean War gradually receded from the forefront of the Joint Chiefs' agenda, where a backlog of other defense and security problems, mainly relating to Europe, clamored for attention. More attuned to the thinking in Washington than MacArthur had been, Ridgway knew that the President and the JCS wanted him to limit the conflict and avoid any actions that might provoke "a worldwide conflagration."[91] Abandoning the quest for Korean reunification, the Joint Chiefs issued new orders on June 1, 1951, that essentially instructed Ridgway to maintain the status quo. Though he remained free to mount operations to protect his forces and to keep pressure on the enemy, he was to restrict his activities to a defensive line in the vicinity of the 38th parallel while military talks explored a ceasefire.[92]

The decision to settle for a stalemate in Korea reflected not only the realities of a war gone sour, but also the deeply held belief of many in the Truman administration, Secretary of State Acheson foremost among them, that vital American interests were more at jeopardy in Europe than in Asia. In Acheson's view, the dynamics of the Cold War centered in Europe; it followed that America's "principal antagonist" was the Soviet Union, not Communist China.[93] The Joint Chiefs believed that Acheson's assessment underestimated China's potential threat and capabilities. But they agreed that, owing to limited resources, the United States should not allow Cold War conflicts in places like Korea and Indochina to become the catalysts for a general war with China.[94] Adopting a frame of reference much like the one that had guided their predecessors in World War II, they accorded the defense of Europe first priority.

Though it predated the Korean War, the European defense buildup had barely begun when fighting broke out in Korea in June 1950. Bureaucratic delays in initiating the Mutual Defense Assistance Program and prolonged debate over NATO's organizing defense plan had slowed European rearmament to a crawl. The basic blueprint was a strategic concept (DC 6/1), adopted by NATO's governing body, the North Atlantic Council (NAC), in January 1950. Written to JCS specifications, DC 6/1 was almost a mirror image of U.S. defense policy at the time, with strategic bombardment provided by the Strategic Air Command (and augmented by British Bomber Command) forming the first line of defense and retaliation. Though the NAC decided against including any specific reference to nuclear weapons, their use was clearly implied. In effect, NATO's members now fell under the extended deterrence protection of the American "nuclear umbrella." The European members' main contribution would be to supply the "hard core" of the Alliance's conventional ground, air, and coastal defense forces. Though the Europeans went along with this division of labor, it was an arrangement that few particularly liked since it made no allowance for them to participate in the command, control, or targeting of the strategic forces that formed their primary protection. Not without justification, some Europeans worried that they were now more than ever the potential target of a Soviet nuclear attack.[95]

Before the Communist invasion of South Korea, the Joint Chiefs had neither the inclination nor the resources to mount an active defense of Europe. Exploratory efforts to incorporate such a defense into U.S. emergency war plans in the spring of 1949 resulted in such high projected costs that the JCS dropped the idea. The war plan they later adopted (OFFTACKLE) called for the evacuation of the two U.S.

divisions on occupation duty in Germany and Austria at the first sign of a large-scale Soviet attack. Aware that the planned withdrawal undercut the U.S. commitment to NATO, Army planners pressed for "retardation bombing" of advancing Soviet forces as part of the strategic air offensive, to give the Europeans a better chance of defending themselves and U.S. forces a better chance of getting out. Air Force and Navy planners viewed the Army's proposal as a diversion of resources from the primary objective of destroying the Soviet Union's war-making capabilities. But through persistence, the Army's position prevailed. Retardation bombing was included, both in the OFFTACKLE plan and in a revised targeting scheme adopted by the Joint Chiefs in August 1950. Even so, the immediate benefits for NATO were uncertain. Retardation bombing remained at the bottom of the JCS priorities list and, because planes and bombs were limited, SAC balked at allocating the necessary assets to anything other than strategic objectives. Bombing military-industrial targets in the Soviet Union, SAC planners insisted, would in the long run retard the Soviet advance as much as anything.[96]

After the outbreak of the Korean War, as funding constraints eased, the JCS reassessed their position and agreed not only to expand the scale and scope of SAC's operations in Europe, but also to bolster NATO's conventional posture by enlarging the U.S. commitment in Germany by up to four divisions. In July 1950, at the same time he ordered the deployment of nonnuclear components to Guam, President Truman approved a similar deployment to facilities in the United Kingdom and accepted a JCS recommendation to send two additional B–29 wings to the UK, tripling the size of the in-country medium bomber force. A secret agreement reached earlier, in April 1950, between the U.S. ambassador to the United Kingdom and Britain's Air Ministry cleared the way for the deployment.[97] By January 1951, JCS planners had earmarked 60 nuclear bombs for NATO retardation purposes. However, SAC commanders winced at even this limited allocation of assets. As one put it, SAC was "not designed for close or general support of ground forces." Rather, it was an organization dedicated to delivering "an atomic offensive against the heart of an enemy wherever that may happen to be."[98]

Having established broad criteria for target selection, the Joint Chiefs left it up to the new NATO Supreme Allied Commander, Europe (SACEUR), General Dwight D. Eisenhower, USA, and his air deputy, General Lauris Norstad, USAF, to develop a working arrangement with the Strategic Air Command. A veteran of the roles and missions quarrels after World War II, Norstad easily perceived that unless the Air Force paid closer attention to retardation bombing and other nonstrategic missions, it would open opportunities for the Army and the Navy to develop their own "tactical" nuclear capabilities and challenge the Air Force's dominant position in atomic warfare. Eventually persuaded to cooperate, the SAC commander, General Curtis E. LeMay, met in late 1951 with Eisenhower and Norstad in Europe to coordinate their respective roles in "retardation operations." The agreement reached allowed SACEUR to determine the military significance and priority of targets, but vested command and control of operations in a new Air Force headquarters element in Europe known as SAC ZEBRA, which dealt only with Norstad and designated U.S. officers. Based on this accord, the Joint Chiefs authorized Eisenhower to prepare atomic annexes for NATO war plans and to carry out independent exercises simulating the use of atomic weapons in support of NATO strategy. In May 1953, SACEUR and SAC conducted the first combined test of their ability to coordinate an atomic operation.[99]

Equally, if not more, frustrating for the Joint Chiefs were the difficulties they encountered in trying to shore up NATO's conventional strength. While atomic weapons and strategic airpower were still the West's most formidable means of retaliation, U.S. nuclear capabilities were as yet too limited to protect Western Europe from an all-out Soviet invasion. As General Bradley put it, "We don't have enough atomic weapons to plaster all of Europe."[100] The initial (pre-Korean) NATO war plan was DC 13, the Medium Term Defense Plan (MTDP), built on the principles in the NAC-approved strategic concept. An ambitious 4-year effort, the MTDP received official sanction in the spring of 1950 and called for the creation of a largely European army of 90 Active and Reserve divisions whose job would be to hold attacking Soviet forces as far to the east as possible in Central Europe. Skeptical whether the plan was economically feasible, the Joint Chiefs urged NATO planners to take a closer look at their requirements and to explore a "radical revision downward" of force goals. But since few NATO leaders took these numbers seriously, treating them instead as a "first approximation," there was little discernible incentive for a more realistic assessment. Planning and preparations for a NATO buildup proceeded at a leisurely pace.[101]

Concern that the Communist attack against Korea might be the prelude to a similar invasion of Western Europe finally prompted a reevaluation of NATO plans and timetables. Not only did it galvanize the European Allies—Britain and France, especially—into stepping up the tempo of their rearmament programs, but it also led them to make new requests for additional military assistance, an increase in U.S. troop strength in Europe, and the creation of an integrated high command. A condition of key importance to the Joint Chiefs in acting on these measures was that the Europeans in return accept the rearmament of West Germany, which the JCS had been studying for some time. Though fully aware that German rearmament was bound to be controversial, the chiefs had come to the conclusion that a

German contribution was unavoidable if NATO was to fill the gaps in its Medium Term Defense Plan and confront the Soviets with a credible defense in Central Europe. Anticipating European resistance, the State Department proposed a North Atlantic or European defense force incorporating German forces under direct Allied command.[102]

Insisting on an all-or-nothing approach, the Joint Chiefs persuaded Secretary of State Acheson to adopt a "one package" negotiating stand that linked the creation of the combined command and increases in U.S. troop strength to European acceptance of German rearmament and progress toward meeting MTDP force goals. Presented to the NAC in September 1950, the U.S. package provoked a livid reaction from the French, who were as irritated by the rigidity of the American proposal as by its contents.[103] Given NATO's need for manpower and materials, German rearmament was only a matter of time. But for many (if not most) Europeans, it was too soon after the War to accept such a prospect. While the French showed a flicker of interest in State's European army concept, the idea needed to gestate and over the next several years it reappeared in several guises, the most well-known being the French-sponsored Pleven Plan, which eventually gave rise to the European Defence Community (EDC). Meanwhile, the only large-scale effort to put Germans back in uniform and under arms was that initiated by the Soviets in the eastern zone.

Unable to achieve a breakthrough on German rearmament, the Joint Chiefs bided their time and turned their attention to the appointment of a supreme Allied commander and the creation of an international command structure. Authorized at the September 1950 NAC meeting, these measures were the first concrete steps toward transforming NATO from a paper alliance into a functioning military organization. The key to the entire enterprise was Eisenhower's willingness to serve as NATO's military head, with Britain's Field-Marshal Bernard Law Montgomery as his deputy. Recommended by the Joint Chiefs in October 1950 and announced that December, Eisenhower's appointment as SACEUR placed him back in a job comparable in many ways to the one he held in World War II, but without the same sweeping authority or resources. From offices hastily constructed on the outskirts of Paris, Eisenhower presided over the Supreme Headquarters, Allied Powers Europe (SHAPE), a multinational headquarters staff charged with planning and coordinating the land and air defense of Western Europe. Though Eisenhower took his orders from the NATO Military Committee via the Standing Group, a select interallied body of senior officers, he also communicated regularly with the Joint Chiefs of Staff and the Secretary of Defense.[104]

Based in Norfolk, Virginia, a separate supreme Allied commander, SACLANT, handled naval planning for the North Atlantic. Though authorized by the NAC in

December 1950, the Atlantic Command did not become active until nearly a year and a half later owing to a bitter contest for control between the British Chiefs of Staff and the JCS. The resolution of this issue in favor of the JCS position was as much a reflection of Britain's demise as a world power as it was NATO's heavy dependence on the United States. Clearly, it was a blow to British pride that needed assuaging. Awarding the Channel Command (ACCHAN) overseeing air and naval operations in the English Channel to the British in February 1952 was meant to serve this purpose. In 1953, the British also received the NATO Mediterranean Command (CINCAFMED), headquartered at Malta. Established as part of SHAPE and not, as the British hoped, as a third supreme command, CINCAFMED had limited assets and authority and exercised no control over the U.S. Sixth Fleet, the most powerful naval force in the area.[105]

Under Eisenhower's guidance and energizing presence, the NATO buildup in Europe gathered momentum quickly. From a force of 15 divisions (in varying degrees of readiness) and fewer than 1,000 aircraft in April 1951, NATO grew to 35 active and reserve divisions and nearly 3,000 planes by the end of the year. During the same time, Congress increased funding for military aid, training for European forces improved, and there were combined field maneuvers to test coordination.[106] Perhaps most important of all, in April 1951, following the "Great Debate" on Capitol Hill, the Senate adopted a resolution sanctioning the deployment of four additional U.S. divisions to Europe, in effect sealing the American commitment under the "transatlantic bargain." Eisenhower had hoped for an infusion of up to 20 American divisions and seemed let down when neither Secretary of Defense Marshall nor the Joint Chiefs would support his request. Aware of Eisenhower's disappointment, the JCS advised him in May 1951 that they were working on plans to make up to 14 divisions available to NATO in an emergency, but cautioned that these numbers were for planning purposes and did not constitute an allocation to SHAPE.[107]

Equally important to NATO's future were Eisenhower's efforts to develop a more coherent strategy for Europe's defense. During his tenure as acting JCS Chairman in 1949, Eisenhower had discussed this problem at length with the Joint Chiefs and, since then, had steadily refined his views. The plan he proposed—a "forward strategy" designated MC 14/1 when formally adopted in December 1952—aimed at blocking invading Soviet forces and stabilizing military ground action as far to the east as possible with a strong conventional defense. NATO's last line of defense would be along the Rhine-Ijssel. Air and naval forces operating from the North Sea and Mediterranean would then hit the invaders "awfully hard from both flanks." The admission into NATO in 1952 of Greece and Turkey—two countries with little in common other than their geographic proximity and antipathy for one another—was meant in large part to bolster this strategy.[108]

The main difference between NATO's initial strategic concept of 1949–1950 and Eisenhower's forward strategy was the increased emphasis on defense by conventional means. Though Eisenhower would not rule out the use of nuclear weapons to augment NATO firepower and delay Soviet forces from advancing, it was well known within the Alliance that the smaller members (Denmark, Norway, and the Benelux countries) were extremely uneasy over the prospect of being caught in a nuclear exchange between the United States and the Soviet Union. For those countries, a war involving the use of nuclear weapons on their territory could mean annihilation. By stressing the role of conventional forces and each country's contributions, Eisenhower sought to ease those anxieties and give the Allies a united frame of reference and stronger sense of common purpose.[109]

In assessing NATO's prospects for implementing the forward strategy, the Joint Chiefs believed that Alliance members possessed adequate actual and potential resources "to discourage, if not deter, aggression in Western Europe."[110] They were less sure, however, whether the Europeans had the political will to support and sustain a rearmament effort much beyond the current level. Studies by various NATO fact-finding and advisory bodies raised similar questions, giving rise to speculation that the Europeans put their economic welfare ahead of security. As a result, the JCS were uneasy over the chances of a successful defense, and toward the end of 1951 they adopted contingency plans separate from NATO's that made provision for a possible retreat by U.S. forces from the Rhine to the Pyrenees and evacuation to the United Kingdom via Cotentin-Cherbourg in the event of a NATO collapse. Though Eisenhower was privy to these plans, the JCS insisted that they not be shown to anyone at SHAPE other than U.S. personnel since they clearly conflicted with NATO strategy.[111]

Whether the Joint Chiefs seriously intended to carry through with the evacuation of U.S. forces in an emergency is unclear. The logistics alone were daunting, and it was unclear what would happen to U.S. dependents. More than likely, these plans were meant to "leak" and serve notice to the Europeans in a subtle yet convincing way that they should not take the United States for granted and expect U.S. forces to carry the main burden of defending Europe. The JCS wanted the Europeans to understand that they needed to shoulder more responsibility for their own security by stepping up their rearmament and by accepting a German contribution to NATO.

Gaining the cooperation of the French was hardest of all. Of France's 15 army divisions, 10 were tied down fighting the Communist Viet Minh insurgency in

Indochina. Implying that what American military planners wanted was excessive, the French government suggested a deal: cooperation on German rearmament in exchange for increased American aid to cover more of the cost of the Indochina war and to guarantee France a military force in Europe on a par with Germany's. Eventually, Washington's acceptance of this offer would lead to a huge jump in U.S. security support assistance to France and additional aid underwriting over half the French war effort against the Viet Minh. But it was a price the Joint Chiefs and the Truman administration were happy to pay if it would bring the German rearmament question to a favorable resolution and bolster the U.S. strategic position in the Far East at the same time.[112]

Matters came to a head in late February 1952 at the North Atlantic Council's Lisbon meeting, which resulted in three major actions: the admission of Greece and Turkey into NATO, thus potentially increasing the conventional force base; the affirmation of NATO force-level objectives for 1954 comparable to those in the MTDP; and a breakthrough in negotiations on a continental European Defense Community under NATO command, with a German contribution of 12 divisions. To ease the financial strain of the buildup, the NAC agreed that less expensive reserve units could make up the bulk of NATO's divisions. Yet even with these relaxed requirements and German rearmament, the Joint Chiefs remained skeptical about the Alliance's capacity to meet its objectives. Within the Office of the Secretary of Defense and the Joint Staff, the operating assumption was that NATO would do well to achieve 80 percent of the Lisbon force goals.[113]

An important postscript to the Lisbon Conference was the signing of the ill-fated Treaty of Paris in May 1952. Symbolic of the evolving Franco-German rapprochement, the treaty's stated purpose was to pave the way for creation of the EDC and, within it, a rearmed West Germany.[114] Though the JCS regarded the treaty as a step in the right direction, they found it to be of no immediate help for filling the gaps in NATO's defenses, which only seemed to widen as the year progressed. Faced with balance of payments deficits, declining industrial production, and rising unemployment, the Europeans treated their economic difficulties as far more urgent and worrisome than falling behind on their defense obligations.

A further blow to NATO's fortunes was Eisenhower's departure as SACEUR in April 1952, and the arrival of his successor, General Matthew B. Ridgway, a month later. Ridgway was the first American officer to serve in what became a routine dual capacity—as the military head of NATO through his role as SACEUR, and as the U.S. Commander in Chief, Europe (USCINCEUR). Though highly regarded as a battlefield commander, Ridgway lacked not only Eisenhower's prestige but also his tact and feel for coalition diplomacy. At SHAPE, he alienated many Europeans by surrounding himself with a mostly American staff. With Eisenhower's departure, Field Marshal Montgomery recalled: "The crusading spirit disappeared. There was the sensation, difficult to describe, of a machine which was running down."[115]

NATO, in brief, was at a crossroads. Despite signs of substantial progress since the Korean War erupted, much remained to be done if the Alliance were to become a credible and effective bulwark against the Soviet Union. According to General Hastings Ismay, NATO's first Secretary General, the Alliance still had only 18 ready divisions by late 1953, half the number called for in the Lisbon goals, facing an estimated 30 Russian divisions in Eastern Europe.[116] Thus far, the burden had fallen most heavily on the United States to provide much of the military power and arms aid to give NATO substance, and to show leadership to set the Alliance on course. While the Joint Chiefs had considerable experience with coalition warfare in World War II, they never had to deal with such problems in peacetime or under an alliance system comprised of so many diverse interests as they faced in NATO. Adjusting took time and would, in fact, prove to be one of the most difficult and continuing Cold War challenges the JCS faced.

The Korean War period was a crucial turning point for the Joint Chiefs of Staff. While it confirmed and strengthened their high-level advisory duties, it also resulted in institutional changes, at the NSC especially, that thrust them and their organization into the mainstream of the policy process. Though not as powerful and influential as they were in World War II, the Joint Chiefs were again at the center of decision. Most important of all was the emergence of the CJCS as their principal representative and spokesman. Functioning in a de facto role that went beyond his official job description, he was a key advisor to the Secretary of Defense, the President, and the NSC in his own right. Much of the enhanced authority and influence that the Chairman—and by extension, the entire Joint Chiefs of Staff—came to enjoy during the Korean War years was the result of General Bradley's presence. Quiet and thoughtful, he projected a common sense approach to problems and a thoroughly professional image that helped overcome the chiefs' reputation for petty quarreling and parochialism in the aftermath of World War II.

Above all, the Joint Chiefs had begun to find their niche and to create for themselves a new institutional role more adapted to Cold War realities. No longer the architects of grand strategy as they had been in World War II, the JCS were part of an interdepartmental "team," functioning within a policy process increasingly dominated by interagency deliberations through the various mechanisms of the National Security Council. Driven by the Soviet A-bomb and the war in Korea, a new consensus had emerged, both at home and abroad, that the containment of communism required a heavier investment in military forces and related programs than anyone had imagined. Not the most efficient organization for dealing with these problems, the Joint Chiefs as a rule worked well enough together, overcoming or papering over their differences as the need arose to keep the military buildup on track. Whether the chiefs would continue to perform at this level once the pressure relaxed and a more "peacetime" atmosphere returned remained to be seen.

NOTES

1. Memo, JCS to SECDEF, April 23, 1948, "Position of the United States with Respect to Support for Western Union and Other Related Free Countries," JCS 1868/1.

2. Steven L. Rearden, *History of the Office of the Secretary of Defense: The Formative Years, 1947–1950* (Washington, DC: Historical Office, Office of the Secretary of Defense, 1984), 471–472; Lawrence S. Kaplan, *NATO 1948: The Birth of the Transatlantic Alliance* (Lanham, MD: Rowman and Littlefield, 2007), 75–103.

3. Though aimed mainly at Europe, MDAP was a comprehensive program and included funding for military assistance to Iran, Korea, Latin America, and the Philippines. See Chester J. Pach, Jr., *Arming the Free World: The Origins of the United States Military Assistance Program, 1945–1950* (Chapel Hill: University of North Carolina Press, 1991), 88–159; and Lawrence S. Kaplan, *A Community of Interests: NATO and the Military Assistance Program, 1948–1951* (Washington, DC: Historical Office, Office of the Secretary of Defense, 1980), 16–34.

4. Rearden, *Formative Years*, 462–463.

5. Kenneth W. Condit, *The Joint Chiefs of Staff and National Policy, 1947–1949* (Washington, DC: Office of Joint History, Office of the Chairman of the Joint Chiefs of Staff, 1996), 204–212.

6. Rearden, *Formative Years*, 477.

7. Forrest C. Pogue, *George C. Marshall: Statesman, 1945–1959* (New York: Viking Penguin, 1987), 64–66.

8. James F. Schnabel, *The Joint Chiefs of Staff and National Policy, 1945–1947* (Washington, DC: Joint History Office, Office of the Chairman of the Joint Chiefs of Staff, 1996), 200–201. Tang Tsou, *America's Failure in China, 1941–50* (Chicago: Chicago University Press, 1963), is the standard treatment of Chiang's demise.

9. Memo, JCS to SECDEF, November 24, 1948, "Strategic Importance of Formosa," U.S. Department of State, *Foreign Relations of the United States, 1949* (Washington, DC: GPO, 1974), IX, 261–262; Memo, JCS to SECDEF, February 10, 1949, "Strategic Importance of Formosa," ibid., 284–86. Hereafter cited as *FRUS*, with year and volume.

10. Memo, JCS to SECDEF, December 16, 1948, "Current Position of U.S. Respecting Delivery of Aid to China," JCS 1721/17.

11. Memcon between Acheson and Truman, November 17, 1949, quoted in Nancy Bernkopf Tucker, "China's Place in the Cold War: the Acheson Plan," in Douglas Brinkley, ed., *Dean Acheson and the Making of U.S. Foreign Policy* (New York: St. Martin's Press, 1993), 110.

12. Memo by Director of Central Intelligence (Hillenkoetter), September 9, 1949, no subject, PSF, Intelligence File, Truman Papers, Harry S. Truman Library (HSTL).

13 Donald P. Steury, "How the CIA Missed Stalin's Bomb: Dissecting Soviet Analysis, 1946–50," *Studies in Intelligence* 49:1 (2005); David Holloway, *Stalin and the Bomb: The Soviet Union and Atomic Energy, 1939–1956* (New Haven: Yale University Press, 1994), 134–149.

14 OSI/SR-10-49, Central Intelligence Agency, "Status of the USSR Atomic Energy Project," July 1, 1949, PSF, Intelligence File, Truman Papers.

15 Letter, Forrestal to Chan Gurney, December 8, 1947, in Walter Millis and E.S. Duffield, eds., *The Forrestal Diaries* (New York: Viking Press, 1951), 350–351.

16 Richard G. Hewlett and Francis Duncan, *A History of the United States Atomic Energy Commission: Atomic Shield, 1947–1952* (Washington, DC: U.S. Atomic Energy Commission, 1972), 394.

17 Memo, JCS to SECDEF, November 23, 1949, "U.S. Military Position with Respect to the Development of the Thermonuclear Weapon," RG 218, CCS 471.6 (12-19-49), sec. 1; portions reprinted in K. Condit, *JCS and National Policy, 1947–49*, 292. See also Memo, JCS to SECDEF, January 13, 1950, FRUS, 1950, I, 503–511.

18 David E. Lilienthal, *The Journals of David E. Lilienthal*, vol. II, *The Atomic Energy Years, 1945–1950* (New York: Harper and Row, 1964), 580–633.

19 Letter, Truman to Acheson, January 31, 1950, FRUS, 1950, I, 141–142.

20 Memo by Acheson, December 20, 1949, FRUS, 1949, I, 612.

21 Dean Acheson, *Present at the Creation: My Years in the State Department* (New York: W.W. Norton, 1969), 348–49.

22 Keith D. McFarland and David L. Roll, *Louis Johnson and the Arming of America: The Roosevelt and Truman Years* (Bloomington: Indiana University Press, 2005), 185–186.

23 Doris M. Condit, *History of the Office of the Secretary of Defense: The Test of War, 1950–1953* (Washington, DC: Historical Office, Office of the Secretary of Defense, 1988), 244.

24 John Lewis Gaddis, *Strategies of Containment: A Critical Appraisal of Postwar American National Security Policy* (New York: Oxford University Press, 1982), 83–90; Steven L. Rearden, "Paul H. Nitze and NSC 68: 'Militarizing' the Cold War," in Anna Kasten Nelson, ed., *The Policy Makers: Shaping American Foreign Policy from 1947 to the Present* (Lanham, MD: Rowman & Littlefield, 2009), 5–28.

25 Quote from Oral History Interview No. 4 with Paul H. Nitze, by Richard D. McKinzie, August 4, 1975, Northeast Harbor, Maine, Oral History Collection, Truman Library; see also Paul H. Nitze, with Ann M. Smith and Steven L. Rearden, *From Hiroshima to Glasnost: At the Center of Decision—A Memoir* (New York: Grove Weidenfeld, 1989), 93–95.

26 NSC 68, "United States Objectives and Programs for National Security," April 14, 1950, PSF, Truman Papers, HSTL. Unfortunately, the version published in *FRUS, 1950*, I, 234–292 is flawed, with unintended deletions.

27 Robert L. Beisner, *Dean Acheson: A Life in the Cold War* (New York: Oxford University Press, 2006), 238–239.

28 Warner R. Schilling, Paul Y. Hammond, and Glenn H. Snyder, *Strategy, Politics, and Defense Budgets* (New York: Columbia University Press, 1962), 321; Nitze, *From Hiroshima to Glasnost*, 96–97.

29. Minutes, 55th Meeting of the National Security Council, April 20, 1950, PSF, Truman Papers, HSTL; Memo, Lay to SecState et al., April 21, 1950, "United States Objectives and Programs for National Security," NSC Records, Truman Papers, HSTL.

30. Memo, BoB to NSC, May 8, 1950, "Comments of the BoB [on NSC 68]," *FRUS, 1950*, I, 298–306.

31. Memo for the Record, May 23, 1950, "Meeting with the President," Papers of Frederick J. Lawton, HSTL; Michael J. Hogan, *A Cross of Iron: Harry S. Truman and the Origins of the National Security State, 1945–1954* (Cambridge, UK: Cambridge University Press, 1998), 304.

32. Memo, SECDEF to SecArmy et al., May 25, 1950, "Military Requirements Under NSC 68," JCS 2101/7.

33. Quote from Matthew M. Aid, *The Secret Sentry: The Untold History of the National Security Agency* (New York: Bloomsbury Press, 2009), 26. For overall estimates of the situation, see John Patrick Finnegan, *Military Intelligence* (Washington, DC: Center of Military History, 1998), 113–14; and James F. Schnabel, *United States Army in the Korean War: Policy and Direction—The First Year* (Washington, DC: Office of the Chief of Military History, 1972), 62–63.

34. ORE 18–50, "Current Capabilities of the Northern Korean Regime," June 15, 1950, in Woodrow J. Kuhns, ed., *Assessing the Soviet Threat: The Early Cold War Years* (Washington, DC: Center for the Study of Intelligence, Central Intelligence Agency, 1997), 390; see also John Lewis Gaddis, *We Now Know: Rethinking Cold War History* (New York: Oxford University Press, 1997), 75.

35. Schnabel, *Policy and Direction*, 64.

36. James F. Schnabel and Robert J. Watson, *History of the Joint Chiefs of Staff: The Joint Chiefs of Staff and National Policy—The Korean War, 1950–1951* (Washington, DC: Office of Joint History, Office of the Chairman of the Joint Chiefs of Staff, 1998), 25–41.

37. Soon Sung Cho, *Korea in World Politics, 1940–1950* (Berkeley: University of California Press, 1967), 61–91.

38. Rearden, *Formative Years*, 255–67.

39. Omar N. Bradley and Clay Blair, *A General's Life* (New York: Simon and Schuster, 1983), 530.

40. Schnabel and Watson, 25–45; D. Condit, 47–55.

41. Acheson, *Present at the Creation*, 405.

42. Matthew B. Ridgway, *The Korean War* (Garden City, NY: Doubleday, 1967), 34.

43. Vandenberg's views in Jessup Memcon, June 25, 1950, *FRUS, 1950*, VII, 159. See also Roger Dingman, "Atomic Diplomacy and the Korean War," *International Security* 13 (Winter 1988–1989), 53–54.

44. Schnabel and Watson, 43; Johnson quoted in D. Condit, 54.

45. Truman, *Years of Trial and Hope*, 336–344; Schnabel and Watson, 36–53.

46. "President's News Conference, June 29, 1950," in *Public Papers of the Presidents of the United States: Harry S. Truman, 1950* (Washington, DC: GPO, 1965), 504. Hereafter cited as *Truman Public Papers*.

47 Minutes, 62d Meeting, National Security Council, July 27, 1950, PSF, NSC Series, Truman Papers; D. Condit, 226–227.

48 John W. Spanier, *The Truman-MacArthur Controversy and the Korean War* (Cambridge, MA: Harvard University Press, 1959), 65–66.

49 Richard F. Hayes, *The Awesome Power: Harry S. Truman as Commander in Chief* (Baton Rouge: Louisiana State University Press, 1973), 177; Spanier, 70–77; Walter S. Poole, *The History of the Joint Chiefs of Staff: The Joint Chiefs of Staff and National Policy, 1950–52* (Wilmington, DE: M. Glazier, 1979/1980), 202–204.

50 Harry S. Truman, *Memoirs*, vol. II, *Years of Trial and Hope* (Garden City, NY: Doubleday, 1956), 355–356.

51 Quoted in D. Clayton James, *The Years of MacArthur*, vol. III, *Triumph and Disaster, 1945–1964* (Boston: Houghton, Mifflin, 1985), 470.

52 Thomas R. Johnson, *American Cryptology during the Cold War, 1945–1989*, Book I, *The Struggle for Centralization, 1945–1960* (Washington, DC: Center for Cryptologic History, National Security Agency, 1995), 43 (redacted); Aid, 28–29.

53 Douglas MacArthur, *Reminiscences* (Annapolis: Naval Institute Press, 2001), 346.

54 Roy E. Appleman, *United States Army in the Korean War: South to the Naktong, North to the Yalu (June–November 1950)* (Washington, DC: Center of Military History, 2000, reprint), 488–502; Schnabel and Watson, 84–89; James, *Triumph and Disaster*, 464–474.

55 Truman, *Years of Trial and Hope*, 358.

56 Truman fired Johnson on September 12, 1950; he officially stepped down on September 19. Bradley recalled that he learned of his promotion at "the beginning of September" but because the action required congressional approval it did not become effective until September 22. See Bradley and Blair, 552–553.

57 Appleman, *South to the Naktong*, 502–606; D. Condit, 66.

58 Minutes, 68th Meeting, National Security Council, September 29, 1950, PSF, NSC Series, Truman Papers; NSC 68/2, "Note by Executive Secretary to the National Security Council on United States Objectives and Programs for National Security," September 30, 1950, *FRUS, 1950*, I, 400.

59 Poole, *JCS and National Policy, 1950–52*, 30.

60 Memo for the President, October 2, 1950, [Summary of Discussion at 68th Meeting, NSC, September 29, 1950], PSF, NSC Series, Truman Papers.

61 NSC 81/1, "United States Courses of Action with Respect to Korea," September 9, 1950, *FRUS, 1950*, VII, 712–721.

62 Johnson, *American Cryptology during the Cold War*, I, 44 (redacted); MacArthur, *Reminiscences*, 360–363; Bradley and Blair, 574–579; Truman, *Years of Trial and Hope*, 364–367; Ferrell, ed., *Off the Record*, 200; "Substance of Statements Made at Wake Island Conference on October 15, 1950," *FRUS, 1950*, VII, 948–960.

63 Memo by Acheson of Meeting with the President, October 19, 1950, Dean Acheson Papers, Truman Library.

64 D. Condit, 244–245; Poole, *JCS and National Policy, 1950–52*, 33–34.

65 NIE–2, November 6, 1950, "Chinese Communist Intervention in Korea," in *Tracking the Dragon: National Intelligence Estimates on China During the Era of Mao, 1948–1976* (Washington, DC: National Intelligence Council, 2004), 69–80.

66 Schnabel, *Policy and Direction*, 233; D. Condit, 74–77; Message, CINCFE to DA, November 6, 1950, quoted in Schnabel and Watson, 126.

67 Schnabel and Watson, 127.

68 Quoted in Eric F. Goldman, *The Crucial Decade and After: America, 1945–1960* (New York: Vintage Books, 1960), 179.

69 Message, CINCFE to JCS, November 28, 1950, *FRUS, 1950*, VII, 1237–1238.

70 D. Condit, 245–246; Truman quoted in Memo, November 24, 1950, [Summary of Discussion at 72d Meeting of the NSC, November 22, 1950], PSF, NSC Series, Truman Papers.

71 Budget and GNP figures from U.S. Department of Defense, *National Defense Budget Estimates for FY88/1989* (Washington, DC: Office of the Assistant Secretary of Defense, Comptroller, 1987), 101, 128; manpower and force levels from Poole, *JCS and National Policy, 1950–52*, 71.

72 Ross, *American War Plans, 1945–50*, 119–132; Poole, *JCS and National Policy, 1950–52*, 88–90.

73 Walter S. Poole, *The Evolution of the Joint Strategic Planning System, 1947–1989* (Washington, DC: Historical Division, Joint Secretariat, Joint Staff, 1989), 2–3.

74 JCS MOP 84, February 1, 1989, "Joint Strategic Planning System," U, JHO 08-0022.

75 See Alfred D. Sander, "Truman and the National Security Council: 1945–1947," *Journal of American History* 59 (September 19, 1972), 369–388.

76 Memo for the President, April 21, 1950, [Summary of Discussion at 55th Meeting of the National Security Council, April 20, 1950], PSF, NSC Series, Truman Papers.

77 Anna Kasten Nelson, "President Truman and the Evolution of the National Security Council," *Journal of American History* 72 (September 19, 1985), 360–378; James S. Lay, Jr., and Robert H. Johnson, *Organizational History of the National Security Council during the Truman and Eisenhower Administrations* (Washington, DC: GPO, 1960), 16–18. The JCS representative to the NSC Senior Staff was Rear Admiral E.T. Wooldridge, Deputy Director, Joint Staff for Politico-Military Affairs, whose previous experience in interagency affairs included having helped develop the 1947 Greek-Turkish aid program.

78 Letter, Truman to Acheson, July 19, 1950, *FRUS, 1950*, I, 348–349.

79 Acheson, *Present at the Creation*, 441.

80 U.S. Department of Defense, *History of the Custody and Deployment of Nuclear Weapons: July 1945 Through September 1977* (Washington, DC: Office of the Assistant to the Secretary of Defense—Atomic Energy, February 19, 1978), 16, and Appendix B (declassified); James, *Triumph and Disaster*, 579–581; Poole, *JCS and National Policy, 1950–52*, 87; Nitze Memcon of Meeting with BG Herbert B. Loper, USA, November 4, 1950, *FRUS, 1950*, VII, 1041–1042.

81 Schnabel and Watson, 168–169.

82 Ian Clark and Nicholas J. Wheeler, *The British Origins of Nuclear Strategy, 1945–1955* (Oxford: Clarendon Press, 1989), 139–145.

83 Interview with MacArthur, published in *U.S. News & World Report*, December 1, 1950, quoted in Schnabel and Watson, 239.

84 Acheson, *Present at the Creation*, 518.

85 Spanier, 197–205; see also Richard F. Haynes, *The Awesome Power: Harry S. Truman as Commander in Chief* (Baton Rouge: Louisiana State University Press, 1973), 232–234.

86 Ferrell, ed., *Off the Record*, 210.

87 Truman, *Years of Trial and Hope*, 447, says Bradley agreed that MacArthur's actions were a "clear cut case of insubordination." Bradley and Blair, 631–632, corrects the record. See also Acheson, *Present at the Creation*, 521–522.

88 Roger M. Anders, ed., *Forging the Atomic Shield: Excerpts from the Office Diary of Gordon E. Dean* (Chapel Hill: University of North Carolina Press, 1987), 127–138; Robert F. Futrell, *The United States Air Force In Korea, 1950–1953* (Washington, DC: Office of Air Force History, 1983, rev. ed.), 287–301; Schnabel and Watson, 246; Bradley and Blair, 629–631.

89 Schnabel and Watson, 247–248; "Statement and Order Relieving GEN MacArthur," April 11, 1951, *Truman Public Papers, 1951*, 222–223.

90 Quoted in Schnabel and Watson, 254.

91 Ridgway, 162.

92 Message, JCS to CINCFE, June 1, 1951, cited in Billy C. Mossman, *United States Army in the Korean War: Ebb and Flow, November 1950–July 1951* (Washington, DC: Center of Military History, 2000), 490.

93 Quoted in Gaddis, *Strategies of Containment*, 115.

94 Memo, JCS to Marshall, November 28, 1950, "Possible Future Action in Indochina," *FRUS, 1950*, VI, 947–948.

95 DC 6/1, "Strategic Concept for the Defence of the North Atlantic Area," December 1, 1949, in Gregory W. Pedlow, ed., *NATO Strategy Documents, 1949–1969* (Brussels: North Atlantic Treaty Organization, 1997), 57–64; Rearden, *Formative Years*, 481–482.

96 K. Condit, *JCS and National Policy, 1947–49*, 159–163. The targeting scheme adopted in August 1950 listed priorities as follows: "blunting" of the enemy's atomic energy capabilities (BRAVO), "destruction" of industrial facilities (DELTA), and "retardation" of advancing forces (ROMEO). See Moody, *Building a Strategic Air Force*, 357.

97 Ken Young, "No Blank Cheque: Anglo-American (Mis)understandings and the Use of the English Airbases," *Journal of Military History* 71 (October 19, 2007), 1143.

98 Poole, *JCS and National Policy, 1950–52*, 79; Moody, *Building a Strategic Air Force*, 342–360; Letter, MG Emmett O'Donnell, Jr., to Gen Hoyt S. Vandenberg, January 26, 1951, Vandenberg Papers, Library of Congress.

99 Moody, *Building a Strategic Air Force*, 366–368; Robert S. Jordan, *Norstad: Cold War NATO Supreme Commander: Airman, Strategist, Diplomat* (New York: St. Martin's Press, 2000), 81–83; Robert A. Wampler, *NATO Strategic Planning and Nuclear Weapons, 1950–1957* (College Park, MD: Center for International Security Studies at Maryland, 1990), 5–6.

100 Quoted in Memo of Discussion of State-Mutual Security Agency—JCS Meeting, January 28, 1953, *FRUS, 1952–54*, V, 714.

101 DC 13, "North Atlantic Treaty Organization Medium Term Plan," April 1, 1950, in Pedlow, 107–177; D. Condit, 311–314; Poole, *JCS and National Policy, 1950–52*, 95.

102 NSC 71, "Views of JCS with Respect to Western Policy Toward Germany," June 8, 1950, *FRUS, 1950*, IV, 686–687; Memo by SecState of Meeting with the President, July 31, 1950, *FRUS, 1950*, III, 167–168; State Department Paper, "Establishment of a European Defense Force," ca. August 16, 1950, ibid., 212–219.

103 Acheson, *Present at the Creation*, 437–440. See also Lawrence S. Kaplan, *NATO and the United States: The Enduring Alliance* (rev. ed.; New York: Twayne Publishers, 1994), 43–46.

104 Robert J. Wood, "The First Year of SHAPE," *International Organization* 6 (May 19, 1952), 175–191; Lord Ismay, *NATO: The First Five Years, 1949–1954* (Paris: North Atlantic Treaty Organization, 1954), 37–38, 70–72; Poole, *JCS and National Policy, 1950–52*, 117.

105 Poole, *JCS and National Policy*, 120–25, 145–47; Ismay, *First Five Years*, 73–77.

106 Ismay, *First Five Years*, 102.

107 D. Condit, 339–341; Poole, *JCS and National Policy, 1950–52*, 126.

108 Notes on White House Meeting, January 31, 1951, *FRUS, 1951*, III, 454; D. Condit, 370; MC 14/1, "NATO Strategic Guidance," December 9, 1952; Pedlow, 193–228.

109 See the discussions between Eisenhower and European leaders resulting from Eisenhower's tour of Europe, January 1951, in *FRUS, 1951*, 402–449.

110 Memo, JCS to SECDEF, January 26, 1951, "Means at the Disposal of the Western Powers to Discourage . . . Aggression," JCS 2073/115.

111 Poole, *JCS and National Policy, 1950–52*, 159.

112 Thomas Alan Schwartz, *America's Germany: John J. McCloy and the Federal Republic of Germany* (Cambridge: Harvard University Press, 1991), 252–253. See also Bruce to Acheson, December 17, 1951, *FRUS, 1951*, IV, 455–459; Bohlen, Memcon with Jean Daridan, Minister Counselor of French Embassy, February 7, 1952, *FRUS, 1952–54*, V, 610–611.

113 Final Communiqué, 9th Session North Atlantic Council, February 26, 1952, *FRUS, 1952–54*, V, 177–179; Poole, *JCS and National Policy, 1950–52*, 151, 157–158.

114 See Edward Fursdon, *The European Defence Community: A History* (New York: St. Martin's Press, 1980), 150–188.

115 Bernard Law Montgomery, *The Memoirs of Field-Marshal the Viscount Montgomery of Alamein* (Cleveland: World Publishing Co., 1958), 462.

116 Ismay to Churchill, February 12, 1954, cited in Kaplan, *NATO and the United States*, 48–49.

Admiral Arthur W. Radford, USN, Chairman, Joint Chiefs of Staff, 1953–1957

Chapter 5

EISENHOWER AND THE NEW LOOK

Dwight D. Eisenhower's election in November 1952 presented the Joint Chiefs of Staff with the prospect of the most radical changes in American defense policy since World War II. A fiscal conservative, Eisenhower saw the heavy military expenditures of the Truman years bankrupting the country. Assuming that the Cold War might go on indefinitely, he sought to develop a sound, yet cheaper, defense posture the United States could maintain over the long haul. The result was a strategic concept known as the "New Look," which incorporated a broader than ever reliance on nuclear weapons and nuclear technology. Indeed, by the time Eisenhower was finished, military policy and nuclear weapons policy were practically synonymous. Some called it simply "more bang for the buck."

The first military professional to occupy the White House since Ulysses S. Grant, Eisenhower was, like Grant, a national hero. Commander of the Allied force that had invaded France and defeated Nazi Germany on the western front in World War II, he had served after the War as Army Chief of Staff, president of Columbia University, unofficial Chairman of the Joint Chiefs of Staff, and NATO Supreme Commander in Europe. To many Americans, he seemed the natural leader to guide them through the increasingly dense thicket of the Cold War.

Eisenhower's advent had a larger and more lasting impact on the JCS than any Commander in Chief until Ronald Reagan in the 1980s. Entering office with unrivaled experience in military affairs and the advantage of personally knowing how the JCS system operated, he knew first-hand how inter-Service competition and parochial interests could thwart agreement among the chiefs on common military policies. Internal differences, he later observed, "tended to neutralize the advisory influence they should have enjoyed as a body."[1] While the JCS had pulled themselves together and worked fairly well as a team during the Korean War, they had functioned more or less as their predecessors had done in World War II—with elastic budgets and under the pressure of events that concealed their internal rivalries and frictions. Anticipating an end to the hostilities in Korea, Eisenhower foresaw a postwar transition period of spending cuts and changes in strategy and force structure leading to renewed inter-Service strife and competition.

THE 1953 REORGANIZATION

In Eisenhower's view, revising the Nation's defense strategy and improving the effectiveness and efficiency of the Joint Chiefs went hand in hand. Knowing that rapid and radical changes could cost him the cooperation of the chiefs and of their supporters on Capitol Hill, he started slowly with modest adjustments. The blueprint he used was a Defense-wide reorganization derived from suggestions offered by former Secretary of Defense Robert A. Lovett and the recommendations of an advisory panel headed by Nelson A. Rockefeller, Eisenhower's protégé. Presented to Congress in April 1953, these changes, known as Reorganization Plan Number 6, took effect under an Executive order in June and required no further legislative action in the absence of congressional objections.

One of Eisenhower's principal objectives was to strengthen the powers of the Chairman, whose de facto role and authority increasingly outweighed the statutory description of his duties. To bring theory and reality more into line, the 1953 reorganization gave the CJCS the beginnings of his own power base by conferring on him authority to manage the work of the Joint Staff and to approve the selection of its members. To get the JCS to concentrate on their advisory and planning functions, the President removed the JCS from the operational chain of command by ending the practice, sanctioned under the 1948 Key West Agreement, that had allowed the Joint Chiefs to name one of their members as the executive agent for each unified or specified command. Henceforth, it would be up to the Service Secretaries to designate these executive agents. The President said that in taking these actions he intended to "fix responsibility along a definite channel of accountable civilian officials as intended by the National Security Act." Eisenhower would have gone further in reforming the JCS, but he recognized that the attempt would have aroused vigorous opposition on Capitol Hill, where the prospect of a more powerful Chairman and a stronger, more independent Joint Staff continued to conjure images of a "Prussian general staff."[2]

The appointment of a new set of Service chiefs and a new Chairman accompanied these structural changes. The "old" chiefs who were in place at the end of the Truman years—Bradley, Army Chief of Staff J. Lawton Collins, Chief of Naval Operations William M. Fechteler, and Air Force Chief of Staff Hoyt S. Vandenberg—were all either close personal friends of Eisenhower or well known to him by reputation. Many of the President's key political supporters, however, accused them of having aided and abetted a no-win strategy in Asia and run-away defense spending at home. Since most of their terms expired in the spring and summer of 1953, it was easy for the President to make a nearly clean sweep. The "new" chiefs included Admiral Arthur W. Radford, previously Commander in Chief, Pacific (CINCPAC), as Chairman, General Nathan F. Twining as Air Force Chief of Staff, General Matthew B. Ridgway, Eisenhower's successor at SHAPE, as Army Chief of Staff, and Admiral Robert B. Carney, formerly the commander of NATO forces in Southern Europe, as Chief of Naval Operations. The only holdover was General Lemuel C. Shepherd, Jr., Commandant of the Marine Corps, who served on the JCS in a limited capacity under legislation enacted in June 1952 allowing the Commandant to participate in JCS deliberations when matters of direct concern to the Marine Corps were under consideration.[3]

Radford's appointment as Chairman sent a powerful political message intended to promote inter-Service unity and cooperation. A naval aviator, Radford had opposed Service unification after World War II and spoken out repeatedly against Louis Johnson's defense policies during the 1949 "Revolt of the Admirals." While selecting a one-time opponent of unification raised more than a few eyebrows, Radford assured the President that his views on defense organization had changed and that he was now fully behind the aims of the National Security Act. Beyond this, he and Eisenhower shared a similar concern for the long-term effects of excessive military spending. Radford's familiarity with the Far East was a further asset at a time when that part of the world seemed to produce one major foreign-policy problem after another. To make the Joint Chiefs into a more effective corporate body, free of Service biases, Eisenhower admonished the admiral to lead the way by divorcing himself "from exclusive identification with the Navy." As an incentive, Eisenhower promised that Radford would have clearer responsibilities and greater authority than his predecessor, General Omar Bradley. Radford would have preferred to be Chief of Naval Operations, and at times he likened his role as CJCS to that of "a committee chairman," as if it were a demotion. But he worked hard on the President's behalf, got along well with Eisenhower's other senior advisors, and did a commendable job of rising above Service interests.[4]

Less successful were Radford's efforts to instill these virtues in his JCS colleagues and forge a consensus among them on basic plans embodying administration policies. During the Indochina and Quemoy-Matsu crises of 1954–1955, he tried to steer the JCS in the direction of military responses that conformed to declared White House positions on the use of nuclear weapons; for his efforts, he wound up being cast in the awkward guise of "party whip."[5] Despite the increased authority the Chairman exercised under Eisenhower, Radford actually had limited influence and control over strategic planning, the Joint Chiefs' key function, which remained a corporate responsibility. Integral to the allocation of resources, strategic planning was a continuing source of inter-Service rivalry. Interminable haggling over phraseology as well as the "force tabs" attached to war plans to lay out the size and composition of forces needed to carry out missions became commonplace.

Unable to agree on a single unified strategy, the JCS resorted to compromises built on broad statements of tasks and objectives that gave something to each Service. Out of this process (known derisively as "log-rolling") the Joint Strategic Capabilities Plan (JSCP) emerged as little more than a yearly inventory of forces available to each joint command in an emergency, while the mid-range Joint Strategic Objectives Plan (JSOP) resembled a compilation of individual Service requirements, assembled in no order of priority. Intended to help the Secretary of Defense and the President project future budgetary needs, the JSOP routinely fell short of its goal and quickly acquired the reputation of being a "wish list" of Service requirements. Occasionally, in this and other areas, Admiral Radford was successful in intervening to mend "splits." But by and large, his most effective weapon in overcoming Service differences was to digest the views of his colleagues and convey them to the President in his own interpretation of JCS advice.[6]

In view of his background and experience, Eisenhower did not hesitate to take matters into his own hands, behaving as Secretary of Defense, Chairman of the Joint Chiefs, and National Security Advisor all in one. Aware of JCS limitations, he frequently took over military planning and issued detailed guidance and direction as the situation warranted. All signs are that he enjoyed these tasks. Yet he still looked to his Secretaries of Defense to attend to day-to-day Pentagon chores and expressed irritation when they failed to measure up.[7] The three who served under him as Secretary of Defense—Charles E. Wilson, Neil H. McElroy, and Thomas S. Gates, Jr.—were business executives in private life and more adept at administration and fiscal management than military affairs. With the exception of Gates, who was Under Secretary and Secretary of the Navy before becoming Secretary of Defense in 1959, their experience in defense matters was exceedingly limited. Wilson, the first, had the hardest time. Formerly the head of General Motors, he was unfamiliar with the ways of the Pentagon and struggled to carry out the President's policies, many of which involved unpopular budget cuts. With Wilson obviously needing help, Eisenhower spent an inordinate amount of time on defense matters to help shore up the Secretary's position, and in the process established a pattern of hands-on involvement that lasted throughout his Presidency.[8]

The Joint Chiefs' most frequent contacts with the President were through the National Security Council, which Eisenhower used as his principal forum for debating and deciding high-level policy. As such, the NSC was a convenient mechanism for double-checking the Chiefs' advice and requirements. The practice that had developed during the Truman years of filtering JCS recommendations through the NSC remained in effect under Eisenhower and became even further institutionalized with the creation of new coordinating mechanisms—an interagency Planning Board, similar to the NSC Senior Staff of Truman's day but with broader powers to review and refine actions going up the "policy hill" to the President and the NSC; and an Operations Coordinating Board (OCB), to deal with intelligence operations and assure the implementation of NSC decisions. All functioned under the discreet and watchful eye of a Special Assistant for National Security Affairs who reported directly to the President. The net effect was a highly structured system of integrated policy review and collective decisionmaking that subjected JCS and Service requests and recommendations to minute scrutiny.[9]

Over time, the Joint Chiefs became highly proficient at working within this system and making it serve their needs. One benefit for them was that it provided reliable lines of communication with other government agencies, especially the State Department. Extremely useful to the chiefs was the administration's practice of conducting annual reviews of basic national security policy, resulting in comprehensive statements of policy that established guidelines and priorities for the development of military and related programs. Exceedingly detailed, these national policy papers emerged only after lengthy discussion and negotiation, with significant inputs from the Treasury and Bureau of the Budget. After laying out the administration's overall policy objectives, these papers virtually guaranteed that once a Service program was adopted, it would enjoy indefinite funding and political support. A major criticism of this system was that it allowed little flexibility in the face of changing international conditions and defense needs. But it suited the Services and the Joint Chiefs by providing them with a predictable platform for assessing requirements and a viable rationale for justifying their claims on resources.

ENDING THE KOREAN WAR

Eisenhower's first order of business as President was to fulfill his campaign promise and bring the Korean War to a swift and honorable conclusion. Stalemated since mid-1951, the war was a growing drain on troops, resources, and the patience of the American people. For the Truman administration, it had become an onerous political liability. Lest the effects linger, Eisenhower wanted an expeditious settlement that would allow the United States to withdraw some, if not most, of its forces. Out of the ensuing efforts to develop a strategy for ending the war emerged many of the key policy strands for the new administration's subsequent basic national security policy—the "New Look."

When Eisenhower took office in January 1953, the principal obstacle to an armistice was the prisoner of war issue. Even though the 1949 Geneva Convention called for mandatory repatriation of POWs, the Truman administration, acting on JCS advice, had embraced a nonforcible repatriation policy. Behind this policy was the chiefs' desire to avoid repeating the unpleasant experience after World War II when the Western allies forcibly repatriated sizable numbers of POWs held by the Germans to the Soviet Union. Reports reaching the West later revealed that Stalin executed many of these POWs and threw others into labor camps. During the Korean conflict, screening done by the UNC confirmed that over 75 percent of the Chinese POWs and a lesser percentage of North Koreans were unwilling to return voluntarily. Having had these figures accidentally revealed to them, Chinese and North Korean negotiators summarily rejected nonforcible repatriation. The armistice talks bogged down and on October 8, 1952, the U.S. chief negotiator, Major General William K. Harrison, Jr., USA, declared an indefinite recess until the Communists tendered a "constructive proposal." Almost immediately, the fighting escalated.[10]

As early as February 1952, the Joint Chiefs had begun to examine alternative courses of action in case the negotiations failed or became prolonged. By the following autumn, the consensus within the JCS organization in Washington and at UNC headquarters in the Far East was that an armistice was unlikely as long as North Korean and Chinese forces continued to occupy the heavily fortified defensive positions they had constructed across the Korean Peninsula. To break the impasse, both the Joint Strategic Plans Committee (JSPC) and General Mark W. Clark, USA, the commander of UN forces in Korea (CINCUNC) recommended a buildup of forces and a large-scale offensive to "carry on the war in new ways never yet tried in Korea."[11] The JSPC's plan incorporated the use of tactical atomic weapons against enemy targets in Korea, China, and Manchuria. Initially, Clark did not include nuclear weapons in his planning. Upon learning of the nuclear provisions in the JSPC's plan, however, he asked for authority to use them if the need arose. In the past, the JCS had shied away from recommending the use of nuclear weapons in Korea for political reasons and because of the limited size of the U.S. nuclear stockpile. But by late 1952, with bomb production up to over 400 assemblies per year, these supply restrictions were less inhibiting.[12]

The Joint Chiefs reviewed General Clark's plans and assured him that they would be given due consideration.[13] The previous summer, anticipating events, the JCS had initiated a buildup of nonnuclear components at storage facilities on Guam and aboard aircraft carriers operating in the Western Pacific.[14] With President Truman's knowledge and approval, the JCS had also taken steps to identify stockpiles of mustard gas and nerve agents at storage depots in the United States for possible use in dislodging the Chinese and North Koreans from their caves and bunkers along the front line in Korea. But with a new administration about to take office, the JCS held further measures affecting a buildup in abeyance.[15]

Meantime, accompanied by General Bradley, Admiral Radford, and Secretary of Defense-designate Wilson, President-elect Eisenhower went on a fact-finding tour of Korea in early December 1952. He returned convinced that stepped-up military pressure held the key to ending the conflict. Soon after the inauguration, he terminated the U.S. naval blockade of Taiwan, ostensibly "unleashing" Chiang Kai-shek to wreak havoc on mainland China, and gave the nod to intensify a conventional bombing campaign against North Korea that the Air Force had launched the previous October. Among the targets the President authorized were hydroelectric power plants on the Yalu River, industrial facilities in congested urban areas, and irrigation dams used in rice production, nearly all of which the previous administration had treated as off limits to bombing for humanitarian reasons.[16]

Between March and May 1953, Eisenhower considered further ratcheting up the military pressure in Korea and asked the Pentagon to come up with plans for a more aggressive campaign involving nuclear weapons, depending "on the advantage of their use on military targets."[17] Uneasy over the direction in which the President seemed headed, the JCS initially hesitated to propose a single course of action and offered instead a choice of six escalating options based on the planning done by the JSPC and CINCUNC. At the low end of the scale was a continuation of the existing level of military activity, followed by successive stages of stepped-up military pressure, culminating in a "major offensive" extending beyond the Korean Peninsula. At this point, all restrictions on the use of chemical and nuclear weapons would be removed.[18] The Planning Board tendered a slightly reworked version of these options (NSC 147) to the NSC in early April, but the Council sent it back with instructions that the JCS provide a specific course of action.[19]

Finally, on May 20, 1953, General Bradley presented an oral report to the NSC that left Eisenhower and the other Council members stunned. Assuming the primary goal to be a military solution, Bradley was convinced that the United States might be "forced to use every type of weapon that we have."[20] Accordingly, he outlined a plan for an all-out offensive in Korea, spearheaded by the use of chemical and tactical nuclear weapons, that would involve taking out targets in China and Manchuria. "We may also," he warned, "be risking the outbreak of global war." In his memoirs, Bradley suggested that the President had known the gist of the chiefs' proposals for some time and that he and Eisenhower had discussed these matters privately on previous occasions. Still, the President seemed taken aback by the aggressive tone of the Chairman's presentation and treated it as a hypothetical inquiry, to be acted upon "if circumstances arose which would force the United States to an expanded effort in Korea." Among the numerous issues yet to be addressed, he mentioned the "disinclination of our allies to go along with any such proposal as this" and the obvious need "to infiltrate these ideas" into their minds.[21]

While Eisenhower elected to hold a major escalation of the Korean War in abeyance, he still believed that military pressure held the key to a truce, and in the weeks following Bradley's presentation to the NSC, conventional air attacks against Communist targets in the north intensified. Irrigation dams received the most attention.[22] Through diplomatic channels, meanwhile, and at the armistice talks in Korea, U.S. representatives served notice that even "stronger" measures were in the offing. These "muffled warnings," as political scientist McGeorge Bundy later characterized them, were an unmistakable threat to use nuclear weapons, but whether they had the impact on the Communist side that Eisenhower claimed remains a matter of conjecture.[23] In any case, the negotiations showed sufficient promise of resolving the POW and other issues for Eisenhower to hold further threats in abeyance and to turn his attention to securing the cooperation of South Korea's recalcitrant President Syngman Rhee.[24] Finally, in July the two sides signed an armistice which avoided the forced repatriation of prisoners and left Korea divided along a demilitarized zone at approximately the same line as where the fighting began in 1950.

The ceasefire brought a respite but did not end JCS involvement in Korean affairs. Although the fighting subsided, tensions between north and south remained high, causing the JCS to keep the situation under constant and close review. For years after the armistice, the United States maintained about 50,000 air and ground forces in Korea under a UN command, while deploying large naval forces nearby and funding a military assistance program to train and equip a South Korean army of 700,000 troops. Next to Western Europe, Korea hosted the largest permanent overseas concentration of U.S. forces during the Cold War. In an increasingly common outcome of Cold War confrontations, neither side scored a clear-cut victory during the Korean conflict, nor did either side suffer a clear-cut defeat.

A NEW STRATEGY FOR THE COLD WAR

Ending the Korean War was the final major task of the "old" chiefs. To the "new" chiefs who succeeded them in the summer of 1953 fell the job of converting the

Armed Forces to a peacetime footing. Despite their ambiguous contributions to ending the Korean War, Eisenhower increasingly viewed nuclear weapons as the key to the country's future security. Stepped-up production of fissionable materials initiated during the Truman years and design improvements leading to new, more purpose-tailored weapons, from high-yield bombs for strategic use to tactical and battlefield weapons, created unprecedented opportunities that Eisenhower proposed to exploit to the fullest. Given the choice, he probably would have preferred a balanced defense posture, in which atomic weapons and conventional forces figured on a roughly equal basis. But from his recent experience in defense matters, as acting chairman of the Joint Chiefs in the late 1940s and as SACEUR, he lacked confidence in being able to overcome the fiscal and political difficulties, either at home or abroad, that raising and maintaining a peacetime conventional force of sufficient size entailed.[25]

Like Truman, Eisenhower viewed a strong defense and a sound economy as the twin pillars of national security. A fiscal conservative, he recoiled at the budget deficits that had accumulated under his predecessor and attributed them in large part to profligate military spending. He pledged to follow "a new policy which would continue to give primary consideration to the external threat but would no longer ignore the internal threat" of an economy weakened by heavy defense expenditures.[26] Assuming a Cold War of indefinite duration, the President rejected the radical changes in national strategy suggested in a high-level study (Project SOLARIUM) carried out during the early months of his Presidency, in favor of continuing the practice of containing Soviet power and influence.[27] Eisenhower also wanted to avoid the "feast or famine" fluctuations in defense programs that the Armed Forces had experienced since the 1920s by establishing a stable level of military spending. To do so, he abandoned the Truman administration's practice of pegging defense programs to a "year of maximum danger," and opted for a military posture that the country could sustain over the "the long pull" without jeopardizing the economy. For this purpose, increased reliance on nuclear weapons was almost ideal.[28]

Eisenhower found the Joint Chiefs to be among the most persistent and irritating obstacles he faced in carrying out his plans. Insisting that the current posture was "sound and adequate," they resisted cuts in conventional strength and argued that uncertainty over the use of nuclear weapons compelled them to retain substantial general purpose forces. Threatening the use of nuclear weapons was one thing; actually carrying through was quite another. The JCS acknowledged the primary importance of nuclear weapons in assuring national security, but wanted a clearer weapons-use policy, removal of the remaining impediments imposed during the Truman years on the military's access to nuclear weapons, and preservation of viable conventional capabilities as backup in the event of a Soviet nuclear attack.[29] While the United States continued to hold a comfortable lead in atomic bombs, intelligence estimates available to the JCS indicated that the Soviets were catching up and that they would have enough weapons by the mid-1960s to match the United States in destructive power.[30] Wholly unexpected was the Soviet detonation in August 1953 of a 400-kiloton thermonuclear device—significantly smaller in explosive power than the U.S. test of the previous November, but with design characteristics that gave the Soviets a deliverable hydrogen bomb (about the same physical size as a "Fat Man" implosion bomb) ahead of the United States.[31]

Eisenhower was well aware that the course he proposed had drawbacks and limitations. But he also knew, as did the Joint Chiefs, that the accuracy of intelligence on Soviet capabilities was questionable and subject to change depending on the available information and how the Intelligence Community interpreted it.[32] Barring an arms control breakthrough, Eisenhower accepted the proliferation of nuclear weapons as essentially unavoidable and sought to turn it to best advantage. He believed the quickest and easiest way was "to consider the atomic bomb as simply another weapon in our arsenal."[33] To those who argued that crossing the nuclear threshold risked all-out war, he replied that applying "tactical" atomic weapons against military targets was no more likely to trigger a "big war" than the use of conventional 20-ton block-busters.[34] Effective deterrence, he believed, meant having not only the capability but also the will to use nuclear weapons. The internal debate surrounding these issues and their impact on defense policy stretched from the summer into the fall of 1953 and revealed sharp differences of opinion. But in the end, the President's views prevailed, at least on paper. The upshot was a new basic national security policy (NSC 162/2) authorizing the Armed Forces to treat nuclear weapons "for use as other munitions" and to plan their force posture accordingly, with "emphasis on the capability of inflicting massive retaliatory damage by offensive striking power."[35]

Admiral Radford publicly described the administration's defense policy as a "New Look" in national security; others, seizing on language used by Secretary of State John Foster Dulles in a 1954 speech, called it "massive retaliation." Eisenhower considered such descriptions misleading because they implied a more sweeping change in the composition of the Armed Forces than he intended.[36] Rather than restructure the military establishment, he wanted to make it more efficient, more up-to-date with the latest technologies, and more economical. "His goal," historian John Lewis Gaddis observed, "was to achieve the maximum possible deterrence of communism at the minimum possible cost."[37]

Most of the savings Eisenhower achieved occurred during his first 2 years in office and came largely from budgets inherited from Truman, whose own plans called for similar reductions at the end of the Korean War. Once the Korean War "bulge" disappeared, Eisenhower faced steadily mounting costs owing to inflation and pressures arising from intelligence estimates pointing to greater-than-expected increases in Soviet strategic air and missile capabilities. Using essentially the same budgeting techniques as the Truman administration, Eisenhower insisted that military requirements fit within fixed expenditure ceilings. To make the money go further, he stretched out procurement and the implementation of approved programs. His major accomplishment was to reduce the rate of growth in military spending, not its overall size. As the largest item in the Federal budget, national security consumed on average about 10 percent of the country's GNP during Eisenhower's presidency. At the end of the administration's 8 years in office, total obligational authority for defense stood at just over $44 billion, roughly the same as when Eisenhower entered the Presidency.[38]

The principal beneficiary under the New Look was the Air Force, whose Strategic Air Command reaped the largest rewards. Force planning for the post–Korean War period done in the waning days of the Truman administration had pointed in this direction.[39] Under Eisenhower's more restricted budgets, the process accelerated. Though Air Force leaders recoiled at some of the funding cuts Eisenhower initially imposed, they soon found themselves enjoying a privileged position. On average, the Air Force received 46.4 percent of the defense budget during the Eisenhower years, compared with 28.3 percent for the Navy and Marines and 25.3 percent for the Army. During this same period, strategic forces (predominantly those under SAC) increased their claim on the total defense budget from 18 percent to nearly 27 percent.[40]

A formidable deterrent, the Strategic Air Command now became the country's undisputed first line of defense and retaliation. Relying primarily on manned bombers during the 1950s, SAC retired its propeller-driven B–29s and B–50s by the middle of the decade in favor of faster jet aircraft: the medium range B–47 and the intercontinental B–52, which replaced the problem-plagued B–36. By the time the Eisenhower administration left office, SAC had an operating force of 1,400 B–47s and 600 B–52s, supported by 300 KC–135 jet tankers for aerial refueling. Early B–52 models (the A through F series) had an unrefueled range of more than 6,000 miles while carrying as many as four gravity-fall atomic bombs; later models (the G and H series) had an unrefueled range of 7,500 to 8,000 miles and could carry up to eight nuclear weapons.[41]

SAC's main weakness during the 1950s was the increasing vulnerability of its bombers to a Soviet surprise attack. Initially, the threat came from the Soviet long-range air force, and later from Soviet intercontinental ballistic missiles (ICBMs). The detonation of the Soviet H-bomb in the summer of 1953 and signs the following year that Moscow might have a larger and more sophisticated strategic bomber program than previously suspected, gave rise to a variety of increased requirements. Based on limited evidence, the Air Force projected a Soviet advantage of up to two-to-one in long-range bombers by the end of the decade. The other Services and the CIA suspected that the Air Force was playing fast and loose with its numbers to pad its budget requests. The give-and-take continued into 1956 when, with the help of U–2 photographs, it became clear that the "bomber gap" grossly exaggerated Soviet capabilities and the matter was laid to rest, but not before the Air Force had acquired additional funding to augment its bomber fleet.[42] At the same time, to reduce SAC's vulnerability to bomber attack, the Eisenhower administration resorted to a series of costly countermeasures, including dispersed basing of SAC's planes, the deployment of an integrated system of missiles and air defense interceptors, extension of the distant early warning (DEW) line, and the creation in 1958 of a combined U.S.-Canadian command and control organization known as the North American Air Defense Command (NORAD). Nonetheless, SAC's vulnerability persisted and gave rise to ever-increasing requirements to allow it to "ride out" an enemy attack, a process that kept alive and aggravated tensions within the Joint Chiefs over the allocation of resources.[43]

The Navy, defying all predictions, adjusted remarkably well to the New Look. While Navy leaders made no secret of their disdain for the pro–Air Force orientation of Eisenhower's defense program, there was no repetition of the nasty sniping after World War II and no second "revolt" of the admirals. Radford's presence as Chairman eased the situation considerably, as did the leeway the Navy received to conduct both a high-profile missile R&D effort, which eventually gave rise to the Polaris fleet ballistic missile system, and a shipbuilding program that included construction of a new generation of heavy carriers. Dating from the Korean War, the carrier program was the brainchild of Chief of Naval Operations Admiral Forrest P. Sherman and initially envisioned replacing the Navy's World War II *Essex*-class carriers with larger *Forrestal*-class ships at a rate of one a year for 10 years. While the pace slowed during the 1950s, the eventual goal remained the same. By the end of the Eisenhower years, the Navy had commissioned four new *Forrestal*-class carriers and had a fifth (the nuclear-powered *Enterprise*) nearing completion. Out of 26 carriers then in service, 15 were large attack carriers (*Essex*-class or bigger), a number that remained nearly constant for the duration of the Cold War.[44]

The size and design of the *Forrestal*-class "super carriers" meant they could embark nuclear-capable aircraft. To avoid renewed accusations of competition with the Air Force, the Navy assigned them a general purpose role. Sherman envisioned these ships serving primarily in the Atlantic or Mediterranean, delivering conventional and atomic attacks against Soviet naval bases and airfields in support of NATO.[45] But because of continuing tensions in the Far East and better port facilities in the Pacific, Admiral Carney persuaded President Eisenhower to modify this strategy. Thus, the carriers came to be concentrated in the Pacific, with the proviso that in a European emergency the Navy would redeploy them as needed to assist NATO. Navy planners were never comfortable with this "swing strategy," and in 1955 Carney's successor, Admiral Arleigh A. Burke, launched a campaign to abolish it. Once in place, however, the swing strategy became a firm fixture of NATO force planning. A symbol of the American commitment, it survived to the dying days of the Cold War, despite one effort after another by the Joint Chiefs to eliminate it.[46]

The JCS member least enamored with the New Look was General Matthew B. Ridgway, who openly disparaged many aspects of the President's defense policy throughout his 2-year term as Army Chief of Staff. Ridgway's main objections to the New Look were that it failed to preserve an adequate mobilization base for rapid Army expansion in an emergency and that it gave undue emphasis to nuclear weapons without fully vetting the concept.[47] His successor, General Maxwell D. Taylor, was, if anything, even more censorious of administration policy. Ridgway knew that the New Look would take a heavy toll on the Army, but he professed to be shocked by the full impact, which involved reducing Army personnel strength by more than 500,000 and slimming down from 20 to 14 Active-duty divisions by 1957.[48] As an economy measure, Eisenhower also wanted the Army to redeploy as many units as possible from NATO and other overseas theaters to the United States, but shelved his plans in the face of strong political and diplomatic objections. As a result, force levels in Europe remained essentially unchanged, while the two divisions left in Korea after the armistice, the one in Hawaii, and those stateside in the Strategic Reserve routinely operated at reduced strength.[49]

Looking ahead, Eisenhower challenged the Army to reconfigure itself around smaller, more mobile divisions designed specifically for the nuclear battlefield. Eisenhower believed that, with the advent of nuclear weapons, no infantry division needed to be bigger than 12,000 men.[50] Studies done at the U.S. Army Infantry School and exercises conducted by the Army Field Forces (later, the Continental Army Command) indicated, however, that combat in a nuclear environment would require divisions to be larger rather than smaller. Efforts to address this problem led in 1956 to the adoption of the "pentomic" division as the blueprint for the Army of the future. Organized into five battle groups, pentomic divisions resembled the structure of the airborne divisions that Ridgway and Taylor commanded in World War II. Each pentomic division had approximately 11,500 men rather than the 17,000 in a post–Korean War "triangular" infantry division. Rated as "dual capable," a pentomic division incorporated conventional firepower and an array of nuclear weapons, from atomic artillery to nuclear-tipped rockets and missiles, and—most unique of all—the "Davy Crockett," a spigot mortar (often erroneously described as a recoilless rifle) adapted to fire a sub-kiloton nuclear warhead. Further study and field tests soon demonstrated that pentomic divisions would lack staying power in a conflict and that much of the hardware and weaponry on which these units depended was not up to the job. By 1960, Army leaders were exploring yet another divisional reorganization scheme.[51]

Even though the pentomic division failed to measure up and soon disappeared, it served a useful purpose by drawing the Army's attention to the impact of new technologies. What the New Look taught Army leaders as much as anything was that, if they were to protect their budget share and remain competitive with the Air Force and the Navy, they had to move beyond an "unglamorous" arsenal of tanks, artillery, and small arms and devote more research and development to guided missiles and other sophisticated weapons. To avoid becoming marginalized, the Army needed to broaden its mission. Taking this lesson to heart as the 1950s progressed, Army leaders used their small but aggressive R&D program to solidify their claim to old functions and lay claim to new ones.[52] In some ways, the Army succeeded too well, for in the process Service roles and missions were again left in disarray. By 1957, the Army was the first Service to test a land-based intermediate-range ballistic missile (IRBM), known as JUPITER, forging ahead of the Air Force's THOR program, and was on the verge of seizing control of the anti-intercontinental ballistic missile (ABM) function with its planned NIKE-ZEUS interceptor missile. This last development was a critical step toward the Army acquiring a major role in strategic warfare and would have reverberations that would echo to the end of the Cold War and beyond.

TESTING THE NEW LOOK: INDOCHINA

While the Joint Chiefs were still digesting the impact of the New Look, events abroad were testing its basic premise that nuclear weapons held the key to the country's future security. No sooner had the dust begun to settle in Korea than the Cold War shifted to Indochina, where the protracted struggle between the French and the Communist Viet Minh appeared to be entering a new and decisive phase. Even though the French had yet to suffer a major setback, war-weariness at home and the inability of French Union troops (predominantly Vietnamese) to sustain the initiative suggested a shift in momentum in favor of the Viet Minh.[53] Finding the war was no longer winnable, the French government notified Washington in July 1953 that it would follow the example of the United States in Korea and end the Indochina conflict as soon as possible, preferably through a negotiated settlement. Expecting the worst, U.S. intelligence sources warned that a Viet Minh victory in Indochina "would remove a significant military barrier" and open the way for communism to "sweep" across Southeast Asia into the Indian subcontinent and beyond.[54]

The Eisenhower administration's initial response was to continue its predecessor's practice of bolstering indigenous forces in Indochina and elsewhere with advice and assistance. Policies adopted in 1953–1954 by the National Security Council, however, indicated a strong willingness to fight to protect U.S. interests in the Western Pacific and to curb the further expansion of Chinese Communist power and influence. JCS contingency planning based on these policies assumed the use of nuclear weapons.[55] But at a five-power staff planners conference hosted by CINCPAC at Pearl Harbor in September–October 1953, the British, French, Australian, and New Zealand military representatives balked at giving prior approval to any military action involving "weapons of mass destruction." A none-too-subtle expression of worry over radioactive contamination from recent atomic weapons tests, the allies' objections also appeared to reflect their growing concern for the potentially adverse impact on Asian opinion that the use of nuclear weapons in that part of the world could have. Even so, the Joint Chiefs reminded CINCPAC after the conference that the exclusion of nuclear weapons, even for planning purposes, was contrary to approved U.S. policy, and directed the Strategic Air Command to develop an atomic attack plan against selected targets in China and Manchuria should Communist Chinese forces intervene in Indochina. Yet, given the reluctance of the other powers in the region to associate themselves with U.S. retaliatory plans, it was likely that in an extreme emergency the United States could find itself acting unilaterally.[56]

While the possible use of nuclear weapons was ever-present throughout the crisis, the larger and more immediate issue facing the Joint Chiefs was whether to get involved at all.[57] Forced to accept sizable budget and troop reductions and having only recently concluded the conflict in Korea, the Service chiefs—Ridgway especially—were uneasy about being drawn into another Asian war. Keeping his options open, President Eisenhower never categorically ruled out direct intervention. But he shared his military advisors' concerns about the costs and consequences, and assured them that he could not imagine putting U.S. ground forces anywhere in Southeast Asia "except possibly in Malaya," where the British and Australians were involved in suppressing a Communist insurgency. Playing down the possibility of U.S. intervention, he likened the American role to fixing "a leaky dike," and in January 1954 he approved policy guidelines limiting retaliation in Indochina to air and/or naval power should the French falter or the Chinese intervene.[58]

Even though the President had seemed to rule out the use of ground troops, events in Indochina conspired to keep the issue alive, and over the next several months Joint Staff and Army planners continued to pay it close attention. The immediate concern was the gathering crisis over Dien Bien Phu, a French redoubt on the Laotian frontier, which had come under siege. By the beginning of 1954, the Viet Minh had the French completely surrounded and wholly dependent on air-delivered reinforcements and supplies. In developing U.S. responses to the ensuing crisis, Eisenhower often bypassed the Service chiefs, finding it more expedient to deal directly with Admiral Radford. Familiar with the Far East, Radford tended to be more open-minded than his JCS colleagues in addressing French requests for assistance and more flexible on the issue of American military intervention, so much so that Indochina was sometimes seen as "Radford's war."

Confirmation that the Chairman was now part of Eisenhower's "inner circle" came from his appointment to the President's Special Committee on Indochina, created in January 1954 to develop a program for aiding the French without overt U.S. participation. The others on the panel were Director of Central Intelligence Allen Dulles, Undersecretary of State Walter Bedell Smith, Deputy Secretary of Defense Roger M. Kyes, and C.D. Jackson, the President's special advisor on psychological warfare. As the composition of the committee suggests, Eisenhower hoped to avoid direct American military involvement in Indochina through the alternative of covert operations, an increasingly common Cold War practice that in this instance had the strong encouragement and endorsement of the Joint Chiefs of Staff.[59] But as the Viet Minh tightened their siege of Dien Bien Phu, doubts grew whether covert operations as planned would be sufficient, causing the Special Committee to speculate that "direct military action" might be required to safeguard U.S. interests.[60]

Toward the end of March 1954, General Paul Ely, chief of the French Armed Forces staff, arrived in Washington appealing for help to stave off a collapse at Dien Bien Phu. Ely estimated the chances of avoiding defeat at fifty-fifty. Convinced that the situation was dire, Radford advised the President that the United States needed "to be prepared to act promptly and in force" to relieve the pressure on Dien Bien Phu.[61] According to Ely's recollections, Eisenhower instructed Radford (in Ely's presence) to make priority responses to all French requests to assist Dien Bien Phu.[62] Ely returned to Paris confident that the United States would provide land- and sea-based air support for a pending operation (code-named *Vulture*) to lift the siege. He apparently believed the United States would employ nuclear weapons. By now, the Joint Staff had several attack plans under consideration—one involving the use of conventional munitions dropped by B-29s, and another plan, derived from Army G-3 staff studies, that envisioned the use of up to six tactical nuclear weapons delivered by Air Force or Navy fighter-bombers. But at a meeting of the Joint Chiefs on March 31, Radford encountered uniformly strong opposition led by Ridgway. Even Twining, who normally backed up Radford in JCS debates, would give only guarded support to the operation, leaving the Chairman isolated as the only JCS member fully favoring armed U.S. involvement.[63]

This meeting was, for all practical purposes, the high-water mark of planning for intervention and for the possible use of nuclear weapons in Indochina. Though Radford continued to promote the project, his better judgment told him it was a lost cause. Unable to carry his JCS colleagues with him, his arguments rang hollow. Indeed, as word "leaked" that the Joint Chiefs were at odds over a plan of action, support for intervention among congressional leaders and within the international community collapsed almost overnight. The French continued to assume that American help was on the way. But as the days passed and no American relief materialized, Dien Bien Phu's fate became certain. On May 7, 1954, after heavy fighting, the garrison capitulated. Later that summer at Geneva, the major powers concluded an agreement ending French rule in Indochina and dividing Vietnam, like Korea and Germany, into Communist and non-Communist states.

Had the Joint Chiefs supported intervention, the course of events assuredly would have been different. But with the long and indecisive involvement in the Korean War a vivid memory and postwar budget cuts eroding force levels, the Service chiefs were averse to embarking on what Ridgway termed "a dangerous strategic diversion of limited United States military capabilities."[64] To them, as to the American public, the use of nuclear weapons, even for limited tactical purposes, still implied a major conflict transcending traditional norms. Administration policy and preferences notwithstanding, the JCS, excluding Admiral Radford, remained uncomfortable with the notion that nuclear weapons were simply another part of the arsenal. Looking back on the crisis, Eisenhower found it "frustrating" that he had not achieved more success in educating the public, his military advisors, or the international community "on the weapons that might have to be used" in future wars.[65]

CONFRONTATION IN THE TAIWAN STRAIT

As the Indochina crisis neared an end, the next test of the New Look was already in the making over a looming confrontation in the Taiwan Strait. At issue was the fate of three small island groups (Tachen, Matsu, and Quemoy) lying a few miles off the China mainland, which the Nationalists had occupied since 1949. The Nationalists used these islands for intelligence gathering, early-warning radar bases, and jumping-off points for commando raids against Communist positions on the mainland. Most U.S. military analysts agreed that the strategic value of these islands was negligible. But after the stalemate in Korea and the French collapse in Indochina, Admiral Radford insisted that the United States could not afford to give up more ground. Indeed, he saw their loss as having "far reaching implications" of a political, psychological, and military nature that could undermine resistance to Communism on Taiwan and throughout the Far East. With air and naval superiority in the area, the Chairman argued, the United States enjoyed distinct advantages that it had not had during the Indochina crisis.[66] Weighing the pros and cons of defending the islands, Eisenhower, though skeptical whether they were of much value militarily, gradually came around to Radford's point of view that their political importance was overriding.[67]

The situation turned critical on September 3, 1954, when the Communist Chinese launched a heavy artillery bombardment of Quemoy. As the confrontation was taking shape, Eisenhower seemed more prepared than ever to entertain a nuclear response and speculated at one point that the People's Republic of China's fleet of junks would make "a good target for an atomic bomb" if the Communists tried to invade Taiwan.[68] Throughout the crisis, he and Radford remained convinced that the use of nuclear weapons in such situations was only a matter of time and that the United States needed to accept the idea in order to be better equipped and ready. But despite tough talk, the administration initially leaned toward a guarded response, and for the first few months of the crisis the United States fell back on more traditional means of applying pressure—the signing of a formal defense treaty with Taiwan in December 1954, obtaining declarations of support for Taiwan from Congress, and a buildup of conventional U.S. air and naval forces in the Taiwan Strait.

The administration's caution and restraint reflected, among other things, the continuing "split" among the Joint Chiefs over the New Look's practical application, reinforced by an underlying worry (common to military and civilian policy-planners alike) that the use of nuclear weapons in Asia could provoke an anti-American backlash and charges of racism. Such thinking may have influenced deliberations during the Indochina crisis, but it was not until the Taiwan Strait episode—when the use of nuclear weapons would undoubtedly have resulted in thousands of Chinese casualties—that the full impact became apparent. Still, it did not stop either Eisenhower or Radford from seriously considering the nuclear option.[69] Having failed to rally JCS support during the Indochina episode, Radford made a determined effort from the outset of the Taiwan Strait crisis to develop a consensus within the JCS that the offshore islands should be defended, leaving aside the question of the means for the time being. A majority of the chiefs, including Twining, Carney, and Shepherd, agreed that the United States had valid security interests at stake and should be prepared to act in their defense. But they refused to hand Radford a blank check and insisted that "available forces," with minor augmentation, could do the job. To Ridgway, however, even a token involvement seemed excessive. Insisting that the offshore islands were of "minuscule importance," he viewed a decision to defend them as folly. If, however, the administration went ahead, it should realize the risks involved and be ready to take "emergency actions to strengthen the entire national military establishment and to prepare for war."[70]

Also weighing into the debate was Secretary of State John Foster Dulles, Eisenhower's most trusted advisor. One of the original architects of the New Look, Dulles was well known for his "hawkish" views on combating communism and as author of the "brinksmanship" concept linking proactive diplomacy to the threatened large-scale use of nuclear weapons. But by the fall of 1954, Dulles was having second thoughts and had come to the conclusion that the use of nuclear weapons short of all-out war would lack popular support at home, alienate U.S. allies in Europe, and hand the Communist Chinese a propaganda issue they could exploit for years to come. The United States, he warned, "would be in this fight in Asia completely alone." Though he supported defending Taiwan, Dulles questioned the strategic value of the offshore islands and persuaded Eisenhower to avoid provocative actions that might turn world opinion against the United States or make it exceedingly difficult to use nuclear weapons later when they might make a difference.[71]

Tensions in the Taiwan Strait, meanwhile, continued to escalate. In January 1955, the Communists began ratcheting up the pressure, first against Tachen, which the Nationalists at U.S. urging evacuated in early February, and then opposite Quemoy and Matsu, where the People's Republic of China (PRC) appeared to be massing troops for an invasion. By March, believing a showdown to be imminent, Radford was laying the groundwork for a nuclear response. "Our whole military structure had been built around this assumption," he told the NSC. "We simply do not have the requisite number of air bases to permit effective air attack against Communist China, using conventional as opposed to atomic weapons." Likely targets identified by the Joint Staff and CINCPAC included Communist Chinese airfields adjacent to Quemoy and Matsu and petroleum storage facilities as far away as Shanghai and Guangzhou (Canton). To minimize collateral damage, Radford insisted that only "precision atomic weapons" would be used.[72]

By then, Radford had a majority of the Service chiefs behind him in support of some form of military action, with Ridgway the lone dissenting voice.[73] Believing that talk of war had gotten out of hand, Eisenhower rejected the JCS majority view favoring "*full-out* defense of Quemoy and Matsu," and sought instead a cooling off period, as much to reassure nervous allies in Europe as to head off a confrontation with China.[74] He seemed to feel, given the uncertainty of the situation, that the use of atomic weapons was becoming more and more a course of last resort.[75] In late April, he sent Admiral Radford and Assistant Secretary of State Walter S. Robertson (both of whom knew Chiang Kai-shek personally) to Taipei to explain the situation.[76] The Communist Chinese also appeared to be having second thoughts, and at the Bandung conference on April 23, Premier Zhou Enlai declared the PRC's readiness to discuss "relaxing tensions" in the Far East, "especially in the Taiwan area."[77]

By the end of May 1955, an informal ceasefire had settled over the offshore islands, causing the issue to drop off the JCS agenda. A revival of tensions in 1958 produced a second offshore islands crisis, replete with renewed bombardment of Quemoy and Matsu, a buildup of forces by both sides, and invasion threats from the PRC. Once again, the Joint Chiefs considered a possible nuclear response but held a decision in abeyance pending a clearer picture of the situation. The crisis ended, like the first, inconclusively and was the last time the Eisenhower administration contemplated the use of tactical nuclear weapons against the People's Republic of China.[78] Only on two further occasions—during the 1961 Laotian crisis and the 1968 siege of Khe Sanh during the Vietnam War—did the JCS again actively consider recourse to nuclear weapons in Asia. In both instances, the advantages to be gained seemed incompatible with the risk.[79] Seizing on the many advances in nuclear technology in the decade following World War II, the New Look gave the Joint Chiefs access to unprecedented power and a wealth of innovative tools for waging war. But it did not do much to clarify how or in what circumstances they might be applied.

THE "NEW APPROACH" IN EUROPE

Doubts and uncertainty among the JCS notwithstanding, the Eisenhower administration remained firmly committed to developing a military posture that stressed nuclear weapons. Nowhere was this commitment more strongly pursued than in Europe where the New Look took the form of the "New Approach," adopted by the North Atlantic Council in December 1954 as MC 48, the new basic blueprint for NATO strategy. While MC 48 affirmed the continuing need for conventional forces, it cited superiority in atomic weapons and the capacity to deliver them as "the most important factor in a major war in the foreseeable future." At the time,

NATO had a mere handful of atomic weapons at its disposal; within a decade, largely as a result of steps taken by the Eisenhower administration, it would have a dedicated arsenal of 7,000 nuclear bombs and warheads.[80]

NATO's embrace of a nuclear response to Soviet aggression in Europe stood in marked contrast to the Allies' opposition to the Eisenhower administration's threatened use of such weapons in Asia. The explanation for this paradox lies in NATO's underlying philosophy of deterrence and defense, and the historic role nuclear weapons had played in NATO strategy. For the European Allies, actually using nuclear weapons and threatening their use were two wholly different matters. Nuclear weapons had been a fundamental part of NATO's politico-military culture since the Alliance's inception in 1949 and had grown steadily in importance. As Stanley R. Sloan and others have shown, not only were nuclear weapons essential for military purposes as NATO's primary deterrent and first line of defense; they were a key ingredient in the political bond holding the Atlantic Alliance together. The U.S. commitment to come to Europe's protection in the event of a Soviet invasion, exposing itself to nuclear retaliation on Europe's behalf, was central to what the American diplomat Harlan Cleveland called the "transatlantic bargain," a community of reinforcing interests. American nuclear weapons, in effect, sealed the deal.[81]

Nonetheless, prior to the Eisenhower administration, the JCS had tried to play down NATO's dependence on nuclear weapons, partly because they remained few in number and out of concern that increased reliance might discourage European conventional rearmament. With the impending advent of nuclear plenty, however, views began to change. The first to acknowledge the opportunities were the British Chiefs of Staff, whose 1952 "white paper" on global strategy offered an alternative course linked directly to the utility of a growing arsenal of nuclear weapons in lieu of conventional capabilities.[82] As British defense planners described it, the aim would be "to increase the effectiveness of existing [NATO] forces rather than to raise additional forces."[83]

With the exception of the Air Force, the military Services in the United States paid little attention to these proposals, believing it premature to write off conventional rearmament efforts. But by the summer of 1953, a combination of factors—the ongoing review of U.S. defense policy that included discussions of withdrawing U.S. troops from Europe, planned cutbacks in U.S. military aid, a slumping European economy, and an embryonic initiative by the British to place their own version of the New Look before NATO—put pressure on the JCS to reexamine their position and to come up with fresh ideas on how to satisfy European security needs. The chiefs agreed that because of the European Allies' economic difficulties, there was little likelihood of NATO meeting declared force goals on time and that a reexamination of NATO strategy would certainly be in order. But there was no unanimity on what the United States ought to suggest.[84] Earlier, as SACEUR, General Ridgway had requested five battalions of the new 280-mm cannon, which could fire either conventional or atomic shells, and had initiated studies on using tactical nuclear weapons to bolster NATO's forward defense strategy and to offset reductions in troop strength. However, the results of these inquiries, based on sketchy data and limited familiarity with nuclear weapons, had disappointed those seeking a relatively cheap and convenient replacement for expensive conventional forces. Now, as Army Chief of Staff, Ridgway shied away from the further nuclearization of NATO and enlisted Admiral Carney in support of keeping the status quo until completion of the U.S. military review then underway. General Twining, the Air Force member, was the only Service chief who ventured to speculate that the solution to NATO's problems might require a sharp departure from current policy and doctrine.[85]

Unable to elicit unanimous advice from the Joint Chiefs, President Eisenhower gave Secretary of State Dulles a free hand to come up with a plan of action. Moving quickly to avoid being preempted by the British, Dulles achieved high-level interagency agreement by late September 1953 on a "new concept" to expand NATO's application of tactical nuclear weapons. At Admiral Radford's request, the State Department postponed a final decision until the JCS had a chance to review the plan.[86] But the chiefs' response, when it came on October 22, skirted the issue by suggesting that the matter be held over for review by the NATO Standing Group, where a final recommendation might have been held up indefinitely.[87] Ignoring the chiefs' proposal, Dulles sounded out his British and French counterparts at the Bermuda conference in December 1953. He then put the issue before the North Atlantic Council, which adopted a resolution instructing NATO's top commanders to review their strategy and force structure, taking account of recent breakthroughs in military technology.[88]

As a result of these actions, the initiative shifted from the JCS in Washington to NATO planners in Paris working under the direction of General Alfred M. Gruenther, USA, Ridgway's successor as SACEUR, and his air deputy, General Lauris Norstad, USAF. Gruenther, a former Director of the Joint Staff, was also one of Eisenhower's closest personal friends. Using fresh intelligence and doctrinal and tactical assumptions in line with the known effects of nuclear weapons, Gruenther and his staff recast the studies Ridgway had done.[89] It was from these "New Approach" studies that MC 48, a 3-year plan for reorganizing NATO's forces, emerged. The key finding was that while the level of M-day forces would remain essentially unchanged, the substitution of nuclear weapons for conventional firepower would cause requirements for follow-on reserve forces to go down.[90] Midway through the

NATO review, in August 1954, the French Assembly voted to defer action on the European Defence Community (EDC), effectively killing the project and throwing the whole question of German rearmament into confusion. Gruenther had always said that, even with an enhanced nuclear capability, NATO would still need a credible conventional "shield" to prevent Western Europe from being overrun. For this reason, he remained a staunch proponent of a full German contribution to NATO and a strong conventional component. But with the EDC a shambles, NATO's credibility now rested more than ever on sharpening its nuclear "sword."[91]

While the Joint Chiefs endorsed the New Approach, two members—Ridgway and Carney—did so with reservations, warning that the collateral damage from nuclear weapons to cities and civilians would be almost catastrophic. They embraced the New Approach and, in the Army's case, the pentomic divisions and other paraphernalia that went along with it, not because they thought these changes would improve European security or save money, but because they were convinced that nuclear weapons would inevitably be used in a major conflict. Plausible deterrence, in the JCS view, therefore dictated that NATO had to be prepared to fight both a conventional and a nuclear war.[92] At the time, deterrence theory rested largely on balancing the raw military power of one side against that of another. Intelligence confirmed that the Soviets continued to devote high priority to their atomic energy program, and following Stalin's demise, there were mounting indications that, like the Eisenhower administration, the Soviet Union's new leaders were shifting the burden of defense from conventional forces to nuclear weapons to cut costs.[93]

The Joint Chiefs expected to be busy for years sorting out how the New Approach should be interpreted and applied. Though convincing the NAC to accept the idea came more easily than expected, there remained a distinct anxiety among the Europeans over who would have the authority to order the use of tactical nuclear weapons if deterrence failed. Since the United States was the only NATO power at the time with a significant nuclear capability (British forces began receiving production nuclear bombs late in 1954), the fate of Western Europe could well rest in U.S. hands. The solution favored by the Joint Chiefs was to give NATO's supreme commanders preexisting approval to carry out agreed defense plans in full.[94] Recognizing the need for greater flexibility but unwilling to go quite so far, the Eisenhower administration in December 1953 liberalized its policy on sharing atomic energy information with other countries where legally permissible.[95] A new Atomic Energy Act, which cleared Congress in the summer of 1954, paved the way for closer collaboration.[96]

The Joint Chiefs allocated nuclear weapons as needed to satisfy NATO requirements and stockpiled them at various locations in Western Europe under the custody and control of the American theater commander (USCINCEUR), also serving as SACEUR.[97] This arrangement allowed Europeans access to U.S. weapons for planning purposes and ostensibly a voice in deciding how and where these weapons would come into play during a conflict. But it did not go far enough to suit some, and by the end of the decade there was growing talk on both sides of the Atlantic of creating a "NATO common stockpile." Proponents contended that the Alliance should have the capacity to operate independently with its own nuclear assets, including not only tactical weapons but also land- and/or sea-based IRBMs that could threaten strategic targets in the Soviet Union.[98] The Joint Chiefs opposed such a move since it would unhinge U.S. war plans from NATO's and duplicate some targeting. But as the Eisenhower administration drew to a close, the momentum within the Alliance was moving toward creation of a NATO-led multilateral nuclear force.[99]

NATO'S CONVENTIONAL POSTURE

Despite the increased emphasis on nuclear deterrence, most of the weapons NATO needed for its new strategy did not reach Europe in appreciable numbers until the late 1950s and early 1960s. Until then, conventional "shield" forces remained the core of NATO's defense posture. Over the course of the decade, the limited introduction of improved tanks, armored personnel carriers, and heavy, self-propelled artillery gradually transformed NATO from a largely foot infantry force into a modern, combined-arms force.[100] Overall, however, these qualitative improvements were insufficient to provide a credible conventional alternative. Nor did they prevent NATO's capabilities from eroding as assets previously allocated for a conventional role (e.g., tactical aircraft) were reconfigured for nuclear missions and as Alliance members unilaterally reduced their contributions in the expectation that nuclear weapons would fill the gaps. The largest and most significant reductions were by the French, whose growing concern over the insurgency in Algeria from 1954 on prompted the eventual transfer of five divisions from NATO to North Africa. By the end of the decade, France had 500,000 troops tied down in Algeria and the equivalent of only one division dedicated to NATO, instead of the 15 to 20 once envisioned.[101]

Other members failed to pick up the slack. As long-time proponents of German rearmament, the JCS expected the admission of the Federal Republic of Germany (FRG) to NATO in 1955 to go far toward solving manpower problems. However, they were taken aback when, a year later, the FRG decided to slow its rearmament program by shortening the length of service for draftees and to seek access to nuclear weapons. The planned structure of the *Bundeswehr* remained 12 divisions and 40 air squadrons, but manpower would be cut by roughly one-third and the target date for completion of the buildup would be moved from 1961 to 1965.[102] Around the same time, bowing to fiscal constraints, Britain, Belgium, and the Netherlands also began to prune their conventional contributions to NATO.[103] In May 1957, the NAC adopted a new formal strategy statement (MC 14/2) confirming that tactical nuclear weapons would be NATO's mainstay against a Soviet invasion, and tentatively set new conventional force goals (formally approved in 1958 as MC 70) of 30 ready divisions in the Central Region for lesser contingencies. But with only about 19 divisions on hand, NATO was still well below its goal.[104]

Even the United States, by far the strongest member of the Alliance, had trouble meeting its commitments. Though U.S. deployments held steady at around six division-equivalents, Army units were often under strength and unevenly equipped. Only the Air Force maintained a level of preparedness consistent with agreed force goals.[105] Uneasy over NATO's prospects, the JCS continued to incorporate provisions in U.S. war plans (as distinct from NATO plans) for a withdrawal of American forces to defensive positions along the Alps and the Pyrenees should the Rhine-Ijssel line be breached.[106] Among Europeans, speculation was rife that the Eisenhower administration had secret plans to reduce its commitment to NATO and that it intended to rely more than ever on Reserve units based in the United States and swing forces in the Pacific to meet its obligations.

Matters came to a head in the summer of 1956 when, in a money-saving move, Admiral Radford attempted to persuade the JCS to accept a radical restructuring of the Armed Forces that included reducing U.S. troops in Europe by 50,000 and reorganizing American ground units into small atomic-armed task forces. The Service chiefs acknowledged the need to reduce overseas deployments, but they could see no place in either the Far East or Europe where this could be done without enormous risks.[107] Secretary of State Dulles and Secretary of Defense Charles E. Wilson backed the plan, and in October 1956 President Eisenhower added his concurrence. However, unauthorized "leaks" making it appear that the United States was preparing to abandon NATO soon followed, provoking an international incident. Embarrassed and chagrined, the administration hastily backtracked, and most of the proposed reductions were restored during the final mark-up of the defense budget at the end of the year, keeping U.S. forces in Europe more or less intact.[108]

Fortunately for NATO, Soviet bloc forces around this time were no better prepared than those in the West, and in certain categories they were probably weaker. Neither side appeared to possess decisive conventional power. Estimates of Soviet capabilities originated within the Services' intelligence offices and focused on counting units and equipment. Owing to a lack of reliable data, these estimates tended to be on the high side and paid little attention to manning levels or the quality, training, and readiness of enemy forces. Studies done in the West routinely depicted the Soviets overrunning NATO even with the Allies using nuclear weapons.[109] The benchmark figure of 175 Soviet divisions remained intact and did not come under close scrutiny until the end of the decade, when the CIA found Soviet divisions to be at various levels of preparedness. Most of those opposite NATO in East Germany proved to be fully ready front-line units. However, only about a third of the Soviet divisions fell into this category, and the rest were either under-strength reserve units or cadres.[110] The creation in 1955 of the Warsaw Pact, an alliance dominated by Moscow, increased the scope of Soviet strategic control in East Europe but probably added little to the Kremlin's immediate capabilities. Political instability within the satellite countries, highlighted by the 1953 East German uprising and the 1956 Hungarian rebellion, cast doubt on the reliability of non-Soviet Warsaw Pact forces.[111]

In sum, NATO planning during the Eisenhower years yielded mixed results. Shaped by essentially the same philosophy and budgetary pressures that were driving defense policy in the United States, the New Approach promised a powerful deterrent against all-out Soviet aggression and a convenient way for NATO's members to save money on defense, but it limited their ability to cope with lesser contingencies. While it did not do away with conventional forces, the New Approach definitely downplayed their role. Missing from this strategy was any provision for a "nuclear pause" or "firebreak" during a crisis to avoid rapid escalation. It was largely for these reasons—the threat of unforeseen consequences and the absence of flexibility—that the Joint Chiefs split over whether a nuclear-oriented strategy significantly improved NATO's defense posture and European security. Radford did his best to promote the President's cause, but he repeatedly ran into strong resistance from the Army and Navy. Though both Services eventually signed off on the New Approach, they did so reluctantly, sensing that they had no choice, and because they knew they would not get the larger conventional forces they wanted.

CURBING THE ARMS RACE

With nuclear weapons in the forefront of American defense policy during the 1950s, the size, composition, and readiness of the U.S. nuclear stockpile became a matter of utmost JCS concern. The Joint Chiefs had conducted a detailed annual review since 1947 to make sure the stockpile was satisfying military requirements. Bowing to JCS requests, the Truman administration in 1949, 1950, and 1952 approved three separate increases in the production capacity for fissionable materials. According to historian David Alan Rosenberg, these decisions recast the country's defense posture by launching the United States into an era of "nuclear plenty" and by generating a construction program capable of providing U.S. forces with nuclear weapons for the duration of the Cold War and beyond.[112] Without the production increases initiated during the Truman years, the New Look would never have been conceivable. From a base of around 1,100 weapons when the Eisenhower administration took office, the nuclear stockpile grew to about 22,000 by the time the President stepped down. Though the Soviets kept the details of their atomic energy program a closely guarded secret, retrospective estimates compiled in the West suggest that their nuclear stockpile increased from a handful of weapons in 1950 to between 1,700 and 4,500 ten years later.[113]

Ironically, this rapid and sustained growth in nuclear weapons production came at a time when the United States and the Soviet Union were taking the first serious steps in nearly a decade to find common ground for resuming arms control and disarmament negotiations. While the JCS had no objection to arms control per se, they were constantly on guard against ill-considered and unenforceable schemes that could compromise national security. With memories of the ill-fated Baruch Plan as a constant reminder, they resisted renewed calls for international control of atomic energy and turned a cold eye on measures that might stifle the development of the nuclear stockpile, like India's 1954 call for a moratorium on atmospheric nuclear testing. Popular and international pressure to curb the "arms race," however, kept the arms control and disarmament issue very much alive and compelled the Joint Chiefs to revisit it more often than they would have preferred.[114]

Eisenhower and the JCS agreed that a significant improvement in the international situation and concrete demonstrations of Soviet goodwill should precede major reductions in either conventional or nuclear arms. Convinced that the Sino-Soviet bloc's vast reservoir of manpower gave it a distinct advantage in a conflict, they believed that the West's most effective counter was its lead in technology—most of all, its superiority in nuclear weapons. The ability of the Soviet Union to duplicate American achievements, including most recently the H-bomb, and to develop delivery systems comparable to those in the U.S. inventory, may have diluted the West's advantage, but it did not, in Eisenhower's or the chiefs' view, negate or lessen the fundamental importance of nuclear weapons to national security. Nuclear weapons, the JCS argued, gave the United States an "indeterminate advantage" over the Soviet Union and its allies that should be nourished and preserved at all costs.[115]

When the Eisenhower administration took office in 1953, ongoing multinational disarmament negotiations before the United Nations still concentrated on sweeping proposals to eliminate conventional and nuclear weapons. While President Eisenhower professed a strong personal interest in arms limitation, the absence of reliable verification measures and the administration's decision to structure the country's defense posture around nuclear weapons raised serious questions of whether the United States should continue to participate in these kinds of negotiations. Lengthy NSC discussions of this issue yielded the affirmation in August 1954 that, from a public relations standpoint, the United States had no choice and needed to be seen as still favoring "a practical arrangement for the limitation of armaments with the USSR."[116] Still, the consensus within the administration was that the time for such agreements had passed and that a more sensible alternative was to pursue limited objectives. While the Joint Chiefs offered no specific opinion during the administration's internal debate, this approach seems to have accorded more closely with their preferences than any other.[117]

Indicative of the administration's shift in focus was the increasing use of "arms control" rather than "disarmament" to describe the goals of American policy. The initial test of the limited-objectives strategy was President Eisenhower's "Atoms for Peace" speech to the United Nations in December 1953. Sidestepping the stalled disarmament debate, the President stressed the peaceful potential of nuclear power and the need to "hasten the day when fear of the atom will begin to disappear." To coordinate peaceful applications, he proposed the creation of an International Atomic Energy Agency (IAEA), a watered-down version of the international control body envisioned under the Baruch Plan. Based in Vienna, Austria, the IAEA began operating in 1957.[118]

The administration's most ambitious initiative, unveiled by the President on July 21, 1955, at the Geneva summit, was the "Open Skies" proposal to allow aerial photography of U.S. and Soviet military installations. Devoid of any direct arms control content, the proposal aimed to build trust and confidence and to improve the prospects for verification, which Eisenhower and his senior advisors regarded as an essential prerequisite to an effective and credible arms control agreement.[119] Though the precise origins of the offer remain vague, Eisenhower claimed that it arose from studies done by his assistant, Nelson A. Rockefeller, in the weeks leading up to the conference, on avoiding a surprise attack through a system of mutual inspections.[120] The threat of a Pearl Harbor with atomic weapons was practically an obsession within the Eisenhower administration, and over the years it had given rise to numerous schemes to penetrate the veil of secrecy surrounding Soviet military programs and possible preparations for a surprise attack. With the exception of an Air Force–run program to monitor Soviet nuclear experiments, none of these efforts had yielded much useful information.[121]

By the mid-1950s, the most promising means of acquiring reliable data on Soviet capabilities and intentions was the U–2, a photo-reconnaissance plane that Lockheed Aircraft was building on a crash basis for the Central Intelligence Agency. Intelligence analysts assumed that orbiting satellites in outer space would someday provide the bulk of the information they needed. But space-based reconnaissance satellites seemed years away and, until then, manned aircraft were the best option. A jet-powered sailplane, the U–2 incorporated design features allowing it to fly above Soviet radar and take pictures with a special high-resolution camera.[122] The Joint Chiefs knew of the U–2, and through the CJCS, who sat on the program's interagency oversight committee, they stayed closely abreast of its progress.[123]

The development of a relatively invulnerable aerial reconnaissance capability was increasingly a source of friction between the CIA and the Air Force. Around the same time as the CIA initiated the U–2, the Air Force came up with a competing proposal using a Bell Aircraft design known as the X–16. Eisenhower, however, opposed putting the military in charge of such a program. His main concern was that if uniformed personnel flew the planes over the Soviet Union, the United States might be committing an act of war. He also suspected that if the Defense Department got involved and tried to manage it, the project would become "entangled in the bureaucracy" and mired in "rivalries among the services."[124] Taking these factors into account, Eisenhower decided in late November 1954 that the CIA would have overall authority and that the Air Force would provide assistance as needed to get the planes operational. Moving quickly to exercise its mandate, the CIA redoubled security on all aspects of the U–2, both because of its sensitivity and to minimize what the agency saw as the danger of Air Force encroachment.[125]

By early spring 1955, the U–2 program was nearing the point of its first flight test. Around this same time, the Technological Capabilities Panel (TCP), a select scientific advisory body chaired by James R. Killian, Jr., president of the Massachusetts Institute of Technology, tendered a new, top secret threat assessment to the NSC. Addressing the problem of surprise attack, the TCP warned of dire consequences should the Soviets launch a preemptive strike. "For the first time in history," it found, "a striking force could have such power that the first battle could be the final battle, the first punch a knockout." To avoid such a calamity, the panel urged the administration to increase its intelligence gathering, expand its early warning capabilities, and accelerate previously planned improvements in offensive and defensive strategic capabilities.[126] In a separate annex on intelligence available only to the President, the CJCS, and a handful of others, the TCP confirmed the U–2 program's progress and the potential of a follow-on system involving space-based satellites.[127] Though Rockefeller appears not to have been privy to this annex, he probably suspected its gist from his access to the main report and from having attended a briefing given by Killian and Edwin H. Land, designer of the U-2's camera, to the NSC on March 17, 1955.[128]

At the President's request, Rockefeller organized a special "vulnerabilities panel" made up of social scientists and intelligence experts to assess the prospects for improved verification. The group met at the Quantico, Virginia, Marine Corps base in early June 1955, and it was from these discussions that the aerial inspections proposal emerged.[129] With time running short, Rockefeller made no attempt to solicit JCS views. Instead, he met a few days prior to the Geneva conference in Paris with Radford and Gruenther to discuss the plan. Both agreed that the United States stood to gain more than it would lose. According to Secretary of State Dulles, who was also present, Radford was "in complete accord and indeed enthusiastic."[130]

With foreknowledge of the U-2, Eisenhower presented the Open Skies proposal, certain that he would have access to information derived from overflights of the Soviet Union with or without Soviet cooperation. The Joint Chiefs' first opportunity to comment as a corporate body did not come until after the conference when Secretary of Defense Wilson asked them for suggestions on how to implement the Open Skies proposal. But by then, Soviet Communist Party leader Nikita Khrushchev had vetoed the plan, making any further action on it rather pointless.[131]

The U-2 made its first test flight in early August 1955 and began reconnaissance of the Soviet Union 11 months later, on July 4, 1956. Though never directly involved in the program, the JCS provided advisory and logistical support and assigned a representative to the Ad Hoc Requirements Committee, chaired by the CIA, which decided the planes' missions.[132] Little more was heard of the Open Skies plan and it survives mainly as a footnote to history. At the time, however, it seemed a daring and ambitious initiative and a possible turning point in the Cold War. "I wonder," recalled Ray S. Cline, a veteran intelligence officer who had been with Eisenhower at Geneva, "if [the Soviets] ever regretted it in the following years as the U-2s began doing unilaterally over the USSR what Eisenhower had proposed they do on a reciprocal basis."[133]

Despite Soviet rejection of the Open Skies proposal, Eisenhower persisted in exploring ways of mitigating the threat of a surprise attack, and by September 1958 he persuaded the Soviets to participate in a conference of technical experts to address the issue.[134] By then, however, the focus of arms control efforts had shifted. Widespread public fear of radioactive fallout from atmospheric nuclear testing now overshadowed the danger of another Pearl Harbor and forced the United States to contemplate a moratorium, resisted by the JCS, on above-ground testing.[135] At the same time, with the information gleaned from U-2 flights over the Soviet Union, the JCS and the President had far better photographic intelligence on Soviet capabilities than ever before. The threat of surprise attack remained, but increasingly it took the form of a Soviet long-range missile program of as yet indistinct proportions, against which existing countermeasures were of questionable value. The strategic environment was again in flux by the mid-to-late 1950s, and as it changed it put renewed pressure on the chiefs to devise appropriate responses.

NOTES

1 Dwight D. Eisenhower, *The White House Years: Mandate for Change, 1953–1956* (Garden City, NY: Doubleday, 1963), 455.

2 Alice C. Cole et al., eds., *The Department of Defense: Documents on Establishment and Organization, 1944–1978* (Washington, DC: Office of the Secretary of Defense, Historical Office, 1978), 149–159.

3 Robert J. Watson, *The Joint Chiefs of Staff and National Policy, 1953–1954* (Washington, DC: Historical Division, Joint Chiefs of Staff, 1998), 14–15.

4 Eisenhower quoted in Michael T. Isenberg, *Shield of the Republic: The United States Navy in an Era of Cold War and Violent Peace* (New York: St. Martin's Press, 1993), 589; Arthur W. Radford, *From Pearl Harbor to Vietnam* (Stanford, CA: Hoover Institution Press, 1980), 305–318; and Richard M. Leighton, *History of the Office of the Secretary of Defense: Strategy, Money, and the New Look, 1953–1956* (Washington, DC: Historical Office, Office of the Secretary of Defense, 2001), 36–37.

5 Maxwell Taylor, *The Uncertain Trumpet* (New York: Harper & Row, 1959), 21; Paul R. Schratz, "The Military Services and the New Look, 1953–1961: The Navy," in David H. White, ed., *Proceedings of the Conference on War and Diplomacy, 1976* (Charleston, SC: The Citadel, 1976), 141.

6 Joint Strategic Capabilities Plan, August 14, 1953, JCS 1844/151, approved and reissued with amendments under SM-285-54, April 2, 1954 as JCS 1844/156; and the first Joint Strategic Objectives Plan for July 1, 1961 (JSOP-61), April 13, 1957, and Decision on January 3, 1958, JCS 2143/69. For the development of these plans, see Watson, *JCS and National Policy, 1953–54*, 94–109; Byron R. Fairchild and Walter S. Poole, *The Joint Chiefs of Staff and National Policy, 1957–1960* (Washington, DC: Office of Joint History, Office of the Chairman of the Joint Chiefs of Staff, 2000), 36–37; and Poole, "Evolution of the Joint Strategic Planning System," 3–5.

7 Fred I. Greenstein, "Dwight D. Eisenhower: Leadership Theorist in the White House," in Fred I. Greenstein, ed., *Leadership in the Modern Presidency* (Cambridge: Harvard University Press, 1988), 91.

8 E. Bruce Geelhoed, *Charles E. Wilson and Controversy at the Pentagon, 1953 to 1957* (Detroit: Wayne State University Press, 1979), 19, 37–39; Leighton, 44–45 and passim; and Dale R. Herspring, *The Pentagon and the Presidency: Civil-Military Relations from FDR to George W. Bush* (Lawrence: University Press of Kansas, 2005), 88–89.

9 Robert R. Bowie and Richard H. Immerman, *Waging Peace: How Eisenhower Shaped an Enduring Cold War Strategy* (New York: Oxford University Press, 1998), 83–95; and Stanley L. Falk, "The National Security Council under Truman, Eisenhower, and

Kennedy," *Political Science Quarterly* 79 (September 1964), 403–434, are excellent overviews of the NSC during the Eisenhower years. See also John Prados, *Keepers of the Keys: A History of the National Security Council from Truman to Bush* (New York: William Morrow and Co., 1991), 57–65; David J. Rothkopf, *Running the World* (New York: Public Affairs, 2005), 65–75; and Henry M. Jackson, ed., *The National Security Council: Jackson Subcommittee Papers on Policy–Making at the Presidential Level* (New York: Praeger, 1965), 33–38.

10 D. Condit, *Test of War*, 139–153.

11 Mark W. Clark, *From the Danube to the Yalu* (New York: Harper & Bros., 1954), 267.

12 James F. Schnabel and Robert J. Watson, *History of the Joint Chiefs of Staff: The Joint Chiefs of Staff and National Policy—The Korean War, 1950–1951* (Washington, DC: Office of Joint History, Office of the Chairman of the Joint Chiefs of Staff, 1998), Pt. 2, 190–193. Atomic weapons production figures derived from U.S. Department of Defense and Department of Energy, "Summary of Declassified Nuclear Stockpile Information," DOE Office of Public Affairs, August 22, 1995.

13 Schnabel and Watson, 193.

14 *History of the Custody and Deployment of Nuclear Weapons: July 1945 Through September 1977* (Office of the Assistant to the Secretary of Defense for Atomic Energy, February 1978), 19 (declassified).

15 Memo, CoS, USA to JCS, March 12, 1953, "Overseas Deployment of Toxic Chemical Agents," JCS 1837/46; Conrad C. Crane, *American Airpower Strategy in Korea, 1950–1953* (Lawrence: University Press of Kansas, 2000), 153–154; Schnabel and Watson, 192–193.

16 Bradley and Blair, *A General's Life*, 658–659; Robert Frank Futrell, *The United States Air Force in Korea, 1950–1953* (Washington, DC: Office of Air Force History, 1983, rev. ed.), 617–629, 666–667; Crane, *American Airpower in Korea*, 157–162.

17 Memo, Cutler to Wilson, March 21, 1953, *FRUS, 1952–54*, XV, Part I, 815.

18 Memo, JCS to SECDEF, March 27, 1953, "Future Course of Action in Connection with the Situation in Korea," JCS 1776/367; Schnabel and Watson, 200–203.

19 Memo of Discussion, 143d Meeting of the NSC, May 6, 1953, *FRUS, 1952–54*, XV, Part I, 975–978.

20 Memo of Substance of Discussion at State-JCS Meeting, March 27, 1953, ibid., 818.

21 Schnabel and Watson, 203–207; Memo, JCS to SECDEF, May 19, 1953, "Courses of Action in Connection with Situation in Korea," *FRUS, 1952–54*, XV, 1059–64; Memo of Discussion, 145th Meeting of the NSC, May 20, 1953, ibid., 1064–1068; Bradley and Blair, *A General's Life*, 658–661.

22 Futrell, 666–670.

23 McGeorge Bundy, *Danger and Survival: Choices About the Bomb in the First Fifty Years* (New York: Random House, 1988), 242–244. See also Rosey J. Foot, "Nuclear Coercion and the Ending of the Korean Conflict," *International Security* 13 (1988–1989), 92–112; and Barry M. Blechman and Robert Powell, "What in the Name of God Is Strategic Superiority?" *Political Science Quarterly* 97 (Winter 1982–1983), 589–602.

24 Clayton D. Laurie, "A New President, a Better CIA, and an Old War: Eisenhower and Intelligence Reporting on Korea, 1953," *Studies in Intelligence* 54, no. 4 (Unclassified Extracts, December 2010), 7–8.

25 Glenn H. Snyder, "The 'New Look' of 1953," in Warner R. Schilling, Paul Y. Hammond, and Glenn H. Snyder, *Strategy, Politics, and Defense Budgets* (New York: Columbia University Press, 1962), is still the best treatment of the origins of Eisenhower's defense policy. See also Gaddis, *Strategies of Containment*, 127–163; Saki Dockrill, *Eisenhower's New Look National Security Policy, 1953–61* (New York: St. Martin's Press, 1996), 19–47; and Samuel F. Wells, Jr., "The Origins of Massive Retaliation," in Robert H. Connery and Demetrious Caraley, eds., *Nuclear Security and Nuclear Strategy* (New York: Academy of Political Science, 1983), 52–73.

26 Notes by L.A. Minnich, Jr., on Legislative Leadership Meeting, April 30, 1953, Eisenhower Papers, Ann Whitman File, DDE Diary, Eisenhower Library.

27 Leighton, 148–151; Watson, *JCS and National Policy, 1953–54*, 11–14.

28 Interview with General Andrew J. Goodpaster, Jr., USA, by Malcolm S. McDonald, April 10, 1982, Eisenhower Library Oral History Collection.

29 Watson, *JCS and National Policy, 1953–54*, 17–20; Leighton, 151–167.

30 SE-46, "Probable Long Term Development of the Soviet Bloc and Western Power Positions," July 8, 1953, in Scott A. Koch, ed., *Selected Estimates on the Soviet Union, 1950–1959* (Washington, DC: History Staff, Central Intelligence Agency, 1993), 159.

31 David Holloway, *Stalin and the Bomb: The Soviet Union and Atomic Energy, 1939–1956* (New Haven: Yale University Press, 1994), 307. The first test of a deliverable U.S. thermonuclear weapon took place on March 1, 1954.

32 See, for example, Memo, CJCS to SECDEF, July 14, 1955, "Comparison U.S. and USSR Technical and Production Capabilities," Nathan F. Twining Papers, box 81, 1955 SecAF (1) folder, Library of Congress.

33 Memo of Discussion, 143d Meeting of the National Security Council, May 6, 1953, Eisenhower Papers, Ann Whitman File, NSC series, Eisenhower Library.

34 Eisenhower quoted in Memo of Conference with the President, May 24, 1956, Eisenhower Papers, Ann Whitman File, Diary Series, Eisenhower Library.

35 NSC 162/2, October 30, 1953, "Review of Basic National Security Policy," *FRUS, 1952–54*, II, Pt. 1, 577–97.

36 Eisenhower, *Mandate for Change*, 449.

37 Gaddis, *Strategies of Containment*, 164.

38 U.S. Department of Defense, *National Defense Budget Estimates for FY 1988/1989* (Washington, DC: Office of the Assistant Secretary of Defense [Comptroller], May 1987), 68–69, 122.

39 See D. Condit, 285–305; and Wells, "Origins of Massive Retaliation," 68–72.

40 *National Defense Budget Estimates FY 1988/89*, 76, 107.

41 Marcella Size Knaack, ed., *Encyclopedia of U.S. Air Force Aircraft and Missile Systems*, vol. II, *Post–World War II Bombers, 1945–1973* (Washington, DC: Office of Air Force History, 1988), 292–293 and passim.

42 John Prados, *The Soviet Estimate: U.S. Intelligence Analysis and Russian Military Strength* (New York: Dial Press, 1982), 38–45; Lawrence Freedman, *U.S. Intelligence and the Soviet Strategic Threat* (Princeton, NJ: Princeton University Press, 1986, 2d ed.), 65–67.

43 David Alan Rosenberg, "The Origins of Overkill: Nuclear Weapons and American Strategy, 1945–1950," *International Security* 7 (Spring 1983), 3–71, gives in-depth treatment of SAC's problems. See also William S. Borgiasz, *The Strategic Air Command: Evolution and Consolidation of Nuclear Forces, 1945–1955* (Westport, CT: Praeger, 1996), 128–133 and passim.

44 Robert W. Love, Jr., *History of the U.S. Navy, 1942–1991* (Harrisburg, PA: Stackpole Books, 1992), 378; Lawrence S. Kaplan, Ronald D. Landa, and Edward J. Drea, *History of the Office of the Secretary of Defense: The McNamara Ascendancy, 1961–1965* (Washington, DC: Historical Office, Office of the Secretary of Defense, 2006), 496.

45 George W. Baer, *One Hundred Years of Sea Power: The U.S. Navy, 1890–1990* (Stanford, CA: Stanford University Press, 1994), 334–335.

46 Love, 372–379.

47 See Matthew B. Ridgway, *Soldier* (New York: Harper, 1956), 286–294.

48 Memo, Discussion at 176th Meeting of the NSC, December 16, 1953, Eisenhower Papers, Ann Whitman File, NSC Series, Eisenhower Library.

49 John B. Wilson, *Maneuver and Firepower: The Evolution of Divisions and Separate Brigades* (Washington, DC: Center of Military History, 1998), 250–254.

50 Goodpaster Memcon, October 2, 1956, *FRUS, 1955–57*, IV, 100.

51 A.J. Bacevich, *The Pentomic Era: The U.S. Army Between Korea and Vietnam* (Washington, DC: National Defense University Press, 1986), 103–127 and passim; Ingo Trauschweizer, *The Cold War U.S. Army: Building Deterrence for Limited War* (Lawrence: University Press of Kansas, 2008), 81–113; and Wilson, *Maneuver and Firepower*, 263–290. Army sources are inconsistent and give various figures for the size of pentomic divisions, ranging from 11,000 to 13,500.

52 Leighton, 478–482; Trauschweizer, 56.

53 NIE 91, "Probable Developments in Indochina Through Mid-1954," June 4, 1953, *FRUS, 1952–54*, XIII, 592–602.

54 Special Estimate prepared by the CIA (SE-52), "Probable Consequences in Non-Communist Asia of Certain Possible Developments in Indochina Before Mid-1954," November 16, 1953, ibid., 866–867.

55 Watson, *JCS and National Policy, 1953–54*, 248–251; and *Joint Chiefs of Staff and the First Indochina War, 1947–1954* (Washington, DC: Office of Joint History, Joint Chiefs of Staff, 2004), 152.

56 "Report by Staff Planners to the Military Representatives of the Five Powers on the Conference Held 21 Sept–2 Oct 1953 at Pearl Harbor," n.d., and Message, JCS 956811 to CINCPAC, February 9, 1954, both in JCS 1992/274; Message, JCS 955782 to COMSAC, January 19, 1954, JCS 2118/62.

57 See John Lewis Gaddis, *The Long Peace: Inquiries Into the History of the Cold War* (New York: Oxford University Press, 1987), 130.

58 Memo of Discussion, 179th Meeting of the NSC, January 8, 1954, *FRUS, 1952–54*, XIII, 947–954; NSC 5405 (formerly NSC 177), January 16, 1954, "U.S. Objections and Courses of Action with Respect to Southeast Asia," ibid., 971–976.

59 *JCS and Indochina War*, 147.

60 "Report by the President's Special Committee on Indochina," March 2, 1954, *FRUS, 1952–54*, XIII, 1116.

61 Memo, Radford to Eisenhower, March 24, 1954, quoted in *JCS and Indochina War*, 155. See also Ronald H. Spector, *Advice and Support: The Early Years, 1941–1960* (Washington, DC: Center of Military History, 1983), 191–194.

62 Laurent Césari and Jacques de Folin, "Military Necessity, Political Impossibility: The French Viewpoint on Operation *Vautour*," in Lawrence S. Kaplan, Denise Artaud, and Mark R. Rubin, eds., *Dien Bien Phu and the Crisis in Franco–American Relations, 1954–1955* (Wilmington, DE: Scholarly Resources, 1990), 108–109. Ely's recollections are the only surviving record of this meeting.

63 Spector, *Advice and Support*, 200–202; Watson, *JCS and National Policy, 1953–54*, 253.

64 Memo, Ridgway to JCS, April 6, 1954, "Indo–China," *FRUS, 1952–54*, XIII, 1270.

65 Memo, Discussion at 209th Meeting of the NSC, August 5, 1954, *FRUS, 1952–54*, II, 706–707.

66 Memo, CJCS to SECDEF, September 11, 1954, "U.S. Policy Regarding Off–Shore Islands," *FRUS, 1954*, XIV, 598–600.

67 Memo of Discussion, 221st Meeting NSC, November 2, 1954, *FRUS, 1952–54*, XIV, 836.

68 Memo of Discussion, NSC Meeting, August 5, 1954, ibid., 519.

69 See Matthew Jones, *After Hiroshima: The United States, Race and Nuclear Weapons in Asia, 1945–1965* (Cambridge, UK: Cambridge University Press, 2010), 240–288.

70 Memo, CJCS to SECDEF, September 11, 1954, "U.S. policy regarding off-shore Islands held by Chinese Nationalist Forces, NSC Action 1206–f," *FRUS, 1952–54*, XIV, 598–609.

71 *FRUS, 1952–54*, XIV, 619, 699. See also Jones, *After Hiroshima*, 269–273; Gaddis, *Long Peace*, 134–135; H.W. Brands, Jr., "Testing Massive Retaliation: Credibility and Crisis Management in the Taiwan Strait," *International Security* 12 (Spring 1988), 124–151; and Bennett C. Rushkoff, "Eisenhower, Dulles and the Quemoy–Matsu Crisis, 1954–1955," *Political Science Quarterly* 96 (Fall 1981), 465–480.

72 Memo of Discussion, 240th Meeting NSC, March 10, 1955, *FRUS, 1955–57*, II, 349; Memo of Discussion, 243d Meeting NSC, March 31, 1955, ibid., 432–433; Kenneth W. Condit, *The Joint Chiefs of Staff and National Policy, 1955–1956* (Washington, DC: Historical Office, Joint Staff, 1992), 205–206.

73 Memo, JCS to SECDEF, March 27, 1955, "Improvement of the Military Situation in the Far East in the Light of the Situation Now Existing in the Formosa Area," JCS 1966/100; *FRUS, 1955–57*, II, 406–408.

74 Diary entry, March 26, 1955, in Ferrell, ed., *Eisenhower Diaries*, 296; Memo, Eisenhower to John Foster Dulles, April 5, 1955, "Formosa," *FRUS, 1955–57*, II, 445–450. Emphasis in original.

75 MFR by Cutler, March 11, 1955, "Meeting in President's Office," ibid., 358–359.

76 Message, Hoover to Robertson and Radford, April 22, 1955, ibid., 501–502.

77 Quoted in Dockrill, *Eisenhower's National Security Policy*, 112. See also Gordon H. Chang, "To the Nuclear Brink: Eisenhower, Dulles, and the Quemoy–Matsu Crisis," *International Security* 12 (Spring 1988), 96–123.

78 Fairchild and Poole, 209–215.

79 See Richard K. Betts, *Soldiers, Statesmen, and Cold War Crises* (Cambridge: Harvard University Press, 1977), 106–107.

80 MC 48, "Report by the Military Committee on the Most Effective Pattern of Military Strength for the Next Few Years," November 22, 1954, approved December 17, 1954, *NATO Strategy Documents, 1949–1969*, 246. For NATO nuclear stockpile figures, see below, chapter 8.

81 Stanley R. Sloan, *NATO's Future: Toward a New Transatlantic Bargain* (Washington, DC: National Defense University Press, 1985), 74–76 and passim.

82 "Defence Policy and Global Strategy: Report by the Chiefs of Staff," June 17, 1952, reprinted in full in Alan Macmillan and John Baylis, eds., *A Reassessment of the Global Strategy Paper of 1952* (College Park, MD: Center for Internal and Security Studies at Maryland, 1994), 19–63.

83 Joint Planners' Minute, February 6, 1953, quoted in Dockrill, *Eisenhower's National Security Policy*, 86.

84 Memo, JCS to SECDEF, September 11, 1953, "Certain European Issues Affecting the United States," JCS 2073/634.

85 Robert A. Wampler, *NATO Strategic Planning and Nuclear Weapons, 1950–1957* (College Park, MD: Center for International Security Studies at Maryland, 1990), 10–11; Watson, *JCS and National Policy, 1953–54*, 290–291, 298; Memo, CoS/AF to JCS, August 31, 1953, "Certain European Issues Affecting the United States," JCS 2073/633.

86 Message, Acting SecState to U.S. Embassy France, October 15, 1953, *FRUS, 1952–54*, V, 444–446.

87 Memo, JCS to SECDEF, October 22, 1953, "U.S. Guidance for 1953 Annual Review," JCS 2073/671.

88 Watson, *JCS and National Policy, 1953–54*, 299–301.

89 Wampler, 12–13.

90 For a summary, see SG 241/3, "Report by the Military Committee to the NSC on the Most Effective Pattern of NATO Military Strength for the Next Few Years," August 19, 1954, enclosure to JCS 2073/900.

91 Robert S. Jordan, ed., *Generals in International Politics: NATO's Supreme Allied Commander, Europe* (Lexington, KY: University Press of Kentucky, 1987), 57–60; Steven L. Rearden, "American Nuclear Strategy and the Defense of Europe, 1954–1959," in David H. White, ed., *Proceedings of the Conference on War and Diplomacy, 1976* (Charleston, SC: The Citadel, 1976), 133–138; Watson, *JCS and National Policy, 1953–54*, 305–306.

92 Memo, JCS to SECDEF, September 24, 1954, "NATO Capabilities Studies," with attachment, JCS 2073/900.

93 NIE 11-5-54, June 7, 1954, "Soviet Capabilities and Main Lines of Policy Through Mid-1959," in Scott A. Koch, ed., *Selected Estimates on the Soviet Union, 1950–1959* (Washington, DC: Central Intelligence Agency, 1993), 209–211.

94 Memo, JCS to SECDEF, June 11, 1954, "Use of Atomic Weapons," JCS 2073/823.

95 NSC 151/2, December 4, 1953, "Disclosure of Atomic Information to Allied Countries," *FRUS, 1952–54*, II, 1256–1284.

96 Richard G. Hewlett and Jack M. Holl, *Atoms for Peace and War, 1953–1961: Eisenhower and the Atomic Energy Commission* (Berkeley, CA: University of California Press, 1989), 113–143.

97 "The Basis for the U.S. Position on the Provision of U.S. Atomic Weapons for the Common Defense of the NATO Area," Appendix to Enclosure A to Memo, CJCS to JCS, November 9, 1957, "Provision of Atomic Weapons to Non–U.S. NATO Forces," JCS 2019/257. For a more detailed discussion, see Fairchild and Poole, 104–105.

98 The most ardent advocate of giving NATO an independent nuclear capability was Gruenther's successor as SACEUR, General Lauris Norstad. See Robert S. Jordan, "Norstad: Can the SACEUR Be Both European and American?" in Jordan, ed., *Generals in International Politics*, 79–82.

99 Fairchild and Poole, 104–112.

100 See Richard L. Kugler, *Commitment to Purpose: How Alliance Partnership Won the Cold War* (Santa Monica, CA: RAND, 1993), 98.

101 John S. Duffield, *Power Rules: The Evolution of NATO's Conventional Force Posture* (Stanford, CA: Stanford University Press, 1995), 141–142; Kugler, 92.

102 Fairchild and Poole, 114–115; James L. Richardson, *Germany and the Atlantic Alliance* (Cambridge: Harvard University Press, 1966), 40–48; Stanley M. Karanowski, *The German Army and NATO Strategy* (Washington, DC: National Defense University Press, 1982), 43–49.

103 Duffield, 137–42.

104 Kugler, 89, 95.

105 K. Condit, 134–135.

106 See Joint Strategic Capabilities Plan July 1, 1955 to July 1, 1956, March 2, 1955, approved April 13, 1955, JCS 1844/178.

107 Leighton, 664–666; Discussion at 307th Meeting of the NSC, December 21, 1956, Whitman File, DDE Papers.

108 *FRUS, 1955–57*, IV, 93–95, 99–102; Robert J. Watson, *History of the Office of the Secretary of Defense: Into the Missile Age, 1956–1960* (Washington, DC: Historical Office, Office of the Secretary of Defense, 1997), 501–502.

109 See for example the 1955 study (WSEG 12) done by the Weapons Systems Evaluation Group, summarized in Wampler, 19–21; and Radford's Briefing on WSEG Report No. 12, undated, available at <http://www.alternatewars.com/WW3/WW3_Documents/JCS/WSEG_12.htm>.

110 Raymond L. Garthoff, "Estimating Soviet Military Force Levels," *International Security* 14 (Spring 1990), 93–116.

111 James D. Marchio, "U.S. Intelligence Assessments and the Reliability of Non–Soviet Warsaw Pact Armed Forces, 1946–89," *Studies in Intelligence* 51 (Extracts, December 19, 2007), 17–18.

112 David Alan Rosenberg, "U.S. Nuclear War Planning, 1945–1960," in Desmond Ball and Jeffrey Richelson, eds., *Strategic Nuclear Targeting* (Ithaca, NY: Cornell University Press, 1986), 41.

113 U.S. stockpile figures from U.S. Department of Defense and Department of Energy, "Declassified Nuclear Stockpile Information," DOE Office of Public Affairs, August

22, 1995. Soviet stockpile figures are those compiled by Albert Wohlstetter, summarized in Thomas B. Cochran et al., *Nuclear Weapons Databook*, vol. IV, *Soviet Nuclear Weapons* (New York: Harper & Row, 1989), 25.

114 Watson, *JCS and National Policy, 1953–54*, 193–194.

115 Memo, JCS to SECDEF, April 30, 1954, "A Proposal for a Moratorium on Future Testing of Nuclear Weapons," JCS 1731/98.

116 NSC 5422/2, August 7, 1954, "Guidelines Under NSC 162/2 for FY 1956," *FRUS, 1952–54*, II, 717.

117 For the gist of JCS thinking, see Memo, JCS to SECDEF, June 23, 1954, "Negotiations with the Soviet Bloc," *FRUS, 1952–54*, II, 680–686; and Watson, *JCS and National Policy, 1953–54*, 197–198.

118 "Address Before the General Assembly of the United Nations on Peaceful Uses of Atomic Energy," December 8, 1953, *Public Papers of the Presidents of the United States: Dwight D. Eisenhower, 1953* (Washington, DC: GPO, 1960), 813–822.

119 John Prados, "Open Skies and Closed Minds," in Günter Bischof and Saki Dockrill, eds., *Cold War Respite: The Geneva Summit of 1955* (Baton Rouge: Louisiana State University Press, 2000), 216–220.

120 Eisenhower, *Mandate for Change*, 519–520.

121 Gaddis, *Long Peace*, 197.

122 Gregory W. Pedlow and Donald E. Welzenbach, *The CIA and the U–2 Program, 1954–1974* (Washington, DC: Central Intelligence Agency, 1998), 24–37.

123 Dwight D. Eisenhower, *The White House Years: Waging Peace, 1956–1961* (Garden City, NY: Doubleday, 1965), 544–545. The other members of the oversight panel were the Secretaries of State and Defense, the Director of Central Intelligence, and Richard M. Bissell, Jr., of the CIA, the program's manager.

124 Eisenhower quoted in James R. Killian, Jr., *Sputnik, Scientists, and Eisenhower* (Cambridge: MIT Press, 1977), 82.

125 Pedlow and Welzenbach, *CIA and the U–2*, 36–37; Richard M. Bissell, Jr., *Reflections of a Cold Warrior* (New Haven: Yale University Press, 1996), 108–110. See also William E. Burrows, "Satellite Reconnaissance and the Establishment of a National Technical Intelligence Apparatus," in Walter T. Hitchcock, ed., *The Intelligence Revolution: A Historical Perspective* (Washington, DC: Office of Air Force History, 1991), 233–250, which traces the evolution of the Air Force–CIA rivalry.

126 Technological Capabilities Panel of the Science Advisory Committee, Office of Defense Mobilization, "Meeting the Threat of Surprise Attack," (MS, February 14, 1955), Vol. I, 5, 14, Eisenhower Papers, Ann Whiteman File, NSC Series, DDEL.

127 Killian, 79–82; R. Cargill Hall, "The Eisenhower Administration and the Cold War: Framing American Astronautics to Serve National Security," *Prologue* 27 (Spring 1995), 62.

128 Memo by J. Patrick Coyne, March 18, 1955, "Discussion at 241st Meeting of the NSC, March 17, 1955," Eisenhower Papers, Ann Whitman File, NSC Series, DDEL.

129 Prados, "Open Skies," 220–221.

130 Message, Dulles to State Department, July 21, 1955, *FRUS, 1955–57*, V, 434.

131 Philip Taubman, *Secret Empire: Eisenhower, the CIA, and the Hidden Story of America's Space Espionage* (New York: Simon & Schuster, 2003), 141–142; K. Condit, 97–98.

132 Anne Karalekas, *History of the Central Intelligence Agency*, S. Report. No. 94–755 (Washington, DC: GPO, 1976), 59.

133 Ray S. Cline, *The CIA Under Reagan, Bush and Casey* (Washington, DC: Acropolis Books, 1981), 181.

134 Held in Geneva between November 10 and December 7, 1958, the conference was largely unproductive. While the Soviets' first concern was the threat of a surprise attack coming from West Germany, the U.S. delegation concentrated on the danger posed by Soviet strategic systems. See Robin Ranger, *Arms and Politics, 1958–1978* (Toronto: Macmillan of Canada, 1979), 31–39.

135 JCSM-337-59 to SECDEF, August 21, 1959, "Study of Nuclear Tests," JCS 2179/183.

General Nathan F. Twining, USAF, Chairman, Joint Chiefs of Staff, 1957–1960

Chapter 6

CHANGE AND CONTINUITY

On October 4, 1957, the Soviet Union stunned the world by sending an artificial satellite, "Sputnik I," into orbit around the Earth. This achievement was the first of its kind and followed the successful launch of a Soviet multistage intercontinental ballistic missile (ICBM) the previous August. It would be more than a year before the United States successfully tested an ICBM.[1] Suggesting a higher level of Soviet technological development than previously assumed, Sputnik I and the Soviet ICBM cast doubt on a key assumption that had shaped U.S. national security policy since World War II—that America's supremacy in science and technology gave it a decisive edge over the Soviet Union. Not since the Soviets tested their first atomic bomb in 1949 had the United States seemed so unprepared and vulnerable. According to James R. Killian, Jr., President Eisenhower's assistant for science and technology, Sputnik I "created a crisis of confidence that swept the country like a windblown forest fire."[2]

A dramatic wake-up call, Sputnik was actually one of several indications of the larger strategic transformation taking place. Around the world, other forces were at work laying the foundations for a new international order in which the underdeveloped countries of the Third World would play a larger and more active part. The most striking changes were those resulting from the end of European colonialism and a rising tide of Third World nationalism and socioeconomic discontent. Starting in Asia, the process had spread to the Middle East and Africa by the mid to late 1950s, creating new security problems as it went along. Meanwhile, a surge of anti-Americanism in Latin America presented fresh challenges there. Most Third World countries were too preoccupied with internal difficulties or regional rivalries to take much interest in the ongoing ideological struggle between East and West. But they were not averse to playing off one superpower against another if they saw it to their advantage.

During this period of transformation, the need for reliable military advice and sound strategic planning continued to place heavy demands on the Joint Chiefs of Staff. Nonetheless, they were slow to rise to the challenge. Quarreling over Service functions and the allocation of resources continued to hobble their ability to address problems of a cross-Service nature and to present consensus recommendations. Rarely did the JCS speak with a single voice on key issues of national strategy and military policy. Despite extensive organizational and administrative reforms introduced in 1958, the JCS system was slow to embrace more efficient and effective ways. While there was some progress toward improving operational planning, clashes and disagreements among the chiefs persisted. Frustrated, the President looked elsewhere for advice in addressing key politico-military problems.

EVOLUTION OF THE MISSILE PROGRAM

The most urgent question raised by Sputnik was whether the United States was as far behind the Soviets as it seemed. When the Eisenhower administration adopted the New Look in 1953, it assumed that while the Soviets would continue to modernize their armed forces, they would be in no position to rival U.S. superiority in nuclear weapons or sophisticated delivery systems for up to 5 years. The initial challenge to this assumption came almost immediately with the detonation of the Soviet H-bomb in the summer of 1953, a smaller-yield but more usable weapon than the H-bomb assemblies in the U.S. arsenal.[3] A year later, the first signs appeared that the Soviets might be developing a long-range heavy bomber force significantly larger than previously believed. Fears of a "bomber gap" eventually proved unfounded. But the episode drew attention to a potentially serious weaknesses in the administration's defense posture and its ability to assess Soviet capabilities. Never again would the Eisenhower administration be quite so sure of its long-term strategic superiority over the Soviets.

By the mid-1950s, concern had shifted from the Soviet Union's long-range bombers to its ballistic missile program. In assessing the Soviet missile effort, the Chairman of the Joint Chiefs, Admiral Radford, warned that it could pose "a major danger" to the continental United States and give Moscow sufficient leverage "to force a showdown" by the end of the decade.[4] To counter that threat, the JCS agreed that the United States needed to step up its development of offensive ballistic missiles, but they were at odds over the objective size and configuration of the U.S. missile force. Seeking to check further growth in the Air Force share of the budget, Army and Navy leaders favored a dispersed missile force tailored to a variety of strategic and tactical missions. For deterrence, they argued, the required force could be kept fairly small as long as it had a high degree of survivability against a Soviet attack and the ability to inflict unacceptable area damage against Soviet cities in retaliation, in effect a posture of "minimum deterrence" resting on "countervalue" targeting.

The Air Force took a more expansive view of missile requirements. Dismissing minimum deterrence as ineffectual, its leaders argued for a "counterforce" posture composed of bombers and missiles that accorded first priority to the destruction of Soviet war-making capabilities. Air Force planners expected manned bombers to remain the principal weapon for this purpose for the foreseeable future, partly owing to a shortage of funds for missile development and also because the size and weight of nuclear weapons limited their application.[5] But with the confirmation in February 1954 by the Teapot Committee, an Air Force scientific advisory panel, that high-yield thermonuclear warheads could be miniaturized, Air Force attitudes began to change in favor of giving ballistic missiles a larger role.[6] Based on the Teapot Committee's findings, Trevor Gardner, the Secretary of the Air Force's special assistant for research and development (R&D), projected an initial operational capability (IOC) of 100 ICBM-type missiles deployed at 20 launch sites around the United States by the end of the decade.[7]

Limited intelligence left U.S. policymakers guessing about the status of Soviet ballistic missile development for most of the 1950s. Citing Moscow's reliance on German scientists to bolster native resources, the Intelligence Community routinely insisted that the Soviets could one day match the United States in missile technology. But lacking hard data, intelligence analysts hedged the date when the Soviet strategic missile program would pose a direct danger. Early estimates, calibrated from the progress in U.S. research programs, placed the IOC for a Soviet ICBM in the 1960–1963 timeframe.[8] But as they gradually pieced together the available information, analysts became concerned that the Soviets might be catching up faster than expected. Prior to the availability of U–2 photographs, practically everything the Joint Chiefs and senior policymakers knew about the Soviet missile program derived from a worldwide complex of seismic and infrared sensors built and maintained by the National Security Agency (NSA).[9] By the mid-1950s, the NSA had detected that the Soviets were testing an intermediate range ballistic missile (IRBM), which many scientists considered the first step in developing an ICBM. In a national intelligence estimate (NIE 11-5-57) issued a few months prior to the Soviet ICBM test of August 1957 and Sputnik, the Intelligence Community predicted that by 1959 the Soviets "probably" would have an IRBM that could strike targets in Western Europe and Japan, and an ICBM prototype for limited operational use against the continental United States by 1960–1961.[10]

Accepting the need to accelerate U.S. missile programs, President Eisenhower decided in September 1955 to make the development of both an ICBM and an IRBM a top priority, but set no target date for acquiring either capability.[11] Later, the Air Force projected that it could have an IRBM ready for deployment in Europe and the Near East by mid-1959 and a small operational force of ICBMs by March 1961.[12] Eisenhower was a committed proponent of ballistic missiles, but not a very enthusiastic one. Hoping to avoid a costly missile competition with the

Soviets, he downplayed the need to preserve strategic superiority and publicly spoke of settling for a posture of "adequacy" or "sufficiency" in overall nuclear capabilities. Though the changes he had in mind were more matters of emphasis than substance, some observers detected the emergence of a "new" New Look that would no longer strive to maintain strategic superiority over the Soviets.[13] On several occasions, Eisenhower denigrated the military value of ballistic missiles and stated that he backed them only for their "psychological and political significance."[14] Other times, he questioned whether much more than a demonstration capability was needed and offered no objection when Secretary of Defense Wilson once estimated that, given the high yield of thermonuclear weapons, "one hundred and fifty well-targeted missiles might be enough." By and large, Eisenhower regarded long-range ballistic missiles as redundant. "We must remember," he told associates, "that we have a great number of bombardment aircraft programmed, and great numbers of tankers that are now being built, and we must consider how to use them."[15]

Eisenhower cautioned against overemphasis on ballistic missiles not only because they were a new and unproven technology, but also because he saw as yet no clear-cut assignment of Service responsibilities for their development and ultimate use. His main regret, he later admitted, was that he had allowed missile development to remain under Service control and had not made it a direct responsibility of the Secretary of Defense.[16] Under the original assignment of functions approved by Secretary of Defense Wilson in November 1955, the Air Force had developmental authority for two first-generation liquid-propellant ICBMs (the Atlas and a backup, the Titan) and an IRBM (the Thor). The Army and Navy were to share responsibility for a fourth missile, a 1,500-mile liquid-propellant medium-range ballistic missile (MRBM) named Jupiter, for launch from land or at sea.[17]

Almost immediately, the Services quarreled over the allocation of resources and access to production facilities. Strife between the Air Force and Army was especially acute. Meanwhile, the Navy lost interest in Jupiter and within a year had shifted its attention to a new missile, the solid-propellant Polaris. More versatile than the Jupiter, the Polaris could be carried aboard submarines and launched from underwater, making the system practically invulnerable. Its principal drawbacks were a limited range (1,000 to 1,500 miles), a relatively small warhead, and questionable accuracy and reliability. Recognizing the advantages of solid-propellant missiles, the Air Force began developing several of its own, including a second-generation ICBM known as Minuteman.[18]

The Joint Chiefs ordinarily confined their participation in R&D to setting general goals and identifying broad categories for exploration. After the 1949 Navy and Air Force clash over airpower, the JCS shied away from participating in decisions assigning specific weapons-development responsibilities to one Service or another. But in

August 1956, with the missile program degenerating into a free-for-all, Secretary Wilson requested JCS help in sorting out functional responsibilities. Not since the Key West and Newport conferences of 1948 had a Secretary of Defense relied so heavily on the JCS to help him resolve a roles-and-missions question of such importance.[19]

Wilson was, of course, asking a lot, since the Joint Chiefs (except the Chairman) served both as military advisors to the Secretary and the President and as the uniformed heads of their Services, in which capacity they were under a moral obligation to defend the interests of their organizations. The ensuing deliberations yielded no consensus that might have pointed to a long-term solution, but they did find the Chairman, the Air Force, and the Navy in basic agreement that three strategic missile programs were too many and that the logical course was to eliminate or curb the Army program. Secretary Wilson agreed and in November 1956 set a range limit (loosely enforced) of 200 miles on future Army missiles and turned the Jupiter over to the Air Force.[20]

While Wilson's clarification of Service functions restored a semblance of order to the ballistic missile program, it left the door open to a resumption of conflict between the Air Force and Navy for control of the strategic bombardment mission. Clearly, the Air Force was in no immediate danger of being displaced. Nor was the Navy's Polaris force, once it became operational in the 1960s, apt to rival the Strategic Air Command's reach and striking power. But as the Services proceeded down the path laid out in the mid-1950s, the country was again heading toward the development of two strategic forces—one run by the Air Force and the other by the Navy—with all the overlapping and duplication of effort separate systems implied. The JCS had yet to address this issue, and, if the past were any guide, they would do everything in their collective power to avoid it. Yet sooner or later the day of reckoning would arrive.

THE GAITHER REPORT

With the U.S. missile program mired in inter-Service rivalry, feuding, and confusion, the task of formulating an effective response to Sputnik became all the more challenging. As it happened, it took an outside inquiry by a group of experts known as the Gaither Committee to break the logjam. The findings, summarized in a top secret report, reached the President and NSC in early November 1957, barely a month after the first Sputnik. Taking a broad-brush approach, the Gaither Committee confirmed the need for vigorous steps to counter Soviet progress in space and ballistic missiles and suggested that U.S. vulnerability might be even greater than previously supposed.

The Joint Chiefs resented intrusions by outsiders like the Gaither Committee but were virtually powerless to do much about it. The panel's origins lay in growing pressure from congressional Democrats who wanted the Eisenhower administration to do more in the area of civil defense against the threat of a Soviet nuclear attack. At issue was a Federal Civil Defense Administration (FCDA) plan, presented to the President in January 1957, urging large-scale civil defense improvements, including a $32 billion nationwide shelter program in lieu of a less expensive evacuation plan.[21] Some administration officials dismissed the shelters as a diversion of resources; others, including former Secretary of the Army Gordon Gray, now the director of the Office of Defense Mobilization (ODM), considered them a valuable contribution to deterrence.[22]

Adopting a neutral position, the Joint Chiefs concurred in the NSC Planning Board's finding that the shelter system needed further study. Seizing on this approach, President Eisenhower arranged in the spring of 1957 with H. Rowan Gaither, a West Coast attorney and chairman of the boards of the Ford Foundation and the RAND Corporation, to conduct an inquiry under ODM auspices. Officially designated the Security Resources Panel (SRP), the group was commonly known as the Gaither Committee. After Gaither fell ill in August, Robert C. Sprague, an electronics company executive who specialized in air and missile defense problems, and former Deputy Secretary of Defense William C. Foster, co-chaired the panel.[23]

Soon after agreeing to head the effort, Gaither persuaded the President's national security advisor, Robert Cutler, to expand the scope of the panel's investigation. Gaither argued that to place civil defense in its proper perspective, he and his committee needed to examine the whole range of the country's preparations for offensive and defensive strategic warfare, much as the Killian Report had done 2 years earlier.[24] Armed with an expanded writ, the SRP launched a wholesale inquiry into the country's strategic posture. Offering limited cooperation, the JCS turned down the committee's request for a list of documents but did provide three briefings—a general review of the Soviet threat, a status report on continental defenses, and an analysis of U.S. retaliatory capabilities.[25] For most of its data, the committee relied on the military Services, the Intelligence Community, and government "think tanks." James Phinney Baxter, the president of Williams College and author of the Pulitzer Prize–winning book, *Scientists Against Time* (1947), the official history of the Office of Scientific Research and Development in World War II, oversaw the preparation of the final report.

Unable to devote full time to the project because of his college duties, Baxter depended on two associates: Colonel George A. Lincoln, USA, a senior planner on General Marshall's staff in World War II and since 1947 a member of the U.S. Military Academy faculty, and Paul H. Nitze, who as director of the State Department's Policy Planning Staff helped write NSC 68 and orchestrate the Truman rearmament program. Lincoln was detached and impartial; Nitze was anything but. An outspoken critic of the Eisenhower administration's heavy reliance on nuclear weapons, he was a leading proponent of the emerging doctrine of flexible response that would reshape American defense policy during the Kennedy-Johnson administrations.[26]

Written in a style reminiscent of NSC 68, the Gaither Report examined the entire panorama of U.S.-Soviet relations. The heart of the report was its assessment of the ominous progress of the Soviet ICBM program, which in the committee's estimation exposed U.S. retaliatory forces to unprecedented risk. "By 1959," the report warned, "the USSR may be able to launch an attack with its ICBMs carrying megaton warheads, against which the Strategic Air Command (SAC) will be almost completely vulnerable under present programs." To address this threat, the committee recommended a $44 billion effort spread over 5 years—$19 billion to expand and upgrade offensive capabilities and $25 billion for active and passive defense programs—with future allocations giving roughly equal priority to offensive and defensive capabilities. Even with these improvements, the committee doubted that the United States could achieve complete security. Looking into the future, it predicted "a continuing race between the offense and the defense" and "no end to the technical moves and countermoves" to gain an advantage. Only through "a dependable agreement" limiting arms and "other measures for the preservation of peace" did the panel see any prospect of ending this vicious cycle.[27]

Despite the Gaither Report's foreboding tone, neither President Eisenhower nor his military advisors saw cause for panic. U–2 photographs (which were off-limits to the Gaither Committee because of their sensitivity) showed a Soviet ICBM capability limited to a single above-ground launch pad at a previously undetected test site near Tyuratam.[28] Whether this information would have changed the Gaither Committee's findings is uncertain. But it made a strong impression on the President's thinking. "Until an enemy has enough operational capability to destroy most of our bases simultaneously and thus prevent retaliation by us," Eisenhower believed, "our deterrent remains effective."[29] Having access to the same intelligence as the President, the Joint Chiefs agreed that the Gaither Committee had exaggerated the threat. Finding little new or unusual in the report, they dismissed its recommendations as excessive, overdrawn, and probably underpriced.[30]

THE "MISSILE GAP" AND BMD CONTROVERSIES

Though classified top secret, key findings of the Gaither Report soon "leaked" to the press, giving rise to speculation that the United States had fallen uncomfortably behind the Soviet Union in missile technology. Under pressure from Congress and the media, President Eisenhower grudgingly requested small increases for missile development and other measures mentioned in the report. Hoping to keep critics at bay, he merely whetted their appetite for more. The ensuing controversy, known as the "missile gap," dogged the Eisenhower administration until it left office. A serious impediment to maintaining stability in military spending, the missile gap also became a major issue in the 1960 Presidential campaign. In fact, Soviet space and missile accomplishments tapered off after a second Sputnik launched in November 1957. However, a well-orchestrated propaganda and deception campaign spearheaded by Soviet leader Nikita S. Khrushchev gave the impression that Soviet missiles were coming off assembly lines "like sausages" and could devastate the United States and Western Europe on a moment's notice. Eisenhower recalled that "there was rarely a day when I failed to give earnest study to reports of our progress and to estimates of Soviet capabilities."[31]

Struggling to hold the line, the White House received relatively little support or cooperation from the two sources—the Intelligence Community and the Joint Chiefs of Staff—that might have given the debate a more rational framework. Closely linked in their day-to-day activities, the JCS organization and the Intelligence Community used much of the same information but tended to interpret the data differently. While all agreed that that the United States still held a commanding lead in strategic nuclear power, there was no consensus on how long it would last. Sputnik had severely rattled the Intelligence Community, and in its aftermath intelligence analysts scrambled to figure out where they went wrong. Generally speaking, their assessment of the Soviet submarine-launched ballistic missile program was always fairly accurate.[32] But having underestimated Soviet ICBM capabilities earlier, they now compensated by overestimating what the Soviets could do. The most excessive estimates were those of Air Force intelligence, which depicted the Soviets as having a more robust missile program than the United States, purposefully designed to produce capabilities for launching a disarming first strike by the early to mid-1960s.[33]

Based in part on these divergent interpretations of intelligence, "splits" persisted among the Joint Chiefs over how the United States should respond in allocating resources. Though hardly conclusive, the best visual evidence the JCS found came from U–2 photographs. For diplomatic reasons, however, President Eisenhower decided in March 1958 to suspend U–2 flights over the Soviet Union, a suspension that lasted until July 1959.[34] Thus, the JCS for all practical purposes were "blind" to the progress in Soviet missile technology for well over a year. Even so, the evidence collected up until the suspension offered uneven support for the Air Force's high-end estimates and its contention that the Soviets were building the infrastructure for a first-strike ICBM force. Not only did launch facilities appear limited to a handful of above-ground pads, but also there was no designated organization to plan and carry out nuclear delivery missions until the formation of the Strategic Rocket Forces (SRF) command in December 1959.[35]

A key figure in eventually settling these debates was General Nathan F. Twining, USAF, who succeeded Admiral Radford as CJCS in August 1957. Twining was not the most forceful or innovative Chairman, but he was well versed in strategic air warfare and did his best to function as an impartial arbiter in settling disputes. The Soviet missile program's ominous potential notwithstanding, Twining believed that the most serious threat to the United States was still the Soviet Union's long-range air force, estimated at 110–115 planes.[36] Looking at these numbers and at the U–2's findings, Twining agreed with his Army and Navy colleagues that there was no need for the "crash" program of ICBM development the Air Force favored. Offering an interim solution, he proposed allowing the missile program to proceed at a measured production rate until the United States had a better picture of the threats it faced and its strategic needs.[37] After further give and take, it was largely on this basis that the Eisenhower administration framed its response to the missile gap.[38]

Meanwhile, an even larger controversy was brewing over the allocation of resources for ballistic missile defense (BMD), one of the programs identified in the Gaither Report as being in urgent need of bolstering. Prior to Sputnik, the Defense Department supported two competing BMD systems: an Air Force program for wide-area defense known as "Wizard" and the Army's Nike-Zeus for point defense, the outgrowth of an earlier antiaircraft missile-radar system. Though both were essentially drawing-board concepts, the Army's was more refined, making it the frontrunner in the competition.[39] Alarmed by the success of Sputnik, Secretary of Defense Neil H. McElroy told President Eisenhower that it might be necessary to launch an initiative comparable to the World War II Manhattan Project to produce an anti-ICBM as quickly as possible.[40] Raising objections, the Air Force and the Navy argued that no program was as yet sufficiently advanced to warrant such action.[41] But with the pressure building, McElroy decided in January 1958 to end further debate by giving the Army primary responsibility for developing an anti-ballistic missile (ABM) system.[42]

Having won the battle for control of the ABM mission, the Army now wanted Nike-Zeus elevated to the same national priority enjoyed by the Air Force and the Navy in offensive missile programs. Projecting deployment by the early 1960s, Army planners sought to move from R&D into full production as quickly as possible. But the high cost of a deployed Nike-Zeus system, estimated at $7 billion to $15 billion, invited further technical analysis which the JCS assigned to the Weapons Systems Evaluation Group (WSEG). While WSEG found Nike-Zeus to have "significant" potential, it also cited the need for more information on technical problems, including the effects of high-altitude nuclear explosions, decoy discrimination, and the vulnerability of incoming nuclear weapons.[43] Bowing to strong congressional pressure to overlook the system's shortcomings, an OSD technical steering group urged the Secretary of Defense in November 1958 to approve a limited production budget.[44] At this stage, a firm, unanimous, and unambiguous response from the Joint Chiefs might have settled the matter. But under the consensus rules that governed JCS deliberations, no such answer emerged. The only area of agreement among the chiefs was that there should be further R&D, a course that McElroy and Eisenhower, hard-pressed to hold down military spending, found more appealing than deployment.[45]

By chance, the President's decision to forego BMD production coincided with a surge in Soviet propaganda and assertions of nuclear superiority. Many Democrats in Congress and some members of the Intelligence Community accepted Soviet claims at face value. An added complication was that the Soviet Union carried out no ICBM tests between May 1958 and March 1959, a hiatus that produced new disputes among intelligence experts. The CIA and most other intelligence organizations interpreted the moratorium on testing as a sign that the Soviet program was having technical difficulties. Air Force intelligence disagreed, however, arguing that an equally plausible explanation was that the Soviets had ceased testing because they had solved their technical problems and were now gearing up for mass production.[46] To settle the matter, McElroy and Twining appealed to the President to resume U–2 overflights of the Soviet Union. At first, the President refused, fearing that the possible loss of a U–2 might provoke a diplomatic incident or worse. Apprised that the reconnaissance satellite project was "coming along nicely" and that the A–12, a faster and more sophisticated spy plane than the U–2, was waiting in the wings, he preferred to wait. But at the urging of both the CIA and State Department, the President changed his mind and in July 1959 authorized a single mission directed against the ICBM test facility at Tyuratam.[47]

The mission found no trace of launch sites other than at the Tyuratam test facility but could neither confirm nor deny whether the Soviets had a large-scale ICBM buildup under way. Still, the absence of new sites was reassuring news and led to a gradual reappraisal of the Soviet missile program. A new NIE, appearing in January 1960, downplayed the likelihood of a Soviet crash program to produce and deploy ICBMs. Based on these findings, George B. Kistiakowsky, the President's special assistant for science and technology, concluded that "the missile gap doesn't look to be very serious."[48]

The new estimate (NIE 11-8-59) projected a deployed Soviet force of 140 to 200 ICBMs by mid-1961, with the Joint Staff endorsing the higher number.[49] At the President's request, General Twining, Secretary of Defense Thomas S. Gates, Jr., and Director of Central Intelligence (DCI) Allen W. Dulles appeared before Congress to explain the new intelligence. All agreed that the fresh data cast doubt on the missile gap. Unfortunately, however, their testimony was poorly coordinated and diverged on critical details, most notably the number of missiles the Soviets might deploy. In closed hearings, Gates and Dulles stressed the lower numbers while Twining stood by the Joint Staff's figures. Seizing on this and other discrepancies, some congressional Democrats questioned the reliability of the administration's assessments, keeping the missile gap controversy alive and well despite growing evidence that the Soviet lead was overblown.[50]

Determined to end the missile gap debate, DCI Dulles persuaded President Eisenhower to increase the frequency of U–2 flights over the Soviet Union. Even before the program began in 1956, Richard M. Bissell, Jr., the coordinator of the effort, had predicted that the U–2 would be able to fly over the Soviet Union with impunity for only about 2 years.[51] Hence, the development of the A–12, a faster plane that could cruise at 90,000 feet. Based on Bissell's estimate, by 1960 the U–2 was living on borrowed time. Increasingly uneasy, Eisenhower reluctantly supported Dulles in hopes of bringing the controversy to a definitive conclusion. The result was a new series of flights, culminating in Francis Gary Powers' ill-fated mission of May 1, 1960, which the Soviets ended abruptly with an SA–2 missile.[52] In addition to wrecking a summit conference between Eisenhower and Khrushchev, the downing of Powers' plane brought an immediate cessation of U–2 flights over the Soviet Union. Thus ended the most reliable source of information the Joint Chiefs and the Intelligence Community had on the Soviet missile buildup until the Discoverer satellite program began to provide detailed pictures later that summer.[53]

Even with the missile gap issue unresolved, the U.S. response was well formed, with much of it in place by the time the Eisenhower administration left office. Unable to agree on an overall strategic blueprint, the Joint Chiefs let the Services pursue their own often overlapping interests and left it up to the Secretary of Defense and the President to resolve conflicts. The result was a fairly predictable allocation of functions that essentially allowed each Service to push its preferred programs—ICBMs and IRBMs for the Air Force, Polaris for the Navy, and Nike-Zeus for the Army. A new strategic buildup driven by dynamic advances in missile technology and energized by arguable claims of Soviet accomplishments had begun.

REORGANIZATION AND REFORM, 1958–1960

The inter-Service rivalry and competition that plagued the missile program left President Eisenhower more convinced than ever that the Department of Defense—and in particular the Joint Chiefs of Staff—needed fundamental organizational reform. Despite the changes made in 1953, Eisenhower was far from satisfied with the results. While the 1953 reforms had streamlined and strengthened the Office of the

Secretary of Defense, they had produced only limited improvements in JCS performance. The central problems, in Eisenhower's view, continued to be the institutional weakness of the Chairman and the influence of "narrow Service considerations" in JCS deliberations. The "original mistake in this whole business," he believed, had been the failure to create a single Service in 1947.[54] Ideally, he wanted the Chairman to have broader powers and the authority to make decisions in the absence of unanimity among the chiefs. He also wanted to simplify lines of command and control, make the JCS members of the Secretary's staff, and turn the Joint Staff into an integrated, all-Service organization similar to the combined staffs he had commanded in Europe in World War II and at SHAPE in the early 1950s.[55]

The Joint Chiefs recognized that their internal differences threatened serious consequences for their role and influence. By failing to reconcile their differences, warned the Air Force Chief of Staff General Thomas D. White, the JCS were placing themselves in jeopardy of ceding important military policy functions to civilians in OSD.[56] Despite the risk, however, none of the chiefs, including White, favored a sharp departure from current practices and procedures; only the Chairman, General Twining, showed significant interest in organizational reform. The most determined of all to preserve the status quo was Chief of Naval Operations Admiral Arleigh A. Burke, who openly denounced "public pressures toward centralization and authoritarianism in defense."[57] To help make their case, the JCS in December 1957 appointed an ad hoc inter-Service panel headed by Major General Earle G. Wheeler, USA, who would later become Chairman of the JCS. Working quickly, the committee came up with an interim report in less than a month, but its findings, which were generally in line with the view that radical changes were to be avoided, proved too little too late to affect the ensuing debate.[58]

The opening salvo in the administration's drive to reform the Pentagon came on January 9, 1958, in the President's State of the Union address. Insisting that defense reorganization was "imperative," he called for "real unity" among the Services, clear subordination of the military to civilian control, improved integration of resources, simplification of scientific and industrial effort, and an end to inter-Service rivalry and disputes.[59] To translate the President's goals into specific recommendations, Secretary of Defense McElroy turned to Charles A. Coolidge, a former assistant secretary, who had worked on defense organizational problems in the past. For assistance, Coolidge formed an advisory group that included General Twining, his two predecessors, Admiral Radford and General Bradley, and General Alfred M. Gruenther, USA (Ret.), the former NATO commander and the first director of the Joint Staff.[60]

Drawing on the findings of the Coolidge group, Eisenhower submitted reform recommendations to Congress on April 3, 1958. Declaring that "separate ground, sea and air warfare is gone forever," the President called for legislation to facilitate closer inter-Service unity and cooperation. Among the changes he sought were authority for the Secretary of Defense to transfer, reassign, consolidate, or abolish military functions; a simplified chain of command; enhanced authority for the Secretary to carry out military research and development through a director of defense research and engineering; removal of the ceiling on the size of the Joint Staff; and stronger powers for the Chairman, allowing him to vote in JCS deliberations and to select (subject to the Secretary's approval) the Joint Staff's director.[61]

Opponents of the President's plan rallied behind Democratic Representative Carl Vinson of Georgia, Chairman of the House Armed Services Committee and a longtime supporter of the Navy.[62] A critic of Service unification, Vinson knew that more power for the Secretary of Defense and the Chairman meant less power and authority for him and his committee. To blunt the President's initiative, he accused the administration of seeking a "blank check" to remake the Joint Staff and revived arguments that the White House was flirting with a Prussian-style general staff. Eventually, he sent proposed legislation to the House floor that fell short of meeting administration requests for changes. Stymied in the House, the administration relied on the Senate to produce a bill more to its liking and trusted a conference committee to iron out the differences in its favor. Although many in Congress shared Vinson's concerns to one degree or another, the overriding sentiment among legislators was that the Commander in Chief should have the latitude to organize the Department of Defense as he saw fit. The resulting compromise, signed into law on August 6, 1958, gave the President nearly everything he sought, but retained a ceiling on the size of the Joint Staff (increased from 210 to 400 officers) and banned its use in any capacity approximating "an overall Armed Forces General Staff."[63]

While most of the President's reforms required enabling legislation from Congress, those affecting the internal organization and operation of the JCS were largely carried out under the existing authority of the Secretary of Defense. Expressing no particular preferences, McElroy left the details to be worked out by the Joint Chiefs themselves. Foremost among the changes thus made was the creation of a conventional military staff structure, which replaced the Joint Staff's committee-group system. In April 1958, Director of the Joint Staff Major General Oliver S. Picher, USAF, suggested establishing functional numbered directorates: J-1 (personnel), J-2 (intelligence), J-3 (operations), J-4 (logistics), J-5 (plans and policy), and J-6 (communications and electronics). The most innovative feature under this arrangement was the creation of the operations directorate, J-3, which had no corresponding organization under the old group system. President Eisenhower had often said that he wanted the JCS more involved in operational matters, but he had never been specific.[64] Arguing that the Joint Staff would be exercising executive authority, Admiral Burke and Commandant of the Marine Corps General Randolph McC. Pate objected to these new arrangements, but offered no alternative other than retention of the status quo. In view of the caveats inserted by Congress into the final legislation, Twining and McElroy agreed that the problems Burke and Pate envisioned appeared highly unlikely, and in late August 1958 they assured Eisenhower that the restructuring of the Joint Staff would proceed as planned.[65]

The 1958 amendments also streamlined relationships under the unified command plan. As the President had stated, a major goal of the reorganization was to establish a more direct chain of command by ending the designation of a military department as the executive agency for each unified command. Under the new law, the chain of command ran from the President to the Secretary of Defense to the unified and specified commanders. The intent was that all combatant forces should operate under the control of a unified or specified commander who would be responsible directly to the Secretary of Defense. The Secretary would exercise control by orders issued through the Joint Chiefs of Staff. In consonance with this intention, the 1958 amendments deleted existing provisions that had authorized a Service chief to command the operating forces of his Service. From this point on, each military department was to organize, equip, train, support, and administer combatant forces but not direct their operations.[66]

Implementing these provisions fell to Secretary McElroy, who issued a revised version of DOD Directive 5100.1, "Functions of the Department of Defense and its Major Components," on December 31, 1958. The directive designated the Joint Chiefs of Staff as the Secretary's "immediate military staff" and described the chain of operational command as extending from the President to the Secretary via the Joint Chiefs to the unified and specified commanders. In effect, the JCS became the conduit through which the National Command Authority, or NCA (i.e., the President, the Secretary of Defense, and the NSC), communicated with the combatant commanders. The new directive also charged the Joint Chiefs with responsibility for recommending to the Secretary of Defense the establishment and force structure of unified and specified commands, the assignment to the military departments of responsibility for providing support to these commands, and the review of the unified commanders' strategic plans and programs.[67]

No less important than the reforms enacted in 1958 was the creation, 2 years later, of the Joint Strategic Target Planning Staff (JSTPS). An administrative extension of the JCS, the JSTPS's function was to plan and coordinate strategic nuclear targeting, a key part of the Joint Chiefs' statutory responsibility for strategic planning. Though the majority of the officers serving on the JSTPS were from the Air Force, it also included naval officers and representatives from each major combatant command allocated nuclear weapons. The origins of the JSTPS lay in the growth of the missile program and the need for better command, control, and coordination of targeting. At issue was how

Figure 6–1.

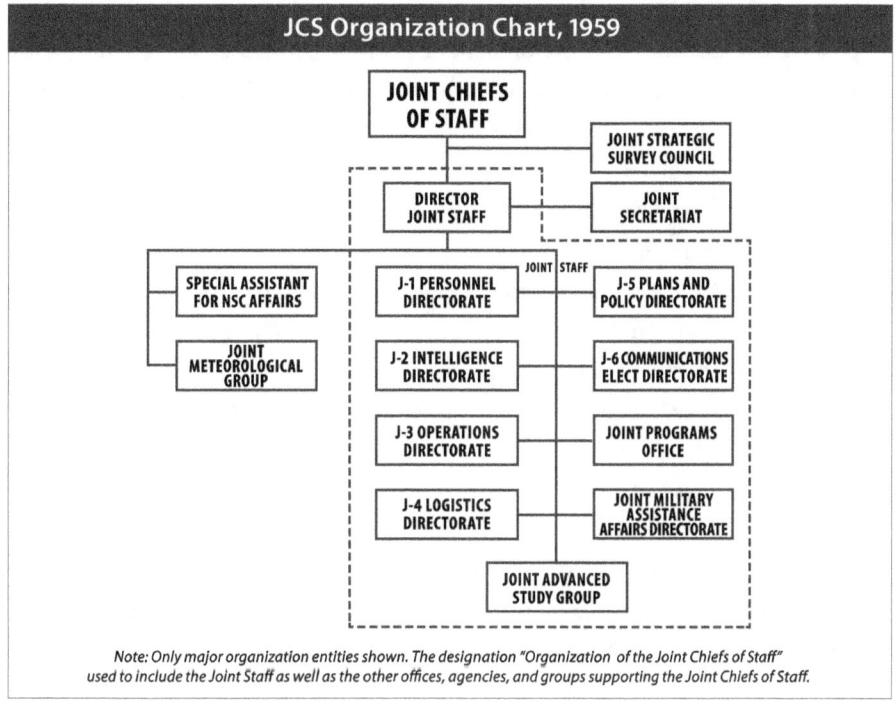

to integrate the Navy's Polaris submarine fleet with other strategic forces when the Polaris boats began deployment in the early 1960s. Initially, there were two competing plans on the table—an Air Force plan to centralize the control of all strategic nuclear forces under an overarching U.S. Strategic Command that would replace SAC, and a Navy plan, supported by the Army and the Marine Corps, to place the Polaris boats under the command and control of unified commanders with major naval forces (Commander in Chief, Atlantic; Commander in Chief, Pacific; and U.S. Commander in Chief, Europe).[68] During the early months of 1959, the debate became, as one senior Air Force planner described it, "an all-out battle" that could shape budget shares and the control of forces and missions for decades to come.[69]

Despite the 1958 reforms, unity among the Joint Chiefs remained more a hope than a reality, frustrating the possibility of an early resolution of the Polaris issue. In May 1959, the Joint Chiefs notified the Secretary that they could only produce a split recommendation on command and control of strategic forces.[70] Absent on medical leave, General Twining had played no part in the chiefs' deliberations. When he resumed his duties that summer he set about finding a solution to the problem, which he identified as essentially the selection of targets, the development of appropriate plans, and the right allocation of resources.[71] Since the Strategic Air Command had most of the assets and experience in these matters, Twining expected any solution to center around SAC. Viewing the creation of a new unified command as the last resort, he preferred to start with the development of a comprehensive target list and a jointly prepared single integrated operational targeting plan. All Polaris submarines would remain under the Navy's tactical control, but the targeting of their weapons would be a joint endeavor, to avoid overlap and unnecessary duplication with other forces. It was from this blueprint that the JSTPS eventually emerged.[72]

Twining urged the Secretary and the President to defer action until they had the results of an ongoing review of targeting priorities by the Net Evaluation Subcommittee (NESC), an inter-Service technical advisory body under the NSC. While the NESC had conducted limited inquiries of this nature before, this was the most in-depth examination of targeting policy since the Joint Chiefs systematized targeting categories in the summer of 1950. Such a review should have been an in-house function, but because of the Joint Staff's limited size, the JCS had yet to develop a war-gaming capability. For technical analysis, they relied on the NESC, WSEG, RAND, the Defense Atomic Support Agency (DASA), and the Services.[73] The key question was whether to concentrate strategic attacks against targets that were primarily military (the preferred Air Force approach), primarily urban-industrial (the Army and Navy view), or an "optimum mix." Toward the end of October 1959, the NESC recommended adopting the latter approach, thereby covering all bases.[74] At this point, a lengthy and acrimonious debate ensued among the Joint Chiefs over the organizational arrangements that should be adopted to implement the NESC report. Resisting pressure from the Air Force, Admiral Burke insisted that there should be no merger of strategic forces and that SAC should have no authority over Polaris.[75] To accommodate Burke's objections, the new Secretary of Defense, Thomas S. Gates, Jr., pushed the idea of a separate joint targeting staff—the JSTPS—responsible to the JCS. Gates told the President that, to reach this point, he had held 15 meetings with the Joint Chiefs.[76]

Patience paid off, and on August 11, 1960, despite continuing objections from Admiral Burke, President Eisenhower gave the go-ahead for the integration of strategic targeting. The decision came at the end of a contentious 2-hour White House meeting involving the President, Gates, and the Joint Chiefs. The most heated exchanges were between Twining, who accused the Navy of habitually operating on its own agenda and flouting the principles of unified command, and Burke, who counterattacked that the proposed targeting system undermined JCS authority and was nothing more than a thinly disguised attempt by the Air Force to seize control of Polaris. Agreeing with Burke that strategic targeting should remain a JCS responsibility, President Eisenhower reminded the chiefs that they should keep the matter under close periodic review. But he found the behavior of all involved in the controversy appalling and admonished them "to try to make arrangements work."[77]

Activated about a month later, the JSTPS operated from Strategic Air Command headquarters at Offut Air Force Base near Omaha, Nebraska, where it had access to SAC's computers and vaults of targeting data. The head of the organization was the commander in chief, Strategic Air Command (CINCSAC), an Air Force four-star general who served as the director of Strategic Target Planning (DSTP). Under him was a Navy vice admiral deputy director in charge of day-to-day management. The JSTPS had an initial strength of just over 200 officers—half the size of the Joint Staff at the time—of which roughly 15 percent were from the Navy.[78] The DSTP communicated directly with the JCS through a liaison office in the Pentagon.[79] Broadly speaking, the JSTPS had two tasks: to maintain and update a comprehensive list of targets, known as the National Strategic Targeting List (NSTL); and to prepare a Single Integrated Operational Plan (SIOP) for the execution of strategic operations against the Soviet Union, Communist China, and the Warsaw Pact countries of Eastern Europe.[80]

By the end of 1960, the JSTPS had produced the first SIOP, designated SIOP–62. A hurry-up job, it contained only one "plan," which was meant for execution as a whole. Though it supposedly conformed to the NESC "optimum mix" philosophy, SIOP–62 was essentially a recapitulation of previous SAC war plans, oriented toward massive retaliation, with the assets of available Polaris boats added in. Eighty percent of the planned attacks were against "military targets." These included not only atomic energy facilities, ICBM sites, air bases, and other military installations, but also factories turning out military equipment located in urban-industrial centers. Planners acknowledged that it was practically impossible to distinguish an attack against a military target from an attack against an urban-industrial target.[81]

Eisenhower's reaction to SIOP–62 was that it did not appear "to make the most effective use of our resources." He said that if the planning had been in his hands, he would have held the Polaris boats in reserve for follow-on attacks. Though Eisenhower still approved the plan, his science advisor, George B. Kistiakowsky, thought the next administration should subject it to a "thorough revision."[82] Herbert F. York, director of Defense Research and Engineering and a key figure in the development of strategic weapons, agreed. York recalled that the programmed attacks were so indiscriminate that their purpose seemed to be "simply to strip-mine much of the USSR."[83]

Eisenhower wanted the targeting controversy settled and the JSTPS up and running before he left office; he did not want to saddle his successor "with the monstrosity we now see in prospect as Polaris and other new weapons come into operating status."[84] But like other organizational reforms initiated toward the end of his Presidency, it was hard to predict how successful the new targeting procedures would be. As the inter-Service quarreling over guided missiles and targeting policy demonstrated, it would take more than an act of Congress to instill unity of spirit and action among the Services. The 1958 reforms had taken the Joint Chiefs of Staff about as far as they could go without discarding the concept of an inter-Service corporate advisory body, creating a full-blown general staff, and giving the Chairman complete control. But at the same time, these reforms had not done much to make the JCS a more efficient and effective entity.

DEFENSE OF THE MIDDLE EAST

As the Joint Chiefs struggled with the impact of guided missiles, new security problems were emerging abroad. At the outset of the Eisenhower administration, the principal Cold War battlegrounds were in Europe and East Asia. But by the mid-1950s, attention turned increasingly to the Middle East, where continuing friction between Israel and the Arab states and a growing Soviet presence created new concerns. To the Joint Chiefs, the strategic importance of the Middle East was self-evident. It contained the largest petroleum reserves in the world, the Suez Canal, and ideal locations for military bases from which to launch strategic air and missile attacks against the Soviet Union in the event of general war. Were the Middle East to become part of the Sino-Soviet block, the results would doubtless have a seriously adverse impact on American interests and the strategic balance.

In considering defense arrangements for the Middle East, the Joint Chiefs moved with caution, partly because of limited resources and partly because British interests and influence predominated there. While the United States had formidable capabilities nearby—the Sixth Fleet in the Mediterranean and air bases in Morocco, Libya, and Turkey—the only U.S. forces assigned to the Middle East were the MIDEASTFOR, a task force of four or five ships in the Persian Gulf under the control of the Commander in Chief, U.S. Naval Forces, Eastern Atlantic and Mediterranean (CINCNELM).[85] Considerably larger, the British presence included a network of military and naval bases, economic holdings, and intelligence assets scattered across the region. Most of the initial defense planning thus occurred in London, where the British Chiefs of Staff took the lead. The organizing concept that emerged from these discussions was the Baghdad Pact, a loose coalition created early in 1955 that included Britain, Turkey, Iraq, Iran, and Pakistan. With NATO to the west and the Southeast Asia Treaty Organization (SEATO) to the east, the Baghdad Pact completed "a globe-girdling wall of containment against communist expansion."[86] The JCS favored full U.S. adherence to the Baghdad Pact, but ran into opposition from the State Department, which worried that U.S. membership would complicate American efforts to ease Arab-Israeli tensions. Eventually, the JCS had to settle for "observer" status, which gave them back-door access to the Pact's military planning.[87]

Conceived as an anti-Communist alliance, the Baghdad Pact's main military function was to block a Soviet invasion of the Middle East and Southwest Asia. At Iran's insistence, the Pact adopted a strategy to defend a line along the rugged Elburz Mountains stretching from the borders of Armenia to the Caspian Sea. JCS planners assessed the concept as "sound" in theory, but found it needing closer coordination than Alliance members seemed prepared to accept.[88] At bottom, the members of the Baghdad Pact had little in common other than their desire for U.S. military assistance, which Iran and Iraq appeared to want to prop up their regimes and preserve internal order rather than to fight the Soviets. Easily destabilized monarchies ruled in both countries, and neither was keen on developing a defense establishment that might become a rival for power. Rating Iran and Iraq of dubious reliability, the JCS viewed a successful defense of the Middle East as resting on Turkey (a NATO ally) and Pakistan, owing to their strategic locations, historic anti-Communism, and commitment to a strong defense posture.[89]

JCS efforts to fashion a credible defense under the Baghdad Pact were further complicated by the rising tempo of anti-Zionism in the Muslim world and the intensification of Arab nationalism. The leading political figure in the region was now Gamal Abdel Nasser, president of Egypt, who had maneuvered his way into power following a 1952 putsch that had toppled the dissolute King Farouk. Nasser aspired to unite the Arab world and mounted unrelenting propaganda campaigns against Israel and the Baghdad Pact. He also aided and abetted Palestinian guerrilla raids into Israel from the Gaza Strip and threatened major military action to wipe out the Jewish state. For support, he turned to the Soviets who obligingly sold him arms through Czechoslovakia. Recognizing Nasser's growing popularity in the Third World, Eisenhower thought it necessary to "woo" him and hesitated to put too much overt pressure on Egypt lest it provoke an anti-American backlash in the Muslim world "from Dakar to the Philippine Islands." Normally, the Joint Chiefs would have agreed. But according to Admiral Burke, the consensus among the chiefs was that one way or another Nasser needed to be "broken."[90]

Nasser's most audacious move was to nationalize the British-owned Suez Canal on July 26, 1956, in retaliation for the withdrawal of American and British financing of the Aswan Dam project. In Admiral Radford's view, Nasser was "trying to be another Hitler."[91] With tensions between Israel and Egypt also escalating, the JCS and the British chiefs quietly began staff talks on possible combined military action in the Middle East in the event of another Arab-Israeli war.[92] After nationalization, the British signaled that they would welcome a collaborative effort along these lines to regain control of the canal.[93] Assuming the President would support the

British, the JCS proposed moving ahead with contingency planning under which the United States would contribute economic and logistical support to a combined operation against Egypt in the event diplomacy failed. Should "third parties" (i.e., the Soviets) intervene, the JCS favored an immediate commitment of U.S. combat forces.[94] Eisenhower, however, refused even to look at such plans. Unless there was a major threat to the Persian Gulf oil fields, he could not perceive U.S. interests to be seriously at risk and had no desire to be accused of coming to the rescue of Anglo-French colonialism. While he acknowledged that "there may be no escape from the use of force" in the current crisis, he did not want the United States directly involved in a confrontation that could draw in the Soviets.[95]

Instead of direct military action, Eisenhower favored weakening Soviet influence and undermining Nasser's regime through covert operations under a combined Anglo-American plan (code-named OMEGA), which he sanctioned in late March 1956. Limited initially to political and economic pressure, the plan's purpose, as Eisenhower described it, was to "help stabilize the situation" in the Middle East and "give us a better atmosphere in which to work."[96] Though the JCS had no direct role in OMEGA, Admiral Radford was in on the planning and aware of the details practically from its inception.[97] OMEGA's chances of success, however, were far from certain, and as planning progressed there were veiled hints that the President's British counterpart (and personal friend) Prime Minister Anthony Eden might take preemptive action on his own. Months before the nationalization, Eden was "quite emphatic that Nasser must be got rid of." But despite their shared antipathy for Nasser, Eden could not persuade the President to participate beyond OMEGA.[98]

Unable to enlist anything other than nominal American support, Eden turned to the French and Israelis and began secretly organizing a military operation against Egypt. Known as MUSKETEER, the British plan called for Israel to feign an invasion of Egypt, giving France and Britain an excuse to intervene, take control of the Suez Canal, and install a new regime in Egypt "less hostile to the West."[99] As preparations for the operation unfolded, the National Security Agency intercepted a new and unfamiliar French code, followed by a "vast increase" in cable traffic between the French and the Israelis.[100] Suspecting something was afoot, President Eisenhower authorized U–2 flights that detected unusual concentrations of British forces on Malta and Cyprus and early signs of Israeli mobilization.[101] An elaborate deception plan mounted by British intelligence sought to convince the CIA and President Eisenhower that the Israeli mobilization was aimed against Jordan, not Egypt, and that the British buildup was to protect Jordan, with whom the UK had a security treaty. Eventually, the Intelligence Community and the Joint Chiefs uncovered the ruse, but by that time it was too late to make much difference.[102]

The invasion began on October 29, 1956, when an estimated six Israeli brigades crossed into the Sinai, breaking through Egyptian defenses. Shortly after hostilities commenced, the Joint Chiefs increased the alert status of selected U.S. forces and deployed additional naval units to the eastern Mediterranean, some to assist in the evacuation of U.S. citizens from Egypt, Israel, Jordan, and Syria. But beyond this, the JCS adopted a low profile and played a limited role in the crisis. In line with declarations from the White House calling on the invaders to cease and desist, the JCS were careful to avoid giving the appearance that the United States was taking sides. Still, the mere presence of increased American forces in the region had the de facto effect of working to the advantage of the Anglo-French-Israeli coalition.[103]

Initially, the attack was a stunning success. Within days, having easily routed the Egyptians, the Israelis were astride the Suez Canal. But after the landing of British and French troops at Port Said on November 6, the invasion began to lose steam. Eden assumed that once the operation was under way, Eisenhower would see the opportunities it presented and throw his support to Britain, France, and Israel.[104] Eden, however, was wrong. The fatal flaw in the allies' plan was that, while the operation seriously crippled Nasser's military machine, it failed to undermine his popularity or bring down his regime. Persuaded that Nasser would survive the setback and that further efforts to unseat him could only harm U.S. interests in the Third World, Eisenhower insisted that the coalition halt its operations, accept an immediate ceasefire, and promptly withdraw. Eden reluctantly agreed, knowing that he would be admitting defeat and have to resign his premiership with no chance of ever regaining control of the canal.

The Suez crisis coincided with two other major events: a popular uprising in Hungary against Soviet domination, which eventually failed to dislodge Communist rule; and the Presidential election in the United States, which Eisenhower won handily. As it turned out, the Hungarian uprising kept the Soviets so preoccupied that they were in no position to provide much help to the Egyptians. Based on the information available at the senior levels in Washington, there was little likelihood that Moscow would intervene on Egypt's behalf. Though Moscow at one point rattled its nuclear sabers against the invaders, Eisenhower dismissed the threats as bluster aimed more at shoring up Moscow's bona fides with Nasser than at influencing decisions in London, Paris, or Tel Aviv. As a precaution, the Joint Chiefs recommended to the President on November 6—election day—that the Strategic Air Command increase its readiness status for an emergency. Eisenhower, however, saw no need.[105]

In the aftermath of the Suez crisis, there emerged a politico-military vacuum in the Middle East which the United States and the Soviet Union rushed to fill—the Soviets by stepping up arms aid and political backing for their major clients, Egypt and Syria, and the United States by offering similar benefits and planning advice to the members of the Baghdad Pact. The operative U.S. policy, unveiled in January 1957, was the "Eisenhower Doctrine," a broad promise of economic and military help for any Middle East country threatened with a Communist takeover.[106] Developed to give the President greater leverage in a future Middle East crisis, the doctrine emerged without even a pro forma review by the Joint Chiefs and had only tepid support in Congress. Even so, it filled an obvious void and gave the JCS a better idea of how far they could go in formulating plans and strategy.[107]

With British influence on the wane, the United States emerged as the de facto leader of the Baghdad Pact. By the summer of 1957, JCS representatives were working directly with pact planners to coordinate defense of the region with assigned taskings for U.S. forces under the Joint Strategic Capabilities Plan. While offering air and naval support, the JCS sought to avoid a commitment of U.S. ground troops and looked to indigenous forces, primarily those of Turkey and Pakistan, to lead the fight on the ground.[108] However, the CINCNELM, Admiral Walter F. Boone, who exercised operational planning responsibility for the region, envisioned a significantly broader U.S. commitment. Citing the Eisenhower Doctrine, Boone requested authority at the first signs of escalating tensions to insert elite combat forces and enlarged military advisory units into the Middle East.[109] In preparation, Boone held exploratory talks with Army commanders at his headquarters in London in September 1957, and in November he hosted a joint conference of Army, Navy, and Air Force representatives to develop joint plans for airborne operations and air transport support in the Middle East.[110]

Several members of the Joint Chiefs expressed concern that Boone was moving too far too fast. Citing the limited availability of resources, the Army and Air Force chiefs of staff questioned the feasibility of Boone's plans and suggested that he had exceeded his authority by presuming to interpret U.S. policy needs under the Eisenhower Doctrine.[111] In February 1958, Boone and the JCS reached an understanding that restricted CINCNELM's planning for intervention to Lebanon and Jordan. Later, the JCS extended this mandate to include the prevention of a coup d'état, rumored to have Egyptian support, aimed at toppling the government of Saudi Arabia.[112]

The first test of these plans came in Lebanon where in May 1958 a Muslim-led revolt broke out against the pro-Western Christian government of Camille Chamoun. Earlier that same year, Egypt and Syria had joined forces to form a United Arab Republic (UAR). Suspecting UAR involvement in the disturbances, President Eisenhower ordered Marines with the Sixth Fleet to be prepared to intervene. But by the end of the month, tensions in Lebanon had eased and U.S. forces stood down. Fearing the unrest would resume and spread, King Hussein of Jordan requested assistance from his cousin, King Faisal II of Iraq. Faisal ordered the Nineteenth Brigade to go to Hussein's aid. Instead of marching on Jordan, dissident units loyal to Brigadier

Abdul-Karim Kassim staged a rebellion in Baghdad against the monarchy. On July 14–15, the insurgents murdered Faisal, his family, and Premier Nuri al-Said and established a military regime allied with Egypt and Syria. Alarmed, Chamoun requested immediate U.S. military intervention under the Eisenhower Doctrine, and on July 15 a Marine battalion landing team went ashore south of Beirut in the first phase of Operation *Blue Bat*. At the same time, demonstrating that it was still a power to be reckoned with, Britain deployed 3,000 paratroopers to Jordan to shore up Hussein's rule.

In contrast to the debates over Korea, Indochina, and the Chinese off-shore islands, the decision to launch Operation *Blue Bat* was relatively quick and easy. Having ironed out most of their differences during the planning phase, the Joint Chiefs were able to move promptly when the time arrived. Though there was some talk of mounting a combined operation with the British, events moved too quickly for the necessary arrangements to be finalized and put into effect. While briefing congressional leaders immediately before U.S. troops landed, General Twining speculated that involvement in Lebanon might require intervention elsewhere in the region.[113] Still, the uneasiness of Congress over an expanded operation, the absence of overt Soviet, Egyptian, or Syrian involvement, and President Eisenhower's own reluctance to make open-ended commitments confined the operation to Beirut. Finding no concrete evidence of Communist involvement, the President declined to justify U.S. intervention as a function of the Eisenhower Doctrine.

The Lebanon incident was the only time during his Presidency, other than during the final months of the Korean War, that Eisenhower resorted to the use of military power. Among other things, Operation *Blue Bat* served to rebut critics (including Army Chief of Staff General Maxwell D. Taylor) who argued that the administration's cutbacks and reallocation of military resources under the New Look had eviscerated the country's conventional forces. To be sure, some of the equipment used in the operation was obsolescent. But within 2 weeks, the JCS were able to deploy the bulk of the Sixth Fleet off-shore and a division-equivalent of Marines and Army troops in and around Beirut, with two more Army divisions standing by in Germany.[114] Initially, Admiral James L. Holloway, Jr., who had succeeded Admiral Boone as CINCNELM in February, directly commanded the entire operation.[115] Evolving quickly into a joint enterprise, the growing scale and scope of the intervention necessitated an expanded command structure, with an Air Force major general in charge of tactical support and air transport operations and an Army major general commanding ground forces ashore. Holloway remained in charge overall.[116] As historian Stephen E. Ambrose later observed: "Lebanon, in short, was a show of force—and a most impressive one."[117]

The Lebanon intervention was the final episode in a fast-paced 2 years since the Suez crisis that witnessed dramatic changes in the political, strategic, and military makeup of the Middle East. From this point until the Six Day War of 1967, the Middle East seemed to quiet down. Even so, the alignment of Egypt and Syria with the Soviet Union, the overthrow of the pro-Western government in Iraq and, with it, the effective collapse of the Baghdad Pact (replaced by a rump alliance calling itself the Central Treaty Organization, or CENTO, in 1959), and continuing tensions between Israel and its Arab neighbors, all made for a sensitive situation that the Joint Chiefs continued to watch carefully. The United States had yet to make a major military commitment to the Middle East. But from the seeds sown in the 1950s, something along those lines seemed unavoidable sooner or later.

CUBA, CASTRO, AND COMMUNISM

Like the Middle East, Latin America experienced growing social, economic, and political turmoil during the 1950s. Building steadily as the decade progressed, these pressures culminated in 1959 in the Cuban revolution, which brought to power a Marxist regime under Fidel Castro. Denouncing the United States, Castro eventually aligned his country with the Soviet Union. At the time these events were taking shape, the Joint Chiefs of Staff had one overriding strategic concern in Latin America—the security of the Panama Canal. They also assisted in training military officers at Defense Department schools and in establishing military advisory programs to assist friendly governments. But as a rule, the JCS dedicated few forces to the region and exercised limited influence over U.S. policy there during the Eisenhower years. If the President needed advice or information, he usually relied on a small circle that included his brother, Milton Eisenhower, a specialist on Latin America, Secretary of State John Foster Dulles, and Allen W. Dulles, the Director of Central Intelligence.

Throughout Eisenhower's years in office, it was axiomatic that a Communist presence in the Western Hemisphere would be intolerable and that the United States should do all it could to prevent Moscow from making inroads. The preferred approach was to use diplomatic channels or covert operations. Prior to the Cuban revolution, the most serious challenge to U.S. policy came from Guatemalan strongman Jacobo Árbenz Guzmán, a former military officer with leftist political sympathies. Installed as president in 1951 after a controversial and violent election, Árbenz adopted tolerant policies toward Communists and made overtures to the Soviet Union, which reciprocated by sending Guatemala a shipload of small arms. Convinced that Árbenz was "merely a puppet manipulated by Communists," Eisenhower gave the Central Intelligence Agency the go-ahead to mount a "black" propaganda campaign against Árbenz's authority and to organize and arm a paramilitary group that ousted Árbenz from power in June 1954.[118]

Following the overthrow of the Árbenz regime, the Eisenhower administration set about bolstering anti-Communist governments in Latin America through, among other things, expanded military training and assistance.[119] The Joint Chiefs supported the administration's overall goal but objected to State Department efforts to micromanage these programs.[120] As time went on, friction over this issue centered increasingly on assistance to Cuba, where dissidents under Fidel Castro, a lawyer turned revolutionary, had been waging a guerrilla war against the country's heavy-handed dictator, Fulgencio Batista, since 1953. Having lost confidence in Batista's honesty and leadership, the State Department charged him with improperly diverting American aid earmarked for hemispheric defense to internal security functions, mainly to fight Castro. In March 1958, without consulting the JCS, State suspended all arms shipments to Cuba.[121] A furious Admiral Burke accused the State Department of committing an "unfriendly act" toward the Cuban government that amounted to aiding the rebels.[122] However, legislators on Capitol Hill supported the State Department, and in the summer of 1958 Congress tightened the terms under which American military assistance could be used for internal security functions in Latin America. The JCS hoped to work around these restrictions, but by the end of the year the tide had so turned in Castro's favor that lifting the arms embargo would have had little effect. On January 1, 1959, Batista fled the country, leaving it in the hands of the rebels.[123]

Castro's almost overnight rise to power ushered in a turbulent era in Cuban-American relations, leading to mutual hostility that would outlive the Cold War. Citing Castro's Marxist rhetoric and anti-American diatribes, the Joint Chiefs were inclined from the beginning to regard him as a Communist who would someday ally himself with the Soviet Union. Others, however, including key figures in the Intelligence Community, found the evidence inconclusive. Not until early 1960, when Cuba and the Soviet Union concluded a series of trade and technical support deals, was Castro's alignment with the Eastern Bloc confirmed beyond all doubt. From that point on, the United States and Castro's Cuba were in a virtual state of war.

In light of the Castro regime's hostility toward the United States and reliance on the Soviet Union, the Joint Chiefs began looking at military options, with Admiral Burke and the Navy in the forefront of advocating a forceful policy. Convinced that a Communist Cuba would be anathema to U.S. interests across the Western Hemisphere, Burke saw military action against Cuba as practically unavoidable, and in February 1960 he suggested that the JCS consider steps to topple Castro's regime. Burke envisioned three possible scenarios: unilateral overt action by the United States; multilateral overt action through the Organization of American States (OAS); and covert unilateral action.[124] The JCS agreed that Burke's suggestions merited a closer look, and by mid-March the Joint Staff had generated preliminary plans to reinforce the defenses around the U.S. naval base at Guantanamo Bay while initiating a naval blockade of Cuba and landing an invasion force of two Army airborne battle groups.[125]

Like the Joint Chiefs, President Eisenhower wanted Castro—a "little Hitler" as he called him—out of the way and was not averse to "drastic" action to achieve his goal.[126] Realizing, however, that Castro appeared as "a hero to the masses in many Latin American nations" and the "champion of the downtrodden," he feared an ugly anti-American backlash across Latin America if U.S. forces became directly involved in Castro's overthrow.[127] Alerted to Eisenhower's concerns, the CIA in January 1960 began assembling plans and supervisory personnel for covert action against Castro, using the Árbenz operation as a model. The original concept envisioned a modest venture in which a small force of Cuban expatriates would invade the island, establish a perimeter, and hold until a provisional government could declare itself and be recognized. Other guerrilla forces would intensify their operations in anticipation that these activities, coupled with unspecified U.S. pressure, would produce a mass uprising leading to Castro's ouster. At a White House conference on March 17, 1960, attended by Admiral Burke, President Eisenhower approved the CIA's plan in principle, noting that he knew of "no better plan for dealing with the situation." It was from this decision that the ill-fated Bay of Pigs operation evolved a little over a year later.[128]

Coordination and oversight for planning Castro's overthrow fell to the supersecret 5412 Committee. Composed of the President's assistant for national security affairs, Undersecretary of State, Deputy Secretary of Defense, and Director of Central Intelligence, the 5412 Committee routinely reviewed and advised on covert operations. Despite earlier discussions of including the JCS in the panel's deliberations, President Eisenhower had seen no need, apparently hoping to keep the committee's activities as closely held as possible.[129] From the outset of planning, the JCS were excluded from direct involvement in the operation. The division of labor that emerged over the summer and autumn of 1960 gave the CIA exclusive jurisdiction over organizing, training, and arming the Cuban exile force, while the Joint Chiefs concentrated on improving security around Guantanamo and in the adjacent airspace. On August 18, 1960, President Eisenhower approved approximately $13 million for the operation and sanctioned the limited use of DOD equipment and personnel for training purposes. At the same time, he reiterated his firm opposition to involving the United States in a combat role.[130]

The Joint Chiefs were finally "read into" the CIA's plans for Cuba on January 11, 1961. Now scheduled for March, the operation had grown from a limited paramilitary venture meant to arouse opposition to the Castro regime into a full-blown invasion involving a "brigade" of 600 to 750 Cuban exiles with their own air support. An ambitious enterprise, the CIA's plan had yet to identify who would take power in Cuba should the invasion succeed, or how to deal with the situation should it fail. At this point, the Joint Chiefs were convinced that a Communist Cuba would pose an intolerable situation and that a failed invasion, leaving Castro in place, would make matters worse. Persuaded that the current plan was seriously flawed, they ordered the Director of the Joint Staff to prepare an alternative course of action. Drawing on Admiral Burke's earlier plan and inputs from the Air Force, the Joint Staff recommended closer politico-military coordination and a reassessment of U.S. military support to assure the operation's tactical success.[131] With a new administration about to take office, however, and with pressure building to move ahead, it was unclear what impact the JCS proposals would have. Eisenhower had set the wheels in motion; it would be up to John F. Kennedy to make the decision to proceed.

BERLIN DANGERS

At the same time the Joint Chiefs were contemplating actions against Cuba, they faced renewed Soviet pressure on Berlin, a source of East-West friction since the city was placed under four-power rule in 1945. The most serious flare-up had been the blockade crisis of 1948–1949, which had nearly provoked a nuclear response from the United States. Since then, even though tensions had eased, the status of the city remained one of the most contentious issues of the Cold War. "Berlin," Nikita Khrushchev reportedly said, "is the testicles of the West. Every time I want to make the West scream, I squeeze on Berlin."[132]

The source of pressure this time was the Soviet Union's demand of November 27, 1958, that the Allies terminate their occupation of Berlin within 6 months and convert the city into a demilitarized zone. If not, the Soviets threatened to conclude a separate agreement with East Germany, end the occupation, and nullify allied access rights to the city. In light of the recent apparent surge in Soviet missiles and nuclear power, it looked as if the Kremlin was trying to flex its muscles and test how far it could go in using its newly found power to exact concessions. Refusing to be blackmailed, the Western powers issued a stiff diplomatic rejection and invited the Soviets to explore a peaceful resolution of the problem through negotiations.[133]

Should diplomacy fail, it would be largely up to the United States to take the lead in formulating a fall-back position. Existing preparations for a replay of the Berlin crisis centered on a set of contingency plans maintained by the U.S. European Command (USEUCOM). Derived from policies adopted in the National Security Council, these plans reflected U.S. thinking at the time that limited wars were to be avoided and that the threat of massive retaliation should be the primary deterrent to aggressive Soviet behavior. Approved by the Joint Chiefs in May 1956, USEUCOM's plans envisioned a narrow range of American and/or allied responses. Assuming that another full-scale airlift would be impractical, USEUCOM proposed to mount a limited resupply by air and initiate a test of Soviet intentions using a platoon of foot soldiers. Rather than risk a firefight that might escalate, the platoon would have orders to withdraw at the first sign of trouble.[134] But by late 1958–early 1959, the Service chiefs regarded these plans as obsolete. Basking in the success of the Lebanon operation, they saw a reemerging role for conventional forces as a means of applying pressure without threatening all-out nuclear war. Urging a policy of firmness in the current crisis, they recommended heightened security along the Autobahn into Berlin and a large-scale mobilization of conventional forces by the Western Allies to demonstrate resolve.[135]

Both the JCS Chairman, General Twining, and President Eisenhower were skeptical of this assessment and did not believe that a conventional buildup would do much to impress Soviet leaders. Both felt that it might instead inadvertently result in a confrontation that could escalate out of control. Convinced that the Service chiefs—Taylor especially—favored a buildup for budgetary reasons, the President dismissed their advice as self-serving and alarmist and told Twining to remind his JCS colleagues that they were "not responsible for high-level political decisions." Adopting a low-key approach, the President authorized limited military preparations, sufficient to be detected by Soviet intelligence but not so great as to cause public alarm, and declared his intention of relying on a combination of diplomacy and deterrence based on "our air power, our missiles, and our allies."[136]

Instead of the JCS, Eisenhower looked to General Lauris Norstad, USAF, the NATO Supreme Commander (SACEUR) since 1956, to handle further military planning. The architect of NATO's air defense system and a key figure in planning NATO's nuclear-oriented New Approach, Norstad stood very high on Eisenhower's list of talented officers. Indeed, when scandal forced his top administrative aide Sherman Adams to resign in September 1958, Eisenhower considered bringing Norstad into the White House as his chief of staff. He realized, however, that Norstad was more valuable in Europe where he enjoyed the absolute trust and confidence of the NATO allies. An ardent proponent of giving NATO its own nuclear stockpile, Norstad treated the New Approach as the first step in that direction. But he also recognized that overreliance on nuclear weapons could have drawbacks and worked assiduously throughout the Berlin crisis to develop and refine other options that would satisfy the both White House and the Joint Chiefs.[137]

Norstad's mechanism for dealing with the crisis was a tripartite (U.S.-UK-French) planning body known as "Live Oak." Established in April 1959, with offices at USEUCOM headquarters outside Paris, Live Oak reported directly to Norstad and operated on its own, separate from NATO, the Joint Chiefs, or any national command structure. Recognizing that the Federal Republic of Germany (FRG) had a major interest in the outcome, Norstad and his Live Oak staff maintained close liaison with West German military planners through the FRG's representative to SHAPE.[138] While Norstad endorsed the concept of a military buildup, he proposed confining it to a token increase of 7,000 troops.[139] Like the Chairman and the President, he was concerned that a large augmentation of allied forces would appear provocative and exacerbate tensions. Above all, he wanted the authority to coordinate the operation as he saw fit and to use minimal conventional force to keep access routes open. He repeatedly cautioned, however, that any military action had to be backed by nuclear weapons and the willingness to use them to be effective.[140]

By late summer 1959, Live Oak's planning was starting to bear fruit. Many of the measures Norstad endorsed avoided the direct use of military power and applied pressure on the Soviets through other means, including covert operations and stepped-up propaganda. Norstad wanted to divert Soviet attention from Berlin by sowing unrest and political instability in the East European satellite countries. Convinced that direct retaliatory measures would only escalate the conflict, he preferred to respond with naval operations that harassed Soviet shipping in a tit-for-tat fashion. Norstad had no doubt that sooner or later a sizable military buildup followed by an "initial probe" might be necessary to determine the extent of Soviet and/or East German resistance should traffic into Berlin be impeded. But he wanted to explore other avenues first to throw the Soviets off balance.[141]

The Service chiefs, meanwhile, continued to take an opposing view. Believing that Eisenhower and Norstad both underestimated the seriousness of the Soviet threat, they were averse to risking nuclear war without a back-up plan. Even though a nuclear confrontation might eventually prove unavoidable, they could see no better way of avoiding one than through a conventional buildup—a concrete demonstration of the West's resolve to defend its rights. But until such time as their advice carried more weight, their only choice was to bide their time and treat Norstad's recommendations as "a suitable basis" for further planning.[142]

Whether the Live Oak plans would be used remained to be seen. Letting the 6-month ultimatum deadline pass without taking action, Khrushchev accepted an invitation to visit the United States, where he and Eisenhower conferred for 2 days in September. While generally unproductive, the meeting seemed to signal a mild improvement in U.S.-Soviet relations and a cooling-off of the Berlin crisis. In January 1960, however, Khrushchev revived his threat to sign a separate peace treaty with the East Germans. A quadripartite summit meeting, held in Paris in May, ended in disarray over the U–2 incident and Khrushchev's tirade denouncing the United States for clandestine overflights of the Soviet Union. Berlin thus became one of a list of high-profile Cold War issues—others being Cuba, the smoldering

Middle East, tensions in Asia, and an escalating competition in ballistic missiles—that the Eisenhower administration passed to its successor.

At the outset of his Presidency, Eisenhower was cautiously optimistic that he could rely on the Joint Chiefs to play a major role in national security affairs, from participating in crisis management to meeting the "long haul" needs of the Cold War by developing a defense posture that would not cripple the economy. But by the end of his administration, he had practically given up using the JCS for those purposes. Increasingly, he turned elsewhere for politico-military advice and assistance that the JCS should have rendered. One side effect was to nudge the administration toward covert operations and the use of surrogates, recruited and organized by the CIA, in lieu of regular military forces and military planners. Though the chiefs had some notable successes (e.g., Lebanon), they were too few and far between to alter the overall picture. Far more typical were the fractious debates that accompanied the Joint Chiefs' deliberations on the guided missile program and related issues like nuclear targeting. The reforms of 1953 and 1958 notwithstanding, there was more dissatisfaction with the Joint Chiefs' performance by the end of the Eisenhower administration than at any time to that point in their history.

NOTES

1. The Air Force flight-tested an Atlas ICBM for the first time in June 1957, but the launch went awry and technicians had to destroy the missile after less than 1 minute in the air. On November 28, 1958, the Air Force finally conducted a fully successful Atlas test that covered a 5,500 nm range.

2. James R. Killian, Jr., *Sputnik, Scientists, and Eisenhower: A Memoir of the First Special Assistant to the President for Science and Technology* (Cambridge: MIT Press, 1977), 7.

3. David Holloway, *Stalin and the Bomb: The Soviet Union and Atomic Energy, 1939–1956* (New Haven: Yale University Press, 1994), 303–312.

4. Radford's comments in Memo of Discussion, June 4, 1954, "200th Meeting NSC, June 3, 1954," Eisenhower Papers, Ann Whitman File, NSC Series.

5. Jacob Neufeld, *The Development of Ballistic Missiles in the United States Air Force, 1945–1960* (Washington, DC: Office of Air Force History, 1990), 95–118. George F. Lemmer, "The Air Force and Strategic Deterrence, 1951–1960" (USAF Historical Division Liaison Office, December, 1967), traces the evolution of Air Force strategic conceptual planning in the 1950s.

6. Max Rosenberg, "USAF Ballistic Missiles, 1958–1959" (USAF Historical Division Liaison Office, July 1960), 2–3; Robert Frank Futrell, *Ideas, Concepts, Doctrine: Basic Thinking in the United States Air Force, 1907–1984*, 2 vols. (Maxwell AFB: Air University Press, 1989), I, 489–490.

7. Memo, Trevor Gardner to Talbot and Twining, March 11, 1954, "Intercontinental Ballistic Missile System Acceleration Plan," Nathan F. Twining Papers, Library of Congress.

8. See Peter Hofmann, "The Making of National Estimates during the Period of the 'Missile Gap,'" *Intelligence and National Security* 1 (September 1986), 336–356.

9   Thomas R. Johnson, *American Cryptology during the Cold War, 1945–1989*, Book I, *The Struggle for Centralization, 1945–1960* (Washington, DC: National Security Agency, 1995), 177 (declassified).

10 John Prados, *The Soviet Estimate: U.S. Intelligence Analysis and Russian Military Strength* (New York: Dial Press, 1982), 35–36; NIE 11-5-57, "Soviet Capabilities and Probable Programs in the Guided Missile Field," March 12, 1957, in Donald P. Steury, ed., *Intentions and Capabilities: Estimates on Soviet Strategic Forces, 1950–1983* (Washington, DC: Central Intelligence Agency, 1996), 59–62.

11 NSC Action No. 1433, September 13, 1955, *FRUS, 1955–57*, XIX, 121.

12 Robert J. Watson, *Into the Missile Age, 1956–1960* (Washington, DC: Historical Office, Office of the Secretary of Defense, 1997), 166–167.

13 See *Eisenhower Public Papers, 1955*, 303; *Eisenhower Public Papers, 1956*, 463–466; NSC 5602/1, March 15, 1956, "Basic National Security Policy," *FRUS, 1955–57*, XIX, 246–247; and Samuel P. Huntington, *The Common Defense: Strategic Programs in National Politics* (New York: Columbia University Press, 1961), 88–105.

14 Memo of Discussion, December 2, 1955, "Meeting of the National Security Council, December 1, 1955," Eisenhower Papers, Whitman File, NSC Series.

15 Memo by Goodpaster, December 15, 1956, "Meeting with SECDEF Wilson and Others, December 7, 1956"; and Goodpaster Memcon, December 20, 1956, "Conference with the President December 19, 1956," both in DDE Diary Series, Ann Whitman File, Eisenhower Papers.

16 Memo by Goodpaster, February 6, 1958, "Conference with the President, February 4, 1958," DDE Diary Series, Ann Whitman File, Eisenhower Papers.

17 Neufeld, 119–147.

18 Harvey M. Sapolsky, *The Polaris System Development: Bureaucratic and Programmatic Success in Government* (Cambridge: Harvard University Press, 1972), 21–34; I.J. Galantin, *Submarine Admiral: From Battlewagons to Ballistic Missiles* (Urbana: University of Illinois Press, 1995), 227–232; Neufeld, 182.

19 K. Condit, *JCS and National Policy, 1955–56*, 71. Advising the Secretary on Service functions was not a JCS statutory responsibility; it was an assigned function under the 1948 Key West agreement, reaffirmed in DOD Directive 5100.1, March 16, 1954, "Functions of the Armed Forces and the Joint Chiefs of Staff." Since Key West, however, Secretaries of Defense had rarely approached the JCS to help resolve roles and missions questions.

20 K. Condit, 66–72; Memo, SECDEF to AFPC, November 26, 1956, "Clarification of Roles and Missions," in Alice Cole et al., *The Department of Defense: Documents on Establishment and Organization, 1944–1978* (Washington, DC: Historical Office, Office of the Secretary of Defense, 1978), 311.

21 Harry B. Yoshpe, *Our Missing Shield: The U.S. Civil Defense Program in Historical Perspective* (Washington, DC: Federal Emergency Management Agency, 1981), 225–234.

22 Memo of Discussion, 318th Meeting NSC, April 4, 1957, *FRUS, 1955–57*, XIX, 460.

23 Morton H. Halperin, "The Gaither Committee and the Policy Process," in Thomas E. Cronin and Sanford D. Greenberg eds., *The Presidential Advisory System* (New York: Harper and Row, 1969), 185–187; David L. Snead, *The Gaither Committee, Eisenhower, and the Cold War* (Columbus: Ohio State University Press, 1999), 43–49.

24 Robert Cutler, *No Time for Rest* (Boston: Little, Brown, 1965), 354–355.

25 Prados, 69; Snead, 97.

26 Paul H. Nitze, with Ann M. Smith and Steven L. Rearden, *From Hiroshima to Glasnost: At the Center of Decision—A Memoir* (New York: Grove Weidenfeld, 1989), 166–167.

27 Security Resources Panel of the Science Advisory Committee, *Deterrence & Survival in the Nuclear Age* (Washington, DC: GPO, 1976), 14, 17, and passim.

28 Gregory W. Pedlow and Donald E. Welzenbach, *The CIA and the U–2 Program, 1954–1974* (Washington, DC: Central Intelligence Agency, 1998), 135–139.

29 Memo by Goodpaster, February 6, 1958, "Conference with the President, 4 February 1958," DDE Diary Series, Ann Whitman File, Eisenhower Papers.

30 Memo, JCS to SECDEF, December 4, 1957, "Report to the President by the Security Resources Panel of the ODM Science Advisory Committee," with appendix, JCS 2101/284.

31 Michael Mihalka, "Soviet Strategic Deception, 1955–1981," *Journal of Strategic Studies* 5 (March 1982), 40–48; Dwight D. Eisenhower, *The White House Years: Waging Peace, 1956–1961* (Garden City, NY: Doubleday, 1965), 390.

32 Peter J. Roman, *Eisenhower and the Missile Gap* (Ithaca, NY: Cornell University Press, 1995), 46–47.

33 Prados, 66–95; Lawrence Freedman, *U.S. Intelligence and the Soviet Strategic Threat* (Princeton, NJ: Princeton University Press, 1986; 2d ed.), 74–77.

34 Pedlow and Welzenbach, 144.

35 Johnson, *Struggle for Centralization*, 175–177.

36 Memo by John S.D. Eisenhower, February 10, 1959, "Conference with the President, February 9, 1959," DDE Diary Series, Ann Whitman File, Eisenhower Papers.

37 CM-407-59 to SECDEF, October 12, 1959, "Minuteman Program," 1st N/H to JCS 1620/277.

38 Watson, 363–379; Fairchild and Poole, *JCS and National Policy, 1957–60*, 45–51; Snead, 137.

39 Benson D. Adams, *Ballistic Missile Defense* (New York: American Elsevier Publishing, 1971), 22–27.

40 Memo by Goodpaster, October 11, 1957, "Conference with the President, October 11, 1957," DDE Diary Series, Ann Whitman File, Eisenhower Papers.

41 Memo, JCS to SECDEF, December 16, 1957, "Anti-Intercontinental Ballistic Missile Developments, JCS 1899/372.

42 *History of Strategic Air and Ballistic Missile Defense: Volume II, 1956–1972* (Washington, DC: Center of Military History, ca. 1975, reprint), 179–182; Adams, *Ballistic Missile Defense*, 27–28; Donald R. Baucom, *The Origins of SDI, 1944–1983* (Lawrence: University Press of Kansas, 1992), 11.

43 Appendix, "WSEG Final Report No. 30: Offensive and Defensive Weapons Systems," [July 15, 1958], 10, JCS 1620/189.

44 Watson, 379.

45   Memo, JCS to SECDEF, November 24, 1958, "Guided Missile Systems," JCS 1620/204; Memo by John S.D. Eisenhower, December 9, 1958, "Conference with the President November 28, 1958," DDE Diary Series, Ann Whitman File, Eisenhower Papers.

46   Freedman, 70.

47   Pedlow and Welzenbach, 159–163. David Robarge, *Archangel: CIA's Supersonic A–12 Reconnaissance Aircraft* (Washington, DC: Central Intelligence Agency, 2007), treats the origins and development of the A–12, precursor of the SR–71. Neither the A–12 nor the SR–71 was ever used for its intended purpose: reconnaissance over the Soviet Union. The A–12 was retired in the mid-1960s; the SR–71 continued to fly until 1997.

48   George B. Kistiakowsky, *A Scientist at the White House: The Private Diary of President Eisenhower's Special Assistant for Science and Technology* (Cambridge, MA: Harvard University Press, 1976), 219.

49   NIE 11-8-59, February 9, 1960, "Soviet Capabilities for Strategic Attack Through Mid-1964," *FRUS, 1958–60*, III, 378; Steury, *Intentions and Capabilities*, 71–107.

50   Watson, 354–355; Roy E. Licklider, "The Missile Gap Controversy," *Political Science Quarterly* 85 (December 1970), 608–609.

51   Pedlow and Welzenbach, 148.

52   Ibid., 165–177. The suspicion at the time was that the Soviets had shot down Powers' plane with a surface-to-air missile. Confirmation came in March 1963 when the U.S. air attaché in Moscow learned that that the Sverdlovsk SA–2 battery had fired a three-missile salvo which, in addition to disabling Powers' aircraft, also scored a direct hit on a Soviet fighter aircraft sent aloft to intercept the U–2.

53   Discoverer was one of two U.S. spy satellite programs at the time. The other was SAMOS (Satellite and Missile Observation System). Between October 1960 and December 1961, there were five SAMOS launchings, only two of which went into orbit. The pictures they provided were of poor quality. See Freedman, 72–73.

54   Memo by John S.D. Eisenhower, July 6, 1960, "Conference with the President: Secretary Gates, July 6, 1960," DDE Diary Series, Ann Whitman File, Eisenhower Papers.

55   Memo for the Record by Goodpaster, November 6, 1957, [Meeting of November 4, 1957, with JCS and Service Secretaries], Eisenhower Papers, Ann Whitman File, Diary Series; Eisenhower, *Waging Peace*, 245–250; Watson, 247–248.

56   CSAFM-306-57 to JCS, December 10, 1957, "Reorganization of the Department of Defense," JCS 1977/24.

57   Watson, 249–251.

58   Report by Ad Hoc Committee to JCS, January 24, 1958, "Organization of the Department of Defense," JCS 1977/26; Fairchild and Poole, 5; and Watson, 250, 252.

59   "Annual Message to the Congress on the State of the Union," January 9, 1958, *Eisenhower Public Papers, 1958*, 7–9.

60   Watson, 251–252. Gruenther served only in a part-time capacity.

61   President's Message, April 3, 1958, in Alice Cole et al., *The Department of Defense: Documents on Establishment and Organization, 1944–1978* (Washington, DC: Historical Office, Office of the Secretary of Defense, 1978), 175–186.

62   Years later, in recognition of Vinson's faithful service, the Navy named a nuclear-powered supercarrier in his honor.

63 DOD Reorganization Act of 1958, August 6, 1958 (PL 253), Cole et al., *Defense Documents*, 218. Watson, 264–275, summarizes the legislative origins of the 1958 amendments.

64 As J-3 evolved, its principal functions were to synchronize operational planning and to monitor the execution and conduct of military operations.

65 Memo by Goodpaster, August 30, 1958, "Conference with the President, August 28, 1958, Following Cabinet Meeting," DDE Diary Series, Ann Whitman File, Eisenhower Papers. See also Fairchild and Poole, 6–7.

66 JCS Chronology, "Substantive Changes to the Unified Command Plan, 1958–1969," undated, 1–2, JHO 7-0014.

67 DOD Directive 5100.1, December 31, 1958, "Functions of the Department of Defense and its Major Components," Military Documents Collection, Pentagon Library.

68 Fairchild and Poole, 51–52; David Alan Rosenberg, "The Origins of Overkill: Nuclear Weapons and American Strategy, 1945–1960," *International Security* 7 (Spring 1983), 60–61.

69 Letter, Maj. Gen. Hewitt T. Wheless to Maj. Gen. Charles B. Westover, May 12, 1959, Thomas D. White Papers, Library of Congress.

70 JCSM-171-59 to SECDEF, May 8, 1959, "Concept of Employment and Command Structure for the POLARIS Weapon System," JCS 1620/257.

71 "History of the Joint Strategic Planning Staff: Background and Preparation of SIOP–62" (History and Research Division, Headquarter, Strategic Air Command, n.d.), 7 (declassified).

72 CM-380-59 to SECDEF, August 17, 1959, "Target Coordination and Associated Problems," JCS 2056/131.

73 Watson, 479. In 1960, to fill the gap, the Joint Chiefs established the Joint War Games Agency which operated outside the Joint Staff. In 1968 it became part of J-5 and in 1970 it merged with the Chairman's Special Studies Group to form the Studies, Analysis, and Gaming Agency (SAGA), reconstituted as the Joint Analysis Directorate (JAD) in 1984.

74 Net Evaluation Subcommittee, National Security Council, "Appraisal of Relative Merits, from the Point of View of Effective Deterrence, of Alternative Retaliation Efforts," October 30, 1959, Enclosure to JCS 2056/145; Fairchild and Poole, 52.

75 Rosenberg, "Origins of Overkill," 4–5, 61; Watson, 485–490.

76 Fairchild and Poole, 52–53; Memo by John S.D. Eisenhower, July 6, 1960, "Conference with the President: Secretary Gates, July 6, 1960," Eisenhower Papers, Ann Whitman File, Diary Series.

77 Memo by Goodpaster, August 13, 1960, "Conference with the President, August 11, 1960," DDE Diary Series, Ann Whitman File, Eisenhower Papers; Rosenberg, "Origins of Overkill," 3–4.

78 "History of the JSTPS and SIOP–62," 14. Considered a separate entity, the JSTPS did not fall under the personnel ceiling that governed the size of the Joint Staff.

79 Memo, SECDEF to CJCS, August 16, 1960, "Target Coordination and Associated Problems," Enclosure A to JCS 2056/164.

80 SM-810-60 to DSTP et al., August 19, 1960, "Implementation of Strategic Targeting and Attack Policy," Enclosure C to JCS 2056/165. See also Desmond Ball, "Development of the SIOP, 1960–1983," in Desmond Ball and Jeffrey Richelson, eds., *Strategic Nuclear Targeting* (Ithaca, NY: Cornell University Press, 1986), 58–61.

81 See "Briefing for the President by CJCS on the Joint Chiefs of Staff Single Integrated Operational Plan 1962 (SIOP-62)," September 13, 1961, JCS 2056/281 (sanitized).

82 Memo by Goodpaster, December 1, 1960, "Conference with the President, November 25, 1960," DDE Diary Series, Ann Whitman File, Eisenhower Papers.

83 Herbert F. York, *Making Weapons, Talking Peace* (New York: Basic Books, 1987), 185.

84 Goodpaster Memcon, August 13, 1960, loc. cit.

85 Michael A. Palmer, *Guardians of the Gulf: A History of America's Expanding Role in the Persian Gulf, 1833–1992* (New York: Free Press, 1992), 46–49. Known as the Middle East Force (MEF), the U.S. flotilla operated out of a British base in Bahrain.

86 Fairchild and Poole, 167.

87 K. Condit, 151–160.

88 Report by JSPC to JCS, June 29, 1956, "Force Requirements for Defense of the Baghdad Pact Area," JCS 1887/220.

89 Memo of Discussion, 231st Meeting of the NSC, January 13, 1955, *FRUS, 1955–57*, XII, 687.

90 Stephen E. Ambrose, *Eisenhower: The President* (New York: Simon and Schuster, 1984), 331; Memo by Goodpaster of Conference with the President, July 31, 1956, *FRUS, 1955–57*, XVI, 64.

91 Memo of Discussion at 292d Meeting of the NSC, August 9, 1956, *FRUS, 1955–57*, XVI, 174.

92 K. Condit, 169–174.

93 Cole C. Kingseed, *Eisenhower and the Suez Crisis of 1956* (Baton Rouge: Louisiana State University Press, 1995), 47.

94 Memo, JCS to SECDEF, July 31, 1956, "Nationalization of the Suez Maritime Canal Company by the Egyptian Government," JCS 2105/38; Memo, JCS to SECDEF, August 3, 1956, "Nationalization of the Suez Canal; Consequences and Possible Related Actions," JCS 2105/39.

95 Letter, Eisenhower to Anthony Eden, September 8, 1956, in Eisenhower, *Waging Peace*, 669–671.

96 Memo, Dulles to Eisenhower, March 28, 1956, "Near Eastern Policies," *FRUS, 1955–57*, XV, 419–421; Diary entry, March 28, 1956, in Robert H. Ferrell, ed., *The Eisenhower Diaries* (New York: W.W. Norton, 1981), 323–324.

97 See *FRUS, 1955–57*, XV, 421–424.

98 W. Scott Lucas, *Divided We Stand: Britain, the U.S. and the Suez Crisis* (London: Hodder & Stoughton, 1991), 109–113 and passim; Diary entry, March 12, 1956, in Evelyn Shuckburgh, *Descent to Suez: Diaries, 1951–56* (London: Weidenfeld and Nicolson, 1986), 346.

99 Lucas, 160.

100 Ambrose, 353; Eisenhower, *Waging Peace*, 82.

101 Pedlow and Welzenbach, 112–117.

102 Ricky-Dale Calhoun, "The Musketeer's Cloak: Strategic Deception During the Suez Crisis of 1956," *Studies in Intelligence* 51, no. 2, 47–58 (unclassified edition); K. Condit, 185.

103 Watson, 60–61; K. Condit, 186–188.

104 Christopher Andrew, *For the President's Eyes Only* (New York: HarperPerennial, 1996), 227.

105 K. Condit, 189.

106 "Special Message to the Congress on the Situation in the Middle East," January 5, 1957, *Eisenhower Public Papers, 1957,* 6–16.

107 Ambrose, 381–383; Fairchild and Poole, 135–136.

108 Report by Joint Middle East Planning Committee, "Strategic Guidance and Concept for the Development and Employment of the Middle East Baghdad Pact Forces," October 18, 1957, and decision on October 23, 1957, JCS 2268/6.

109 CINCNELM, "Annual Report 1 January 1959 to 31 December, 1959," 2–4, copy in JHO Collection, summarizes the scale and scope of planned operations. CINCNELM was the Navy component commander to USCINCEUR; he functioned also as Commander in Chief, U.S. Specified Command, Middle East (CINCUSSPECOMME). In February 1960 CINCNELM became Commander in Chief U.S. Naval Forces Europe (CINCNAVEUR) and CINCSPECOMME was discontinued.

110 "The U.S. Army Task Force in Lebanon" (Headquarters, U.S. Army, Europe, 1959), 9.

111 Memo, CNO to JCS, November 7, 1957, "Draft CINCSPECOMME Operation Plan 215-58," JCS 2034/34; CSAM 183-57 to JCS, November 14, 1957, same subject, JCS 2034/36; CSAFM 291-57 to JCS, November 20, 1957, same subject, JCS 2034/38.

112 Report by JSPC to JCS, January 31, 1958, "Military Planning for the Middle East," and decision on February 5, 1958, JCS 1887/433; Message, JCS to USCINCEUR et al., March 22, 1958, enclosure A to JCS 1887/438; Fairchild and Poole, 140–142.

113 Memo of Conference with the President, July 14, 1958, *FRUS, 1958–60,* XI, 225.

114 H.H. Lumpkin, "Operation Blue Bat" (Paper Prepared by Office of Command Historian, USEUCOM, November 4, 1958), 2–3, JHO Collection.

115 Holloway's son, James L. Holloway III, also participated in the Lebanon operation as a naval aviator. Later, from 1974 to 1978, he served as Chief of Naval Operations and a member of the JCS.

116 Fairchild and Poole, 156–158; Bernard C. Nalty, "The Air Force Role in Five Crises, 1958–1965," USAF Historical Division Liaison Office, June 1968, 13.

117 Ambrose, 472.

118 Eisenhower, *Mandate for Change,* 421–427; Richard H. Immerman, "Guatemala as Cold War History," *Political Science Quarterly* 95 (Winter 1980–81), 629–653.

119 NSC 5432/1, September 3, 1954, "United States Objectives and Courses of Action With Respect to Latin America," *FRUS, 1952–54,* IV, 81–88.

120 *FRUS, 1955–57,* VI, 266–268, 270, 294–295, 297–298; *FRUS, 1958–60,* VI, 117–119.

121 Jules R. Benjamin, *The United States and the Origins of the Cuban Revolution* (Princeton, NJ: Princeton University Press, 1990), 151–153.

122 Memo, CNO to JCS, April 19, 1958, "Suspension of Delivery of Military Equipment and Military Sales to Cuba," JCS 1976/243; Memo of Discussion at DoS–JCS Meeting, June 27, 1958, *FRUS, 2958–60*, VI, 118.

123 Fairchild and Poole, 179–180.

124 Memo, CNO to JCS February 19, 1960, "U.S. Action in Cuba," JCS 2304/2.

125 Memo of Discussion 437th Meeting of the NSC, March 17, 1960, *FRUS, 1958–60*, VI, 857; Watson, 763–764.

126 Ferrell, ed., *Eisenhower Diaries*, 379; Andrew, *For the President's Eyes Only*, 252.

127 Eisenhower, *Waging Peace*, 525.

128 Goodpaster Memo of Conference with the President, March 17, 1960, *FRUS, 1958–60*, VI, 861–863. For the approved plan, see 5412 Committee Paper, March 16, 1960, "A Program of Covert Action Against the Castro Regime," ibid., 850–851.

129 Ambrose, 507.

130 Memo of Meeting with the President, August 18, 1960, *FRUS, 1958–60*, VI, 1057–1060; Watson, 764.

131 Report by Director, Joint Staff to JCS, January 24, 1961, "U.S. Plan of Action in Cuba," JCS 2304/19.

132 Khrushchev quoted in Gaddis, *We Now Know*, 140.

133 Jack M. Schick, *The Berlin Crisis, 1958–1962* (Philadelphia: University of Pennsylvania Press, 1971), 10–27 and passim.

134 Fairchild and Poole, 125–126.

135 See JCSM-16-59 to SECDEF, January 13, 1959, "Berlin Situation," JCS 1907/162; JCSM-82-59 to SECDEF, March 11, 1959, "U.S. Position on Berlin," JCS 1907/175; Talking Paper Approved by JCS for Conference on March 14, 1959, JCS 1907/179; and JCSM-93-59 to SECDEF, March 16, 1959, "Statement of Policy on Berlin and Germany," JCS 1907/180. See also Fairchild and Poole, 126–129.

136 Eisenhower, *Waging Peace*, 341; Memo by John S.D. Eisenhower of Conference with the President, March 9, 1959, quoted in Ambrose, *Eisenhower the President*, 515–516. See also Fairchild and Poole, 128.

137 Eisenhower, *Waging Peace*, 318. Robert S. Jordan, *Norstad: Cold War NATO Supreme Commander: Airman, Strategist, Diplomat* (New York: St. Martin's Press, 2000), 103–136.

138 H.H. Lumpkin, "Live Oak: A Study in Thirty Months of Combined Planning: February 1959–September 1961" (Study prepared at Headquarters U.S. European Command, August 19, 1961), 25–26; Jordan, *Norstad*, 157.

139 *FRUS, 1958–60*, VIII, 423.

140 Jordan, 155–166.

141 Live Oak Planning Group Paper, July 24, 1959, "Berlin Contingency Planning: More Elaborate Military Measures," JCS 1907/231.

142 JCSM-355-59 to SECDEF, August 21, 1959, "Berlin Contingency Planning: More Elaborate Military Measures," JCS 1907/237.

Confrontation in Berlin, 1961

Chapter 7

KENNEDY AND THE CRISIS PRESIDENCY

For an organization that did not adapt easily to change, John F. Kennedy's Presidency was one of the most formidable challenges ever to face the Joint Chiefs of Staff. Representing youth, enthusiasm, and fresh ideas, Kennedy entered the White House in January 1961 committed to blazing a "New Frontier" in science, space, and the "unresolved problems of peace and war."[1] As a Senator and Presidential candidate, Kennedy had been highly critical of the Eisenhower administration's defense program, faulting it for allowing the country to lag behind the Soviet Union in missile development and for failing to develop a credible conventional alternative to nuclear war. "We have been driving ourselves into a corner," Kennedy insisted, "where the only choice is all or nothing at all, world devastation or submission—a choice that necessarily causes us to hesitate on the brink and leaves the initiative in the hands of our enemies."[2] Instead of threatening an all-out nuclear response, Kennedy advocated graduated levels of conflict tailored to the needs of the situation and the degree of provocation, in line with the "flexible response" doctrine put forward by retired General Maxwell Taylor, former Secretary of State Dean Acheson, and others.

Refining and implementing the President's concepts fell mainly to the new Secretary of Defense, Robert S. McNamara, who served both Kennedy and his successor, Lyndon B. Johnson. President of the Ford Motor Company before coming to Washington, McNamara had no prior experience in defense affairs other than his service as a "statistical control" officer in the Army Air Forces during World War II. Applying an active management style, McNamara soon became famous for his aggressive, centralized administrative methods and sophisticated approach toward evaluating military programs and requirements. To assist him, McNamara installed a management team that mixed experienced officials with younger "whiz kids" adept at "systems analysis," a relatively new science based on complex, computerized quantitative models. The net effect by the time McNamara stepped down in 1968 was a veritable revolution in defense management and acquisition and an unprecedented degree of civilian intrusion into military planning and decision-making. "I'm here to originate and stimulate new ideas and programs," McNamara declared, "not just to referee arguments and harmonize interests."[3]

As the McNamara revolution unfolded, the Joint Chiefs looked on with a mixture of awe and apprehension. Made up initially of holdovers appointed by Eisenhower, the JCS were generally older than McNamara and his entourage and skeptical of making abrupt changes to practices and procedures built on years of experience, painstaking compromise, and meticulous planning. To the incoming Kennedy administration, the JCS seemed overly cautious, tradition-bound, and impervious to new ideas. Inclined to give McNamara the benefit of the doubt, Chief of Naval Operations (CNO) Admiral Arleigh A. Burke at first lauded the Secretary's "sharp, decisive" style and expected him to be "extremely good." By the time he retired as CNO in August 1961, however, Burke saw McNamara and the JCS as working at cross purposes. Air Force Chief of Staff General Thomas D. White agreed. In White's opinion, McNamara and his staff were "amateurs" who had little or no appreciation of military affairs. Most uneasy of all was the Chairman of the Joint Chiefs, General Lyman L. Lemnitzer, USA, an Eisenhower appointee steeped in "old school" ways. Though McNamara promised not to act on important matters without consulting his military advisors, he offered no assurances that he would heed their views. All too often, Lemnitzer recalled, the JCS would deliberate "long and hard" to resolve a problem and reach a consensus, only to have McNamara turn their recommendations over to a systems analysis team "with no military experience" to reshape their advice.[4]

In addition to their difficulties with McNamara, the JCS faced an uphill struggle to retain influence at the White House. Believing the National Security Council system had become unwieldy and unresponsive under Eisenhower, Kennedy opted for a simplified organization and a streamlined NSC Staff with enhanced powers. The principal architect of the new system was Kennedy's assistant for national security affairs McGeorge Bundy, who believed that simplified methods would give the President a broader range of views. "[T]he more advice you get," he assured the President, "the better off you will be."[5] Soon to go were the Planning Board, the Operations Coordinating Board, and the other support machinery created by Eisenhower that had given the JCS direct and continuous access to the top echelons of the policy process. As one sign of their diminished role, the Joint Chiefs closed their office of special assistant for national security affairs, which they had maintained in the White House since the early 1950s, and conducted business with the NSC through a small liaison office located next door in the Old Executive Office Building.[6]

Under Kennedy, the NSC became a shadow of its former self. Cutting staff by one-third, he abandoned the practice of developing broad, long-range policies in the NSC and used it primarily for addressing current problems and crisis management. Meetings followed an irregular schedule and were informal compared with the two previous administrations. In addition to the statutory members, regular participants at NSC meetings came to include the President's brother, Attorney General Robert F. Kennedy, and White House political consultant Theodore C. Sorensen. By law, the JCS remained advisors to the council, but under the new structure and procedures they were further removed than ever from the President's "inner circle." Still, whatever problems or weaknesses Kennedy's deconstruction of the NSC may have introduced, there was no rush to correct them under the succeeding Johnson administration, which seemed content with the status quo.[7]

A further blow to the Joint Chiefs' influence was Kennedy's decision in the aftermath of the Bay of Pigs fiasco in April 1961 to give retired Army Chief of Staff General Maxwell D. Taylor an office in the White House as the President's Military Representative (MILREP). The President originally had Taylor in mind to succeed Allen Dulles as Director of Central Intelligence, but after the Bay of Pigs embarrassment, Kennedy wanted an experienced military advisor close at hand to avoid another "dumb mistake."[8] Taylor's position was analogous in some ways to Admiral Leahy's during World War II, though Taylor did not participate in the Joint Chiefs' deliberations or represent their views. Upon taking the job as the President's MILREP, Taylor assured Lemnitzer that he did not intend to act as a White House "roadblock" to JCS recommendations.[9] His assigned tasks were to provide the President with an alternative source of military advice, to review recommendations from the Pentagon before they went to the Oval Office, and to serve as the President's liaison for covert operations.[10] Taylor's appointment actually worked out better than the JCS expected because they now had someone between McNamara and the President. According to Henry E. Glass, who served as special assistant to the Secretary of Defense, McNamara resented having his advice second-guessed and eventually persuaded Kennedy that Taylor would be more valuable at the Pentagon as Chairman of the Joint Chiefs (where McNamara would have control over him) than at the White House. Arthur M. Schlesinger, Jr., historian-in-residence at the White House, had a different view. He characterized Taylor's appointment as a temporary measure until General Lemnitzer's term expired and Kennedy could move Taylor to the Pentagon as CJCS. In any event, on October 1, 1962, Taylor became Chairman, replacing Lemnitzer, who went to Europe as NATO Supreme Commander.[11]

THE BAY OF PIGS

Kennedy's early months in office were the formative period in his relationship with the Joint Chiefs and left an indelible impression on all involved. His primary aim in defense policy was to move away from Eisenhower's heavy reliance on nuclear weapons by developing a more balanced and flexible force posture. Most of the JCS at the time—the Army and Navy especially—agreed with Kennedy's basic objective and welcomed his efforts to make changes. However, the JCS soon found McNamara's methods of carrying out the President's orders heavy-handed and counterproductive to the development of smooth and efficient civil-military relations. Efforts to convince McNamara and his staff that it would take time and patience to implement the changes the President wanted initially met with strong quizzical objections. The honeymoon between the administration and the JCS was brief. Rumors of growing tensions and discontent at the Pentagon surfaced within weeks after the inauguration.

No episode more aptly captured these difficulties of adjustment than the Joint Chiefs' role in the Bay of Pigs operation, the ill-fated attempt by the Central Intelligence Agency (CIA), using Cuban expatriates, to invade Cuba and overthrow Fidel Castro in the spring of 1961. By the time Kennedy took office, the Bay of Pigs invasion had been in gestation for nearly a year, though few outside the CIA knew of the program's existence. Not until early January 1961 did the Joint Chiefs officially became privy to the details, though even then, by Admiral Burke's account, they were "kept pretty ignorant" and told only "partial truths." All the same, what the CIA revealed of its preparations up to that point was far from reassuring and left the Joint Chiefs and their special operations staff decidedly uneasy over achieving stated goals.[12]

Similar misgivings had raced through President-elect Kennedy's mind when he first learned of the operation during a CIA briefing on November 18, 1960.[13] On the eve of the inauguration, realizing that Kennedy had doubts, Eisenhower assured him that nothing was firm and that it would be up to the new administration to decide whether to proceed. Taking Eisenhower at his word, Kennedy gave the matter top priority and during his early days in office he held a round of meetings with the Director of Central Intelligence, Allen Dulles, the Joint Chiefs, and other senior advisors to examine the details and explore options. Much to the President's surprise, the CIA described plans and preparations that were substantially farther along than Eisenhower had let on, leaving the distinct impression that it might be too late to turn back.[14]

Indications are that, at this stage, Kennedy looked to the Joint Chiefs to provide him with ongoing analysis of the invasion plans and to apply a brake on any ill-conceived actions by the CIA. With the new administration still organizing itself, Kennedy had practically nowhere to turn other than the JCS for the professional expertise and insights he needed. Somewhere along the way, however, lines of communication broke down. Having had limited involvement in the operation from its inception and knowing only what the CIA chose to disclose to them about the invasion force, the Joint Chiefs were uncomfortable offering much more than a general assessment. Weighting one thing against another, Joint Staff planners (J-5) rated the chances of success as "very doubtful."[15] But in their formal submission to the Secretary of Defense and the

President, the JCS appeared to offer a more upbeat evaluation and suggested that the operation as originally conceived stood a "fair chance of ultimate success." The chiefs neglected to mention, however, that "fair chance" meant one in three.[16]

Perhaps sensing his military advisors' uneasiness over the operation, Kennedy continued to evince misgivings. The initial plan presented to him by the CIA called for the exiles to land in force near the town of Trinidad, a popular seaside resort on Cuba's south-central coast. But at the State Department's urging, the President agreed to tone down the operation lest it provoke adverse reactions in Latin America and the United Nations. Terming the Trinidad plan too "spectacular," he directed the CIA to find a "quiet" site for the landing. The upshot was the selection of the Bay of Pigs, a swampy but relatively secluded area in Cuba's Zapata region to the west.[17] After examining the amended plan, Admiral Burke upped the odds for success slightly and told the President he thought they were about fifty-fifty. Burke, however, was offering a personal opinion. Later, Kennedy complained that the JCS had let him down by not giving him better warning of the risks and pitfalls.[18]

The landing, which took place on April 17, 1961, was probably doomed before the invaders hit the shore. Inadequately equipped, ill-trained, and ineptly led, the 1,400 Cuban expatriates in the invasion force were no match for Castro's larger veteran army. Poorly coordinated air attacks launched from bases in Central America failed to suppress the Cuban air force. The action was over in 3 days. Whether a more hospitable landing site and/or stronger air support would have changed the outcome is a matter of conjecture. The Joint Chiefs had taken a dim view of moving the landing from Trinidad to the Bay of Pigs and had considered effective air support the key to the entire operation. But they had never pressed their views in the face of the President's obvious determination to minimize overt U.S. involvement. Nor had McNamara, still new to dealing with the military, insisted that the JCS be more forthcoming and specific. Never again would he hesitate to second-guess the chiefs or to offer an opinion on their advice.

To sort out what went wrong, President Kennedy persuaded General Taylor to oversee an investigation. Assisting him were Attorney General Kennedy, Director of Central Intelligence Dulles, and Chief of Naval Operations Admiral Burke. During the inquiry, General Taylor and Robert Kennedy developed a close and lasting friendship. Taylor and the study group took extensive testimony from those who had been in on the planning and decisionmaking. On June 13, 1961, they presented their findings to the President. Written almost exclusively by Taylor, the group's final report took the Joint Chiefs to task for not critiquing the CIA's plan more closely and for not being more forthcoming in offering the President options. "Piecing all the evidence together," Taylor recalled, "we concluded that whatever reservations the Chiefs had about the Zapata plan . . . they never expressed their concern to the President in such a way as to lead him to consider seriously a cancellation of the enterprise or the alternative of backing it up with U.S. forces."[19]

Despite the study group's findings, Kennedy never publicly blamed anyone other than himself for the debacle. Seeking to avoid similar incidents, he told the chiefs that in the future he expected them to provide "direct and unfiltered" advice and to act like "more than military men."[20] All the same, it was Taylor's impression that the whole experience "hung like a cloud" over Kennedy's relations with the JCS. Attempting to clear the air, Kennedy met with them in the Pentagon on May 27, 1961. Though no detailed records of the meeting survived, Kennedy at one point apparently lectured the chiefs on their responsibility for providing him with unalloyed advice, drawing on a paper Taylor wrote earlier. But the response he got was "icy silence."[21] Henceforth, Kennedy remained respectful but skeptical of JCS advice. "They always give you their bullshit about their instant reaction and their split-second timing," he later remarked, "but it never works out. No wonder it's so hard to win a war."[22]

BERLIN UNDER SIEGE

No sooner had the fallout from the Bay of Pigs begun to settle when a more ominous crisis arose over access rights to Berlin. Kennedy knew that the city was a frequent flashpoint and had named former Secretary of State Dean Acheson as his special advisor on NATO affairs in February 1961, with the Berlin question part of his mandate.[23] Existing plans for defending Western access rights to the city rested on NATO doctrine of the 1950s, stressing the early use of nuclear weapons, and bore the strong imprint of the NATO Supreme Commander General Lauris Norstad, USAF, a leading proponent of deterrence through the threat of massive retaliation. In a preliminary assessment that reached the Oval Office in early April 1961, Acheson dismissed these plans as dangerous and ineffectual and urged Kennedy to call the Soviets' bluff by pursuing a combination of diplomatic initiatives and nonnuclear military options that involved, among other things, sending a heavily armed convoy down the Autobahn to Berlin.[24]

The Joint Chiefs recommended a more cautious response. Given the limitations of U.S. conventional forces at the time, they would not rule out possible recourse to nuclear weapons if the crisis escalated, though as a practical matter they seemed to feel that with skillful diplomacy the situation need not go that far. Treating Acheson's proposals as overly provocative, they assured the President that they had already explored the convoy idea and similar military actions and had reached the conclusion that the use of substantial ground forces "even if adequately supported by air is not militarily feasible." A smaller probe, they argued, would serve just as well as a test of

Soviet intentions and would be far less confrontational than a heavily armed convoy. Two years earlier, with strengthening deterrence their main objective, the JCS had recommended a large-scale conventional buildup in Europe, both to impress the Soviets with the West's resolve and to be better prepared if a showdown did occur. This continued to be the Joint Chiefs' preferred approach to addressing the crisis.[25]

To be effective, the Joint Chiefs' recommended strategy would have required a mobilization of forces, increased defense spending, and an acceptance that, should all else fail, recourse to nuclear weapons might be unavoidable. As yet, President Kennedy was unprepared to go quite that far. With memories of the Bay of Pigs still fresh, he was doubly cautious in listening to JCS advice or endorsing a course of military action. But after his disastrous Vienna summit meeting with Khrushchev in early June, he steadily revised his thinking. Hoping the Vienna meeting would lay the groundwork for a peaceful settlement, Kennedy was instead taken aback by Khrushchev's bullying and refusal to engage in serious negotiations. When Khrushchev finished brow-beating Kennedy, he placed another ultimatum on the table, threatening to sign a treaty with the East Germans by the end of the year. "I've got a terrible problem," Kennedy observed afterwards. "If [Khrushchev] thinks I'm inexperienced and have no guts, until we remove those ideas we won't get anywhere with him."[26] In late June 1961, convinced that a showdown was coming, Kennedy created an interdepartmental Berlin Task Force to coordinate overall policy and directed McNamara to take a closer look at military preparations to counter Khrushchev's ultimatum.[27]

The ensuing review confirmed that the United States had yet to achieve a credible flexible-response force posture. In a rough estimate of requirements, the Joint Chiefs recommended a supplemental appropriation of $18 billion, mobilization of Reserve units, and an increase in the size of the Armed Forces by 860,000. Yet even with these increases in strength, "main reliance" would still come down to a nuclear response.[28] Meeting with McNamara, Lemnitzer, Taylor, and Secretary of State Dean Rusk at his Hyannis Port home on July 8, Kennedy declared the chiefs' recommendations to be unacceptable and said he wanted a "political program" backed by enhanced conventional military power "on a scale sufficient both to indicate our determination and to provide the communists time for second thoughts and negotiation."[29] With the President's goals further clarified, the Joint Staff assembled revised estimates that became part of the discussion at a series of ad hoc meetings involving McNamara, Acheson, Rusk, and senior White House staff.[30] The upshot was the President's nationally televised speech on July 25 warning of grave dangers over Berlin and calling for a supplemental budget increase of $3.2 billion to augment the Armed Forces by 217,000, with most of the increase in ground troops.[31]

Congress acted quickly to give the President practically everything he wanted. But the need to develop an agreed position with U.S. allies and problems associated with mobilizing the Reserves posed unexpected delays. Moreover, from conversations between McNamara and Norstad in Paris in late July, it was clear that SACEUR lacked a workable plan for assuring access to Berlin using solely or even primarily conventional forces.[32] Until these problems were resolved, the administration had no choice but to fall back on the nuclear-oriented posture it inherited from Eisenhower, a decision that became almost automatic after the increase in tensions precipitated by erection of the Berlin Wall on August 13. Seeing the wall as a major escalation of the crisis, Kennedy resolved to meet the challenge head on, telling McNamara that the time had come to adopt "a harder military posture."[33]

Behind the President's decision to toughen his stance over Berlin was a growing body of credible evidence debunking the missile gap and the artificial constraints it imposed on the administration's behavior. As early as February, a skeptical McNamara had acknowledged that the missile gap was probably more myth than reality during a background briefing for reporters. But he retracted his statement under pressure from the White House.[34] Based on information provided by Colonel Oleg Penkovskiy, the CIA's "mole" inside the Soviet General Staff, and photos from the Discoverer satellite program, it became apparent over the summer that earlier intelligence estimates had overstated Soviet long-range missile capabilities and that the United States retained overall strategic nuclear superiority. Though Kennedy refused to treat the new evidence as conclusive, there was no denying that the gap, if it existed at all, was far less extreme than previously assumed.[35]

On September 13, 1961, the Joint Chiefs gave President Kennedy his first formal briefing on SIOP-62, the current war plan for strategic bombardment of the Sino-Soviet bloc. Afterwards proclaiming the plan to be overly rigid, he ordered changes (already initiated by McNamara) that would allow greater choice in the selection of targets and the timing and sequence of attacks.[36] At the same time, however, knowing that the United States retained the edge in strategic power, Kennedy and key aides adopted a significantly tougher line toward the Berlin crisis, both to reassure U.S. allies and to pressure the Soviets. Thus, in the weeks following the SIOP briefing, McNamara, Rusk, and Deputy Secretary of Defense Roswell L. Gilpatric all made high-profile public appearances in support of administration policy. Echoing the policies of the previous 8 years, they reaffirmed the President's determination to stand fast and their certainty that the United States had the resources to prevail. "Our nuclear stockpile," McNamara confirmed, "is several times that of the Soviet Union and we will use either tactical weapons or strategic weapons in whatever quantities wherever, whenever it's necessary to protect this nation and its interests."[37]

The blueprint for carrying out these declarations was National Security Decision Memorandum (NSDM) 109, a compendium of phased responses for the defense of Western rights to Berlin, also known as the "poodle blanket" paper. The first three phases involved pressure through diplomatic channels, economic sanctions, and maritime harassment, followed by or in conjunction with military pressures and escalation to the full use of nuclear weapons. Adopted by the NSC in late October 1961, NSDM 109 was largely the product of the Office of International Security Affairs in OSD, headed by Assistant Secretary of Defense Paul H. Nitze. "In case one response failed," Nitze recalled, "we would go to the next and then the next, and so on." Many of the proposed measures in the "preferred sequence," such as the use of diplomatic protests, small unit probes, and the coercion of Soviet shipping in retaliation for obstruction of access to Berlin, resembled the options compiled the year before by Norstad's Live Oak planners in Paris. But as far as Nitze and his staff were concerned, Live Oak had barely scratched the surface. Early drafts of NSDM 109 listed so many possible courses of action that the joke around Nitze's office was that it would take a piece of paper the size of a horse blanket to list them all. A condensed version reduced the horse blanket to the size of a "poodle blanket." Hence the paper's nickname.[38]

Although the Joint Chiefs belatedly offered their own "preferred sequence" paper, it was almost entirely oriented toward military sanctions and too detailed, in the opinion of Deputy Secretary Gilpatric, to serve as policy guidance.[39] Moreover, during a meeting with the President on October 20, it slipped out that the JCS had yet to reach full agreement on how their preferred sequence plan should be implemented. In a scene reminiscent of their internal quarrels over Laos (see below), Lemnitzer and Army Chief of Staff General George H. Decker wanted to move quickly with the deployment of forces once mobilization reached its peak, while Air Force Chief of Staff General Curtis LeMay and Chief of Naval Operations Admiral George W. Anderson, Jr., urged patience and delay. Assured by McNamara that a final decision need not be taken until November, Kennedy politely shrugged off the matter as an honest difference of professional opinion.[40]

A week later, on October 27–28, the crisis peaked with the dramatic confrontation between U.S. and Soviet tanks at "Checkpoint Charlie," a key transit point between the Soviet and U.S. sectors in Berlin. Anticipating trouble, the Joint Chiefs had taken steps to bolster the city's garrison but had warned the President that there was little chance allied forces could hold against a determined Soviet attack. Taylor agreed, describing it as "a hell of a bad idea" to try to defend the city.[41] Despite the face-off, however, neither side seemed eager for a fight and the incident ended peacefully, with Soviet tanks the first to withdraw. From that point on, though the wall remained, tensions gradually relaxed.

Exactly why the Soviets backed down may never be known. But thanks to the limited opening of Soviet and East European archives following the end of the Cold War, the explanation that suggests itself is that the Warsaw Pact high command lacked confidence in the ability of its forces to prevail in a showdown. On September 25, 1961, with the crisis gathering momentum, the Warsaw Pact announced that over the next few weeks it would conduct a command post exercise called BURIA. The Warsaw Pact's largest exercise to date, BURIA simulated a military conflict arising from ongoing tensions over Berlin and tested the Eastern Bloc's ability to conduct unified operations. With the exercise under way, the CIA assessed BURIA's purpose as two-fold: to convince the West of the Soviet bloc's military strength, readiness, and determination in the current crisis, and to increase pressure on the West to make concessions or to acquiesce to Communist demands.[42]

BURIA lasted from September 28 to October 10, 1961, and proved a disappointment to the Warsaw Pact high command. Once fighting erupted, the Soviets and their East European allies were supposed to shift quickly from a defensive to an offensive posture. Using tactical nuclear weapons and fast-moving tank divisions to spearhead the assault, Warsaw Pact forces planned to smash through NATO defenses and occupy Paris within a fortnight. But as the exercise unfolded, it encountered unexpected command and control, mobilization, transportation, and logistical problems. Assuming nuclear retaliation by the West, Soviet army doctors reckoned a 50 percent loss of strength in front line units. A shortage of interpreters and faulty radio equipment crippled coordination among East German, Soviet, Polish, and Czech commanders. Communications between land and sea forces off the north German coast were practically nonexistent. Soviet maps provided to East European forces proved largely useless because they were written in Russian.[43]

Whether Kennedy and the Joint Chiefs paid much attention to BURIA is unclear. Even though Western intelligence monitored the exercise, there are few references to it in subsequent estimates. Still, those in Washington with access to the intelligence on BURIA knew that Warsaw Pact forces were poorly organized and in a relatively weak position to risk a military confrontation with the West. About their only option would have been to use nuclear weapons, a dangerous course that the Joint Chiefs expected the Soviets to avoid unless they felt seriously threatened. Precipitating a nuclear conflict was never the U.S. intention in any event. Despite their differences over the scale and scope of the Western military buildup, Kennedy and the JCS agreed that its fundamental purpose was to pressure the Soviets into respecting the status quo. With the exception of the Berlin Wall, which remained in place for nearly three decades, they by and large succeeded. "It's not a very nice solution," Kennedy conceded, "but a wall is a hell of a lot better than a war."[44]

LAOS

At the same time President Kennedy and the Joint Chiefs were wrestling with Soviet threats to Berlin, there loomed an equally grave crisis on the other side of the world, in the small, remote kingdom of Laos, formerly part of French Indochina. Like Cuba and Berlin, Laos was another of the unresolved problems passed from Eisenhower to Kennedy. At issue was a steadily escalating political and military conflict between the Communist Pathet Lao, supported by neighboring North Vietnam, Communist China, and the Soviet Union, and the U.S.-backed Royal Lao Government (RLG) dominated by General Phoumi Nosavan. By the beginning of 1961, the two sides were locked in a see-saw battle for control of the Laotian administrative capital of Vientiane. In alerting Kennedy to the situation as he was leaving office, Eisenhower warned of larger implications: "If Laos is lost to the Free World, in the long run we will lose all of Southeast Asia." By comparison, the gathering conflict in neighboring South Vietnam was a mere sideshow.[45]

The Joint Chiefs initially advised the incoming administration to do all it could to keep Laos from going Communist, up to and including unilateral U.S. intervention with "sizable" military forces.[46] Even though the Laotian army (Forces Armées de Laos, or FAL) had seldom made effective use of U.S. assistance, Kennedy agreed to consider increasing American help. But he strongly opposed the go-it-alone approach and leaned toward a negotiated settlement that would neutralize the country under a coalition government. Above all, he wanted it understood that intervention with U.S. combat troops was a last resort. Apparently not expecting the President to take such a firm stand, General Lemnitzer assured him that the JCS did not advocate the deployment of "major U.S. forces" and that their main concern was to bolster "indigenous" capabilities. Guarding his options, Kennedy directed the JCS to continue to study U.S. intervention but indicated he would hold a decision in abeyance until efforts to reach a diplomatic solution ran their course.[47]

Throughout the crisis, the Joint Chiefs and their superiors had less than reliable intelligence on the situation inside Laos. SIGINT was virtually nonexistent and U–2s had limited applicability.[48] The information the JCS received came mainly from the U.S. Embassy in Vientiane and U.S. military advisors working with the FAL. In early March 1961, with the military balance tipping in favor of the Communists, the President approved an interagency plan (MILL POND) for limited overt and covert assistance to the RLG and its allies.[49] Over the next several weeks, Pacific Theater Commander (CINCPAC) Admiral Harry D. Felt stepped up the delivery of arms and equipment to the FAL. At the same time, he began assembling a command staff and earmarking U.S. units for a joint task force (JTF 116) that would form the nucleus of a Southeast Asia Treaty Organization (SEATO) Field Force should he be ordered to intervene. Estimates assembled by the Joint Staff projected an intervention force of some 60,000 U.S. troops, augmented by token units from nearby SEATO countries. Anything smaller, JCS planners insisted, would fail to impress or pressure either the Soviets or North Vietnamese and could draw the United States into an open-ended war on the Asian mainland.[50]

Despite preparations to intervene, the preferred U.S. solution remained a diplomatic settlement. As Secretary of State Dean Rusk described the administration's strategy, "Even if we move in, the object is not to fight a big war but to lay the foundation for negotiation."[51] During talks with British Prime Minister Harold Macmillan toward the end of March, Kennedy acknowledged that he did not accord much strategic importance to Laos and was prepared to accept "anything short of the whole of Laos being overrun." Should intervention become unavoidable, he was thinking of deploying four or five U.S. battalions to hold Vientiane and a few bridges across the Mekong River long enough to reach an agreement. But he had yet to settle on a specific course of action and looked to the British to help find a solution through diplomacy.[52]

Convinced that the President was underestimating the seriousness of the situation and Laos' importance, the Joint Chiefs of Staff continued to favor a strong show of force as the only way of avoiding a larger conflict. But as time passed with no new decisions from the White House and as the FAL suffered one setback after another, the JCS saw the opportunity for effective action slipping away. With large-scale intervention appearing unlikely, they advised staying out. At a pivotal meeting on April 29 with the Secretary of Defense and Attorney General Kennedy, they made their concerns known and urged shelving plans for intervention, provoking McNamara to remark snidely that "we had missed having government troops who were willing to fight." Most cautious of all was Army Chief of Staff Decker, who considered a conventional war in Southeast Asia a losing cause. Decker offered one reason after another why going into Laos at this point had drawbacks. Ultimately, in his view, it came down to a question of whether the results would be worth the cost. "[I]f we go in," he said, "we should go in to win, and that means bombing Hanoi, China, and maybe even using nuclear bombs."[53]

Decker's reference to the use of nuclear weapons was not the first time the subject came up with respect to Laos, but it put the potential consequences of an escalating and widening conflict in Southeast Asia into sharper focus than ever before. Whether the JCS had a specific plan for mounting nuclear operations in Laos is unclear. Detailed planning for a Laotian operation was a function of Admiral Felt's staff, which produced several operational and concept plans during the crisis, none involving nuclear weapons other than against the threat of large-scale Chinese intervention.[54] But in light of the nuclear-oriented tactics and strategy introduced during the Eisenhower years, it was practically routine for the use of nuclear weapons to be considered at one point or another in the planning process. The Kennedy administration had vowed to change that practice, but its preferences had yet to affect the planning guidance employed by the Joint Staff and the combatant commanders.[55]

The use of nuclear weapons was thus present, if not explicit, in policymakers' and military planners' minds throughout the Laos crisis. Yet the decisive factors that steered Kennedy away from military intervention were the absence of congressional support for military action and his own concern, in the aftermath of the Bay of Pigs, about the quality and soundness of JCS advice. In early May, Kennedy polled the chiefs for their views. All, to one degree or another, still favored the application of some form of military power, but speaking individually, they offered no coherent courses. Instead, they described a series of separate measures which, taken together, might invite a full-scale war with North Vietnam and Communist China.[56] With the Joint Chiefs unable to offer a credible military option, Kennedy continued to rely on diplomacy to yield a settlement. "Thank God the Bay of Pigs happened when it did," he later remarked. "Otherwise we'd be in Laos by now—and that would be a hundred times worse."[57]

Meanwhile, a fragile ceasefire descended on Laos, opening the way by mid-May for the 14 nation Geneva Conference to reconvene work on a negotiated settlement. Without the continuing threat of U.S. and/or SEATO military intervention, the Joint Chiefs doubted that there could ever be an agreement that did not favor the absorption of Laos into the Sino-Soviet bloc. W. Averell Harriman, the senior U.S. representative to the Geneva talks and, in President Kennedy's eyes, a highly respected authority on negotiating with the Communists, took a similar view.[58] Consequently, as the talks went forward, the Joint Staff, with White House approval, continued to review plans and preparations to insert U.S. or SEATO forces into Laos. But by September, the administration's preoccupation with Berlin and the diversion of military assets to meet the crisis there left the Joint Chiefs skeptical of achieving a favorable outcome at the bargaining table. JCS efforts to interest senior policymakers in a limited SEATO buildup in the region, backed by U.S. air, sea, and logistical support, met with the cold rebuff from OSD that such actions might "dilute other deployments."[59]

The Laotian situation heated up again in the spring of 1962. Blatantly disregarding the ceasefire, Pathet Lao and North Vietnamese troops laid siege to the provincial capital of Nam Tha. As the crisis unfolded, General Lemnitzer and Secretary McNamara were in Athens for a NATO ministerial meeting. Ordered by Kennedy to take a first-hand look at the situation, they arrived in Southeast Asia soon after Nam Tha had fallen to the Communists, with the remnants of the FAL in full retreat. An aerial inspection confirmed that the Mekong River offered little or no defense against a Communist invasion of either Thailand or South Vietnam. Arriving back in Washington on May 12, McNamara and Lemnitzer immediately debriefed the President and the NSC and urged a prompt but restrained show of force in line with "precautionary steps" recommended by Admiral Felt. This time Kennedy agreed, giving CINCPAC the go-ahead to move a Marine battalion with its helicopters and other air support to Thailand and to shift a U.S. Army battle group already there for maneuvers to the strategically important town of Ubon.[60] Should the Communist advance fail to stop, Kennedy sanctioned planning for a larger intervention, mainly to protect South Vietnam. The JCS and CINCPAC were still working out the details when, in mid-June 1962, the warring parties in Laos announced agreement on a coalition government, ending the crisis but leaving Laos effectively partitioned along lines that gave the North Vietnamese avenues to infiltrate troops and weapons into South Vietnam and to threaten Thailand as well.[61]

The battle for Laos was essentially over, and for all practical purposes the Communists had won. Gaining what they had wanted all along, they now had unfettered access into South Vietnam and beyond. Once again, the Joint Chiefs and President Kennedy had failed to see eye-to-eye on a crucial issue. After the Bay of Pigs Kennedy never fully trusted JCS advice. As a result, JCS efforts to persuade Kennedy to take a strong stand on Laos fell largely on deaf ears until it was too late. Unlike the President, the JCS never regarded Laos as expendable. Rather, they saw it as a small but strategically important country whose fate would determine that of its neighbors. In the chiefs' view, once Laos was lost it was only a matter of time before the United States faced larger conflicts in South Vietnam, Thailand, and beyond.

ORIGINS OF THE CUBAN MISSILE CRISIS

The last major foreign crisis of Kennedy's presidency was the October 1962 confrontation with the Soviets over their deployment of strategic nuclear missiles in Cuba. By then, Kennedy had replaced the military advisors he inherited from Eisenhower with people of his own choosing. Two of these personnel changes came on October 1, when General Earle G. Wheeler replaced Decker as Army Chief of Staff and Maxwell Taylor returned to active duty, succeeding Lemnitzer as Chairman. Earlier, Anderson had replaced Burke as CNO and General Curtis E. LeMay had succeeded Thomas D. White as Air Force Chief of Staff. In Taylor's view, LeMay was a superb operational commander, as demonstrated by his accomplishments in World War II and during the years he ran the Strategic Air Command. But his appointment as Air Force Chief of Staff was a "big mistake." Kennedy, on the other hand, felt he had no choice. Though he found LeMay coarse, rude, and overbearing, he felt he had to promote him in view of the general's seniority and strong popular and congressional following.[62]

In contrast, President Kennedy regarded Taylor as "absolutely first-class." Indeed, he was one of the few military professionals he respected and felt comfortable with.[63] To his JCS colleagues, however, Taylor's return to the Pentagon was less than welcome owing to the political overtones surrounding his appointment, his identification with administration policies, and his criticism of the Joint Chiefs following the Bay of Pigs. As Chairman, he saw himself mainly as the agent of his civilian superiors and tried to craft military recommendations that harmonized with civilian views and administration programs. Aware that the JCS were losing influence, he attributed this situation in part to the Joint Staff, which he characterized as only "marginally effective" because of its "inherent slowness" in addressing issues and providing timely responses.[64] Some of the Service chiefs believed they could not always count on Taylor to convey their views fairly and accurately to the President. Nor could they rely on him to report precisely what the President or other senior officials said, a problem that Taylor's hearing difficulties may have exacerbated.[65]

Taylor was still in the White House as the President's military representative when the Cuban missile crisis unfolded. Its origins went back to the spring of 1961, in the aftermath of the Bay of Pigs episode, when the Kennedy administration resolved to isolate Castro's Cuba and to undermine its authority and influence. The Joint Chiefs' contribution was a set of plans for a swift and powerful U.S. invasion of Cuba to overthrow Castro's government in an 8-day campaign.[66] Meeting with Secretary McNamara and Admiral Burke on April 29, 1961, President Kennedy concurred in the general outline of the plan.[67] But after further review, the NSC decided against military intervention at that point and elected to put pressure on Castro through diplomatic and economic means and a covert operations program known as MONGOOSE. To coordinate the effort, the President turned to his brother, Robert, who preferred to draw on Taylor—a family friend—rather than the JCS for military advice.[68]

Like the struggle for Laos, the Kennedy administration's growing obsession with Cuba reflected a fundamental shift in the focus of Cold War politics. During the late 1940s and 1950s, Europe and Northeast Asia had been at the center of the Cold War. But by the early 1960s, despite occasional flare-ups over Berlin and along the demilitarized zone between North and South Korea, the contest for control in these areas was essentially over and a stalemate had settled in. Realizing that further gains in the industrialized world were unlikely, the Soviets turned their attention to the emerging Third World countries of Asia, Africa, and Latin America where Khrushchev in a celebrated speech of January 6, 1961, proposed to unleash a wave of Communist-directed "liberation wars." President Kennedy referred to

Khrushchev's speech often and considered it clear evidence that the United States needed to pay more attention to the Third World. In particular, he stressed the development of aid programs to improve living conditions and the acquisition of more effective tools for counterinsurgency warfare.[69]

Khrushchev found the temptation of establishing a strong Soviet presence in Cuba, 90 miles from the southern coast of the United States, irresistible. Not only would these weapons counterbalance the deployment of American forces in Europe and the Near East, but Cuba would also serve as a hub for spreading Communism throughout Latin America. Less clear is why Khrushchev risked losing his foothold in Cuba by placing strategic nuclear missiles there, a provocation that was almost certain to draw a sharp U.S. response. In his memoirs, Khrushchev justified his actions as providing Castro with deterrence against American attack. "Without our missiles in Cuba, the island would have been in the position of a weak man threatened by a strong man."[70] The missiles in question, however, were strategic offensive weapons, not defensive ones, which would have afforded Cuba better protection. Though there may also have been a handful of Soviet tactical nuclear weapons in Cuba at the time, the evidence of their presence is sketchy and has never been positively confirmed. Nor is it clear who, if anyone, had authority to use them.[71] The most plausible explanation for Khrushchev's actions is that he was trying to bolster the Soviet Union's strategic posture and overplayed his hand. The consensus among Kennedy loyalists like diplomat George Ball was that Khrushchev was a "crude" thinker who miscalculated that he could push the President around with impunity. According to Ball, Khrushchev's decision to place offensive missiles in Cuba resulted from his desire to "bring the U.S. down a peg, strengthen his own position with respect to China, and improve his standing in the Politburo with one bold stroke."[72]

Whatever the reasons, Khrushchev was adept at refining and carrying out his plan. The decision to deploy missiles in Cuba emerged from an informal meeting in the spring of 1962 between Khrushchev and Marshal Rodion Malinovskiy, the minister of defense, at Khrushchev's *dacha* in the Crimea. Malinovskiy complained about the presence of 15 U.S. Jupiter medium-range ballistic missiles (MRBMs) in Turkey and the need to redress this situation. The Jupiters had been operational for about a year. While not in Malinovskiy's view a serious military threat, they were an irritant requiring a diversion of resources. One thing led to another and it was from these conversations that Khrushchev seized on the idea of putting strategic missiles in Cuba.[73]

To implement his policy, Khrushchev relied on the Soviet General Staff to concoct an elaborate deception scheme. Code-named ANADYR, the operation involved assembling and outfitting in total secrecy over 50,000 soldiers, airmen, and sailors, calculating the weapons, equipment, supplies, and support they would need for a prolonged stay in Cuba, finding 85 freighters for transportation, and completing the mission in 5 months.[74] Apparently, senior members of the Soviet Defense Council initially resisted the idea, but as a practiced expert in bullying people, Khrushchev got his way.[75] Toward the end of May, a high-level Soviet military delegation, posing as engineers, visited Havana and secured Castro's agreement to the plan. Preparations continued over the summer, and on September 8, 1962, the first SS–4 MRBMs were unloaded in Cuba. Their nuclear warheads began arriving a month later, though their presence went undetected by U.S. intelligence.[76]

Despite tight security and elaborate deception measures, the Soviets could not fully conceal their activities. By summer, rumors were rife within intelligence circles and the Cuban exile community in south Florida that the Soviets were up to something. Attention focused on an apparent buildup of conventional arms, which the CIA confirmed in July and August through U–2 photographs, HUMINT sources, and NSA surveillance of Soviet ships passing through the Dardanelles.[77] The CIA also detected increased construction activity for SA–2 antiaircraft missile installations (the same weapon used to shoot down Gary Powers's U–2 in 1960) and a partially finished surface-to-surface missile complex at the Cuban coastal town of Banes, reported to President Kennedy on September 7. The Banes installation was for short-range anti-ship cruise missiles and did not pose a serious threat to U.S. vessels, but the discovery caused President Kennedy to impose tight compartmentalization on all intelligence dealing with offensive weapons. Earlier, he had imposed similar constraints on the dissemination of SA–2 surface-to-air missile (SAM) information. These precautions severely limited the distribution of intelligence data, even among high-level officials and senior intelligence analysts. Whether they prevented critical intelligence from reaching the JCS is unclear.[78]

As part of the deception operation, the Soviets maintained that they had no plans to deploy offensive nuclear missiles in Cuba. Until U–2 pictures proved otherwise, the Intelligence Community accepted these assurances at face value.[79] Monthly U–2 overflights of Cuba had been routine since the Bay of Pigs and by September 1962, with reports of increased Soviet activity, the Kennedy administration fell under growing pressure to step up surveillance. But as more SA–2 sites became operational, the U–2s were increasingly vulnerable, raising fears of a repetition of the Powers incident. Over CIA objections, National Security Advisor McGeorge Bundy and Secretary of State Dean Rusk persuaded President Kennedy in mid-September to suspend U–2 flights across Cuba and to approve new routes along the periphery of the island. To gloss over the loss of coverage, the White House termed these "additional" flights, which technically they were. But the overall result, as one CIA analyst characterized it, was "a dysfunctional surveillance regime in a dynamic situation."[80]

These procedural changes took place at the very time Soviet offensive missiles were starting to arrive in Cuba and delayed their discovery by a full month. As late as September 24, however, General Lemnitzer still considered U.S. surveillance of Cuba to be "adequate" in light of current policy and military requirements.[81] Though the JCS were well aware of the danger posed by the growing Soviet presence in Cuba, it was Castro's stubborn hold on power despite ongoing economic, diplomatic, and covert efforts to loosen his grip that concerned them even more. Convinced that the time was fast approaching when only a military solution would suffice, the JCS continued to focus on various contingency plans to cripple or topple Castro's regime. By the end of September, their attention had settled on three concepts: a large-scale air attack (OPLAN–312–62); an all-out combined arms invasion (OPLAN–314–61) that would take approximately 18 days to organize; and a quick reaction version of the invasion plan (OPLAN–316–61) that could be launched with immediately available forces in 5 days.[82] Also on the table was a Joint Strategic Survey Council proposal to impose a naval blockade of Cuba. However, the JCS paid less attention to this option than the others because there was no guarantee it would assure Castro's downfall.[83]

Treating these plans as exceedingly sensitive, the Joint Chiefs did not discuss them in any detail with senior administration figures outside the Pentagon. Consequently, their possible political and diplomatic impact remained unassessed. The President's views, insofar as they were known to the JCS, favored continuing surveillance of the island and avoidance of a military confrontation.[84] As a concession to preparedness, Kennedy asked Congress in September for authority to call up 150,000 Reservists, and in early October he and McNamara discussed the possibility of an air strike to take out the SA–2 sites.[85] But before taking further action, the President wanted better information. On October 12, with the SA–2 threat still his uppermost concern, he transferred operational command and control of U–2 flights over Cuba from the CIA to the Strategic Air Command and authorized the resumption of direct overflights, limited to the western tip of the island for the time being. Two days later, SAC's first U–2 mission confirmed that the Soviets were deploying SS–4 medium-range surface-to-surface missiles on the island. Subsequent flights revealed that the Soviets were also constructing SS–5 intermediate-range ballistic missile (IRBM) sites.[86]

SHOWDOWN OVER CUBA

The discovery that the Soviets were deploying offensive strategic missiles in Cuba and that the weapons were on the verge of activation presented Kennedy with the most serious foreign policy crisis of his Presidency. Militarily, the MRBMs and IRBMs the Soviets were deploying in Cuba were comparable to the Thor and

Jupiter missiles the United States had deployed to Britain, Italy, and Turkey the previous few years. With ranges of up to 1,200 miles for the MRBMs and 2,500 miles for the IRBMs, the Soviets could threaten most of the eastern half of the United States with nuclear destruction. By themselves, these weapons may have done little or nothing to change the overall strategic balance since the United States continued to hold a substantial lead in ICBMs and long-range strategic bombers. All the same, the threat was much too large and close to home to ignore. With the congressional mid-term elections looming, a decisive response became all the more certain.

To manage the crisis, Kennedy improvised through an ad hoc body known as the Executive Committee, or ExCom. Hurriedly assembled, ExCom operated for security reasons with no pre-set agenda and initially consisted of Cabinet-level officials, a handful of their close aides, and a few outside advisors.[87] As time passed, the list of attendees steadily grew to more than seventy people, mostly civilians. Even though the Joint Chiefs were actively engaged in contingency planning throughout the crisis, they were not directly privy to ExCom's deliberations or even much of the information that passed through it. General Taylor was the sole JCS member on the ExCom and one of its few members with significant military experience. During the crisis, the Joint Chiefs met privately with the President only once—on October 19. The rest of the time, Taylor or McNamara acted as intermediary. In his memoirs, Taylor acknowledged that some of the chiefs distrusted him. He added, however, that over the course of the crisis he repeatedly volunteered to arrange more meetings with the President, but that none of the Service chiefs showed any interest.[88]

The main advantage of a larger and more conspicuous JCS presence in the ExCom would have been closer coordination. Policymakers would have had a clearer understanding of the military options and the Joint Chiefs a fuller appreciation of the political and diplomatic dimensions of the problem.[89] In the JCS view, the deployment of offensive missiles in Cuba was a serious provocation that more than justified Castro's removal from power by force if necessary. Thus, from the onset of the crisis, the JCS (including Taylor) favored a direct and unequivocal military response to eliminate all Soviet missiles from Cuba and, in the process, to "get rid" of Castro.[90] It was a position Kennedy found both too extreme and too risky. During the Bay of Pigs, he had wanted the Joint Chiefs to speak out more. By the time of the Cuban missile crisis, he had little interest in what they had to say. By keeping them at arm's length, he could acknowledge their suggestions but ignore them as well. "The first advice I'm going to give my successor," he later observed, "is to watch the generals and to avoid feeling that just because they are military men their opinions on military matters are worth a damn."[91]

The Joint Chiefs came to their position during the early days of the crisis and stuck to it. Throughout their deliberations, there was little repetition of the squabbling that had exposed their disunity and marred their effectiveness during the Berlin and Laos episodes. Treating military action as inevitable, their initial preference was for a strong air attack to take out all known IR/MRBM sites, SA–2 installations, and other key military facilities, followed by implementation of the quick-reaction invasion plan (OPLAN–316). From mid-October on, the JCS carried out a steady buildup of airpower in Florida, reaching a strength of over 600 planes, and positioned supplies and ammunition for an invasion. They also designated Admiral Robert L. Dennison, Commander in Chief, Atlantic, a unified command, to exercise primary responsibility for Cuban contingencies. Facing a shortage of conventional munitions, McNamara authorized U.S. combat aircraft to fly with nuclear weapons.[92]

While treating an invasion as unavoidable, the Joint Chiefs accepted McNamara's advice and confined their presentation to the President on October 19 to the air attack phase. Predictably, the most ardent advocate of this course was LeMay, the Air Force chief, who doubted whether a naval blockade or lesser measures would permanently neutralize the missile threat. Kennedy seemed to like the idea of a "surgical" air strike against the IR/MRBM sites alone. However, a large-scale air campaign (especially one that might involve tactical nuclear weapons) was another matter, and in exploring options with the JCS, he expressed concern that it might invite Soviet reprisals against Berlin. "We would be regarded," he said, "as the trigger-happy Americans who lost Berlin." And, he added: "We would have no support among our allies." Kennedy also feared that an American attack of any sort on Cuba with the Soviets there could escalate into a nuclear exchange. "If we listen to them and do what they want us to do," Kennedy later said of the Joint Chiefs, probably with LeMay in mind more than any of the others, "none of us will be alive later to tell them that they were wrong."[93]

If it resolved anything, the President's meeting with the Joint Chiefs left Kennedy more convinced than ever that he urgently needed to find an alternative to direct military action. The next day, after a rambling 2-hour ExCom session, the President decided to put both an air campaign and an invasion on hold and to impose a blockade, or "quarantine" as he publicly called it since a blockade amounted to a declaration of war in international law. During the ExCom debate, General Taylor strenuously defended the JCS position in favor of air strikes and played down the possibility that the use of nuclear weapons against Cuban targets would invite nuclear retaliation from the Soviets.[94] Afterwards Taylor returned to the Pentagon to brief his JCS colleagues. "This was not," he told them, "one of our better days." In explaining the President's blockade decision, Taylor said that the decisive votes had come from McNamara, Rusk, and UN Ambassador Adlai E. Stevenson, all of whom strongly opposed air attacks. Pulling Taylor aside as the meeting broke up, the President had added: "I know that you and your colleagues are unhappy with the decision, but I trust that you will support me in this decision." The Chairman assured him that the JCS would back him completely.[95]

Kennedy and the Joint Chiefs were not, in fact, as far apart as it seemed. Even though the President preferred the quarantine, he had not categorically ruled out either an air attack or an invasion, and over the next several days, while the Navy was organizing the quarantine, he directed the Joint Chiefs to proceed with the military buildup opposite Cuba. As part of the show of force, the Joint Chiefs ordered the Commander in Chief, Strategic Air Command, to begin generating his forces toward DEFCON 2 (maximum alert) and to launch SAC bombers up to the "radar line" where the Soviets would detect them. Shelving OPLAN–314 for a large-scale invasion, the Joint Chiefs instructed Admiral Dennison on October 26 to concentrate his preparations on OPLAN–316, which he could execute on shorter notice. By leaving the invasion and other military options open, McNamara told the ExCom, the United States would "keep the heat on" the Russians. Kennedy thus found military power indispensable, even if at times he felt events were taking over. But to go beyond a show of force, as he demonstrated time and again during the crisis, was out of the question without the most extreme provocation.[96]

As the showdown approached, the accompanying tensions further exacerbated the already strained relationship between the Joint Chiefs and their civilian superiors. The most serious clash was between McNamara and Chief of Naval Operations Admiral George Anderson. Though Anderson professed the utmost respect for civilian authority, he vehemently objected to the intrusion of civilians into the management of naval operations, as evidenced by the run-in he had with McNamara on October 24. The night before, the Office of Naval Intelligence (ONI) had received unconfirmed reports that, rather than risk inspections under the quarantine, many Soviet merchant ships heading for Cuba, including some suspected of carrying missiles, had slowed, changed course, or turned back. However, ONI insisted on visual verification from U.S. warships and reconnaissance aircraft before giving the information wide distribution. As a result, it was not until noon the next day that Secretary McNamara and the White House finally received the information. Furious at the delay, McNamara confronted Anderson that evening in the Navy's Flag Plot command center in the Pentagon where, according to one account, he delivered "an abusive tirade." Anderson declined to explain why it had taken so long for the information to reach McNamara and took umbrage at the Secretary's manner. Tempers flared and the Secretary of Defense stalked out, resolving as he left to be rid of Anderson at the earliest convenient opportunity.[97]

A similar communications lapse took place a few days later, on October 27, during the height of the crisis, as chances for a negotiated settlement seemed to dwindle. At issue was a truculent letter from Khrushchev linking the removal of the U.S. Jupiter MRBMs from Turkey to the removal of Soviet offensive missiles from Cuba.[98] Deployed above ground at "soft" fixed sites, the Jupiters were vulnerable to a preemptive attack and had a low level of readiness because they used nonstorable liquid fuel. Kennedy had never attached much military value to them and, treating them as "obsolete," was inclined to deal. But there was little support in the ExCom, where the prevailing opinion held that such a trade could seriously harm U.S. relations with Turkey and perhaps drive a wedge between the United States and NATO.[99] That evening back at the Pentagon, Taylor briefed the chiefs on the stalemate regarding the Jupiters and added: "The President has a feeling that time is running out." At this point the Joint Chiefs began making preparations to go to the White House the next morning to bring the President up to date on the status of war plans and to secure his approval to initiate direct military action.[100]

Unknown to Taylor and the Service chiefs, Secretary of State Rusk had come up with a scheme to break the impasse, and early that evening he and the President held a short meeting in the Oval Office. Others present were McGeorge Bundy, McNamara, Gilpatric, Robert Kennedy, George Ball, Theodore Sorensen, and Llewellyn E. Thompson, the former U.S. Ambassador to Moscow. It was at this gathering that Kennedy approved a secret initiative, which his brother Robert conveyed to Soviet Ambassador Anatoly Dobrynin a short while later.[101] The offer was in two parts. The first was a pledge by the United States not to invade Cuba or to overthrow Castro in exchange for removal of the Soviet missiles; the second, at Rusk's instigation, was an informal assurance that in the not-too-distant future the United States would quietly remove its Jupiter missiles from Turkey. The concession on the Jupiters appears to have been unnecessary since an offer to discuss the matter at a later date probably would have sufficed. But in his eagerness to avoid coming to blows, Kennedy chose to sweeten the deal and give Khrushchev fewer grounds for objecting.[102]

The Joint Chiefs were never consulted, nor were they given an opportunity to comment on the strategic implications of this settlement. General LeMay was disappointed that the President, with a preponderance of strategic and tactical nuclear power on his side, had not demanded more concessions from the Soviets. "We could have gotten not only the missiles out of Cuba," LeMay insisted, "we could have gotten the Communists out of Cuba at that time."[103] The first inkling the chiefs had of the deal ending the Cuban missile crisis came the next morning from a ticker tape news summary announcing Moscow's acceptance of the American no-invasion pledge in exchange for the withdrawal of Soviet offensive missiles.[104] Little by little over the next few days the Joint Chiefs learned more about the deal and about "a proposal" to withdraw the Jupiters from Turkey and to assign Polaris boats in their place. The consensus on the Joint Staff was that the United States had come out on the poorer end of the bargain. Not only did the Jupiters make up one-third of SACEUR's Quick Reaction Alert Force, they also carried a much larger payload than Polaris and were more reliable and accurate. Believing withdrawal of the Jupiters to be ill-advised, the Joint Chiefs considered sending the Secretary of Defense a memorandum recommending against it. But upon discovering that it was a done deal, they let the matter drop. Kennedy had what he wanted most of all—removal of the Soviet missiles from Cuba—and the crisis was winding down.[105]

AFTERMATH: THE NUCLEAR TEST BAN

By the time the Cuban missile crisis ended, relations between the Kennedy administration and the Joint Chiefs of Staff (Taylor excepted) were at an all-time low. In contrast, Kennedy's public stature and esteem had never been higher. Lauded by his admirers and critics alike for showing exemplary statesmanship, fortitude, and wisdom in steering the country through the most dangerous confrontation in history, the President emerged with his credibility and prestige measurably enhanced. But to end the crisis he made compromises and concessions that his military advisors considered in many ways unnecessary and excessive. Worst of all, in the chiefs' view, the United States had left Castro's regime in place. The presence of an outpost of communism in the Western Hemisphere left the JCS no choice but to continue allocating substantial military and intelligence resources for containment purposes. Looking back, McGeorge Bundy acknowledged that Kennedy had kept the Joint Chiefs "at a distance" throughout the crisis, sensing that their perception of the problem "was not well connected with his own real concerns." "The result," Bundy added, "was an increased skepticism in his view of military advice which only increased the difficulty of exercising his powers as commander in chief."[106]

Despite the estrangement between Kennedy and his military advisors, the only member of the Joint Chiefs to become a casualty of the episode was Admiral Anderson, whose 2-year term as Chief of Naval Operations expired in August 1963 and was not extended. Sending Anderson to Portugal as U.S. Ambassador, Kennedy selected the more even-tempered David L. McDonald to be CNO. Well liked and highly respected among his peers, McDonald was serving with NATO at the time of his selection and would have preferred to stay in London.[107] Kennedy and McNamara might have gone further in purging the chiefs, but they knew that LeMay, the other candidate for removal, had strong support in Congress and was virtually untouchable. Furthermore, in the aftermath of the missile crisis, the administration's foreign policy agenda began to move away from the confrontational approach that had characterized its first 2 years, toward a rapprochement with the Soviets based on the negotiation of outstanding differences. The Cuban missile crisis settlement was the opening wedge.

To realize his policy goals, Kennedy knew he would need the agreement if not the outright support of the JCS. Central to Kennedy's quest to improve relations with the Soviet Union was the nuclear test ban, a measure that had been on the back burner since the waning days of the Eisenhower administration. Before winning the White House, Kennedy had spoken in favor of curbs on nuclear testing and in his inaugural address he listed "the inspection and control of [nuclear] arms" as a major objective of his Presidency.[108] But at his meeting with Khrushchev in Vienna in June 1961, he had been unsuccessful in enlisting the Soviet leader's cooperation. The United States was then observing a voluntary moratorium on nuclear testing both above and below ground that Eisenhower had introduced in October 1958. Without progress in negotiations, however, Kennedy knew that at some point he would face concerted pressure from Congress, the Atomic Energy Commission, and the JCS to resume testing.

The Joint Chiefs had been urging Kennedy to resume testing almost from the moment he took office, if not in the atmosphere then underground, underwater, and in outer space. Some of their arguments were highly technical, but their overall position was relatively simple and straightforward: without testing they could neither verify the effectiveness of the existing nuclear deterrent nor be assured of new weapons to protect future security.[109] After the Soviets resumed atmospheric testing in September 1961, Kennedy gave in.[110] One of the experiments the Soviets conducted, on October 30, 1961, was a colossal "super bomb" nicknamed *Tsar Bomba* (King of Bombs) that had an explosive yield of 58 megatons, the largest nuclear device ever detonated. Seeing no practical military requirement for a bomb that size, the Joint Chiefs dismissed the test as a stunt, designed for propaganda purposes and to intimidate other countries.[111]

The U.S. testing program resumed in a less flamboyant fashion, getting off to a shaky and slower start. Owing to the moratorium, U.S. expertise in conducting nuclear experiments had "gone to pot," as one of those in charge put it, causing delays and difficulties during the first round of underground tests (Operation *Nougat*) in Nevada during the fall of 1961. Problems persisted into the spring of 1962, when the AEC and the Defense Atomic Support Agency (DASA), the organization in charge of proof-testing weapons, resumed atmospheric testing in the Pacific (Operation *Dominic*). Near the outset of the series, several important experiments connected to the development of an antiballistic missile system went awry. Subsequent tests were notably more successful. For the first time, a Polaris submarine launched one of its missiles and detonated the nuclear warhead. Other experiments demonstrated the feasibility of increasing the yield-to-weight ratio and the shelf life of warheads. From these data eventually emerged a new generation of more advanced nuclear weapons.[112]

Ending in November 1962, with its final experiments carried out during the Cuban missile crisis, *Dominic* was the last series of atmospheric tests the United States conducted. As the missile crisis wound down, Kennedy and Khrushchev expressed interest in reducing international tensions, starting with a renewed effort to reach a nuclear test ban. A major stumbling block then and for years to come was the need for reliable and effective verification. Khrushchev's agreement to permit aerial inspections by the United Nations to verify the removal of the missiles from Cuba was for some in the Kennedy administration a promising sign that the Soviets were becoming more open-minded about accepting reliable verification measures.[113] The Joint Chiefs were less optimistic, and in formulating a negotiating position they raised numerous objections.[114] While he went along with his colleagues' recommendations, Taylor felt increasingly frustrated and wanted to do more to further the President's agenda. Seeking to put a positive face on the chiefs' approach to the problem, he asked the Joint Staff what would constitute an "acceptable" agreement to the JCS. But to his disappointment, the Joint Staff found each option to contain shortcomings "of major military significance."[115]

Uncertain whether the Joint Chiefs would support a test ban, Kennedy worked around them as he did during the Cuban missile crisis. Conspicuously absent from the 13-member U.S. delegation that went to Moscow in July 1963 to do the negotiating was a JCS representative.[116] Kennedy would have preferred a comprehensive agreement barring all forms of testing. But he realized that there was insufficient support for such an accord either at home or in the Kremlin. A complete ban would have been tantamount to proscribing new nuclear weapons. Curbing his expectations, he authorized his chief negotiator, W. Averell Harriman, to pursue a treaty banning atmospheric, outer space, and underwater explosions.[117] With the negotiations entering their final stage, Kennedy summoned the Joint Chiefs to the White House on July 24, 1963, to urge their cooperation. As Taylor recalled, the Service chiefs reacted with "controlled enthusiasm."[118] At the time, the Joint Chiefs were considering a draft memorandum to the Secretary of Defense urging rejection of the accord unless "overriding nonmilitary considerations" dictated otherwise. Yielding to pressure from Taylor and the President, the chiefs shelved their objections and during Senate review of the treaty they grudgingly endorsed it.[119]

Signed in Moscow on August 5, 1963, the Limited Test Ban Treaty entered into force the following October. A major breakthrough in arms control, it helped set the stage for the Strategic Arms Limitation Talks (SALT) later in the decade. Weak as it was, JCS support was crucial to the treaty's passage and rested on acceptance by Congress and the President of four safeguards: an aggressive program of underground testing; maintenance of up-to-date research and development facilities; preservation of a residual capability to conduct atmospheric testing; and improved detection capabilities to guard against Soviet cheating. Had the Joint Chiefs opposed the treaty, it almost certainly would have failed of adoption.[120]

Taylor's role, both personally and as Chairman, was crucial to the treaty's approval. Without his persistence in nudging the Service chiefs along and keeping them in line, the outcome almost certainly would have been different. Institutionally, the test ban episode demonstrated that power and influence within the JCS organization were moving slowly but surely into the hands of the Chairman, as Eisenhower's 1958 amendments had largely intended. No longer merely a presiding officer or spokesman, the Chairman emerged from the treaty debate as a key figure in interpreting the chiefs' views and in shaping their advice and recommendations. Henceforth, the Chairman would become more and more the personification of the military point of view, and thus his interpretation of his colleagues' advice would be the final word.

In contrast, the overall authority, prestige, and influence of the Joint Chiefs of Staff as a corporate advisory body had never been lower than by the time the test ban debate drew to a close. Though JCS views still carried considerable weight on Capitol Hill, the same was not true at the White House and elsewhere in the executive branch. Having lost faith in the Joint Chiefs after the Bay of Pigs, Kennedy never regained confidence in his military advisors. Except for Taylor, a trusted personal friend, he kept the JCS at arm's length. Rarely ever openly critical of their superiors, the Joint Chiefs accepted these ups and downs in their fortunes as part of the job. Reared in a tradition that stressed civilian control of the military, they instinctively deferred to the Commander in Chief's lead and were not inclined to challenge his decisions lest it appear they were impugning his authority. But in so doing, it became increasingly difficult for them to maintain their credibility and to provide reliable professional advice.

NOTES

1 Kennedy first used the phrase "New Frontier" in his acceptance speech at the 1960 Democratic National Convention. See Theodore C. Sorensen, *Kennedy* (New York: Bantam Books, 1966), 187–189.

2 John F. Kennedy in Allan Nevins, ed., *The Strategy of Peace* (New York: Harper, 1960), 184.

3 Quoted in Joseph Kraft, "McNamara and His Enemies," *Harper's Magazine* (August 1961), 42.

4 Lawrence S. Kaplan, Ronald D. Landa, and Edward J. Drea, *History of the Office of the Secretary of Defense: The McNamara Ascendancy, 1961–1965* (Washington, DC: Historical Office, Office of the Secretary of Defense, 2006), 10–11; Charles A. Stevenson, *SECDEF: The Nearly Impossible Job of Secretary of Defense* (Washington, DC: Potomac Books, 2006), 30; and L. James Binder, *Lemnitzer: A Soldier for His Time* (Washington, DC: Brassey's, 1997), 279.

5 Andrew Preston, *The War Council: McGeorge Bundy, the NSC, and Vietnam* (Cambridge: Harvard University Press, 2006), 41.

6 Steven L. Rearden, "The Secretary of Defense and Foreign Affairs, 1947–1989" (MS, Study Prepared for the Historical Office, Office of the Secretary of Defense, December 1995), chap. IV, 3–5; *Chronology of JCS Organization, 1945–1984* (Historical Division, Joint Secretariat, Joint Chiefs of Staff, December 1984), 178.

7 Sorensen, 284; John Prados, *Keepers of the Keys: A History of the National Security Council from Truman to Bush* (New York: William Morrow, 1991), 102; and Stanley L. Falk, "The National Security Council Under Truman, Eisenhower, and Kennedy," *Political Science Quarterly* 79 (September 1964), 428–433.

8 Edgar F. Raines, Jr., and David R. Campbell, *The Army and the Joint Chiefs of Staff* (Washington, DC: Center of Military History, 1986), 106–108; H.R. McMaster, *Dereliction of Duty: Lyndon Johnson, Robert McNamara, the Joint Chiefs of Staff, and the Lies That Led to Vietnam* (New York: Harper Perennial, 1997), 8–9.

9 Binder, 285.

10 Memo, Bundy to CJCS, June 28, 1961, "Functions of the Military Representative to the President," and Letter, Kennedy to Taylor, June 26, 1961, both attached to JCS 1977/141. See also Maxwell D. Taylor, *Swords and Plowshares* (New York: W.W. Norton, 1972), 195–203.

11 Interview with Henry E. Glass by Maurice Matloff et al., October 28, 1987, OSD Historical Office, cited in Kaplan et al., 6, 556 (note no. 18); Arthur M. Schlesinger, Jr., *A Thousand Days: John F. Kennedy in the White House* (Cambridge, MA: Houghton, Mifflin, 1965), 297.

12 Memo of ADM Burke's Conversation with CDR Wilhide, April 18, 1961, in Mark J. White, ed., *The Kennedys and Cuba: The Declassified Documentary History* (Chicago: Ivan R. Dee, 1999), 33; Walter S. Poole, *The Joint Chiefs of Staff and National Policy, 1961–1964* (Washington, DC: Office of Joint History, Office of the Chairman of the Joint Chiefs of Staff, 2011), 109–111.

13 Schlesinger, *A Thousand Days*, 232–233. Also see *An Analysis of the Cuban Operation* (Study prepared by the Deputy Director [Plans], Central Intelligence Agency, January 18, 1962), sec. IV, 4 (declassified).

14 Sorensen, 331; Stephen E. Ambrose, *Eisenhower: The President* (New York: Simon and Schuster, 1984), 615; Kaplan et al., 176.

15 Enclosure C to Report by the DJS to JCS, January 24, 1961, "U.S. Plan of Action in Cuba," JCS 2304/19.

16 JCSM-57-61 to SECDEF, February 3, 1961, "Military Evaluation of the CIA Para–Military Plan, Cuba," *FRUS, 1961–63*, X, 67–78; Peter Wyden, *Bay of Pigs: The Untold Story* (New York: Simon and Schuster, 1979), 89.

17 Haynes Johnson, *The Bay of Pigs* (New York: W.W. Norton, 1964), 65–66; "Narrative of the Anti–Castro Operation Zapata," June 13, 1961, 8–10, JHO 16–0061, JHO Collection.

18 Kaplan et al., 176–177, 183; Sorensen, 341–342; Taylor, 189.

19 Taylor, 188.

20 NSAM 55, June 28, 1961, "Relations of the JCS to the President in Cold War Operations," Kennedy Papers, National Security Files, JFK Library.

21 Taylor, 189; Mark Perry, *Four Stars* (Boston: Houghton Mifflin, 1989), 115.

22 Quoted in Richard Reeves, *President Kennedy: Profile of Power* (New York: Simon & Schuster, 1993), 363.

23 Douglas Brinkley, *Dean Acheson: The Cold War Years, 1953–71* (New Haven: Yale University Press, 1992), 117.

24 Memo, Acheson to Kennedy, April 3, 1961, "Berlin," JCS 1907/293.

25 JCSM-287-61 to SECDEF, April 28, 1961, "Berlin," JCS 1907/295; Kaplan et al., 148–49.

26 Kennedy quoted in John Lewis Gaddis, *We Now Know: Rethinking Cold War History* (Oxford: Clarendon Press, 1997), 146.

27 Note by Secretaries to JCS, July 1, 1961, "Berlin Planning," JCS 1907/315; Report by J5 to JCS, July 9, 1961, "Berlin Contingency Planning," JCS 1907/318; Paul H. Nitze, with Ann M. Smith and Steven L. Rearden, *From Hiroshima to Glasnost: At the Center of Decision—A Memoir* (New York: Grove Weidenfeld, 1989), 196–202.

28 JCSM-464-61 to SECDEF, July 1, 1961, "Berlin Planning," JCS 1907/316; JCSM-476-61 to SECDEF, July 12, 1961, "Partial Mobilization," JCS 1907/321; Kaplan et al., 152–153.

29 Schlesinger, *Thousand Days*, 388–389.

30 J-5 Report on Partial Mobilization, July 12, 1961, JCS 1907/321.

31 Poole, *JCS and National Policy, 1961–64*, 145–147; "Radio and Television Report to the American People on the Berlin Crisis," July 25, 1961, *Public Papers of the Presidents of the United States: John F. Kennedy, 1961* (Washington, DC: GPO, 1962), 535.

32 Nitze, *From Hiroshima to Glasnost*, 200–201.

33 Memo, Kennedy to McNamara, August 14, 1961, Kennedy Papers, NSF, quoted in Fred Kaplan, *The Wizards of Armageddon* (New York: Simon and Schuster, 1983), 298.

34 Kaplan et al., 12–13.

35 NIE 11-8/1-61, [September 21, 1961], "Strength and Deployment of Soviet Long Range Ballistic Missile Forces," in Donald P. Steury, ed., *Intentions and Capabilities: Estimates on Soviet Strategic Forces, 1950–1983* (Washington, DC: History Staff, Center for the Study of Intelligence, Central Intelligence Agency, 1996), 121–138; Jerrold L. Schecter and Peter S. Deriabin, *The Spy Who Saved the World* (New York: Charles Scribner's Sons, 1992), 271–283.

36 "Briefing for the President by the CJCS on the JCS Single Integrated Operational Plan 1962 (SIOP–62)," September 13, 1961, JCS 2056/281 (sanitized). Scott D. Sagan, "SIOP-62: The Nuclear War Plan Briefing to President Kennedy," *International Security* 12 (Summer 1987), 22–51, reproduces the briefing in full.

37 "Interview with Peter Hackes," September 28, 1961, in Alice C. Cole, ed., *Public Statements of the Secretaries of Defense: Robert S. McNamara, 1961* (Washington, DC: Historical Office, Office of the Secretary of Defense, n.d.), III, 1443. See also Kaplan et al., 161–162; and McGeorge Bundy, *Danger and Survival* (New York: Random House, 1988), 378–383.

38 NSDM 109, October 23, 1961, "U.S. Policy on Military Actions in a Berlin Conflict," Kennedy Papers, NSF, JFK Library; Letter, VADM John Marshall Lee, USN (Ret.) to author, May 18, 1984, Nitze Papers, Library of Congress; Nitze, *From Hiroshima to Glasnost*, 202–204.

39 JCSM-728-61 to SECDEF, October 13, 1961, "Preferred Sequence of Military Actions in a Berlin Conflict," JCS 1907/433; Memo, DepSECDEF to CJCS, October 17, 1961, "Preferred Sequence," JCS 1907/435.

40 Bundy, Memo of State-Defense Meeting, October 20, 1961, *FRUS, 1961–63*, XIV, 517.

41 Quoted in John Newhouse, *War and Peace in the Nuclear Age* (New York: Alfred A. Knopf, 1989), 156.

42 SNIE 11-10/1-61, October 5, 1961, "Soviet Tactics in the Berlin Crisis," in Donald P. Steury, ed., *On the Front Lines of the Cold War: Documents on the Intelligence War in Berlin, 1946 to 1961* (Washington, DC: CIA History Staff, Center for the Study of Intelligence, 1999), 621–622.

43 Matthias Uhl, "Storming on to Paris: The 1961 *Buria* Exercise and the Planned Solution of the Berlin Crisis," in Vojtech Mastny, Sven G. Holtsmark, and Andreas Wenger, eds., *War Plans and Alliances in the Cold War* (London: Routledge, 2006), 52–71. Uhl's analysis of the exercise makes no mention of the role of airpower in Warsaw Pact plans.

44 Quoted in Gaddis, 149.

45 Memo, McNamara to Kennedy, January 24, 1961, [Meeting with Eisenhower, January 19, 1961], *FRUS 1961–63*, XXIV, 41–42. Eisenhower may also have advised intervention with ground troops, but the various records of the meeting are unclear on this point. See Fred I. Greenstein and Richard H. Immerman, "What Did Eisenhower Tell Kennedy about Indochina? The Politics of Misperception," *Journal of American History* 79 (September 1992), 568–587.

46 JCSM-34-61 to SECDEF, January 24, 1961, "Courses of Action in Laos," JCS 1992/903.

47 Memo, Nitze to McNamara, January 23, 1961, "White House Meeting on Laos, January 23, 1961," *FRUS 1961–63*, XXIV, 26–27; Goodpaster Memo, January 25, 1961, "Conference with Kennedy," ibid., 42–44.

48 Pedlow and Welzenbach, *CIA and the U-2 Program*, 221.

49 Memo, CJCS to JCS, March 16, 1961, "Funding for Mill Pond Operations," JCS 1991/932; Kaplan et al., 237.

50 *CINCPAC Command History 1961*, Pt. II, 83–84; Schlesinger, *A Thousand Days*, 332.

51 MFR by Nitze, March 21, 1961, "Discussion of Laos at White House Meeting," *FRUS 1961–63*, XXIV, 95.

52 Harold Macmillan, *Pointing the Way, 1959–1961* (New York: Harper and Row, 1972), 337–338.

53 Memcon, April 29, 1961, "Laos," *FRUS, 1961–63*, XXIV, 150–154; Kaplan et al., 242; Richard K. Betts, *Soldiers, Statesmen, and Cold War Crises* (Cambridge: Harvard University Press, 1977), 178.

54 See "Concept for the Recapture of the Plaine des Jarres," March 9, 1961, JCS 1992/929; and *CINCPAC Command History 1961*, Pt. II, 85–89.

55 Victor B. Anthony and Richard R. Sexton, *The War in Northern Laos, 1954–1973* (Washington, DC: Center for Air Force History, 1993), 46; a previously classified USAF study released in 2008 asserts that the JCS had plans to use nuclear weapons, but it cites only unclassified sources.

56 JCS views summarized in editorial note, *FRUS 1961–63*, XXIV, 169–170. See also Poole, *JCS and National Policy, 1961–64*, Pt. II, 104–109; and Kaplan et al., 242–243.

57 Quoted in Sorensen, 726.

58 See Kaplan et al., 243.

59 JCSM-688-62 to SECDEF, September 29, 1961, "Military Intervention in Laos," JCS 2344/14; Memo, DepSECDEF to CJCS, October 3, 1961, "Planning for Southeast Asia," JCS 2344/17.

60 *Chronological Summary of Significant Events Concerning the Laotian Crisis, Sixth Installment: 1 May 1962 to 31 July 1962* (Historical Division, Joint Secretariat, Joint Chiefs of Staff, n.d.), 33–34, 53–54.

61 Poole, *JCS and National Policy, 1961–64*, Pt. II, 126–140.

62 Taylor quoted in Dino A. Brugioni, *Eyeball to Eyeball: The Inside Story of the Cuban Missile Crisis*, Robert F. McCort, ed. (New York: Random House, 1990), 266; Reeves, 182–183.

63 Benjamin C. Bradlee, *Conversations with Kennedy* (New York: Pocket Books, 1976), 117.

64 Letter, Taylor to McNamara, July 1, 1964, *FRUS, 1964–68*, X, 97–98.

65 Betts, *Soldiers, Statesmen, and Cold War Crises*, 68–69; and McMaster, 22–23. Jeffrey G. Barlow, "President John F. Kennedy and His Joint Chiefs of Staff" (Ph.D. Diss., University of South Carolina, 1981), 202, points out Taylor's hearing problems.

66 Memo, SECDEF to CJCS, April 20, 1961, no subject, JCS 2304/29; JCSM-278-61 to SECDEF, April 26, 1961, "Cuba," JCS 2304/30.

67 Memo, SECDEF to JCS, May 1, 1961, "Cuba Contingency Plans," JCS 2304/34.

68 Memo, SECDEF to SecArmy et al., May 9, 1961, "U.S. Policy Toward Cuba," JCS 2304/36; Arthur M. Schlesinger, Jr., *Robert Kennedy and His Times* (Boston: Houghton, Mifflin, 1978), 476–477.

69 Schlesinger, *A Thousand Days*, 302–304.

70 Nikita Khrushchev, *Khrushchev Remembers: The Last Testament*, trans. and ed. by Strobe Talbott (Boston: Little, Brown, 1974), 511.

71 The notion that the Soviets had tactical nuclear weapons in Cuba comes from General Anatoli I. Gribkov, head of the operations directorate of the Soviet General Staff in 1962, speaking at a 1992 conference in Havana. For an elaboration of his views, see Anatoli I. Gribkov and William Y. Smith, *Operations ANADYR: U.S. and Soviet Generals Recount the Cuban Missile Crisis* (Chicago: Edition Q, Inc., 1994), 4–7, 63–68. Mark Kramer, "Tactical Nuclear Weapons, Soviet Command Authority, and the Cuban

Missile Crisis," *Cold War International History Project Bulletin*, no. 3 (Fall 1993), 40–46, debunks the notion that the Soviets were preparing to use tactical nuclear weapons. Other than Gribkov, no Soviet official ever mentioned the presence of these weapons in Cuba. Nor do the documents Gribkov cites confirm that tactical nuclear weapons had actually reached Cuba at the time of the crisis.

72 James G. Blight and David A. Welch, eds., *On the Brink: Americans and Soviets Reexamine the Cuban Missile Crisis*, 2d ed. (New York: Noonday Press, 1990), 33–34.

73 Steven Zaloga, "The Missiles of October: Soviet Ballistic Missile Forces During the Cuban Crisis," *Journal of Soviet Military Studies* 3 (June 1990), 307–308. Aleksandr Fursenko and Timothy Naftali, *"One Hell of a Gamble": Khrushchev, Castro, and Kennedy, 1958–1964* (New York: Norton, 1997), 171, suggest that the idea of putting missiles in Cuba was Khrushchev's all along.

74 Gribkov and Smith, 23–24. Anadyr was the name of a river in the extreme northeastern Soviet Union.

75 Fursenko and Naftali, 180–182.

76 James H. Hansen, "Soviet Deception in the Cuban Missile Crisis," *Studies in Intelligence* 46, no. 1 (2002), 49–58 (unclassified edition).

77 Matthew M. Aid, *The Secret Sentry: The Untold History of the National Security Agency* (New York: Bloomsbury Press, 2009), 65.

78 Pedlow and Welzenbach, *CIA and the U-2 Program*, 200–201; Current Intelligence Memorandum, September 13, 1962, "Analysis of the Suspect Missile Site at Banes, Cuba," in Mary S. McAuliffe, ed., *CIA Documents on the Cuban Missile Crisis, 1962* (Washington, DC: History Staff, Central Intelligence Agency, 1992), 71–73; Dino A. Brugioni, *Eyeball to Eyeball: The Inside Story of the Cuban Missile Crisis* (New York: Random House, 1990), 120–126.

79 See SNIE 85-3-62, September 19, 1962, "The Military Buildup in Cuba," in McAuliffe, ed., *Cuban Missile Crisis, 1962*, 91–93.

80 Max Holland, "The 'Photo Gap' that Delayed Discovery of Missiles in Cuba," *Studies in Intelligence* 49, no. 4 (2005), 21 (unclassified edition).

81 CM-977-62 to SECDEF, September 24, 1962, "A U.S. Position for Presentation to the Foreign Ministers of the American States," JCS 2304/61.

82 Report by J3 to the JCS, October 14, 1962, "Cuba," JCS 2304/69.

83 Memo, JSSC to JCS, September 19, 1962, "Cuban Situation," JCS 2304/58; 1st N/H to JCS 2304/58, October 3, 1962. Also see Poole, *JCS and National Policy, 1961–64*, Pt. II, 231–232.

84 JCS Meeting, October 15, 1962, Notes Taken by Walter S. Poole on Transcripts of Meetings of the JCS, October–November 1962, dealing with the Cuban Missile Crisis, JHO 07-0035 (declassified and released 1996), hereafter cited as "Poole Notes." Why the President specified 3 months is unknown, but it would have taken him beyond the November elections and into the new Congress.

85 Letter to President of the Senate and Speaker of the House Transmitting Bill Authorizing Mobilization of Ready Reserve, September 7, 1962, *Kennedy Public Papers, 1962*, 665; Memo, McNamara to Kennedy, October 4, 1962, "Presidential Interest in SA-2 Missile System and Contingency Planning for Cuba," Kennedy Papers, National Security File, Kennedy Library.

86 Pedlow and Welzenbach, *CIA and the U–2 Program*, 207–208.

87 NSAM 196, October 22, 1962, "Establishment of an Executive Committee of the National Security Council," Kennedy Papers, National Security Files, JFK Library. Though not officially constituted until October 22, 1962, ExCom began functioning on October 16.

88 Taylor, 269.

89 See the roundtable discussion of this issue in Blight and Welch, eds., *On the Brink*, 85–86.

90 See JCSM-828-62 to SECDEF, October 26, 1962, "Nuclear-Free or Missile-Free Zones," JCS 2422/1, a good overall summary of the JCS position.

91 Quoted in Bradlee, *Conversations with Kennedy*, 117.

92 Kaplan et al., 206 and 582, fn. 48; Bernard C. Nalty, "The Air Force Role in Five Crises, 1958–1965" (Study done for USAF Historical Division Liaison Office, June 1968), 39–40 (declassified).

93 "Meeting with the JCS on the Cuban Missile Crisis," October 19, 1962, in Timothy Naftali and Philip Zelikow, eds., *The Presidential Recordings: John F. Kennedy—The Great Crises, Volume Two* (New York: W.W. Norton & Co.: 2001), II, 580–599; Reeves, 379.

94 "NSC Meeting on the Cuban Missile Crisis," October 20, 1962, Naftali and Zelikow, eds., *Presidential Recordings*, II, 605.

95 JCS Meeting, October 20, 1962, Poole Notes. See also Walter S. Poole, "The Cuban Missile Crisis: How Well Did the Joint Chiefs of Staff Work?" (Paper presented to the Colloquium on Contemporary History, Washington, DC, September 22, 2003); and editorial comments in Naftali and Zelikow, eds., *Presidential Recordings*, II, 614.

96 Poole, *JCS and National Policy, 1961–64*, Pt. II, 281–296; McNamara quoted in "ExCom Meeting of the NSC on the Cuban Missile Crisis," October 27, 1962, in Philip Zelikow and Ernest May, eds., *The Presidential Recordings: John F. Kennedy—The Great Crises, Volume Three* (New York: W.W. Norton, 2001), 360.

97 Thomas R. Johnson, *American Cryptology during the Cold War, 1945–1989: Book II—Centralization Wins, 1960–1972* (Washington, DC: National Security Agency, 1995), 329 (declassified); Aid, *Secret Sentry*, 73–76; Brugioni, *Eyeball to Eyeball*, 398–400; and Deborah Shapley, *Promise and Power: The Life and Times of Robert S. McNamara* (Boston: Little, Brown, 1993), 176–178.

98 Letter, Khrushchev to Kennedy, October 27, 1962, *FRUS 1961–63*, XI, 257–260.

99 Memo by Bromley Smith, "Summary Record of NSC Executive Committee Meeting No. 7, October 27, 1962, 10:00 AM," Kennedy Papers, National Security File, Kennedy Library.

100 JCS Meeting, October 27, 1962, Poole Notes.

101 Memo, Robert F. Kennedy to Rusk, October 30, 1962, *FRUS 1961–63*, XI, 270–271.

102 Bundy, 432–434.

103 LeMay Interview in Richard H. Kohn and Joseph P. Harahan, eds., *Strategic Air Warfare* (Washington, DC: Office of Air Force History, 1988), 114.

104 JCS Meeting, October 28, 1962, Poole Notes.

105 Draft Memo, JCS to SECDEF, enclosure to Report, J5 to JCS, October 29, 1962, "Proposal for Substitution of Polaris for Turkish Jupiters," JCS 2421/73. In mid-November 1962, Kennedy sought to reopen the deal he had cut with Khrushchev by including Soviet IL–28 fighter-bombers as among the weapons banned from Cuba. The Soviets agreed to remove their IL–28s, but offered no pledge not to reintroduce them or some similar plane.

106 Bundy, 458.

107 Interview with McDonald by Harold Joiner, June 1, 1963, for *Atlanta Journal-Constitution*, available at <http://victorianmaysville.com/history/notables/dlm/1963_story.htm>.

108 "Inaugural Address," January 20, 1961, *Kennedy Public Papers, 1961*, 2.

109 JCSM-99-61 to SECDEF, February 21, 1961, "Resumption of Nuclear Testing," JCS 2179/230; JCSM-182-61 to SECDEF, March 23, 1961, "Nuclear Arms Control Measures," *FRUS, 1961–63*, VII, 21–27; JCSM-445-61 to SECDEF, June 26, 1961, "Proposed Letter to the President on Resumption of Nuclear Weapons Testing," JCS 1731/466.

110 "Statement by the President on Ordering Resumption of Underground Nuclear Tests," September 5, 1961, *Kennedy Public Papers, 1961*, 589–590. On November 30, 1961, Kennedy approved the resumption of atmospheric testing but delayed announcing the decision until March 1962 just as tests were scheduled to begin.

111 SNIE 11-14-61, November 1961, "The Soviet Strategic Military Posture, 1961–1967," in Gerald K. Haines and Robert E. Leggett, eds., *CIA's Analysis of the Soviet Union, 1947–1991: A Documentary Collection* (Washington, DC: Center for the Study of Intelligence, 2001), 234–235; Glenn T. Seaborg, *Kennedy, Khrushchev, and the Test Ban* (Berkeley, CA: University of California Press, 1981), 114; Poole, *JCS and National Policy, 1961–64*, Pt. I, 371.

112 Christian Brahmstedt, *Defense's Nuclear Agency, 1947–1997* (Washington, DC: Defense Threat Reduction Agency, 2002), 160–163.

113 Seaborg, 176.

114 See JCSM-160-63 to SECDEF, February 22, 1963, "Draft Treaty Banning Nuclear Weapon Tests in All Environments," JCS 1731/672; JCSM-234-63 to SECDEF, March 19, 1963, "Draft Treaty Banning Nuclear Weapons Tests," JCS 1731/684; JCSM-241-63 to SECDEF, March 21, 1963, "US/USSR Weapons Capabilities," JCS 1731/668; and JCSM-327-63 to SECDEF, April 20, 1963, "Nuclear Test Ban Issue," JCS 1731/696.

115 CM-643-63 to DJS, June 4, 1963, "Test Ban Hearings before the Stennis Subcommittee," JCS 1731/707; Report, Special Assistant to JCS for Arms Control to JCS, June 11, 1963, "JCS Views on Important Aspects of the Test Ban Issue," JCS 1731/707-3.

116 Ronald J. Terchek, *The Making of the Test Ban Treaty* (The Hague: Martinus Nijhoff, 1970), 22.

117 "Instructions for Hon. W. Averell Harriman," July 10, 1963, *FRUS 1961–63*, VII, 786.

118 Taylor, 285–287.

119 Poole, *JCS and National Policy, 1961–64*, Pt. I, 399–402.

120 "Statement of Position of the Joint Chiefs of Staff on the Three–Environment Nuclear Test Ban Treaty," August 12, 1963, Enclosure A to JCS 1731/711-30. See also Seaborg, 269–270.

Robert S. McNamara, Secretary of Defense, 1961–1968

Chapter 8

THE MCNAMARA ERA

The assassination of President Kennedy on November 22, 1963, shook the Joint Chiefs as much as the country at large and left a void that the new President, Lyndon B. Johnson, moved quickly to fill. To reassure the Nation and to promote stability, he pledged continuity between his administration and Kennedy's. "I felt from the very first day in office," he recalled, "that I had to carry on for President Kennedy. I considered myself the caretaker of both his people and his policies."[1] One of those who stayed on was Secretary of Defense Robert McNamara. The dominant figure at the Pentagon before Kennedy's death, McNamara would exercise even more power and authority during Johnson's Presidency.

By the time Johnson became President, the Joint Chiefs were grudgingly coming to terms with McNamara's policies and methods. Under Kennedy, McNamara had firmly established his authority, using it to carry out two major revolutions within the department—one, to redesign the military strategy and Armed Forces of the United States to achieve greater flexibility and effectiveness; the second, to install new methods of analysis and decisionmaking in the areas of planning, management, and acquisition.[2] Like Kennedy, McNamara grew to be skeptical of JCS advice and once characterized the Joint Chiefs as "a miserable organization" hamstrung by collegial and parochial interests.[3] For analytical support, he established his own group of advisors known as the "whiz kids," made up predominantly of young and eager civilians who routinely checked and double-checked the programmatic recommendations of the military Services and the Joint Chiefs of Staff. Out of this process emerged both a new approach to solving defense problems and a significant expansion of the power and authority vested in the Office of the Secretary of Defense.

THE MCNAMARA SYSTEM

By late 1963, when the Johnson administration assumed office, McNamara had largely accomplished what he initially set out to do—transform the Department of Defense into a more tightly knit and efficient organization. The original impetus for these changes lay in the increased defense spending during the final years of the Eisenhower administration in response to Sputnik and the perceived Soviet lead in missile technology. To cope with these issues, Eisenhower had backed away from the rigid budget ceilings he had imposed earlier and began accepting increases in military spending that soon threatened to get out of hand. Further additions to defense spending at the outset of the Kennedy administration exacerbated the situation. In assessing the underlying causes of the problem, McNamara and his staff identified two principal culprits: a compliant Congress, which was prone to overspend on defense programs; and the Joint Chiefs of Staff, whose elaborate but ineffectual strategic planning process had failed to apply the necessary discipline in determining military requirements, curb excessive expenditures, and eliminate unnecessary duplication in Service programs. Though there was not much McNamara could do to reform Congress, he used his newly minted authority under the 1958 amendments to the National Security Act to bring defense planning and programming under his direct control.

The system McNamara imposed during the first 2 years of the Kennedy administration came into effect with limited consultation between the Secretary's office and the Joint Chiefs of Staff. Kennedy wanted a more robust defense posture, which he expected McNamara to achieve, in part, through better management. At the heart of McNamara's reforms was the use of computer modeling techniques known as "systems analysis," which the Secretary and his staff used to develop 5-year projections of military spending based on "program packages" in various functional areas such as strategic nuclear forces, general purpose forces, continental defense, and airlift and sealift forces. The organizing mechanism was the planning, programming, budgeting system (PPBS), a sophisticated decisionmaking apparatus for integrating Service requirements and national objectives. Recommendations to the President took the form of draft Presidential memorandums (DPMs) detailing force levels and their funding for the upcoming fiscal year and projections for the next 4 years. Initially in 1961, McNamara submitted two DPMs for the President's consideration; by 1968, when he stepped down, he was submitting 16.[4]

Through refinements to the Joint Program for Planning, the JCS had tried, with mixed success, to develop something similar in the 1950s. The focus of the JCS effort had been the Joint Strategic Objectives Plan (JSOP), a mid-range projection of military requirements.[5] When the Joint Chiefs adopted the JSOP format in the early 1950s, they envisioned it serving as a statement of integrated requirements that would be updated annually to assist in smoothing out the ups and down of the budget cycle. But because of disagreements over basic strategy—especially the relative balance between strategic and general purpose capabilities—the Services were constantly at odds over force-level recommendations, known in JCS parlance as "force tabs." By the end of the decade, the Joint Chiefs had given up trying to produce an integrated plan and had turned the JSOP into a compilation of unilateral Service estimates, organized in no particular order of priority, as their projection of future needs. Invariably, these estimates exceeded available funding.[6]

As a rule, McNamara and his staff paid little serious attention to the JSOP, which they and other critics of the JCS system dismissed as a "wish list." In the spring of 1962, McNamara introduced an alternative means of calculating requirements known as the Five-Year Defense Program (FYDP), a mission-oriented projection of future costs and manpower. To justify the estimates in the FYDP, the Office of the Secretary of Defense prepared a lengthy and detailed analysis known as the Secretary's "posture statement." Under Lemnitzer's leadership, the Joint Chiefs sought more generalized guidance and supported the adoption of a broad basic policy paper, similar to those generated under the Truman and Eisenhower administrations. However, after General Maxwell Taylor's arrival as Chairman of the Joint Chiefs, the JCS position changed. Finding little practical value in such papers, Taylor prevailed in getting the project cancelled.[7] In January 1963, President Kennedy confirmed that the project was dead and indicated that the Secretary's posture statement, along with other "major policy statements" by senior officials, would constitute the country's basic national security policy.[8]

McNamara hoped that, using his posture statement as guidance, the Joint Chiefs would turn the JSOP into "a primary vehicle for obtaining the decisions on force structure necessary for validating the ensuing budget."[9] The first Chairman to take up the task was General Taylor, whose efforts yielded mixed results. Knowing how intractable the JCS system could be, Taylor had no illusions and told Joint Staff officers assigned to preparing the JSOP that reaching a consensus on the rationale for force requirements was imperative, no matter how difficult the task. As a first step, he ordered the JSOP redesigned to incorporate some of the same supporting rationales as in the Secretary's DPMs. But even though there was some progress toward harmonizing Service interests, neither he nor his successor, General Earle G. Wheeler, USA, could ever totally eliminate Service "splits" and present the Secretary with a fully integrated statement of military requirements. On the contrary, instead of going down, the number of splits went up, from 13 in 1962 to 43 in 1963 and 47 a year later.[10] From the mid-1960s on, with Wheeler in charge and attention focused on meeting requirements in Vietnam, the Joint Chiefs lost interest in trying to reform the JSOP and left it as it was—a compilation of Service estimates in no particular order of priority that routinely averaged 25 to 35 percent above authorized levels.[11]

RECONFIGURING THE STRATEGIC FORCE POSTURE

McNamara's most ambitious reforms were in reconfiguring the size and composition of the strategic nuclear deterrent. Kennedy wanted a more flexible force posture with less emphasis on nuclear retaliation, but he had also campaigned for the Presidency on claims that the United States had fallen behind the Soviet Union in

ICBMs. Though McNamara suspected soon after taking office that the infamous "missile gap" was overstated, it was not until August–September 1961 that Kennedy came to a similar view.[12] In consequence, during the administration's early months, McNamara was under heavy pressure from the White House to make "quick fixes" to bolster strategic forces that would shore up the defense posture. Paying little attention to the slow-moving Joint Chiefs, he turned to his systems analysis experts to produce the program he needed. Drawing heavily on work done earlier at the RAND Corporation and in the Weapons Systems Evaluation Group, McNamara and his staff promptly assembled a list of remedial measures—acceleration of the Polaris missile submarine program, increased production of Minuteman ICBMs, and improved alert measures for portions of the manned-bomber fleet.[13] To help offset the cost of these improvements, McNamara accelerated the phase-out of older systems (notably the Atlas and Titan I ICBMs, the Snark intercontinental cruise missile, and the B-47 bomber, long the work horse of the Strategic Air Command) and ordered a closer look at several other high-profile programs. Most prominent among the latter was the B-70 supersonic bomber, the planned follow-on to the B-52, which McNamara canceled over strenuous objections from Air Force Chief of Staff General Curtis E. LeMay.[14]

McNamara was concerned that "some in the U.S. Air Force" were striving under the massive retaliation doctrine for nothing less than a first-strike capability that would completely disarm the Soviet Union.[15] Persuaded that such a force posture was neither sound nor practicable, he asked the Joint Chiefs to develop a "doctrine" ending reliance on massive retaliation and establishing in its place a set of controlled responses allowing for pauses to negotiate an end to nuclear exchanges.[16] The JCS cautioned that acquiring the requisite capabilities would be expensive; in April 1961 they added a further caveat that to pursue the matter at the present time could "gravely weaken" nuclear deterrence.[17] These replies seem only to have whetted McNamara's interest all the more, and for the next several years he and the Joint Chiefs engaged in a running battle to redefine U.S. strategic doctrine. Much of the conflict centered on the particulars of the SIOP—the Single Integrated Operational Plan for nuclear retaliation against the Sino-Soviet bloc—but there were also broader considerations affecting the worldwide disposition of forces and the design and acquisition of new weapons systems. By the time all was said and done, the United States had adopted a new principle as the basis for its nuclear strategy. Known as "assured destruction," the new concept rested on a "triad" of land-based ICBMs, submarine-launched ballistic missiles (SLBMs), and long-range bombers.

Assured destruction was part massive retaliation and part controlled response. As McNamara described it to President Johnson in December 1963, assured destruction was "our ability to destroy, after a well planned and executed Soviet surprise attack on our Strategic Nuclear Forces, the Soviet government and military controls, plus a large percentage of their population and economy (e.g., 30 percent of their population, 50 percent of their industrial capacity, and 150 of their cities)." Damage beyond those levels, McNamara believed, would be gratuitous and not cost-effective.[18] McNamara would have preferred a more controlled and measured execution of strategic options, and on two occasions—at a closed meeting of the NATO ministers in Athens early in 1962 and publicly at Ann Arbor, Michigan, that spring—he lofted the trial balloon of a "counterforce/no-cities" doctrine that downplayed attacks on urban-industrial areas in favor of retaliation against high-priority politico-military targets. Yet the counterforce doctrine, as McNamara conceived it, failed to catch on. Considered impractical by the Joint Chiefs, it received an even cooler reception in Europe, where many leaders viewed it as weakening deterrence by relieving the Soviets of the threat of wholesale nuclear destruction.[19] Khrushchev, for his part, suspected a ruse. Upon learning of the Ann Arbor speech, he thought McNamara was trying to conceal a secret expansion of America's nuclear arsenal.[20]

JCS skepticism rested on the high demands that the counterforce/no-cities doctrine would place on strategic assets. Except for LeMay, a die-hard proponent of massive retaliation, the Joint Chiefs were amenable to adding flexibility to the SIOP and to strategic plans in general.[21] But they insisted on firm assurances of having the time and money to make the necessary changes in plans and force structure. To execute something as complex as the no-cities strategy, the JCS estimated, would involve expanded requirements for weapons and supporting command, control, communications, and intelligence (C3I) that would necessitate funding well above current and foreseeable levels. Though McNamara and his systems analysts routinely picked apart the JCS numbers, they were never able to overcome the chiefs' fundamental argument that it would take an unstinting dedication of resources extended over a period of years, if not decades, to achieve reasonable confidence of success. In consequence, McNamara gave up on seeking sweeping revisions in the SIOP and settled for piecemeal changes resulting in the gradual introduction of greater flexibility and more selective targeting options.[22]

To meet the Secretary's targeting requirements, the Commander in Chief, Strategic Air Command (CINCSAC), General Thomas S. Power, estimated that SAC would need 10,000 ICBMs by the end of the decade.[23] Favoring quantity over quality, Power wanted as many weapons as possible with which to threaten the Soviets.[24] Taking a more reserved approach, the Joint Chiefs recommended between 1,350 and 2,000 deployed ICBMs.[25] Comparing JCS estimates with the intelligence on existing and projected Soviet capabilities, McNamara concluded that the decisive factor was the number of targetable bombs and warheads, not delivery vehicles. Operating on this premise, he persuaded President Johnson to accept the eventual leveling-off of strategic programs at 41 ballistic missile submarines with a total of 656 launchers, 1,054 ICBMs, and approximately 600 long-range (B–52) bombers. According to those familiar with McNamara's thinking, the numbers he chose were arbitrary, but to give them greater credibility he paired them with the concept of assured destruction.[26]

Air Force leaders, who were most directly affected by the new force structure, were dismayed and openly critical. Having struggled for years to gain a decisive advantage over the Soviet Union, they saw their efforts coming to naught. As one later put it, McNamara and his OSD staff "did not understand what had been created and handed to them. SAC was about at its peak. We had, not supremacy, but complete nuclear superiority over the Soviets."[27] Yet to McNamara, nuclear superiority was an ephemeral thing, perhaps attainable for a short while but difficult if not impossible to perpetuate without an open-ended commitment of resources. Convinced that neither side could ever "win" a nuclear war, he opted for lesser capabilities, which he thought would do more to save money, promote deterrence, and achieve a stable strategic environment in the long run.

The impact of these decisions extended well beyond restructuring the strategic deterrent. First, it ended the Joint Chiefs' exclusive monopoly on strategic nuclear planning, a function they had exercised without serious challenge as one of their statutory responsibilities since World War II. Henceforth, insofar as basic policy and targeting doctrine were concerned, strategic nuclear planning became a shared responsibility of the JCS and civilian analysts in OSD. Only the actual preparation of SIOP remained firmly under JCS control. Second, it brought about a reordering of spending priorities that dramatically reshaped both the military budget and the Pentagon's claim on resources. From consuming nearly 27 percent of defense spending when the Kennedy administration took office in 1961, strategic forces declined to slightly over 9 percent by the end of the decade. During this same time, national defense (comprising the Department of Defense and related security programs) dropped from 9.1 percent of the country's gross national product, to 7.8 percent. McNamara hoped that, out of the savings realized from cuts in strategic programs, he could bolster conventional capabilities. Yet the demands of the Vietnam War and competition for funds from President Johnson's "Great Society" and other civilian-sector programs disrupted his plans. As a result, spending on general purpose forces increased only slightly over the decade, from 33 percent to just under 37 percent of the military budget.[28]

Still, when McNamara was finished, the country's defense posture was vastly different from when he became Secretary. Most notably, strategic doctrine placed less emphasis on carrying out preemptive attacks than at any time since the end of World War II. In terms of size, composition, and destructive power, U.S. strategic nuclear forces functioned largely as a second-strike deterrent geared toward inflicting punishing retaliation. Meanwhile, as the United States was reining in its strategic programs, the Soviet strategic buildup was starting to surge with the deployment of a second generation of ICBMs (see below). In consequence, by the early 1970s the two sides had reached approximate parity in strategic nuclear power. With the United States then preoccupied in Vietnam, the loss of strategic superiority was barely noticed at the time other than by the Joint Chiefs, a few astute Members of Congress, and a small coterie of academics and strategic analysts. But from that point on, the Joint Chiefs' confidence in being able to confront and deal with the Soviets would never be the same.

NATO AND FLEXIBLE RESPONSE

The quest for greater choice and flexibility that drove the Joint Chiefs to accept changes in U.S. strategic doctrine also inspired the Kennedy and Johnson administrations to seek a reordering of military priorities in Europe. During the 1961 Berlin crisis, President Kennedy had set great store in a nonnuclear buildup, both to impress upon the Soviets the seriousness of Western resolve and to expand the range of plausible military options to lessen the need for early recourse to nuclear weapons. But he had had trouble explaining to the Joint Chiefs and to the European Allies what he wanted to do and how. While flexible response was well formed in theory, it was less refined in practice. As a deterrent, its reliability and effectiveness were untested. In contrast, the concept of nuclear deterrence was widely known and accepted, and while it entailed great risks, it was also far more affordable to Europeans than a conventional defense, since the United States shouldered most of the costs of nuclear forces. Despite its ominous implications and potential dangers, a nuclear-oriented defense posture continued to enjoy strong support in Europe.

At the outset of the 1960s, NATO strategy (MC 14/2) rested on the Eisenhower era concept of massive retaliation and made no allowance for trying to defend Europe by fighting a large-scale conventional war.[29] The product of painstaking negotiation and delicate compromise, MC 14/2 embodied the "trip wire" theory, under which the primary function of conventional forces was to delay a Warsaw Pact invasion until NATO could mount a nuclear response. Some of the nuclear weapons at NATO's disposal were British, in accordance with a pledge made by Prime

Minister Clement R. Attlee in December 1950 dedicating his country's nuclear arsenal (once it came into being) to NATO.[30] But the bulk of the Alliance's atomic capabilities consisted of American bombs, warheads, and delivery systems assigned and/or deployed to Europe under bilateral agreements with the host countries and targeted by SACEUR in collaboration with the Joint Strategic Target Planning Staff in Omaha.[31] At the time, the United States still had Thor IRBMs in the United Kingdom and Jupiter MRBMs in Italy and Turkey. While the missiles were in the operational hands of the host governments, their warheads were under a "dual key" system, according to which the United States and the host country shared custody and control.[32] Looking beyond the current situation, General Lauris Norstad, USAF, who served as SACEUR until 1962, and his successor, General Lyman Lemnitzer, USA, both subscribed to the view that NATO should eventually have its own organic nuclear capability. With this end in view, the United States had come up with the idea of a multilateral nuclear force (MLF) in the late 1950s.[33]

Under the revised (flexible response) strategy that McNamara proposed, the trip wire would give way to a conventional defense as far forward as possible to meet and defeat Soviet aggression near the point of attack. McNamara could see no practical application for tactical nuclear weapons and lacked confidence in using them without risking escalation to a full-scale nuclear exchange. Rather than relying on tactical nuclear weapons, he stressed the security NATO enjoyed under the protective "umbrella" of the U.S. strategic deterrent, a concept known as "extended deterrence." But with U.S. policy and doctrine in flux, European leaders wanted more concrete assurances of nuclear support. Hence their continuing interest, now stronger than ever, in achieving something along the lines of the MLF. On both sides of the Atlantic, military planners continued to believe that tactical nuclear weapons were NATO's first line of defense and that selective use of atomic firepower, even though it might heighten risk, would not necessarily result in total war. Weighing one thing against another, the Joint Chiefs urged McNamara to move cautiously in making changes in NATO's strategy and defense posture and to observe "a proper balance between nuclear and non-nuclear forces."[34]

Had this been the Truman or Eisenhower administration, there probably would have been an in-depth interagency study, with detailed inputs from the Joint Chiefs and other agencies, coordinated through the NSC, to develop a master plan of action. But neither President Kennedy nor those close to him, including McNamara, had the patience for what they considered the tedious consensus-building and hair-splitting staff work of the past. As usual, McNamara paid less attention to professional military advice than to his civilian systems analysts. First up for review was the Soviet Order of Battle. Using computerized models, OSD analysts concluded that it was practically impossible for the Soviet economy to train, equip, and sustain, along with other forces, 175 front-line divisions, the benchmark figure applied by Western intelligence since the late 1940s for "sizing" the Soviet army. Applying different methods, the CIA reached a similar conclusion. Based on a reexamination of evidence accumulated over the past decade, the CIA calculated that instead of 175 divisions, the Soviets had closer to 140, with at least half at reduced strength, some no more than cadres. Persuaded that previous estimates had exaggerated the Soviet threat, McNamara became convinced that with modest increases in Alliance spending (about $8.5 billion spread over 5 years) and technical improvements, NATO could carry out a "forward strategy" and hold its own in a conventional confrontation with the Warsaw Pact.[35]

While McNamara had some valid points, he and his staff presented their arguments clumsily to the Europeans. In so doing they antagonized the NATO allies and made them more resistant than ever to change. As a result, the Europeans became suspicious of the Kennedy administration's whole approach to nuclear deterrence, from its contemplated shift at the strategic level to a counterforce/no-cities doctrine, to its proposed curbs on theater and tactical weapons.[36] What most European political leaders and military planners wanted was more nuclear support, not less, and greater control over the assets at hand in case the United States reneged on its commitments. McNamara, conversely, was set on limiting both.

THE SKYBOLT AFFAIR

Throughout the debate over Europe's nuclear future, the Joint Chiefs found themselves increasingly marginalized as McNamara and his whiz kids took matters more and more into their own hands. While it was one thing for the JCS to be ignored, it was quite another to have a majority recommendation blatantly overruled as happened with the "Skybolt" program, which McNamara decided to cancel in November 1962. Initiated under Eisenhower, Skybolt was a strategic air-to-surface ballistic missile being developed by the Air Force in collaboration with the British, who planned to use it to prolong the active service life of their obsolescent Vulcan bombers. Late in 1959, seizing on an offer from Washington to codevelop Skybolt, the British shelved a similar program ("Blue Streak"), which they had been pursuing on their own.[37] With a planned range of over 1,000 miles, Skybolt's mission was to carry out stand-off attacks against targets inside the Soviet Union. By the end of the Eisenhower administration, however, technical problems and rapidly escalating costs threatened the program's future. Aware that these issues could scuttle Skybolt, President Kennedy saw an opportunity to pressure the British into phasing out their nuclear deterrent, in keeping with the administration's policy of curbing nuclear proliferation. In April 1961, he authorized McNamara to explore such a possibility if the missile failed to measure up.[38] McNamara continued to nurse the project along, but in August 1962 both the OSD Comptroller, Charles J. Hitch, and the Director of Defense Research and Engineering, Harold Brown, advised McNamara to terminate Skybolt. From the technical data laid before him, McNamara concluded that Skybolt was "a pile of junk."[39]

Despite his growing skepticism concerning Skybolt, McNamara hedged a final decision on the program's future. On November 8, he told the British ambassador that the United States was "reconsidering" the program, but conveyed the impression that any action was conditional upon the receipt of JCS views.[40] As part of their annual review of the Secretary's budget submission, the Joint Chiefs weighed in with a split recommendation. Insisting that Skybolt was necessary to maintain a "clear margin of superiority," the Service chiefs unanimously favored retaining the program. However, the Chairman, General Taylor, disagreed. The newest member of the JCS, Taylor had no vested interests to protect and felt that he could view the situation more objectively. Terming Skybolt "a relatively marginal program," he shared the prevailing view in OSD that the money could be better spent on other systems.[41]

In the past, upon receiving a split recommendation, the Secretary of Defense invariably sided with the majority or sought a compromise. But in this case, Taylor's lone dissent prevailed. While taken on technical and cost-effectiveness grounds, McNamara's decision to cancel Skybolt nevertheless had strong geopolitical overtones. Without Skybolt, McNamara knew that the only readily available alternative the British had was the Blue Steel Mk I, an air-launched missile with limited range and penetration capabilities. Lacking a more up-to-date and effective system, Britain's entire nuclear weapons program would face an uncertain future.[42] In early December, McNamara flew to London and presented British Minister of Defence Peter Thorneycroft with three options—continue Skybolt as a solely British program, adopt a less capable U.S. weapon, the "Hound Dog," or participate in whatever arrangements emerged from ongoing discussions to create an MLF. Lurking in the background was a fourth possibility—British acquisition of Polaris technology. But to McNamara's surprise, Thorneycroft did not raise it. In fact, the British Ministry of Defence had explored this option earlier but considered it too costly and incompatible with Britain's overall weapons and shipbuilding program.[43]

Just before Christmas, at a mini-summit between President Kennedy and Prime Minister Harold Macmillan at Nassau, a solution emerged. In preparation for the meeting, Taylor asked McNamara whether he should attend to assure the availability of a senior military advisor should the need arise. McNamara told the Chairman to stay home since "substantive discussions" appeared unlikely.[44] Whether McNamara simply misread the situation or was purposefully excluding the CJCS is unclear. Yet even if Taylor had been there, the outcome doubtless would have been the same. Conceding that Skybolt was a lost cause, Macmillan agreed that acquiring Polaris was the only choice that made sense if Britain were to remain a strategic nuclear power. While the British would supply their own nuclear warheads, the boats and missiles would conform to U.S. design. The decision was technically without prejudice to the future of the UK's independent deterrent, but it came with strings attached that severely limited British freedom of action. Most constraining of all was the requirement that all forces acquired by the UK under the agreement be "assigned and targeted" as part of a NATO nuclear force in keeping with current practice. Only if Britain's "supreme national interests" were at stake could it withdraw its Polaris boats from NATO command and control.[45]

DEMISE OF THE MLF

Over the long run, the Nassau agreement probably created more problems than it solved. Not only did it show the Kennedy administration backtracking from its declared policy of curbing nuclear proliferation; it also resurrected the notion of a U.S.-UK "special relationship," which the French, Germans, and other Europeans resented. In fact, the close links forged between Washington and London in World War II were long gone. But by agreeing to share some of its most sensitive military technology with the British—technology to which no other Alliance member had comparable access—the Kennedy administration had left itself vulnerable to charges of favoritism.

Earlier, anticipating that problems of this sort might arise, the Kennedy White House had endorsed a variation on the theme of a NATO-wide multilateral force. The original MLF concept of the late 1950s had the strong personal imprint of the Supreme Allied Commander, General Lauris Norstad, who envisioned a mix of land- and sea-based medium-range ballistic missiles to replace aging aircraft and the obsolescent Thor and Jupiter missiles in the United Kingdom, Italy, and Turkey.[46] Had Norstad's conception of the MLF prevailed, NATO would have become, in effect, the fourth nuclear power, alongside the United States, the United Kingdom, and the Soviet Union. But with the advent of the Kennedy administration, Norstad's vision faded almost immediately. Determined to reduce nuclear proliferation, President Kennedy discouraged the creation of an autonomous nuclear force under NATO and proposed in May 1961 that NATO concentrate on strengthening its conventional forces rather than its nuclear posture. To minimize the risk, he reaffirmed an offer made by President Eisenhower that the United States would dedicate five U.S. Polaris submarines to NATO and move toward the next stage—the creation of a full-blown MLF—"once NATO's nonnuclear goals have been achieved."[47]

Though McNamara and the Joint Chiefs saw no compelling military need for the MLF, they went along with the idea largely in deference to the enthusiastic backing it had among Kennedy loyalists in the State Department. To this group, the MLF was a crucial component of the President's "Grand Design" for European political, military, and economic integration, and another step toward eventually achieving European union. The Pentagon's main contribution was to push the concept in the direction of a predominantly, if not exclusively, sea-based system to expedite the project and to minimize costs through the use of existing technologies. The proposed force would comprise 25 surface vessels armed with 200 Polaris A–3 missiles, manned by multinational crews and funded collectively by contributions from NATO members. Costs would be limited to 1 to 5 percent of a nation's military budget. Though SACEUR would have operational command and control of the ships and their missiles, the United States would retain custody of the warheads and exercise ultimate veto power over their use. Many Europeans disparaged these arrangements as being not much better than the current system.[48]

The Kennedy administration's recasting of the MLF concept encountered no strong objections from the JCS. Lukewarm toward the MLF from the start, the chiefs supported it as long as it posed no excessive drain on American resources and caused no major diversion of assets from SAC or other major commands. Their most serious concerns had to do with the composition of the force. Siding with Norstad, they repeatedly urged McNamara to include mobile land-based MRBMs in the MLF, along with Polaris. More accurate and reliable than sea-based missiles, land-based MRBMs would give NATO a broader range of capabilities and options and help dissuade the FRG and others from developing independent nuclear capabilities outside of NATO.[49] McNamara, however, believed that a European land-based missile force would drive up costs and duplicate functions already assigned to U.S. strategic forces. Still, in deference to growing pressures, he agreed to think about it and acknowledged to NATO leaders in May 1962 that a land-based MRBM might be acceptable to the United States under the right conditions.[50]

Despite efforts by Washington to come up with an acceptable plan, European opponents of the MLF, led by the French, continued to make headway. Turning their back to a multilateral solution, the French remained focused on acquiring an independent nuclear *force de frappe*. A low-key affair for much of the 1950s, the

French nuclear program grew out of theoretical studies dating from 1951 and gathered momentum quickly in the aftermath of the Suez affair when the United States failed to support Britain, France, and Israel in their attack on Egypt. Convinced that the Americans were capricious friends, the French sought a "trigger" that was certain to bring U.S. nuclear power to bear regardless of American policy. Denied American assistance, France pursued collaboration with Italy and West Germany. With the return to power of General Charles de Gaulle in 1958, this brief partnership ended and France embarked on unilateral development. In February 1960, France detonated its first atomic explosion, a plutonium bomb, in the Sahara desert. At first, the French relied on air-delivered weapons using Mirage IV bombers. But as the 1960s progressed, they expanded their arsenal to include silo-based IRBMs and submarine-launched ballistic missiles.[51]

For de Gaulle, the force de frappe was part of a larger effort to restore France's faded power, glory, and international prestige. Leader of the Free French in World War II, de Gaulle had emerged from his wartime experience feeling that the British and Americans had slighted him. According to diplomatic historian Erin R. Mahan, de Gaulle carried "a smoldering animosity toward *les Anglo-Saxons*" practically his entire adult life.[52] Dismissing the MLF as "a web of liaisons," he opposed any measure that did not give France a veto over the use of nuclear weapons. In place of the MLF, he wanted a tripartite (Anglo-French-American) directorate, with each country having an equal voice in decisions on when and where to use nuclear weapons. After the Nassau conference, he became convinced that NATO was in the hands of an Anglo-Saxon cabal and redoubled his efforts to assure France its independence in foreign and defense affairs, a process leading eventually to the announcement in February 1966 that French forces would cease to operate under NATO's integrated command.

Never excessively strong to begin with, the momentum behind the MLF slowed to a crawl in the face of unrelenting French resistance and the lukewarm support of other NATO members. Scrambling to salvage what he could, President Kennedy sent veteran diplomat Livingston Merchant to Europe in the spring of 1963 to mobilize British and West German support for the beleaguered MLF. Despite Merchant's upbeat reports, he achieved no major breakthroughs.[53] About the same time, under pressure from McNamara, the Joint Chiefs offered a tepid endorsement of the MLF, not for its military value (which they assessed as negligible) but as a brake on nuclear proliferation. The chiefs' support, however, made little difference. By the time of Kennedy's assassination, the MLF was practically moribund, the victim of its own muddled objectives and shortcomings and waning interest on both sides of the Atlantic.[54]

Learning from his predecessor's experience, President Johnson distanced himself from the MLF and never seriously pursued it.[55] Nevertheless, on the off chance that there might be a revival of the idea, McNamara asked the Joint Chiefs in the summer of 1964 for a fresh analysis of the MLF's command and control procedures, with particular attention given to the prevention of an unauthorized or accidental detonation. McNamara wanted to reassure anxious Members of Congress that a "pilot project" involving a NATO crew operating a U.S. guided-missile destroyer, USS *Claude V. Ricketts*, would not compromise the custody and control of any U.S. nuclear weapons.[56] But even though the Navy rated the *Ricketts* experiment a success, it failed to generate any appreciable renewed support for the MLF. Dropped from further discussion at NATO meetings, the MLF passed into the history books sometime in late 1964 or early 1965, with the exact date of its obsequy still unknown.

A NEW NATO STRATEGY: MC 14/3

Following the MLF's demise, the Johnson administration sought other arrangements for nuclear sharing and coordination. The JCS wanted to explore closer cooperation through military channels between U.S. and French nuclear forces, with the goal of eventually integrating the force de frappe into NATO.[57] But as it became apparent that Paris was determined to pursue an independent course, not only in nuclear affairs but in all aspects of military planning, the United States dropped efforts to placate de Gaulle and refocused on strengthening neglected ties with the FRG and other NATO members. Meanwhile, with McNamara in the forefront, the Johnson administration continued to push for formal adoption of a forward defense strategy resting on flexible response. The upshot during 1966–1967 was the creation of a new high-level consultative body, the Nuclear Planning Group (NPG), to guide the Alliance in nuclear matters, and a reconfiguration of basic NATO strategy around a new policy directive (MC 14/3) that finally brought the era of massive retaliation to a close.

Overshadowing these accomplishments was a perceptible diminution of American power and influence within the Alliance, accelerated by the American preoccupation with Vietnam and the attendant diversion of resources. Unable to give NATO the time and attention accorded it in the past, the Johnson administration struggled to preserve U.S. leadership. The most serious challenger remained de Gaulle, whose assault on the U.S. dollar and unrelenting criticisms of American foreign policy left the Alliance in tension and disarray until February 1966, when France announced the withdrawal of the last of its forces from NATO command.

By any measure, de Gaulle's decision to secede from the NATO military structure (the first and only defection of its kind until Greece withdrew its forces in 1974 over the Cyprus issue) was a severe blow to Alliance solidarity and to American prestige. Summarily evicted from its facilities in France, NATO's weighty military and civilian bureaucracy had to scramble to find new offices and headquarters. Though relocating to Belgium proved less difficult than the Joint Chiefs expected, it was still a major disruption that left the Alliance dependent in the short term on hastily organized and largely untested lines of support and communication.[58]

The erosion of the American presence in Europe was especially apparent from the shrinking size and quality of the U.S. forces committed to NATO. As of the mid-1960s, just as the Vietnam buildup was beginning, the United States had almost 5 Army divisions, 3 regimental combat teams, and 28 combat air squadrons assigned to Europe. But because of rising costs, the French drain on U.S. gold reserves, and growing requirements in Southeast Asia, it was only a matter of time before the United States reassessed its military role in Europe. The Joint Chiefs invariably opposed cutbacks in U.S. forces. Arguing that it would weaken NATO's defenses, they saw any lessening of the U.S. presence as setting a poor example and making it harder for the United States to elicit troop contributions from the European Allies. As time passed, however, and as the requirements for Vietnam grew, the chiefs' position became increasingly untenable. The solution pushed by McNamara and his systems analysts was "dual-basing"—the prepositioning of supplies and equipment in Europe and the rotation of selected units between there and the United States. Initially opposed to the idea, the JCS became more amenable when Presidential preferences for McNamara's approach left them no choice. The near-term practical results were a 10 percent troop reduction and the withdrawal from Europe of two combat brigades of the 24th Mechanized Division and three tactical air squadrons. Pleading financial difficulties, the British, Dutch, and Belgians soon followed suit with similar troop reductions.[59]

Pressures on the force structure complicated the work of NATO planners in translating flexible response into concrete plans. Once skeptical of the whole idea, the Joint Chiefs had gradually come to accept it as long as it did not rule out recourse to nuclear weapons should a defense with conventional firepower falter.[60] Accordingly, throughout the 1960s, the Joint Chiefs continued to stockpile tactical-sized nuclear weapons and delivery systems in Europe. By the end of the decade, the nuclear arsenal earmarked for NATO had doubled in size to more than 7,000 bombs and warheads.[61] At the same time, because of overriding priorities in Vietnam, U.S. reserves available to NATO declined drastically. In 1961 when Kennedy and McNamara began talking about flexible response, the United States had a strategic reserve force of one infantry and two airborne divisions earmarked for immediate deployment to Europe. By 1968, the NATO-committed reserve was down to two airborne brigades available by M+30 and one airborne, one mechanized, and two infantry brigades by M+60. Time and again, from the mid-1960s on, the Joint Chiefs urged a call-up of Reservists and an increase in Active-duty strength to overcome the shortfall. For fiscal and political reasons, President Johnson turned them down.[62]

Meanwhile, efforts to achieve a nonnuclear defense continued to meet strong resistance from NATO's European members. The most difficult to convince (once the French took themselves out of the debate by withdrawing from NATO's Military Committee in 1966) were the West Germans, who feared that flexible response would increase the risk of a conventional conflict. Clinging to the defense doctrines of the 1950s, West German military leaders contended that threatening the Soviets with the early use of nuclear weapons constituted "the very nature of the strategy of deterrence." Operating on this premise, they insisted that nuclear weapons continued to be "the most significant political instrument for the defense of NATO Europe."[63] But under persistent American pressure, their resistance gradually wore down, paving the way for NATO planners to reconcile differences and adopt the new flexible response strategy in December 1967.

A tribute to McNamara's hard work and determination, MC 14/3 was the most far-reaching revision of NATO strategy since adoption of the original strategic concept in 1950. Ending primary reliance on nuclear weapons, it mandated an initial defense "as far forward as is necessary and possible," supported by "sufficient ground, sea and air forces in a high state of readiness." While MC 14/3 did not dictate exclusive reliance on conventional arms, it clearly stated that the "first objective" should be to "counter the aggression without escalation."[64] In interpreting these instructions, the rule of thumb for Alliance planners was that NATO should be capable of mounting sustained conventional operations for up to 30 days.[65] According to Sir Michael Quinlan, Britain's leading nuclear strategist and a key participant in the debate leading up to the adoption of MC 14/3, one of the purposes behind the new strategy was to send a clear signal to the Soviets. "We rightly believed," Quinlan later related, "[that] Soviet Intelligence would obtain accounts of the policy discussions that had taken place behind closed doors, so we tried to ensure that two key messages got through to Moscow—first, NATO had faced up to the tough issues of nuclear use; and second, NATO would not take provocative or hasty action."[66]

A companion document—MC 48/3—dealt with implementation measures. Framed in broad language, MC 48/3 called for improved intelligence, coordination, readiness, and logistical support to increase NATO's capacity for flexibility in response to aggression. Unlike earlier exhortations, however, this one fell mostly on deaf ears and remained unapproved in NATO's military committee system for the next several years.[67] A more accurate barometer of NATO sentiment was the Harmel Report, adopted in conjunction with MC 14/3. Named for Belgian Foreign Minister Pierre Harmel, the chairman of the committee that produced it, the report addressed "future tasks which face the Alliance" and reflected a distinctly European perspective in urging a dual policy of defense and détente. As part of this process, it suggested exploring confidence-building measures to improve East-West relations and stepped up efforts toward arms control and disarmament. "Military security and a policy of détente," the report argued, "are not contradictory but complementary." Given the overall tenor of the Harmel panel's findings, it was clear that, while the European Allies accepted flexible response in principle, they viewed it as a less than credible form of deterrence unless accompanied by a fundamental change in the East-West political climate. Before proceeding much further in implementing flexible response, they wanted to explore relaxing tensions and improving relations with the Soviet bloc.[68]

Whether flexible response would reduce the dangers of a nuclear war never ceased to be a hotly contested issue. With the United States preoccupied in Vietnam and with many European Allies skeptical of the American commitment, the link between the security of NATO territory and nuclear weapons was as strong and as close as ever, the adoption of MC 14/3 notwithstanding. Acknowledging as much, McNamara told President Johnson that, despite "years of effort," NATO still had a long way to go "to deal successfully with any kind of nonnuclear attack without using nuclear weapons ourselves." Concurring with this assessment, the Joint Chiefs continued to see no other choice than "early selective employment of nuclear weapons" to counter even a limited Warsaw Pact attack. An agreed concept on paper, flexible response still had a long way to go before becoming an attainable objective on the battlefield.[69]

THE DAMAGE LIMITATION DEBATE

NATO's adoption of flexible response marked a major turning point in Alliance strategy. In theory, it moved away from dependence on massive retaliation and, by positing a broader range of conventional responses, lessened the dangers of a nuclear war in Europe. But by the mid-1960s, the larger and more urgent problem facing the Joint Chiefs and other Western military planners was the relentless expansion of the Soviet ICBM force. As these deployments continued, they threatened to negate the U.S. advantage in strategic nuclear power and, with it, the concept of extended deterrence on which transatlantic security ultimately rested. With their own strategic force levels effectively frozen, the JCS sought qualitative enhancements to U.S. capabilities, largely in two areas. One was a new system of multiple independently targetable reentry vehicles, or MIRVs, which enhanced the capabilities of a single long-range missile by increasing the number of warheads it could carry. The other was the advent of improved interceptors and tracking radars for ballistic missile defense, which made an American ABM a more credible and attractive option for countering the growing Soviet missile force. Out of the debate over these issues, summarily referred to by the Joint Chiefs as "damage limitation" measures, emerged not only a series of fateful decisions affecting refinements in the strategic posture, but also a new realm of negotiations with the Soviets—the Strategic Arms Limitation Talks (SALT).

MIRV appealed to McNamara and the JCS alike, but for different reasons. For the Joint Chiefs, MIRV was a way of upgrading strategic capabilities while staying within the limits of the programmed missile force, which the Secretary of Defense had capped at 1,054 ICBMs and 41 ballistic missile submarines. For McNamara, it was a convenient way of fending off JCS requests for new systems—an advanced manned strategic bomber (the AMSA, later the B–1) to replace the obsolescent B–52 and a larger, more powerful ICBM (the MX)—on the grounds that, with MIRV factored in, programmed delivery systems would more than satisfy targeting requirements. As McNamara saw it, in other words, MIRV enhanced the Services' capabilities, but it was also a mechanism for imposing restraint on the acquisition process.

Proposals for deploying multiple warheads on a missile dated from the late 1950s. The earliest missile that actually incorporated a multiple warhead design was the Navy's Polaris A–3, first tested in 1962 and declared operational aboard submarines 2 years later. Capable of carrying three 200 kiloton warheads, the A–3 employed a system of multiple reentry vehicles (MRVs) that, instead of being independently targeted, applied a "shotgun" pattern against a single target. Since SLBMs were less accurate and reliable than land-based ICBMs, targeting planners in Omaha generally held them in reserve for follow-on attacks against "soft" targets like troop concentrations and urban-industrial facilities.[70]

Fully developed MIRV systems came along later, emerging from design studies done by the Air Force's Ballistic Systems Division in the early 1960s and the Navy's Special Projects Office. More sophisticated and versatile than the A–3, a MIRVed reentry vehicle (known as a "bus") could attack several separate targets simultaneously or one target redundantly. The Joint Chiefs considered it imperative to develop a submarine-launched MIRV missile (the Poseidon), and indicated that they would also welcome MIRVed versions of the Minuteman ICBM (known as Minuteman III), which the Air Force planned to deploy in the late 1960s.[71] To increase the versatility and effectiveness of programmed forces, the JCS sought and obtained penetration aids, improvements in command and control, and increased missile accuracy. However, they were unsuccessful in persuading McNamara to accept higher-yield warheads and other qualitative improvements that would have further boosted counterforce potential by threatening "hardened" targets like Soviet missile silos and command bunkers. Although McNamara conceded that these measures would limit damage to the United States, he refused to embrace damage-limitation as his overriding priority.[72]

McNamara believed that were he to accept the full range of the Joint Chiefs' proposed enhancements and make damage limitation a high-priority objective, he would be signaling to the Soviets that the United States was striving for a first-strike capability. The result, he feared, would destabilize relations with Moscow and increase the risk of a Soviet preemptive attack in a crisis. Thus, as plans and preparations for MIRV deployment went forward, McNamara continued to think in terms of assured destruction. For Poseidon, he rejected a counterforce MIRV package consisting of warheads in the three-megaton range, and opted instead for the C-3 reentry vehicle, which could deliver a large number of relatively small warheads and was best suited for urban-industrial attacks. He likewise insisted that the Air Force's Minuteman III use a three-warhead "light" version of the MK-12 RV, a configuration the Air Force considered best suited for attacking soft targets, rather than the MK-12 "heavy" design (also known as the MK-17), which could have delivered a larger payload.[73]

Unable to make much headway with McNamara in configuring offensive forces for damage-limitation purposes, the Joint Chiefs eyed recent advances in ballistic missile defense technology to help achieve their goals. By 1965, the JCS had changed their minds about ABM and now embraced it as an essential strategic requirement in the JSOP, their annual mid-range estimate of military programs.[74] What sparked the shift in the JCS position is not clear. British historian Lawrence Freedman explains it as a reaction within the military to McNamara's policies, a feeling that the time had come to challenge his whole strategic philosophy.[75] Morton H. Halperin, who served on McNamara's staff, remembered it more as the product of tradeoffs between the Services and bureaucratic politics.[76] Personalities also played a part. As Air Force Chief of Staff, LeMay had never had much confidence in ABM being able to cope with a large-scale enemy attack. Preferring to invest in offensive weapons, LeMay had probably done as much as anyone other than McNamara to block JCS endorsement of the ABM program. However, his successor, General John P. McConnell, USAF, who joined the JCS in February 1965, was more open-minded and flexible on the missile defense issue.[77] Also, with Taylor's departure in July 1964 to become Ambassador to South Vietnam and General Earle G. Wheeler's appointment as Chairman, the JCS were again under "one of their own," in whom they had greater confidence to present their views and argue their case with the Secretary and the President.

Whatever their motivations, the Joint Chiefs had a strong incentive, based on intelligence reports, to review and change their position on missile defense. From about 50 ICBM launchers in mid-1962 and a handful of ballistic missile submarines, Soviet capabilities had increased to an estimated strength to 350–400 ICBMs and 36 ballistic missile submarines by the mid-1960s. As part of this buildup, the Soviets had phased out their first generation SS–6 ICBMs, and were proceeding posthaste with the deployment of a more effective and easier-to-use second generation (the SS–7, the SS–9, and the SS–11). Though about 40 percent of the Soviet ICBM force remained above ground in "soft" configurations, all new deployments were in hardened underground silos. The Intelligence Community projected Soviet capabilities of approximately a thousand ICBM launchers by the end of the decade (equal in number to the programmed U.S. deployment) and 40–50 ballistic missile submarines.[78]

No less unsettling was evidence that the Soviets were pursuing a well-defined ballistic missile defense R&D program, which could complicate U.S. targeting and reduce the attainment of assured destruction goals. Like the Soviet ICBM program a few years earlier, Soviet BMD development had become a source of intense controversy within the Intelligence Community. The Army and Air Force saw the Soviets engaged in a massive BMD effort, while the CIA, State Department, National Security Agency, and Naval Intelligence reserved judgment.[79] Under study were the characteristics and capabilities of three known systems: one around Leningrad, apparently started as an air-defense system, which the Soviets suddenly dismantled in 1964 prior to completion; a second, known as the "Tallinn Line," also for air defense with discernible ABM capabilities; and a third, known as "Galosh," the most advanced and sophisticated, under construction around Moscow. All three exhibited design features seen at the Soviet ABM development and test center at Shary Sagan.[80]

Worried that the Moscow system might give the Soviets a critical advantage, the Joint Chiefs recommended in early December 1966 that Secretary McNamara and President Johnson begin full-scale ABM production and deployment without delay.[81] The ABM the Joint Chiefs proposed to field was the Nike-X, successor to the Army's earlier Nike-Zeus, which offered initial protection for up to 25 American cities. In contrast to the point defense concept used in Nike-Zeus, Nike-X was a layered defense with area-wide applications. Employing the basic Zeus missile (later renamed Spartan) for long-range interception, it would use a second interceptor, the Sprint, to destroy whatever leaked through the first line of defense. The principal advantage of Nike-X over any of its predecessors was its phased array radar, a major breakthrough in battle management pioneered by the Advanced Research Projects Agency (ARPA; later, the Defense Advanced Research Projects Agency) under a set of studies known as Project Defender. Faster and more accurate than the manually operated Nike-Zeus radar system, phased array radar used solid-state electronics and high-capacity computers to process large amounts of data quickly, track multiple reentry vehicles, and guide interceptor missiles to their targets all at the same time. While there were still serious "bugs" in the system, not the least of which was its limited capacity to distinguish decoys from real warheads, Nike-X seemed a giant stride toward more effective missile defense.[82]

Nike-X had originally been McNamara's idea, an outgrowth of his efforts during the Kennedy administration to find a more reliable and cost-effective alternative to Nike-Zeus.[83] But as he drifted away from the counterforce/no-cities doctrine and became increasingly committed to the assured destruction concept, he lost interest in pursuing strategic defense and related damage-limitation options such as civil defense.[84] "It is our ability to destroy an attacker," he argued, ". . . that provides the deterrent, not our ability to partially limit damage to ourselves."[85] For advice, he turned to scientists who opposed the whole notion of ABM and contractors who doubted whether Nike-X was sufficiently advanced for deployment.[86] Contrary assessments, like the 1964 Betts Report, an internal DOD study that endorsed missile defense as both feasible and compatible with the preservation of mutual nuclear deterrence, had no apparent impact on his thinking.[87] Indeed, McNamara became firmly convinced that the pursuit of BMD by both sides was provocative and destabilizing and that it represented an open-ended invitation to a costly escalation of the arms race. He seemed to feel also that the responsibility for showing restraint fell more on the United States than the Soviet Union. "Were we to deploy a heavy ABM system throughout the United States," he maintained, "the Soviets would clearly be strongly motivated to so increase their offensive capability as to cancel out our defensive advantage."[88]

The showdown over Nike-X came during the final markup of the FY 1968 defense budget in early December 1966, shortly after the JCS recommended proceeding with deployment. By now, the Moscow "Galosh" ABM was public knowledge, and there was growing support for missile defense among key Democrats in Congress. Among these were some of the President's closest friends, including Senators Richard Russell, Jr., Henry M. Jackson, and John C. Stennis, whose continuing cooperation the White House needed in the face of mounting opposition to the administration's Vietnam policies.[89] While the President shared McNamara's concerns over an expensive and dangerous arms race with the Soviets, he leaned toward the JCS position that the time had come to settle the ABM debate.

Matters reached a head at a budget review meeting attended by McNamara, the Joint Chiefs, and the President in Austin, Texas, on December 6, 1966. Though aware that the decision could go either way, the JCS had good reason to be confident that the momentum was moving in their favor. Taking steps to outmaneuver them, McNamara offered a compromise that consisted of two components—a token deployment of Nike-X in the mid-1970s against an as-yet nonexistent (but expected) Chinese ICBM threat, to show the Soviets and critics alike that the administration was serious about missile defense, in tandem with exploratory talks to see if Kremlin leaders would be interested in a negotiated "freeze" on future ABMs. The deal was too good for Johnson to pass up. Not only would it confirm the administration's determination to respond to an increasingly dangerous situation, it would also save money at a time when the costs for the Vietnam War were becoming onerous. Above all, both elements of the compromise would shore up the President's image as a peacemaker. Making no secret of their disappointment, the Joint Chiefs acquiesced.[90]

The President's decision proved neither firm nor final. Returning to Washington, he met in early January 1967 with a group of distinguished scientists who convinced him (apparently without much difficulty) that even a limited ABM deployment would accelerate the arms race, undermine the chances for arms control, and be "extremely dangerous."[91] Johnson accepted the scientists' advice and in his budget message to Congress toward the end of the month he announced his intention to continue "intensive development" of Nike-X, but to hold production and deployment in abeyance pending exploratory talks with the Soviets to curb or freeze ABMs.[92]

Whether the talks with the Soviets would be productive remained to be seen. Until then, arms control negotiations involving the United States and the Soviet Union had yielded only two agreements—the 1963 Limited Test Ban Treaty negotiated under Kennedy, and a pending Nuclear Non-Proliferation Treaty (NPT), with key clauses still in draft form. Doubtful whether the NPT would significantly improve U.S. security, the JCS hoped the administration would not "aggressively pursue" it.[93] In the case of the proposed freeze on ABMs, as with practically all other arms control matters, the Joint Chiefs' uppermost concern was the adequacy and effectiveness of verification measures. To avoid any misunderstanding of their position, they notified McNamara that they would resist "any proposal" that might foreclose deployment of missile defenses or prevent planned improvements to offensive forces.[94]

Seeking a breakthrough, President Johnson and Secretary McNamara arranged a meeting with Soviet Premier Alexei N. Kosygin at Glassboro, New Jersey, in late June 1967, while Kosygin was in the United States addressing the UN General Assembly. A showy, impromptu affair, the Glassboro summit benefited from none of the detailed staff work and prior exchanges that might have narrowed differences and paved the way for an agreement. Neither side brought along any senior military representatives. Despite a vigorous presentation of his views, McNamara failed to convince Kosygin that a freeze on ABM deployments was in the best interests of all concerned. By stressing curbs on missile defense, he apparently misled Kosygin into believing that the United States was indifferent toward restraints on offensive arms. McNamara and Johnson rushed to assure Kosygin that this was not so. But the damage was irremeable. As McNamara recalled the scene, Kosygin "absolutely erupted." Turning red in the face, he pounded the table. "Defense is moral," he declared, "offense is immoral." Concluding that the Soviet leader probably lacked the authority to make a deal, Johnson and McNamara shrugged off their disappointment and returned to Washington empty-handed.[95]

SENTINEL AND THE SEEDS OF SALT

If it accomplished nothing else, the Glassboro summit confirmed that the administration had no choice but to move ahead on ABM. Indeed, once Kosygin rejected American overtures for a freeze, deployment by the United States became virtually certain. In addition to the Soviet threat, there was now the danger posed by the Communist Chinese, who had detonated a thermonuclear device the week before the Glassboro summit. Speculation was rife that even if the administration opted against a "heavy" ABM aimed against the Soviets, it would still deploy a "thin" defense against the Chinese.[96] A heated debate developed in Congress, while at the Pentagon the Joint Chiefs put renewed pressure on McNamara to lift the prohibition on deployment and approve the heavy system they had recommended earlier. The JCS rarely prioritized military or Service programs. But in this instance, they told McNamara that they could think of "no other action ... more necessary" to the Nation's security than full production and deployment of Nike-X. Not only would a firm decision remove all doubt about American resolve, they maintained, but also it "would either stimulate Soviet participation in meaningful negotiations or disclose their lack of serious interest in this matter." Here in a nutshell was the conundrum of Cold War arms control: to convince the other side to curb or eliminate weapons, one had first to demonstrate one's readiness to bear the risk and expense of acquiring them, if only to see them later negotiated away as a bargaining chip.[97]

With the JCS and powerful figures in Congress pushing for deployment, McNamara launched a feverish search for a credible alternative. Above all, he wanted to preserve the arms control option and avoid giving the Soviets an excuse to increase their offensive arsenal. For budgetary planning purposes, he notified the JCS in early August 1967 that he was still leaning toward a limited ballistic missile defense to deal with the emerging Chinese threat and, as a bonus, to provide a small degree of protection for Minuteman missile fields.[98] The JCS did not doubt the potential threat posed by the Chinese, but they saw that threat as rather remote and could find no urgent need for the protection of missile fields, given the imperfect accuracy of Soviet missiles and the hardness of U.S. silos. They continued to believe that the first order of business should be a full-scale nationwide ABM deployment.[99] As far as McNamara was concerned, however, the matter was closed. In a well publicized speech in San Francisco on September 18, 1967, he confirmed that the United States was going ahead with a limited ABM deployment aimed against Communist China rather than the Soviet Union. While this was a volte-face from his previous position on missile defense, McNamara insisted that his strategic objectives were unchanged and that preserving assured destruction ("the very essence of the whole deterrence concept") remained his paramount concern.[100]

Attempting to put the best possible interpretation on the Secretary's decision, the Joint Chiefs treated it not as the beginning of the end for ABM, but as the end of the beginning.[101] Nonetheless, ABM faced an uncertain future and over the ensuing year it remained the subject of intense legislative debate, diplomatic maneuvering, and Pentagon infighting. As he was leaving office in February 1968, McNamara was still cautioning against an anti-Soviet ABM and insisting that assured destruction constituted the only reliable and effective form of deterrence. Skeptical, the Joint Chiefs in April 1968 urged McNamara's successor, Clark M. Clifford, and his deputy, Paul H. Nitze, to approve a nationwide ABM system (now dubbed Sentinel) for full deployment by FY77. Their efforts, however, were no more successful with Clifford and Nitze than they had been with McNamara.[102] Even though the Army began acquiring Sentinel deployment sites during this time, it remained to be seen whether the incoming Nixon administration would carry the program forward.[103]

As the Johnson Presidency drew to a close, it was increasingly likely that the fate of ABM would be decided at the negotiating table, as McNamara had hoped. Though the Glassboro summit had failed to achieve a breakthrough, behind-the-scenes talks held afterwards in conjunction with the final negotiation of the NPT yielded broad agreement between Washington and Moscow that the time was ripe to address larger arms controls issues. On July 1, 1968, in conjunction with the signing of the NPT, the two sides announced their intention to discuss limiting offensive and defensive strategic weapons systems.[104] The date and place of these talks were about to be announced when, on August 20, Warsaw Pact forces invaded Czechoslovakia, crushing that country's nascent democracy and causing the Johnson administration to postpone arms control negotiations indefinitely. But as the Nixon administration was taking office on January 20, 1969, the Soviet Foreign Ministry expressed renewed interest in limitations on strategic arms. The long-anticipated SALT negotiations were soon to begin.

For the Joint Chiefs, as for others in the military establishment, McNamara's departure and the end of the Johnson administration constituted a watershed. In the corridors of the Pentagon it was said that the history of the Defense Department fell into two periods—before McNamara and after. Not only did the administrative and managerial reforms he instituted reshape Pentagon business practices; they also had profound effects in the areas of weapons procurement, force structure, and military doctrine. More than any other Secretary of Defense, he fundamentally transformed the way the country thought about and approached armed conflict. Prior to McNamara, the decisions affecting the force structure, its composition, and the strategic concepts under which it operated had been largely in the hands of military professionals—the Joint Chiefs of Staff—who worked under broad guidance from the President, the Secretary of Defense, and the National Security Council. But during McNamara's tenure, such decisions became a joint function of the JCS organization and analysts in the Office of the Secretary of Defense, with the latter often having the final word. Combined with the handling and political repercussions of the war in Vietnam, the net effect was a dramatically reduced role and influence for the military in national security affairs, to a level not seen since the 1930s.

NOTES

1 Lyndon Baines Johnson, *The Vantage Point: Perspectives of the Presidency, 1963–1969* (New York: Holt, Rinehart and Winston, 1971), 19.

2 William W. Kaufmann, *The McNamara Strategy* (New York: Harper & Row, 1964), 2–3.

3 Interview with Stephen Ailes by Maurice Matloff, June 6, 1986, 22–23, OSD Oral History Collection, OSD Historical Office.

4 Alain C. Enthoven and K. Wayne Smith, *How Much Is Enough? Shaping the Defense Program, 1961–1969* (New York: Harper & Row, 1971), 54.

5 Report, J-5 to JCS, May 29, 1959, "Joint Program for Planning," JCS 2089/13. Limited to 3 years at first, the JCS extended the timeframe of the JSOP to 5 years because, by the late 1950s, that was about the length of time it took to develop and field a new weapon. In the mid-1960s, it was extended to 8 to make it more comprehensive.

6. See Memo, JCS to SECDEF, June 6, 1958, "JSOP for July 1, 1962," JCS 2143/78; and Byron R. Fairchild and Walter S. Poole, *The Joint Chiefs of Staff and National Policy, 1957–1960* (Washington, DC: Office of Joint History, 2000), 37–42.

7. Walter S. Poole, *The Joint Chiefs of Staff and National Policy, 1961–1964* (Washington, DC: Office of Joint History, Office of the Chairman of the Joint Chiefs of Staff, 2011), 20–23.

8. Lawrence S. Kaplan, Ronald D. Landa, and Edward J. Drea, *The McNamara Ascendancy, 1961–1965* (Washington, DC: OSD Historical Office, 2006), 296–297. From FY72 on, the term "posture statement" applied to the annual threat assessment submitted to Congress by the CJCS.

9. CM-109-62 to DJS, November 14, 1962, "Planning Directive for Reorienting JSOP-68," JCS 2143/177.

10. JCS Decision Statistics, 1958–1982, JCS "Splits" folder, JHO 14-003.

11. Walter S. Poole, *The Evolution of the Joint Strategic Planning System, 1947–1989* (Washington, DC: Historical Division, Joint Secretariat, Joint Staff, 1989), 7–8; Enthoven and Smith, 94–95.

12. See Lawrence Freedman, *U.S. Intelligence and the Soviet Strategic Threat*, 2d ed. (Princeton, NJ: Princeton University Press, 1986), 73, 205 fn. 36.

13. Memo, SECDEF to SecArmy et al., February 10, 1961, "Military Budgets and National Security Policy," JCS 1800/401; Letter, McNamara to Kennedy, February 20, 1961, *FRUS, 1961–63*, VIII, 35–48.

14. Memo, SECDEF to Director BoB, March 10, 1961, "Revisions to Defense FY 1962 Budget," *FRUS, 1961–63*, VIII, 56–65.

15. Robert S. McNamara, *Blundering into Disaster* (New York: Pantheon Books, 1987), 51.

16. Item No. 2, "Projects Within the Dept of Defense Assigned 8 March 1961," enclosure to Memo, SECDEF to Secretaries Military Departments et al., March 8, 1961, JCS 2101/413.

17. JCSM-153-61 to SECDEF, March 11, 1961, "Foreign Policy Considerations Bearing on the U.S. Defense Posture," JCS 2101/412; CM-190-61 to SECDEF, April 18, 1961, "Doctrine on Thermonuclear Attack," JCS 1899/640.

18. DPM, December 6, 1963, "Recommended FY 1965–1969 Strategic Retaliatory Forces," *FRUS, 1961–63*, VIII, 549. See also Edward J. Drea, *McNamara, Clifford, and the Burdens of Vietnam, 1965–69* (Washington, DC: Historical Office, Office of the Secretary of Defense, 2011), 347.

19. See Kaplan et al., 305–309; and Janne E. Nolan, *Guardians of the Arsenal: The Politics of Nuclear Strategy* (New York: Basic Books, 1989), 74–77.

20. Aleksandr Fursenko and Timothy Naftali, *Khrushchev's Cold War: The Inside Story of an American Adversary* (New York: W.W. Norton, 2006), 442.

21. See JCSM-252-61, April 18, 1961, "Doctrine on Thermonuclear Attack," JCS 1899/640.

22. Desmond Ball, "The Development of the SIOP, 1960–1983," in Desmond Ball and Jeffrey Richelson, eds., *Strategic Nuclear Targeting* (Ithaca, NY: Cornell University Press, 1986), 57–70; David Alan Rosenberg, "U.S. Nuclear Strategy: Theory and Practice," *Bulletin of the Atomic Scientists* (March 1987), 23–26; and Kaplan et al., 310–311, 316–319.

23. McNamara Interview in Michael Charlton, *From Deterrence to Defense: The Inside Story of Strategic Policy* (Cambridge: Harvard University Press, 1987), 9–10. The 10,000 missile figure was apparently Powers' personal estimate and had no official Air Force standing.

THE MCNAMARA ERA

24 Bernard C. Nalty, "USAF Ballistic Missile Programs, 1962–1964" (Study Prepared for the USAF Historical Division Liaison Office, April 1966), 10–11.

25 JSOP FY 68–70, Vol. I, Pt. I, April 19, 1963, JCS 2143/201; Kaplan et al., 123.

26 Robert S. McNamara, *The Essence of Security* (New York: Harper & Row, 1968), 51–67; author's conversations with Paul H. Nitze, Henry Glass, and others.

27 Interview with General David A. Burchinal, USAF, in Richard H. Kohn and Joseph P. Harahan, eds., *Strategic Air Warfare* (Washington, DC: Office of Air Force History, 1988), 113.

28 U.S. Office of the Assistant Secretary of Defense (Comptroller), *National Defense Budget Estimates for FY 1985* (Washington, DC: DOD, March 1984), 80, 128.

29 See Fairchild and Poole, *JCS and National Policy, 1957–60*, 96–97.

30 Humphrey Wynn, *The RAF Strategic Nuclear Deterrent Forces: Their Origins, Roles, and Deployment, 1946–1969* (London: HMSO, 1994), 104.

31 Report, J-5 to JCS, April 24, 1961, "NATO-U.S. Targeting," JCS 2305/462.

32 Office of the Assistant to the Secretary of Defense (Atomic Energy), "History of the Custody and Deployment of Nuclear Weapons: July 1945 through September 1977" (February 1978), 59–61, 83–91 (declassified). Available at http://www.dod.gov/pubs/foi/reading_room/306.pdf.

33 Kaplan et al., 386–387.

34 JCSM-175-61, with Letter, JCS to Acheson, March 20, 1961, JCS 2305/414; Poole, *JCS and National Policy, 1961–64*, 188; Kaplan et al., 360–362.

35 Enthoven and Smith, 132–142; Raymond L. Garthoff, "Estimating Soviet Military Force Levels," *International Security* 14 (Spring 1990), 93–116; Poole, *JCS and National Policy, 1961–64*, 189–190.

36 See Jane E. Stromseth, *The Origins of Flexible Response: NATO's Debate over Strategy in the 1960s* (New York: St. Martin's, 1988), 48–53; Kaplan et al., 362 and passim; and Edward Drea, "The McNamara Era," in Gustav Schmidt, ed., *A History of NATO: The First Fifty Years* (London: Palgrave, 2001; 3 vols.), III, 183–195.

37 Wynn, 373–402.

38 Memo, Acting Exec Sec NSC to NSC, April 24, 1961, "NATO and the Atlantic Nations," JCS 2305/466.

39 Richard E. Neustadt, "Skybolt and Nassau: American Policy-Making and Anglo-American Relations" (Report to the President, November 15, 1963, sanitized), 4–9 (copy in Kennedy Library); McNamara quoted in Kaplan, 376, from an Oral History Interview by Alfred Goldberg and Maurice Matloff, April 3, 1986, 25–26, OSD Oral History Collection.

40 Kaplan et al., 380; Neustadt, 18.

41 JCSM-907-62 to SECDEF, November 20, 1962, "Recommended FY 1964–FY 1968 Strategic Retaliatory Forces," JCS 1800/644; CM-128-62 to SECDEF, November 20, 1962, same subject, JCS 1800/649. See also Neustadt, 22.

42 See Wynn, 403–441.

43 Peter Nailor, *The Nassau Connection: The Organization and Management of the British POLARIS Project* (London: HMSO, 1988), 3–7.

44 Poole, *JCS and National Policy, 1961–64*, 201.

45 Harold Macmillan, *At the End of the Day, 1961–1963* (New York: Harper & Row, 1973), 356–361; Kaplan et al., 402–403; Wynn, 420–421.

46 All Thors and Jupiters were retired in 1963. Removal of the Jupiters from Turkey was part of the settlement ending the Cuban missile crisis. See above, chap. 7.

47 "Address Before the Canadian Parliament in Ottawa," May 17, 1961, *Kennedy Public Papers, 1961*, 385.

48 SM-763-63, June 13, 1963, "Basic Elements of Future MLF Agreement," enclosure to JCS 2421/434; Lawrence S. Kaplan, "The MLF Debate," in Douglas Brinkley and Richard T. Griffiths, eds., *John F. Kennedy and Europe* (Baton Rouge: Louisiana State University Press, 1999), 52–55; Stromseth, 79–81.

49 See JCSM-407-61 to SECDEF, June 15, 1961, "NATO Force Requirements for 1966," JCS 2305/498; and Appendix to JCSM-594-61 to SECDEF, August 30, 1961, "NATO Force Requirements for End–1966," JCS 2305/578.

50 Kaplan et al., 393–397. The missile in question did not actually exist. Barely a drawing-board concept, it became known in NATO circles as "Missile X." Later, McNamara backtracked and told the JCS that he would resist a new MRBM for Europe. See Memo, SECDEF to CJCS, September 11, 1965, "DPM on NATO," JCS 2450/77.

51 Georges-Henri Soutou, *The French Military Program For Nuclear Energy, 1945–1981*, trans. by Preston Niblack (College Park, MD: Center for International Security Studies at Maryland, 1989), 1–5; Wilfrid L. Kohl, *French Nuclear Diplomacy* (Princeton: Princeton University Press, 1971), 84; Beatrice Heuser, *NATO, Britain, France and the FRG: Nuclear Strategies and Forces for Europe, 1949–2000* (London: Macmillan, 1997), 93–123.

52 Erin R. Mahan, *Kennedy, de Gaulle and Western Europe* (London: Palgrave Macmillan, 2002), 14.

53 Kaplan, "MLF Debate," 60–62.

54 JCSM-350-63 to SECDEF, May 2, 1963, "Multilateral Seaborne Ballistic Missile Force," JCS 2421/434; Kaplan, "The MLF Debate," 62–65.

55 Philip Geyelin, *Lyndon B. Johnson and the World* (New York: Praeger, 1966), 159–180.

56 Kaplan et al., 414.

57 JCSM-1014-64 to SECDEF, December 4, 1964, "French Nuclear Weapons Program," JCS 2278/78-4.

58 Walter S. Poole, *The Joint Chiefs of Staff and National Policy, 1965–1968* (Washington, DC: Historical Division, Joint Secretariat, Joint Chiefs of Staff, May 1985), Pt. I, 285–317.

59 Poole, *JCS and National, 1965–1968*, Pt. I, 323–337; Johnson, *Vantage Point*, 308–309; Richard L. Kugler, *Commitment to Purpose: How Alliance Partnership Won the Cold War* (Santa Monica, CA: RAND, 1993), 207, 220.

60 See Stromseth, 71–72.

61 Stockpile figures from McNamara's unclassified statement accompanying his FY69 budget submission in U.S. Congress, Senate, Committee on Armed Services, *Hearings on Authorization for Military Procurement, Research and Development, Fiscal Year 1969, and Reserve Strength*, 90th Cong., 2d Sess., 1968, 183.

62 Poole, *JCS and National Policy, 1965–68*, Pt. I, 111–198, 395–396.

63 Letter, Heinz Trettner (Inspector General of the German Army) to Wheeler, May 13, 1966, JCS 2124/370.

64 See MC 14/3 (Final), January 16, 1968, "Overall Strategic Concept for the Defense of the North Atlantic Treaty Organization Area," in Gregory W. Pedlow, ed., *NATO Strategy Documents, 1949–1969* (Brussels: North Atlantic Treaty Organization, 1999), 345–370.

65 Ivo H. Daalder, *The Nature and Practice of Flexible Response: NATO Strategy and Theater Nuclear Forces since 1967* (New York: Columbia University Press, 1991), 74–76.

66 Quoted in Gordon S. Barrass, *The Great Cold War: A Journey Through the Hall of Mirrors* (Stanford, CA: Stanford University Press, 2009), 194.

67 MC 48/3 (Final), December 8, 1969, "Measures to Implement the Strategic Concept for the Defence of the NATO Area," in Pedlow, ed., *NATO Strategy Documents*, 371–400.

68 "The Future Tasks of the Alliance (The Harmel Report)," December 1967, reprinted in Stanley R. Sloan, *NATO's Future: Toward a New Transatlantic Bargain* (Washington, DC: National Defense University Press, 1985), 219–222.

69 McNamara quoted in Drea, *McNamara, Clifford, and the Burdens of Vietnam*, 391; JCS assessment from SM-863-67, December 12, 1967, enclosing Joint Strategic Capabilities Plan, July 1, 1968–June 30, 1969 (JSCP–69), JCS 1844/488.

70 Graham Spinardi, *From Polaris to Trident: The Development of U.S. Fleet Ballistic Missile Technology* (Cambridge, UK: Cambridge University Press, 1994), 66–67.

71 JCSM-131-65 to SECDEF, March 1, 1965, "Joint Strategic Objectives Plan for FY 1970–74 (JSOP–70)," JCS 2143/248.

72 Kaplan et al., 319–320; Drea, *McNamara, Clifford, and the Burdens of Vietnam*, 346–356; Jerome H. Kahan, *Security in the Nuclear Age: Developing U.S. Strategic Arms Policy* (Washington, DC: Brookings, 1975), 105–106.

73 Spinardi, 89–94; Ted Greenwood, *Making the MIRV: A Study of Defense Decision Making* (Cambridge, MA: Ballinger Publishing, 1975), 3–11, 64–65; Bernard C. Nalty, "USAF Ballistic Missile Programs, 1964–1965" (Study Prepared by the USAF Historical Division Liaison Office, March 1967), 34–37 (declassified).

74 See JCSM-131-65 to SECDEF, March 1, 1965, loc. cit.

75 Lawrence Freedman, *The Evolution of Nuclear Strategy* (New York: St. Martin's Press, 1983), 253.

76 Morton H. Halperin, *The Decision to Deploy the ABM: Bureaucratic Politics in the Johnson Administration* (Washington, DC: Brookings, 1973), 67–69, 83–84, 93.

77 Ernest J. Yanarella, *The Missile Defense Controversy: Strategy, Technology, and Politics, 1955–1972* (Lexington: University Press of Kentucky, 1977), 92–93.

78 See NIE 11-8-62, [July 6, 1962], "Soviet Capabilities for Long Range Attack," in Donald P. Steury, ed., *Intentions and Capabilities: Estimates on Soviet Strategic Forces, 1950–83* (Washington, DC: Center for the Study of Intelligence, 1996), 181–183; and NIE 11-8-66 [October 20, 1966], amended by M/H NIE 11-8-66, "Soviet Capabilities for Strategic Attack," ibid., 209–224.

79 Prados, *Soviet Estimate*, 155–157; Freedman, *U.S. Intelligence and the Soviet Strategic Threat*, 86–94; NIE 11-3-65, November 18, 1965, "Soviet Strategic Air and Missile Defense," *FRUS, 1964–68*, X, 330–332.

80 Donald R. Baucom, *The Origins of SDI, 1944–1983* (Lawrence: University Press of Kansas, 1992), 27–30; NIE 11-3-66, November 17, 1966, "Soviet Strategic Air and Missile Defenses," *FRUS, 1964–68*, X, 446–550.

81 JCSM-742-66 to SECDEF, December 2, 1966, "Production and Deployment of the NIKE–X," JCS 2012/283-1 and *FRUS, 1964–68*, X, 458.

82 Baucom, 15–20; Richard J. Barber Associates, "The Advanced Research Projects Agency, 1958–1974" (Unpublished Study Prepared for the Advanced Research Projects Agency, December 1975), chaps. IV–VI (unclassified).

83 Kaplan et al., 123–125.

84 Harry B. Yoshpe, *Our Missing Shield: The U.S. Civil Defense Program in Historical Perspective* (Washington, DC: Federal Emergency Management Agency, 1981), 362–375.

85 McNamara statement to House Armed Services Committee, January 23, 1967, quoted in David Goldfischer, *The Best Defense: Policy Alternatives for U.S. Nuclear Security from the 1950s to the 1990s* (Ithaca, NY: Cornell University Press, 1993), 181.

86 John Newhouse, *Cold Dawn: The Story of SALT* (New York: Holt, Rinehart and Winston, 1973), 85.

87 Yanarella, 104–106; Baucom, 22–23.

88 McNamara, *Essence of Security*, 64.

89 Morton H. Halperin, *Bureaucratic Politics and Foreign Policy* (Washington, DC: Brookings Institution, 1974), 298.

90 "Notes on Meeting with the President," December 6, 1966, *FRUS, 1964–68*, X, 459–464; JCSM-804-66 to SECDEF, December 29, 1966, "Production and Deployment of Nike–X," JCS 2012/283/2 and *FRUS, 1964–68*, X, 510–511.

91 James Killian quoted in Memo by Walt Rostow of Meeting with President Johnson, January 4, 1967, "ABMs," *FRUS, 1964–68*, X, 526–531.

92 "Annual Budget Message to the Congress, Fiscal Year 1968," January 24, 1967, *Public Papers of the Presidents of the United States: Lyndon B. Johnson, 1967* (Washington, DC: GPO, 1968), Pt I, 48.

93 JCSM-602-65 to SECDEF, August 5, 1965, "ACDA Memorandum for the Committee of Principals, Paper on Nonproliferation Agreement, 16 July 65," JCS 1731/878-2.

94 JCSM-143-67 to SECDEF, March 14, 1967, "Proposal on Strategic Offensive and Defensive Missile Systems," JCS 1731/966-2.

95 Johnson, *Vantage Point*, 485; Kosygin's comments paraphrased by McNamara in Michael Charlton, *From Deterrence to Defense: The Inside Story of Strategic Policy* (Cambridge: Harvard University Press, 1987), 27. Memorandum of Conversation, June 23, 1967, "Luncheon given by President Johnson for Chairman Kosygin," *FRUS, 1964–68*, XIV, 528–531, gives a less vivid account of the discussion.

96 Baucom, 34.

97 JCSM-425-67 to SECDEF, July 27, 1967, "Initiation of Nike-X Production and Deployment," JCS 2012/308-1; *FRUS, 1964–68*, X, 562–564.

98 Draft Memo, SECDEF to President, August 1, 1967, "Strategic Offensive and Defensive Forces," enclosure to Memo, SECDEF to CJCS, August 2, 1967, same subject, JCS 2458/272.

99 JCSM-481-67 to SECDEF, August 28, 1967, "DPM on Strategic Offensive and Defensive Forces," JCS 2458/272-2.

100 Robert S. McNamara, "The Dynamics of Nuclear Strategy," September 18, 1967, in Department of State, *Bulletin* 57, no. 1476 (October 9, 1967), 443–451.

101 See JCSM-481-67 to SECDEF, August 28, 1967, loc. cit.

102 Poole, *JCS and National Policy, 1965–68*, Pt. I, 104–109; and Drea, *McNamara, Clifford, and the Burdens of Vietnam*, 371–373.

103 Baucom, 39–40.

104 "Remarks at Signing of the Nuclear Nonproliferation Treaty," July 1, 1968, *Public Papers of the President: Lyndon B. Johnson, 1968-69* (Washington, DC: GPO, 1970), Pt. II, 349.

Chopper Pick Up, by Brian H. Clark (Courtesy of the Center of Military History)

Chapter 9

VIETNAM: GOING TO WAR

The scene was reminiscent of many amphibious operations of World War II. On the morning of March 8, 1965, with a light mist reducing visibility, elements of 9th Marine Expeditionary Brigade landed on a sandy beach near Da Nang, South Vietnam. Wading ashore with their gear, they encountered reporters, photographers, the mayor of Da Nang, and their commander, Brigadier General Frederick J. Karch, whom local school girls had laden with garlands in celebration of the occasion. A cordial welcome, it belied the presence of Viet Cong guerrillas a few miles away. Climbing into waiting trucks, the Marines were transported to the nearby American air base to take up security duties. The vanguard of a larger U.S. presence yet to come, these Marines were the first American combat troops to arrive in Vietnam. "Americanization" of the Vietnam War had begun.[1] It was a policy the Joint Chiefs of Staff had helped to shape, but not one that gave them much satisfaction or sense of confidence. The war in Vietnam was entering a new phase, and with it came growing uncertainty among the JCS whether they would have the tools and resources at their disposal to make that policy succeed.

THE ROOTS OF AMERICAN INVOLVEMENT

By the time U.S. combat troops began to deploy to Vietnam in 1965, the United States had been involved there fighting Communism for more than a decade and a half. With the escalation of Cold War tensions brought on by the Korean War, the Truman administration funneled massive support to the French effort in Indochina (Vietnam, Laos, and Cambodia) against the Communist Viet Minh. In 1954, after the Viet Minh victory over the French at Dien Bien Phu, an international conference in Geneva agreed to a settlement that resulted in the division of Vietnam between a Communist regime in the North and a non-Communist one in the South. The Joint Chiefs of Staff viewed the Geneva accords as a major setback for U.S. interests in Southeast Asia. But given the American public's war-weariness in the wake of the Korean conflict and the Eisenhower administration's reallocation of resources limiting the size and capabilities of general purpose forces, they ruled out recommending direct military involvement to change the outcome. Elections leading to unification were never held owing to chronic political instability in the South (much of it instigated by agents from the North) and the intransigence of South Vietnamese Prime Minister Ngo Dinh Diem, a stalwart anti-Communist, whose rejection of the vote the United States fully supported.

After the French withdrawal in 1954–1955, the United States assumed major responsibility for South Vietnam's economic welfare, political stability, and military security. Expecting continuing pressure from the North, the Joint Chiefs saw a Korean War-style invasion, assisted by the Chinese, as the most serious threat that South Vietnam might face. Since the Joint Staff lacked the requisite personnel and resources at the time, the JCS relied on ad hoc fact-finding committees or the Army General Staff for assessments and recommendations. The results of one such inquiry in 1955 credited the South Vietnamese with a limited capacity for offering resistance and estimated that it would take up to eight U.S. divisions, two to three tactical air wings, a carrier task force, and a Marine landing force to defeat a full-scale North Vietnamese invasion.[2] The Eisenhower administration had no desire to become involved in Vietnam on such a scale and turned instead to heavy infusions of political, economic, and military assistance to buttress South Vietnam's position. But by the end of the decade an increase in assassinations, terrorism, and guerrilla activity by the Viet Cong (successor to the Viet Minh) pointed to the need for stronger measures to avert a Communist takeover. In April 1960, at JCS instigation, the Commander in Chief, Pacific (CINCPAC) assembled a group of senior U.S. officers on Okinawa to take a fresh look at the problem. Based on supposed lessons learned in the recent insurgencies in Malaya and the Philippines, the conference recommended a counterinsurgency plan (CIP) that included increases in military strength for the South Vietnamese armed forces and paramilitary units, and major political and administrative reforms in the Diem government.[3]

Action on the CIP was still pending when the Kennedy administration took office in January 1961. By then, insurgency and terrorism had grown into the most ubiquitous forms of conflict worldwide. In the aftermath of Khrushchev's speech of January 6, 1961, welcoming "wars of national liberation," the new President had all the more reason to be concerned. The development of countermeasures, however, was still in its infancy. Among the JCS and elsewhere within the military there was considerable debate over strategy and doctrine. One of the leading figures in counterinsurgency warfare was Brigadier General Edward G. Lansdale, USAF, who had been instrumental in defeating the Communist Hukbalahap in the Philippines after World War II. Turning to Lansdale for advice and guidance, President Kennedy decided to expand the use of covert operations and to increase the size of U.S. Army Special Forces (the "Green Berets"). The JCS alternative for dealing with the crisis at the time in neighboring Laos seemed to be a costly and politically risky large-scale military buildup, the prelude to possible intervention. The Chairman of the Joint Chiefs, General Lyman L. Lemnitzer, USA, lacked the President's confidence in special forces and disputed the notion that current programs in Vietnam were insufficient and ineffective against the guerrilla threat. But in the aftermath of the Bay of Pigs episode, Kennedy paid little attention to JCS advice. In April-May 1961, he approved a series of counterinsurgency measures using the Green Berets to spearhead the effort.[4]

Along with increased military activity, Kennedy sought political and economic reforms from Diem to bolster his regime's credibility and popularity. This process of attempting to develop a "balanced" policy lasted, with mixed success, from Kennedy's Presidency into Lyndon Johnson's. But as early as the autumn of 1961 it was clear that without a major improvement in the security situation, efforts to achieve political and economic reform would fall short of the goal. Military power by itself might not determine the outcome of the struggle for Vietnam, but the side without it in preponderance was unlikely to prevail.

The catalyst for the rapid and sustained expansion of the American military presence in South Vietnam was the Taylor-Rostow report, the product of a fact-finding mission jointly headed by the President's MILREP, General Maxwell D. Taylor, USA (Ret.), and Walt W. Rostow, an economist on the NSC Staff who specialized in underdeveloped countries. Delivered to President Kennedy in early November 1961, the report painted a bleak picture of the situation in South Vietnam and recommended an "emergency program" of additional assistance, to include allowing U.S. trainers and advisors to "participate actively" in planning and executing operations against the Viet Cong. The most controversial part of the report was its call for the introduction of an 8,000-man "task force" to boost security while ostensibly assisting in flood repair and other civic action projects in the Mekong Delta. Later, as Ambassador to South Vietnam in 1964–1965, Taylor was apprehensive about the introduction of U.S. combat troops, arguing that it could undermine the government's commitment to the war. In 1961, however, he saw things differently and insisted that "there was a pressing need to do something to restore Vietnamese morale and to shore up confidence in the United States."[5]

The Joint Chiefs agreed that the situation was critical, but they believed that if the United States intervened, it should do so wholeheartedly and without illusion. In General Lemnitzer's view, the "8000-man force," once in place, would be too thinned out to make much difference.⁶ Working in collaboration with CINCPAC, the JCS came up with an alternative contingency "Win Plan" that would involve the use of up to six divisions and put heavy military pressure directly on North Vietnam with air and naval power.⁷ Initially, McNamara seemed to prefer the JCS Win Plan to the limited course outlined in the Taylor-Rostow report. But upon further reflection, he and Secretary of State Dean Rusk concurred that even though the introduction of U.S. combat troops might someday become unavoidable, there was no immediate need to go quite so far, a conclusion Kennedy gladly embraced.⁸ On November 15, 1961, leaving the question of combat troops in abeyance, Kennedy approved a revised Vietnamese assistance policy (characterized as a "first phase" program), which authorized an increase in the number of U.S. advisors and specialized support units and an expansion of their role.⁹

Kennedy's decision entirely reshaped the U.S. commitment in Vietnam. From a strength of around 1,000 advisors in 1961, the U.S. military advisory presence grew to over 5,000 by the end of the following year. To increase the mobility of government troops, the United States also sent nearly 300 helicopters and transport planes to Vietnam.¹⁰ In February 1962, to oversee the expanded effort, President Kennedy authorized a new command structure—the U.S. Military Assistance Command, Vietnam (COMUSMACV)—a subordinate unified command which reported through CINCPAC to the Joint Chiefs, the Secretary of Defense, and the President.¹¹ Officially, U.S. policy drew the line at the direct involvement of American advisory personnel in combat operations. The reality, however, was different. Having previously served in rear echelon training areas and command posts, U.S. advisors now fanned out into the countryside, operating at the battalion level or lower. Some advisors actually fought alongside government troops; others flew combat missions.¹² But with Berlin, Cuba, and other hot spots capturing the headlines, Vietnam remained a remote and distant war for policymakers and the American public alike.

By the start of 1963 the surge of American advisors and assistance appeared to be having the desired effects of reinvigorating the South Vietnamese armed forces and placing the Viet Cong on the defensive. By now there were over 11,000 U.S. military personnel in Vietnam. Confident of ultimate success, McNamara told the JCS to plan on U.S. advisors being out of the country in 3 years.¹³ But just as the war appeared to be looking up, it took a turn for the worse, owing to unexpected setbacks suffered by the South Vietnamese Army (ARVN), increased political protests against Diem by the Buddhists and other noncommunist groups, and stepped up infiltration of men and supplies from the North. By the summer of 1963, the progress of the previous year was a fading memory. Knowing the President's aversion to the use of combat troops, the Joint Chiefs, CINCPAC, and the CIA came up with a plan (later designated OPLAN 34A) to bring the war home to North Vietnam through a campaign of sabotage and covert operations.[14] However, it was too late for any improvement in the course of the war to save Diem's crumbling regime, which fell victim in early November 1963 to a bloody coup d'état fomented, with American encouragement, by disgruntled South Vietnamese generals. Weapons, tactics, and equipment meant to fight the Viet Cong were used instead to settle old scores and to prop up the new military junta.

Shortly before his death, President Kennedy said publicly that he was confident most U.S. advisors could leave Vietnam in the foreseeable future and turn the war over to the ARVN.[15] But he had no fall-back strategy in case he found withdrawal ill advised and remained averse to putting pressure on North Vietnam, other than through limited, indirect means, to cease and desist its support of the Viet Cong. Though the Joint Chiefs grudgingly accommodated themselves to the President's wishes, they had yet to be convinced that a policy of restraint would succeed. What they saw evolving was an ominous repetition of the stalemate in Korea—a remote war, offering no sign of early resolution, consuming precious resources, and diverting attention from larger threats. Hence their support for a more aggressive, immediate strategy to confront the enemy directly with strong, decisive force. Militarily, the chiefs' solution had much to recommend it. The United States still possessed overwhelming strategic nuclear superiority and could have used that power as an umbrella for large-scale conventional operations against North Vietnam. But it was a strategy fraught with enormous political risks that Kennedy was unwilling or unprepared to take. It would be up to his successor to try to find a more durable solution.

THE ROAD TO AN AMERICAN WAR

By the time Lyndon Johnson entered the Oval Office in November 1963, the situation in South Vietnam had clearly deteriorated to the point that a Communist takeover seemed more probable than ever. Remembering the backlash against Truman over the "loss" of China after World War II, Johnson was determined not to become tagged as the President who "lost" Vietnam. While professing continuity with Kennedy's policy, he quietly abandoned his predecessor's timetable for the withdrawal of U.S. advisors and told General Maxwell Taylor, the new Chairman of the Joint Chiefs, to treat Vietnam as "our most critical military area right now." Identifying the problem as one of insufficient will and commitment, he exhorted Taylor and the JCS to pay close attention to the selection of personnel and to send only "our blue ribbon men" to Vietnam as advisors.[16]

By early 1964, it was apparent from the continuing political turmoil in Vietnam and a surge in Viet Cong activity that reducing the U.S. presence could have adverse consequences. General Wallace M. Greene, Jr., Commandant of the Marine Corps, believed the situation had reached the point where the United States needed "a clear-cut decision either to pull out of South Vietnam or to stay there and win."[17] Embracing the latter course, the Joint Chiefs offered a ten-point program of "increasingly bolder actions in Southeast Asia" that amounted to a virtual take-over of the war. Among the recommended measures were overt and covert bombing of the North, increased reconnaissance, large-scale commando raids, the mining of North Vietnamese harbors, operations in Laos and Cambodia, and the commitment of U.S. forces "as necessary" in direct actions against North Vietnam.[18] An expansion of the war at this time, however, was the last thing President Johnson wanted. Meeting with the Joint Chiefs on March 4, 1964, he stated that he remained committed to keeping South Vietnam out of Communist hands, but would do nothing that might involve the country in a war before the November elections. "We haven't got any Congress that will go with us," he told them, "and we haven't got any mothers that will go with us in the war."[19]

Until the election, then, Johnson all but ignored JCS advice on Vietnam, finding it excessively focused on applying overwhelming military power.[20] Limiting his contacts with the chiefs, he saw only Taylor on a regular basis and turned to a small circle of civilian advisors for guidance on the war. Increasingly preeminent within this group was McNamara, who remained confident that the careful and selective application of military power (as opposed to the sweeping intervention favored by the JCS) could produce the desired results. Applying the lessons he had drawn from the Berlin and Cuban Missile crises, McNamara viewed a successful outcome in Vietnam in relatively narrow terms that involved applying precisely the right amount of pressure to achieve the withdrawal of North Vietnamese support for the Viet Cong without escalating the war into a superpower confrontation. With this strategy in mind, he returned from a fact-finding trip to Saigon around mid-March 1964, cautioning against large-scale U.S. military action against North Vietnam and favoring only a limited buildup of American airpower, "tit for tat" reprisal air strikes by the South Vietnamese, and stepped-up commando raids against the North.[21]

A key figure in developing the flexible response doctrine, Taylor shared McNamara's view that the graduated application of finely tuned military pressure would produce the desired results in Vietnam and avoid the need for large-scale intervention. Urging his JCS colleagues to support the Secretary's plan, Taylor defended it as a suitably aggressive, yet measured, response. But to the Service chiefs it smacked of more of the same and did not go nearly far enough to satisfy them.[22] "We are swatting flies," complained Air Force Chief of Staff General Curtis LeMay, "when we ought to be going after the manure pile."[23] Intelligence reports supported the Service chiefs' contention that North Vietnam would be largely impervious to the limited raids and retaliatory attacks McNamara had in mind. Yet despite the drawbacks, President Johnson preferred McNamara's plan over a full-blown war, and on March 17, 1964, he decided to put it into action.[24]

Shortly after the President's decision, in April 1964 the Joint Chiefs conducted a wargaming exercise (SIGMA I–64) to test McNamara's hypothesis that a strategy of graduated pressure against the enemy would turn the war around. Organized under the JCS Joint War Games Agency, SIGMA I involved military officers from the lieutenant colonel to the brigadier general level, their civilian equivalents, and representatives of the Intelligence Community. Described by historian H.R. McMaster as "eerily prophetic," the exercise's main finding was that steadily escalating military pressure failed to have any significant deterrent effect on North Vietnamese behavior.[25] On the contrary, as the game progressed, it led to both a stiffening of North Vietnamese resistance and a worsening of the political-military situation in the South that narrowed American options to two unappealing alternatives—a greatly expanded war against the North that risked Chinese intervention, or a humiliating withdrawal with a marked loss of U.S. credibility and prestige worldwide. As one participant in the game later observed: "The thesis of escalated punishment of North Vietnam had again been tested by interagency experts and found wanting."[26]

With their doubts about a strategy of graduated pressure steadily growing, the Joint Chiefs, less General Taylor, continued to urge the use of large-scale military force to thwart the North Vietnamese and to curb the insurgency. But without Taylor's support and endorsement, their ideas and recommendations stood little chance of having much impact.[27] On July 1, 1964, Taylor stepped down as Chairman to take up new duties as Ambassador to Saigon. His departure came as a relief to the Service chiefs who believed, almost without exception, that he could have done a more effective job representing them and conveying their views to the Secretary of Defense and the President.

Whether his successor, General Earle G. Wheeler, USA, would be a more forceful spokesman for JCS views remained to be seen. The third army officer in a row to serve as CJCS, Wheeler came to the job largely on Taylor's recommendation. Having once been Director of the Joint Staff, he knew the ins and outs of the JCS system as well as anyone. As Army Chief of Staff immediately prior to becoming Chairman, Wheeler had been critical of the administration's emerging strategy of graduated response in Vietnam, but he had been far less outspoken than the other chiefs.[28] Throughout his years at the Pentagon prior to becoming CJCS, he had always gotten along well with his superiors. Though he might not always agree, they could count on him, once a decision was taken, to implement it without complaint. Perhaps the most obvious shortcoming in his otherwise distinguished résumé was his limited combat experience (confined to a few months as chief of staff to an infantry division in Europe in World War II), a drawback in the eyes of some of his peers, but not a great concern to either McNamara or President Johnson.[29]

THE GULF OF TONKIN INCIDENT AND ITS AFTERMATH

Wheeler was still settling into his job as Chairman when in early August 1964 the fateful Gulf of Tonkin incident occurred. At the time, it seemed that North Vietnamese torpedo boats had launched two separate attacks 2 days apart against two U.S. destroyers operating in international waters off North Vietnam. The first attack, against USS *Maddox*, occurred August 2; the second, involving both the *Maddox* and USS *Turner Joy*, appeared to follow 2 days later. Both ships were part of the Desoto Patrol Program, a JCS-authorized effort conducted by the Seventh Fleet to collect intelligence on Sino-Soviet bloc electronic and naval activity. Since mid-December 1962, Desoto Patrols had paid regular visits to the Gulf of Tonkin. Despite a loose system of coordination, Desoto Patrols and the covert missions mounted by Commander, U.S. Military Assistance Command, Vietnam (COMUSMACV) against North Vietnam under OPLAN 34A were separate and independent of one another. Thus, the possibility of one set of operations overlapping or interfering with the other was ever-present. Matters came to a head in late July 1964 when South Vietnamese commandos, part of the 34A program, carried out a pair of raids along the North Vietnamese coast. Apparently in response to these raids, the North Vietnamese attacked the *Maddox*, mistaking it for part of the raiding force.[30]

The role of the Joint Chiefs in this episode was relatively minor and consisted mainly of drawing up a list of targets for retaliatory air strikes following reports of the second attack. As was increasingly the custom, the only member of the Joint Chiefs to attend face-to-face meetings with the President was the Chairman, General Wheeler. To expedite matters, McNamara at several critical points bypassed the Joint Chiefs and dealt directly with CINCPAC. On the morning of August 4, while McNamara was attending an emergency NSC meeting with the President, the JCS prepared their recommendations and forwarded them to the White House, urging severe retaliation against North Vietnamese naval bases and petroleum, oil, and lubricants (POL) storage in the Vinh area. That afternoon, McNamara returned to the Pentagon and told the JCS that the President had approved their recommendations, with several notable modifications. In a foretaste of the micromanagement of the air war yet to come, the President had added two base areas to the target list but had decided that, except for striking the storage tanks, the U.S. attacks would be mounted against boats only, not against the bases or port facilities. The next day carrier-based aircraft executed the mission.[31]

Soon after the Tonkin Gulf incident, questions arose over whether the second attack had actually taken place. The issue was especially relevant since it was on the basis of the second attack that President Johnson had decided not only to order retaliatory air strikes against North Vietnam, but to seek authorization from Congress for further military action in the event of additional provocations. Had there been only one attack, the President said, he was prepared to dismiss the incident with a diplomatic protest.[32] Years later, a reexamination of the evidence confirmed suspicions that the North Vietnamese never mounted a second attack, though it may have appeared so at the time to Sailors aboard the *Turner Joy*. According to a detailed study by Robert J. Hanyok, a historian for the National Security Agency, errors in the translation of North Vietnamese radio traffic and the Navy's mishandling of SIGINT led to the misidentification of a North Vietnamese salvage operation as a second attack. Hanyok found nothing to indicate that the Navy, the National Security Agency, or the White House had manipulated the data or acted improperly. But under the pressure of events, those monitoring the situation interpreted the evidence as pointing to two separate incidents.[33]

The most important long-term consequence of the Gulf of Tonkin episode was a joint congressional resolution giving the White House practically carte blanche in Southeast Asia. The idea of seeking such authority had apparently originated with Walt W. Rostow, then serving in the State Department, who began discussing the matter with members of the NSC Staff as early as December 1963. By June 1964, Rostow's suggestion had attracted the attention of McGeorge Bundy, the President's National Security Advisor, who felt that a congressional resolution would "give additional freedom to the Administration in choosing courses of action." President Johnson agreed, but with the election looming, he was reluctant to tarnish his image as the "peace" candidate unless the situation warranted.[34]

The Tonkin Gulf episode had a galvanizing effect on administration policy toward Vietnam. With the White House unsure how far it could go in Vietnam, it became the rallying point for testing support of the war and mobilizing congressional backing. Leading the charge in the Senate was J. William Fulbright, Chairman of the Foreign Relations Committee. Later, as the war degenerated into a stalemate, Fulbright became one of the administration's harshest critics and a key figure in the antiwar movement on Capitol Hill. But at the time of the Tonkin Gulf incident, he was still a strong advocate of taking firm action to curb the "aggressive and expansionist ambitions" of the North Vietnamese. The upshot was a unanimous vote in the House and overwhelming support in the Senate to give the White House a free hand to retaliate—the closest the United States came to a formal declaration of war.[35]

Following the Gulf of Tonkin episode, the Johnson administration launched yet another review of its Vietnam policy. In light of the recent congressional resolution and the stepped-up pace of military activity, the Joint Chiefs now viewed direct U.S. intervention as inevitable, though they were split over the form it should take. Confident that airpower could be decisive, LeMay downplayed the need for large-scale troop deployments and urged an intensive bombing campaign against 94 high priority military and industrial targets across North Vietnam. "All of his experience," one of LeMay's colleagues recalled, "taught him that such a campaign would end the war."[36] The intent, as the Joint Chiefs described it to the Secretary of Defense, would be to deal the enemy "a sudden sharp blow." If it failed, the United States could reconsider whether to commit a large ground force.[37] However, the new Army Chief of Staff, General Harold K. Johnson, doubted whether the increased use of airpower, without accompanying increases on the ground, would have the desired impact on the insurgency in the South. In Johnson's view, expanding the war in the air and on the ground should go hand in hand.[38] Unable to achieve a full reconciliation of their differences, the chiefs papered them over and in late August recommended a program of "prompt and calculated responses" emphasizing "air strikes and other operations" against enemy targets in North Vietnam, Laos, and Cambodia.[39]

The JCS found their advice for expanding the scale and scope of the war no more welcome now than earlier. Having decided to cast his Republican Presidential opponent, Senator Barry Goldwater, in the role of warmonger, Johnson often went out of his way to avoid making it appear that he was under the military's spell or influence. The result, however, was a policy that seemed to straddle two stools. "I haven't chosen to enlarge the war," the President declared publicly. "Nor have I chosen to retreat and turn [Vietnam] over to the Communists."[40] Gathering his key advisors at the White House on September 9, he heard a report by Ambassador Taylor on the unsettled political situation in Saigon and a reiteration of JCS views on the air campaign—"this bombing bullshit," the President called it.[41] The next day he approved increasing the military pressure against North Vietnam but limited it to low-profile activities that included the resumption of Desoto naval patrols in the Gulf of Tonkin and covert operations by the South Vietnamese against the North. He also approved discussions with the Laotian government to allow South Vietnamese air and ground operations in the Lao panhandle, and preparations for an "appropriate" response (i.e., a further build-up of air power in the South and off the coast) should the North Vietnamese resume attacks on U.S. forces.[42]

With "graduated response" becoming the accepted strategy, the Joint Chiefs decided to take another look at its probable effects. The upshot was a second round of war games known as SIGMA II–64. Conducted in mid-September 1964, SIGMA II–64 occurred at the same time the President was reviewing proposals to step up operations in Vietnam. Organized this time to include senior officials, SIGMA II produced about the same results as SIGMA I. Not only was the graduated application of military power, including bombing of the North, unlikely to stop the North Vietnamese; it was also apt to draw the United States more deeply into an inconclusive war. But despite the exercise's disturbing findings, McNamara paid little attention and later dismissed SIGMA II as further evidence that the JCS were looking for an excuse to ramp up the war. Interpreting the findings somewhat differently, he chose to see them as confirmation that an expanded and more intensified bombing effort would be a largely pointless waste of lives and resources.[43]

Increasingly frustrated and troubled, the Joint Chiefs made no attempt to conceal their dissatisfaction with the current policy or the limited influence of their advice. Soon, reports of "considerable unhappiness" among the JCS over their exclusion reached McGeorge Bundy and were a source of concern to the President's staff. In mid-November, with the election now out of the way, Jack J. Valenti, a White House aide who handled liaison with Congress, urged Johnson to have the Joint Chiefs "sign on" before taking further actions in Vietnam because their inclusion in presidential decisions would help to shield the administration from possible congressional recriminations. If the Joint Chiefs participated at pertinent NSC meetings, Valenti believed, "they could have their views expounded to the Commander-in-Chief face to face." He added, "That way, they will have been heard, they will have been part of the consensus, and our flank will have been covered in the event of some kind of flap or investigation later." Johnson agreed and at a November 19 White House meeting he informed his top civilian advisors that in the future no decisions on Vietnam "would be made without participation by the military."[44]

While the President was willing to give the chiefs the opportunity to say their piece, he was no more inclined than before to accept their advice that the strategy of graduated response was flawed. Johnson had no interest in a full-scale war. But as the situation in Vietnam deteriorated, with the Viet Cong escalating attacks against Americans, he knew it was only a matter of time before the United States moved in with more of its military power. Exactly when the President came to this realization is unclear, but between the election in November 1964 and the Viet Cong attack on Pleiku in early February 1965, deliberations with his top advisors were almost nonstop. What he wanted from them was a consensus recommendation. The options under consideration fell into three general categories: 1) continuation of the present policy of support for counterinsurgency in the South and limited pressure on the North; 2) a graduated increase in military pressure on the North Vietnamese meshing at some point with negotiations; and 3) an intensive bombing campaign of the North as recommended earlier by the JCS, known variously as the "hard knock" or "fast, full squeeze" option, which might or might not include the use of nuclear weapons.

The ensuing debate followed the "Goldilocks principle" that if the first and third choices appeared either inadequate or too extreme, the middle course was just right.[45] Assistant Secretary of State William P. Bundy, an ardent advocate of graduated response, denounced the JCS position as an "almost reckless" invitation to Chinese intervention.[46] Arguing that it would keep the commitment of U.S. prestige and resources from getting out of hand, Bundy and likeminded others, including Walt Rostow, now director of State's Policy Planning Council, Assistant Secretary of Defense John T. McNaughton (McNamara's most trusted advisor), and White House assistant Michael Forrestal, all insisted that the graduated application of military power would give the United States the flexibility to negotiate or withdraw should things go sour.[47] Anticipating the debate's outcome, McNamara ordered Wheeler to have the Joint Staff draw up a military plan to support a graduated bombing campaign. Wheeler complied, but in submitting the plan, the JCS expressed little confidence in it and urged the Secretary to develop a "clear set of military objectives before further military involvement in Southeast Asia is undertaken."[48]

McNamara refused the chiefs' request to pass their views to the President. The reason he gave at the time was that their recommendations would become known at the White House in due course as part of an interagency review.[49] Later, however, he acknowledged that he had lost confidence in JCS advice, feeling that it was too extreme. "The president and I were shocked," McNamara recalled, "by the almost cavalier way in which the chiefs . . . referred to, and accepted the risk of, the possible use of nuclear weapons."[50] Be that as it may, the inclusion of nuclear weapons in contingency planning, especially in connection with large-scale operations, was then still a well-known routine practice, so it seems odd that McNamara and the President were somehow surprised. The Joint Chiefs, as they saw it, were merely doing their job and presenting the available options.

Still, the Joint Chiefs must have known that they were engaged in a losing cause. Arrayed against them were the President's best and brightest senior advisors, nearly all of whom—McNamara, McGeorge Bundy, and Maxwell Taylor—favored some form of the graduated response option. So, too, did the COMUSMACV in Saigon, General William C. Westmoreland, USA. Unprepared to take on a full-scale war, Westmoreland hoped that with a modest increase in pressure, he could buy time until the South Vietnamese were better able to hold their own.[51] Practically the only support for the JCS position was that of the CINCPAC in Hawaii, Admiral U.S. Grant Sharp, who thought the time was ripe to "hit hard" and turn the war around. But like the Joint Chiefs, his views made little difference.[52] Compelled to retreat, the JCS grudgingly concurred in what they characterized as a "controlled program" of "intense military pressure" against North Vietnam, "swiftly yet deliberately applied." A lukewarm endorsement, it left the door open to the proposal of stronger measures should the need or opportunity arise.[53]

President Johnson had yet to be convinced that bombing, controlled or otherwise, would produce the desired results, and after listening to Secretary of State Rusk and George W. Ball, the veteran diplomat, he decided in early December 1964 to postpone overt military action against North Vietnam for at least 30 days to give the State Department time to explore the possibility of negotiations and to round up contributions of troops and support from other countries. Depending on the responses, decisions could be taken to conduct U.S. and South Vietnamese air strikes against North Vietnam during the next 2 to 6 months, starting with targets south of the 19th parallel and working northward. Mining of North Vietnamese ports and a naval blockade could follow in due course. The approved policy made no mention of inserting U.S. combat units, but neither did it rule out such a possibility. A partial victory for the Joint Chiefs, the President's decision acknowledged that military power remained a key component of American policy in Southeast Asia. But it further postponed the "hard knock" that the JCS believed to be necessary, sooner or later, to win the war.[54]

While the 30-day period specified by the President elapsed in mid-January, new decisions on military action were held in abeyance owing to political instability in Saigon. Then, on February 7, 1965, the Viet Cong attacked the U.S. military advisory compound near Pleiku in the Central Highlands of South Vietnam, killing 8 U.S. Servicemen, wounding more than 100 others, and destroying 20 U.S. aircraft. The next day President Johnson ordered reprisal raids (code-named FLAMING DART) and gave the Joint Chiefs the go-ahead to prepare an 8-week bombing campaign of the North.[55] For reprisal purposes, the Joint Chiefs recommended immediate large-scale air attacks against seven enemy targets which, after review, the President whittled to two. Both were army barracks complexes used by the North Vietnamese to resupply the Viet Cong. Initial reports indicated the effects of the bombing as "moderate to good" in destroying enemy facilities. Upon closer inspection, however, it became clear that FLAMING DART had fallen short of expectations, and within days, enemy operations in the targeted areas were back to normal.[56]

The modest success of the FLAMING DART raids left the Joint Chiefs more persuaded than ever that if airpower were to be effective, it needed to be concentrated in repeated heavy doses. Hoping to move policy in that direction, the JCS secured the Secretary's approval to transfer an additional 325 aircraft, including 30 B–52s, to the Western Pacific. Sustained bombing of the North (Operation *Rolling Thunder*), initially disguised as retaliation, began on March 2, but followed no coherent strategy or consistent political objectives. Seeing an opportunity to revive the hard-knock strategy, the new Air Force Chief of Staff, General John P. McConnell, proposed a 28-day campaign to destroy all 94 targets on the Joint Chiefs' earlier target list. At the same time, Admiral Sharp recommended an "eight week pressure program" against the enemy's logistical lines.[57] Putting these proposals together, the Joint Staff came up with a revised bombing plan for a four-phase, 12-week air campaign for the systematic destruction of North Vietnam's rail network, ports, and war-production facilities, culminating in heavy attacks on key military-industrial targets in the vicinity of Hanoi and Haiphong.[58] Sharp and McConnell were convinced that over time a concerted bombing campaign would significantly degrade North Vietnam's capacity and willingness to support the Viet Cong. However, Wheeler and Johnson (the Army Chief of Staff) were skeptical and would sanction only the program's initial phases, which were underway by early April. Straddling two stools, Wheeler told McNamara that while the bombing thus far had not reduced North Vietnam's military capabilities in "any major way," he was confident that eventually it would cause a "serious stricture."[59]

An expansion of the U.S. ground role in the South accompanied the enlarged bombing campaign against the North. The heralded arrival in early March of the Marines at Da Nang was in response to General Westmoreland's request the month before for additional security around U.S. air bases and coincided with Army Chief of Staff General Harold K. Johnson's fact-finding visit to Saigon, instigated at the request of the White House to "get things bubbling." Clearly, the momentum was building for a larger commitment of U.S. forces. By far the most cautious member of the JCS at the time, Johnson was also the least enthusiastic about the air war and further U.S. involvement in general. A survivor of the Bataan death march in World War II and a veteran of Korea, he knew the rigors and pitfalls of waging war in the Far East as well as anyone. Johnson had been in Vietnam 3 months earlier and was astonished by the rapid deterioration of security at the local level. Persuaded that the situation was critical, he dismissed as "fictional" General Omar Bradley's admonition against U.S. involvement in wars on the Asian mainland. Upon returning to Washington, he secured prompt endorsement from the President for 21 stop-gap measures ("band aids of a sort," the general called them) aimed at strengthening the existing advisory and support effort.[60] For the longer term, he believed it imperative that U.S. combat forces assume major responsibility for defending towns and installations and for operating offensively against the Viet Cong. Ultimately, he speculated, it might take as many as 500,000 troops and 5 years to complete the mission. "None of us," McNamara recalled, "had been thinking in anything approaching such terms."[61]

General Johnson's unsettling assessment seemed to confirm what the Joint Chiefs had been saying all along—that without a wholehearted U.S. commitment, Vietnam was lost. Even so, his predictions of what would be required and the length of time it would take to turn the situation around exceeded anything the Joint Chiefs had thus far envisioned. Despite their tough talk about a buildup of forces and delivering "hard knocks" to the enemy, the JCS had not looked much beyond a 3- or 4-month campaign. If the Chief of Staff was right, the United States faced a long, expensive, and arduous war. With that possibility in mind, General Wheeler began laying the groundwork the day after Johnson's return for an expanded conflict by having the Joint Staff initiate studies of the various administrative, funding, and logistical adjustments that would have to be made.[62]

Among the Chairman's JCS colleagues, however, there was not much inclination to look beyond the immediate crisis. As sobering as Harold Johnson's warnings of an open-ended conflict may have been, they were slow to sink in. Indeed, not even Johnson himself had thought far beyond the current situation, except in highly generalized terms. As a result, instead of trying to devise a long-range strategy, the JCS turned to hashing out differences among themselves over near-term solutions—the size and composition of the ground force, where to insert it, and whether it or the air war should have priority. Resorting eventually to compromise, they agreed that stepping up the air war and deploying forces on the ground (one full Marine division, one Army division, and one division from the Republic of Korea, if it could be arranged) should proceed in tandem and be aimed at achieving "an effective margin of combat power."[63]

Earlier studies done by the Joint Staff estimated a minimum requirement of six divisions to defend Southeast Asia, so the deployment of two to three divisions would not be much more than a foot in the door.[64] Nonetheless, the decision to intervene in force, even at this critical stage of the conflict, was far from automatic. While he supported graduated bombing of the North, Ambassador Taylor resisted the introduction of U.S. combat troops, arguing that it would shift the burden on to the United States and weaken South Vietnamese resolve. Others, including McNamara, Secretary of State Rusk, and the NSC's McGeorge Bundy, increasingly believed that the United States had no choice, though in making their case they urged the President to show restraint and hold down the number of committed troops. Knowing that he would be hard pressed to mobilize public and congressional backing for an immediate deployment of the size the Joint Chiefs of Staff proposed, President Johnson opted in early April for a lesser figure of 20,000 logistical troops and two Marine battalions with tactical air support—a token commitment that barely disguised the fundamental shift in administration policy. Even more significant, he broadened the Marines' mission "to permit their more active use" against the enemy.[65] General Wheeler promptly advised Admiral Sharp and General Westmoreland that this decision meant a change in employment from "static defense" to "counterinsurgency combat operations."[66] By the end of April 1965, U.S. forces were engaging the Viet Cong in firefights; by June they were regularly conducting offensive operations around their bases. For the United States, the advisory phase of the war was essentially over and a new, more deadly combat phase was beginning.

INTO THE QUAGMIRE

The President's decision of April 1965 committing organized units of U.S. ground troops to combat ushered in a rapid expansion of the American role in the war. Shortly after the President's action, General Wheeler accompanied Secretary McNamara to Honolulu for a 1-day conference on April 20 to take stock of the situation and to discuss future deployments. The other key participants were Admiral Sharp, Ambassador Taylor, and General Westmoreland. It was at this meeting that the broad outlines of basic strategy for the next 3 years emerged. If the meeting accomplished nothing else, McNamara wanted to win over Taylor's support for a stepped-up air and ground war in the South, on the assumption that this was where the war would be decided. Dominating the discussion, McNamara sought to impress upon the others his view that destruction of the Viet Cong, rather than pressure on the North, was crucial to a successful outcome and that land-based tactical air should be completely at Westmoreland's disposal for this purpose. *Rolling Thunder*, the coercive air campaign against the North, assumed a secondary role.[67] Afterwards, without much discussion, the JCS recommended eight U.S. battalion equivalents, with appropriate air and logistical support, for immediate reinforcement of the ground effort, with an additional twelve battalions earmarked for deployment at a later date.[68]

As it happened, these decisions coincided with the onset of a smaller but still alarming crisis in the Dominican Republic, brought on by a long-simmering power struggle between rival political factions. Convinced that the threat of a Communist coup loomed large, President Johnson in late April directed U.S. military intervention to restore order.[69] As General Wheeler explained to his immediate staff, the President had made up his mind to use "the force necessary" to prevent another Cuba in the Caribbean.[70] Moving quickly, the Joint Chiefs deployed nearly 24,000 troops (Marines and Army airborne) in a matter of days and by late May the situation was under control. A quick, hard-hitting operation, mounted as a joint effort, the American show of force in the Dominican Republic seemed to do its job with relative ease and barely a whiff of inter-Service friction.[71]

Whether the U.S. venture in Vietnam would enjoy the same success as the operation in the Dominican Republic remained to be seen. One skeptic, Army Lieutenant General Bruce Palmer, Jr., commander of U.S. forces in the Dominican Republic, hinted darkly that American intervention in Vietnam, undeniably a far bigger affair, might be too little too late.[72] However, very few, if any, of his colleagues agreed. Indeed, by now, a race was on between the United States and North Vietnam to see who could put the most troops into Vietnam in the shortest possible time to gain the advantage. Gathering momentum over the summer of 1965, the U.S. buildup accelerated rapidly as logistical capabilities improved. From around 60,000 troops in mid-1965, American military strength in Vietnam increased to 185,000 by the end of the year. A year later, it had grown to 385,000, and by the end of 1967, it reached 490,000. American combat casualties also mounted—28,000 killed in action by the time the Johnson administration left office and 18,000 more before the ceasefire took effect in 1973.[73]

Contrary to what the JCS expected or hoped to see, the American buildup came with vague war aims and constrained methods of achieving them. Since the 1950s, the stated goal of American involvement in South Vietnam had been to preserve the country's independence and prevent it from falling into Communist hands. A widely accepted hypothesis held that an enemy victory would set off a chain reaction of Communist takeovers across Asia (the "domino theory"). The Joint Chiefs fully subscribed to the domino theory and under the Johnson administration it became the most often cited rationale for U.S. involvement in Vietnam.[74] Given the high stakes involved, however, the White House remained uncommonly restrained in authorizing the application of military power, in contrast to the JCS position that the United States should hit hard and fast. In the summer of 1965, as the administration was contemplating how to manage the buildup, the Joint Chiefs assumed that the President would order a national emergency, mobilize the Reserves and National Guard, and seek supplemental appropriations. Nothing less, General Wheeler argued, would convince the American people "that we were in a war and not engaged in some two-penny military adventure."[75] For political reasons, however, the President decided otherwise. Treating social reforms at home (the "Great Society") as his first priority, he believed that a declared emergency and a call-up of the Reserves would divert attention and resources from his domestic agenda.[76] He thought he could downplay the war, juggle funds from current appropriations, and rely on volunteer enlistments and the draft to supply the necessary manpower. A "guns and butter" approach, the President's decision effectively stripped the war effort of experienced noncommissioned officers (NCOs) and over the long run played a large part in turning public opinion against the conflict by focusing anti-war sentiment on the draft.

The strategic concept governing the deployment of U.S. forces further underscored the restrained nature and limited aims of the American commitment. As described by Secretary McNamara at a Cabinet meeting in June 1965, a military victory in the traditional sense was not the U.S. objective. Rather, the function of American forces was to produce a "stalemate" that would convince the Viet Cong and the North Vietnamese that even if they continued fighting, they could never win. "We think that if we can accomplish that stalemate," McNamara contended, "accompanied by the limited bombing program in the North, we can force them to negotiations, and negotiations that will lead to a settlement that will preserve the independence of South Vietnam."[77] Translating these broad objectives into a military strategy, Westmoreland came up with what amounted to a war of attrition which he formally presented to the Secretary of Defense and the Chairman in July 1965. While the ARVN protected the population centers, U.S. forces would conduct "search and destroy" missions to take back captured territory, restore government authority, and wear down the enemy.[78]

The Joint Chiefs endorsed this strategy, but pointed out (largely at the insistence of the Air Force and the Navy) that the only way it could achieve significant results was in conjunction with heavy pressure from air and naval power on North Vietnam to cease directing and supporting the Viet Cong.[79] From the start, however, the White House insisted that operations on the ground be confined as much as possible to the South. The only exceptions were occasional commando raids against the North and into neighboring Cambodia and Laos where the Viet Cong and North Vietnamese Army (NVA) had their supply lines and base camps. Although CINCPAC had contingency plans for invading North Vietnam, they were rarely mentioned in high-level discussions and never used. On the contrary, as Undersecretary of State George W. Ball, the President's friend and confident, acknowledged, the administration went out of its way to send signals "that we do not seek to bring down the Hanoi regime or to interfere with the independence of Hanoi."[80]

The air campaign was the most guarded of all. Part of the reason was the persistent lack of a consensus among the Joint Chiefs over whether the air war or the ground war should have priority. These differences had hobbled the Joint Chiefs in developing clear-cut positions during the advisory phase and continued into the combat phase, with the Army favoring emphasis on land operations, the Air Force arguing for an intensive air campaign, and the Navy and Marine Corps somewhere in between.[81] Yet even if the JCS had been united, it probably would have made little difference. While CINCPAC coordinated Navy and Air Force attacks against the North under the *Rolling Thunder* campaign, COMUSMACV controlled tactical air operations over South Vietnam and had first call on air assets under the allocation of resources decided at the April 1965 Honolulu conference. Dominated by Army officers, even with the presence of an Air Force deputy, Westmoreland's command in Saigon regarded airpower as the handmaiden of the ground forces and used it for close air support, escort operations, and interdiction of infiltration routes.[82]

The principal impediment to a more effective air war remained the President himself. Near the outset of the buildup, President Johnson made a conscious decision not to exploit the full potential of the air campaign against the North lest it invite Soviet or Chinese intervention, alienate opinion abroad, or encourage further dissent at home. "In Rolling Thunder," observes Air Force historian Wayne Thompson, "the Johnson administration devised an air campaign that did a lot of bombing in a way calculated *not* to threaten the enemy regime's survival."[83] By avoiding certain targets while delaying or moderating attacks on others, the administration allowed the initiative to pass to the enemy. NVA air defenses quickly became a formidable obstacle, costing the United States dearly in pilots and planes. A few years earlier, when the Joint Chiefs had begun urging stronger measures, the United States had had undisputed superiority in strategic nuclear power over the Soviet Union and might have carried out operations against North Vietnam with minimal worry for the wider consequences. But by 1965-1966, U.S. nuclear superiority was on the wane, leaving both McNamara and the President convinced that if they pushed too hard against North Vietnam, they would invite serious trouble with China or the Soviet Union.[84]

The most controversial aspect of the air war was the choice of targets for U.S. planes to bomb. A professional function customarily the domain of the Joint Chiefs, target selection came to be closely controlled and managed by the President in collaboration with McNamara, Rusk, and his other top civilian advisors at his "Tuesday lunch." As the Joint Chiefs became accustomed to the process, their targeting recommendations came to hinge as much on arbitrary assessments of what the President might accept as on what was needed to achieve military results.[85] For reasons never fully explained, the CJCS did not become a member of the Tuesday lunch group until late 1967. Until then, Secretary of Defense McNamara was the military's sole voice at these sessions, at which the President would go over JCS-proposed target sets in minute detail, approve, disapprove, or amend the selections, schedule attacks, and review the results of previous raids. Until the summer of 1966, Hanoi and Haiphong were off limits to bombing and U.S. planes were prohibited from approaching any closer than 30 miles to the Chinese border. Believing that attacks on the North by B-52s would appear provocative, President Johnson limited their use to bombing in the South and along the demilitarized zone (DMZ) separating North from South Vietnam.[86] "This piecemeal application of airpower," one senior Air Force commander recalled, "was relatively ineffective because it still avoided many of the targets that were of most value to the North Vietnamese."[87]

The restraint shown by Washington in prosecuting the war contrasted sharply with the all-out commitment and well-honed objectives of the Viet Cong and the North Vietnamese and the support they received from Communist Bloc countries. As revealed in documents captured by U.S. forces in Cambodia in 1970, the North Vietnamese Communist Party made a binding decision in December 1963 to do whatever it took to "liberate" the South and to reunify it with the North under a Communist regime. Militarily, this meant increasing assistance to the Viet Cong, transitioning from guerrilla warfare to "big unit" tactics involving regimental-sized operations, and sending regular NVA units into the South along the Ho Chi Minh Trail. Thus, while Washington thought it was still dealing with a guerrilla war, North Vietnam was gearing up for a full-scale conflict which it intended to win at any cost. Implementation of this strategy started slowly owing to disagreements within the party over tactics and the reluctance of Soviet Premier Nikita Khrushchev to sanction and assist the intensification of the conflict. But after Khrushchev's ouster from power in October 1964, Moscow became more amenable to providing stepped-up assistance to the Communist insurgency in the South and weapons, including sophisticated air defense systems, to protect North Vietnam against U.S. retaliation.[88]

While U.S. intelligence detected NVA formations in South Vietnam as early as April 1965, the battle of the Ia Drang Valley that November was the first solid confirmation of large-unit North Vietnamese involvement. Like the Chinese intervention in Korea in late 1950, the bloody combat in the Ia Drang Valley that left nearly 300 Americans dead was a shocking experience. Gilding over the losses, Westmoreland treated the battle as a major victory. Yet it should have been a wake-up call for the Joint Chiefs to push for a reexamination of U.S. tactics and strategy, to assess whether a war of attrition was realistic and feasible against a well-armed enemy increasingly composed of highly trained and disciplined North Vietnamese regulars. But by then, with Westmoreland and McNamara fully in control of military strategy, the Joint Chiefs were in no position to raise such questions or make many demands. Projecting a self-assured air, Westmoreland insisted he would prevail and took the President at his word that he could have all the resources he needed. A strategic review of sorts did take place, in mid-January 1966 in Hawaii, with President Johnson himself chairing some of the sessions. But it treated an inordinately broad range of topics, from combat operations to agricultural reform under the pacification program, and was so large (over 450 U.S. and South Vietnamese military and civilian participants) that it more properly resembled a pep rally. The outcome was a resounding reaffirmation of the current course and a full endorsement of Westmoreland's plan to add 102 maneuver battalions (79 of them American) to his force structure over the coming year.[89]

Immediately following the Hawaii Conference, the Joint Chiefs resumed their efforts to convince the President and the Secretary of Defense to mobilize the Reserves, all to no avail. By now, there was serious concern among the JCS that they were losing control over the strategic direction of U.S. military forces, not only in Southeast Asia but worldwide, as the burgeoning demands of Vietnam were beginning to erode force levels everywhere. Should a crisis erupt in Europe or Korea, the JCS warned, the United States would be hard put to mount an effective response.[90] Though fully aware of the situation, McNamara and the President regarded it as an acceptable risk. The mobilization of the Reserves would have required approval by Congress, where anti-war sentiment was on the rise. Even though the measure doubtless would have passed, it probably would have fallen well short of the resounding support shown for the Gulf of Tonkin resolution a year and a half earlier.

While deteriorating support for the war at home was rarely an explicit factor in JCS decisions and recommendations, it was ever-present in the background of their deliberations and impossible to ignore. Indeed, antiwar demonstrations soon became an almost daily occurrence on the steps of the Pentagon. Meanwhile, as the war dragged on, it seemed to acquire a life of its own, an open-ended conflict with no clear resolution in sight. Westmoreland's strategy of attrition may have looked sound on paper, but it was costly, time consuming, and hard to assess in terms of its success. One method of evaluation was a controversial practice known as the "body count" of enemy dead, which the command in Saigon published weekly, claiming it to be evidence of progress in destroying the enemy. The numbers ran into the thousands, though whether they were accurate became a matter of some dispute. The theory was that eventually the VC and NVA would tire of taking heavy losses and cease their aggression. But with U.S. losses averaging around 1,300 per week killed and wounded, the evidence was mixed as to whether the American effort was making much headway toward its goal. The war of attrition, in other words, could cut both ways.[91]

A further complication was the continuing indifference of both Secretary McNamara and President Johnson toward JCS advice and their preference for dealing directly with Westmoreland in managing the conflict. A subunified command to CINCPAC, the COMUSMACV was several steps down the chain of command. Yet almost from the start, McNamara and the White House treated Westmoreland as being on a par with his superiors and normally put greater credence in COMUSMACV's assessments than those of the theater commander, Admiral Sharp, or the JCS. In fact, Westmoreland's views and those of the JCS were often practically identical, with the JCS sometimes even coaching Westmoreland on what to say or how to present it. Yet in the day-to-day handling of the war, Johnson and McNamara seemed to believe that because Westmoreland was closer to the situation, he was more familiar with the nuances and tempo of the conflict, making his advice more authoritative. Whether Westmoreland's reportage and evaluations were in fact accurate and reliable became one of the most hotly debated issues of the conflict. Looking back, McNamara acknowledged that some of their discussions and the information he received were "superficial." But he never suggested that he considered Westmoreland's advice unsound or that he made a mistake by not paying more attention to the JCS.[92]

In these circumstances, it was almost inevitable that the Joint Chiefs of Staff would exercise limited influence on high-level decisions and the strategy and tactics used in the war. That they stuck it out, refusing to resign in protest as some have argued they should have, underscores their willingness to persevere (their "can do" spirit, as General Bruce Palmer, Jr., called it), and their sense of duty in the face of mounting adversity.[93] A shrewd politician, Lyndon Johnson thought he could handle the JCS like he had handled his political competitors over the years, by offering them compromises and meeting their proposals halfway. But in facing up to a confrontation with the North Vietnamese and, by extension, their Soviet and Chinese allies, the Joint Chiefs realized something the President did not: that halfway measures would never suffice and that waging a war against such an enemy meant accepting great risks or getting out. From the outset, the JCS had wanted a more vigorous response than President Johnson was willing to contemplate; for that reason, he and McNamara elected to ignore the Joint Chiefs and to follow a different path. Though it led ultimately to the same destination—a massive military commitment in Southeast Asia—it had more twists and turns and brought power to bear in increments that the enemy had less trouble absorbing.

NOTES

1. Jack Shulimson and Charles M. Johnson, *U.S. Marines in Vietnam: The Landing and the Buildup, 1965* (Washington, DC: History and Museums Division, Headquarters, U.S. Marine Corps, 1978), 9–15.

2. Ronald H. Spector, *Advice and Support: The Early Years, 1941–1960* (Washington, DC: Center of Military History, 1983), 270.

3. Jack Shulimson, *The Joint Chiefs of Staff and The War in Vietnam, 1960–1968* (Washington, DC: Office of Joint History, 2011), Pt. 1, 25; Lawrence S. Kaplan, Ronald D. Landa, and Edward J. Drea, *History of the Office of the Secretary of Defense: The McNamara Ascendancy, 1961–1965* (Washington, DC: Historical Office, Office of the Secretary of Defense, 2006), 262; Spector, 357–368.

4. Kaplan et al., 266–269; Shulimson, Pt. 1, 33–38.

5. Maxwell D. Taylor, *Swords and Plowshares* (New York: W.W. Norton, 1972), 238–242. The full Taylor–Rostow Report is in *FRUS, 1961–63*, I, 477–532.

6 MFR February 6, 1961, "Meeting to discuss the recommendations of the Taylor Mission to South Viet–Nam," ibid., 534.

7 *CINCPAC Command History, 1961*, 20–22, 170–171; Memo, McNamara to Kennedy, November 8, 1961, "South Vietnam," *FRUS, 1961–63*, I, 559–561.

8 Memo, Rusk and McNamara to Kennedy, November 11, 1961, *Pentagon Papers*, Book II, 359–366; Robert S. McNamara with Brian VanDeMark, *In Retrospect: The Tragedy and Lessons of Vietnam* (New York: Times Books, 1995), 38–39.

9 NSAM 111, "First Phase of Viet–Nam Program," February 22, 1961, John F. Kennedy Papers, National Security Files, JFK Library.

10 Taylor, *Swords and Plowshares*, 288; Shulimson, 193–196.

11 Graham A. Cosmas, *MACV: The Joint Command in the Years of Escalation, 1962–1967* (Washington, DC: Center of Military History, 2006), 21–29.

12 Robert F. Futrell, with Martin Blumenson, *The United States Air Force in Southeast Asia: The Advisory Years to 1965* (Washington, DC: Office of Air Force History, 1981), 135–149.

13 Cosmas, *MACV, 1962-67*, 79–80.

14 Richard H. Shultz, Jr., *The Secret War Against Hanoi* (New York: Perennial, 1999), 35. For CIA cooperation, see Memo, Krulak to Taylor, June 6, 1963, "Conversation with Mr. John A. McCone," Taylor Papers, box 7, NDU.

15 "The President's News Conference," November 14, 1963, *Public Papers of the Presidents of the United States: John F. Kennedy, 1963* (Washington, DC: GPO, 1964), 846.

16 Memo, Johnson to Taylor, December 2, 1963, no subject, Johnson Papers, LBJ Library; Futrell and Blumenson, 195.

17 CMCM 12-64 to JCS, February 24, 1974, "Situation in Vietnam," JCS 2343/326-2.

18 JCSM 46-64 to SECDEF, January 22, 1964, "Vietnam and Southeast Asia," JCS 2339/117-2; Futrell and Blumenson, 198.

19 Johnson quoted in H.R. McMaster, *Dereliction of Duty* (New York: Harper Perennial, 1997), 70. See also Memo by Taylor of Conversation Between JCS and the President, March 4, 1964, *FRUS, 1964–68*, I, 129–130.

20 See Doris Kearns, *Lyndon Johnson and the American Dream* (New York: Harper & Row, 1976), 242.

21 Memo, McNamara to Johnson, March 16, 1964, "South Vietnam," *FRUS, 1964–68*, I, 153–167; Johnson, *Vantage Point*, 66–67.

22 McMaster, *Dereliction of Duty*, 76–77. See also CSAFM 164-64, February 21, 1974, "Revitalized South Vietnam Campaign," JCS 2343/326/1; CMCM 12-64, February 24, 1974, "Situation in Vietnam," JCS 2343/326-2; and CNOM 59-64, February 24, 1974, "U.S. Policy Toward Southeast Asia," JCS 2443/326/3.

23 LeMay quoted in Futrell and Blumenson, 201.

24 Memo on Vietnam by DCI, March 3, 1964, *FRUS, 1964–68*, I, 120–127; NSAM 288, March 17, 1964, "Implementation of South Vietnam Programs," ibid., 172–173.

25 H.R. McMaster, *Dereliction of Duty*, 90. See also H.R. McMaster, "The Human Element: When Gadgetry Becomes Strategy," *World Affairs* (Winter 2009).

26 Harold P. Ford, *CIA and the Vietnam Policymakers: Three Episodes, 1962–1968* (Washington, DC: Central Intelligence Agency, 1998), 57.

27 CM-1450-64 to SECDEF, June 2, 1964, "Transmittal of JCSM-471-64,"; and JCSM-471-64 to SECDEF, June 2, 1964, "Objectives and Course of Action—Southeast Asia," both derived from JCS 2343/394-1.

28 See statement of Army views in Enclosure B to Report by the SACSA to JCS, February 11, 1974, "Revitalized South Vietnam Campaign," JCS 2343/317-1.

29 See, for example, the observations on Wheeler in Lewis Sorley, *Honorable Warrior: General Harold K. Johnson and the Ethics of Command* (Lawrence: University Press of Kansas, 1998), 182–183; and McMaster, *Dereliction of Duty*, 108–111.

30 Edward J. Marolda and Oscar P. Fitzgerald, *The United States Navy and the Vietnam Conflict*, vol. II, *From Military Assistance to Combat, 1959–1965* (Washington, DC: Naval Historical Center, 1986), 393–436.

31 Joint Chiefs of Staff, *The Joint Chiefs of Staff and the War in Vietnam, 1960–1968* (Washington, DC: Historical Division, Joint Secretariat, Joint Chiefs of Staff, July 1970), Pt. I, chap. 11, 22–23.

32 Johnson, *Vantage Point*, 113. See also Edward J. Drea, "'Received Information Indicating Attack'": The Gulf of Tonkin Incident 40 Years Later," *Military History Quarterly* (Summer 2004).

33 Robert J. Hanyok, "Skunks, Bogies, Silent Hounds, and the Flying Fish: The Gulf of Tonkin Mystery, 2–4 August 1964," *Cryptologic Quarterly* 19–20 (Winter 2000/Spring 2001), 1–55.

34 Quotation from Draft Memo by McGeorge Bundy, June 10, 1964, "Alternative Public Positions for U.S. on Southeast Asia," *FRUS, 1964–68*, I, 493–496. On the internal debate over seeking a congressional resolution, see Andrew Preston, *The War Council: McGeorge Bundy, the NSC, and Vietnam* (Cambridge: Harvard University Press, 2006), 149–150.

35 Joint Chiefs, *JCS and the War in Vietnam*, 24–27.

36 William W. Momyer, *Air Power in Three Wars* (Washington, DC: Office of Air Force History, 1978), 13.

37 JCSM-460-64 to SECDEF, May 30, 1964, "Air Campaign Against North Vietnam," JCS 2343/383; and JCSM-729-64 to SECDEF, August 24, 1964, "Target Study—North Vietnam," JCS 2343/383-2. For the origins and development of LeMay's proposal, see CSAFM 459-64 to JCS, May 28, 1964, "Objectives and Course of Action—Southeast Asia," JCS 2343/394; CSAFM 665-64, August 10, 1964, "Recommended Course of Action—Southeast Asia," JCS 2343/438; and CSAFM 667-64 to JCS, August 10, 1964, "Air Strikes Against NVN," JCS 2343/442.

38 CSAM 417-64 to JCS, August 6, 1964, "Planning Guidance for Outline Plan," JCS 2343/439.

39 JCSM-746-64 to SECDEF, August 26, 1964, "Recommended Courses of Action—Southeast Asia," JCS 2343/444-1. See also McMaster, *Dereliction of Duty*, 142–150, 225–226; and Bruce Palmer, Jr., *The 25-Year War: America's Military Role in Vietnam* (Lexington: University Press of Kentucky, 1984), 34–35.

40 "Remarks at a Barbecue in Stonewall, Texas," August 29, 1964, *Johnson Public Papers, 1963–1964*, 1022.

41 See David Halberstam, *The Best and the Brightest* (New York: Random House, 1969), 488, 500.

42 NSAM 314, September 10, 1964, "U.S. Actions in South Vietnam," *FRUS, 1964–68*, I, 758–760.

43 SIGMA II–64 Collection (declassified) in War Games, vol. 2 folder, box 30, National Security File, Agency File, JCS, LBJ Library; McMaster, *Dereliction of Duty*, 156–158; McNamara, *In Retrospect*, 153.

44 Memo, Valenti to LJB, February 14, 1964, Folder F6 115-4 JCS, box 21, Confidential File FC 115 (1966), LBJ Library.

45 See Preston, 161.

46 Futrell and Blumenson, 255.

47 McMaster, *Dereliction of Duty*, 184–187.

48 JCSM-967-64 to SECDEF, February 18, 1964, "Courses of Action in Southeast Asia," JCS 2339/157-1; also in *The Pentagon Papers (Gravel Edition), The Defense Department History of United States Decisionmaking on Vietnam*, 4 vols. (Boston: Beacon Press, 1971–1972), III, 639–640.

49 Memo, McNamara to CJCS, February 21, 1964, "Courses of Action in Southeast Asia," 1st N/H to JCS 2339/157-1.

50 McNamara, *In Retrospect*, 160.

51 Message, COMU.S.MACV to JCS, February 27, 1964, cited in *JCS and War in Vietnam, 1960–1968*, Pt I, chap. 14, 13–14.

52 U.S. Grant Sharp, *Strategy for Defeat: Vietnam in Retrospect* (Novato, CA: Presidio Press, 1978), 52. See also *JCS and War in Vietnam, 1960–68*, Part I, chap. 14, 14.

53 JCSM-982-64 to SECDEF, February 23, 1964, "Courses of Action in Southeast Asia," JCS 2339/161-2; also in *FRUS, 1964–68*, I, 932–935.

54 "Position Paper on Southeast Asia," December 2, 1964, *FRUS, 1964–68*, I, 969–974; *JCS and War in Vietnam, 1960–68*, Part I, chap. 14, 33–35.

55 *JCS and War in Vietnam, 1960–68*, Part II, chap. 18, 5.

56 Summary Notes of 547th Meeting of the NSC, February 8, 1975, *FRUS, 1964–68*, II, 188; *JCS and War in Vietnam, 1960–68*, Part II, chap. 17, 17–25.

57 Mark Clodfelter, *The Limits of Air Power: The American Bombing of North Vietnam* (New York: Free Press, 1989), 79–81.

58 JCS and War in Vietnam, Pt. II, chap. 18, 21–22; JCSM-221-65 to SECDEF, March 27, 1965, "Air Strike Program Against North Vietnam," with Annex containing synopsis of Joint Staff bombing program, JCS 2343/551.

59 *JCS and the War in Vietnam*, 18, 22–26.

60 Cosmas, 202–205; Sorley, 197.

61 McNamara, *In Retrospect*, 177. General Wallace Greene, Commandant of Marines, gave a similar estimate during congressional testimony in the summer of 1965. General Johnson's prediction is in the recollections of General Andrew Goodpaster, Wheeler's assistant, following Johnson's return from Vietnam and his meeting with the President on March 15, 1965. See Senate Report No. 100-163, Pt. 3, 165–166; and John P. Burke and Fred I. Greenstein, *How Presidents Test Reality: Decisions on Vietnam, 1954 and 1965* (New York: Russell Sage Foundation, 1989), 160–161.

62 *JCS and War in Vietnam, 1960–68*, Pt. II, chap. 20, 14–15.

63 Jacob Van Staaveren, "U.S.AF Plans and Operations in Southeast Asia, 1965" (U.S.AF Historical Liaison Office, October 1966), chap. III, 21–22; JCSM-204-65 to SECDEF, March 20, 1965, "Deployment of U.S./Allied Combat Forces to Vietnam," *FRUS, 1964–68*, II, 465–467.

64 Poole, *JCS and National Policy, 1965–68*, Pt. I, 114–115.

65 NSAM 328, April 6, 1965, "Decisions with Respect to Vietnam," *FRUS, 1964–68*, II, 537–539.

66 *JCS and War in Vietnam, 1960–68*, Pt. II, chap. 21, 2–3.

67 Sharp, 77–90; McNamara, *In Retrospect*, 182–183; Taylor, *Swords and Plowshares*, 342–343; John Schlight, *The War in South Vietnam: The Years of the Offensive, 1965–1968*, in *The United States Air Force in Southeast Asia* series (Washington, DC: Office of Air Force History, 1988), 31.

68 *Commander in Chief Pacific: Command History, 1965*, vol. II, 288; JCSM-321-65 to SECDEF, April 30, 1965, "Program for the Deployment of Additional Forces into South Vietnam," JCS 2343/564-7.

69 Johnson, *Vantage Point*, 201.

70 Bruce Palmer, Jr., *Intervention in the Caribbean: The Dominican Crisis of 1965* (Lexington: University Press of Kentucky, 1989), 4.

71 Drea, *McNamara, Clifford, and the Burdens of Vietnam*, 289–315.

72 Walter S. Poole, *The Joint Chiefs of Staff and National Policy, 1965–1968* (Washington, DC: Office of Joint History, Joint Chiefs of Staff, publication forthcoming), chap. 9.

73 Military strength figures from Richard W. Stewart, ed., *American Military History*, Vol. II, *The United States Army in a Global Era, 1917–2003* (Washington, DC: Center of Military History, 2005), 306. Combat deaths from U.S. Bureau of the Census, *Statistical Abstract of the United States: 1969*, 90th ed. (Washington, DC: GPO, 1969), 256; and U.S. Bureau of the Census, *Statistical Abstract of the United States: 1976*, 97th ed. (Washington, DC: GPO, 1976), 337. Veterans organizations generally cite a figure of 58,000 killed in the Vietnam War, which includes noncombat fatalities.

74 Johnson, *Vantage Point*, 50, 232, and passim.

75 Wheeler quoted in Poole, *JCS and National Policy, 1965–68*, Pt. I, 117.

76 See Kearns, *Lyndon Johnson and the American Dream*, 251–253.

77 Minutes of Meeting of the President's Cabinet, June 18, 1965, Johnson Papers, Cabinet Papers File, LBJ Library.

78 Briefing for SECDEF, July 16, 1965, JCS 2343/636; William C. Westmoreland, *A Soldier Reports* (Garden City: Doubleday, 1976), 144–161; McNamara, *In Retrospect*, 211–212.

79 JCSM-652-65 to SECDEF, August 27, 1965, "Concept for Vietnam," JCS 2343/646-1.

80 Quoted in Clodfelter, 117, from George W. Ball, "How Valid Are the Assumptions Underlying Our Viet–Nam Policies?" *Atlantic Monthly*, October 5, 1964, 38.

81 See McNamara, *In Retrospect*, 175; and Palmer, Jr., *The 25-Year War*, 33–34.

82 Schlight, *Years of the Offensive, 1965–68*, 30–44; Cosmas, 134–135.

83 Wayne Thompson, *To Hanoi and Back: The United States Air Force and North Vietnam, 1966–1973* (Washington, DC: Air Force History and Museums Program, 2000), 31. Emphasis in original.

84 McNamara, *In Retrospect*, 160–161, 211.

85 Thompson, 42.

86 David C. Humphrey, "Tuesday Lunch at the Johnson White House: A Preliminary Assessment," *Diplomatic History* 8 (Winter 1984), 81–101; Clodfelter, 118–122.

87 William W. Momyer, *Air Power in Three Wars* (Washington, DC: Department of the Air Force, 1978), 23.

88 Cosmas, 120–122.

89 John M. Carland, *Stemming the Tide: May 1965 to October 1966*, in *United States Army in Vietnam* series (Washington, DC: Center of Military History, 2000), 155–157.

90 JCSM-130-66 to SECDEF, March 1, 1966, "CY 1966 Deployments to SE Asia and World-wide U.S. Military Posture," JCS 2343/760-5.

91 U.S. casualty figures from Drea, *McNamara, Clifford, and the Burdens of Vietnam*, 173.

92 McNamara, *In Retrospect*, 203.

93 Palmer, *25-Year War*, 46.

General Earle G. Wheeler, USA; General Creighton W. Abrams, USA; and Secretary of Defense Melvin R. Laird, Vietnam, ca. 1969

Chapter 10

VIETNAM: RETREAT AND WITHDRAWAL

On March 31, 1968, President Lyndon B. Johnson announced over national television that he would not seek reelection and would instead devote the remainder of his tenure in the White House to finding a peaceful settlement in Vietnam. At home, the President faced a rising crescendo of protests against the war, mounting economic difficulties brought on by war-induced inflation, and challenges to his political leadership from Senators Eugene McCarthy and Robert F. Kennedy. Meanwhile, in Vietnam, recent heavy fighting—the Communist Tet offensive and the ongoing battle for Khe Sanh—had shattered administration predictions that the United States was winning and that the war would soon be over. With an American-imposed solution appearing less and less feasible, the President ordered a halt to the bombing of North Vietnam above the twentieth parallel. Henceforth, the United States would concentrate on strengthening the South Vietnamese armed forces to resist Communist aggression on their own.[1] Since committing U.S. combat forces to Vietnam 3 years earlier, the United States had yet to suffer a major defeat. But it had also been unable to score a decisive victory. As a practical matter, President Johnson's announcement was the first step toward U.S. disengagement from Vietnam, a process that would still take 5 more years to yield what his successor, Richard M. Nixon, termed "peace with honor."

STALEMATE

Long before President Johnson announced his decision not to stand for reelection, the war in Vietnam had degenerated into a stalemate. At the outset of large-scale U.S. intervention in the summer of 1965, Secretary of Defense McNamara had wanted to demonstrate to the North Vietnamese that their aggression would never succeed and that their only choice was to withdraw their forces and accept a negotiated settlement.[2] The stalemate that McNamara envisioned had indeed come to pass, but it had not worked as he had predicted. Even though American intervention had thwarted a Communist takeover and bolstered the South Vietnamese government and its armed forces, the U.S. presence had failed to intimidate the enemy. Trained and equipped for a war in Europe, American forces initially found themselves awkwardly adjusting to unfamiliar tactics and terrain. Dominant in mobility and firepower, they repeatedly inflicted heavy losses on the Viet Cong and North Vietnamese, but could not achieve decisive results. The longer the war went on, the more resilient the Viet Cong and North Vietnamese became. Rather than wearing down the enemy's will and ability to fight, General William C. Westmoreland's strategy of attrition was having the opposite effect. By demonstrating the limits of American military power, it strengthened Viet Cong and North Vietnamese resolve. While they might not prevail in every engagement, they fought with growing confidence that they could stand up to the Americans, inflict enough casualties to turn public opinion in the United States against the war, and eventually win.[3]

Efforts by the Johnson administration to rally support for its involvement in Vietnam yielded disappointing results. At home, a growing and increasingly strident antiwar movement challenged the administration's policies with mass protests, acts of civil disobedience, and draft card burnings. In Europe and elsewhere overseas, opposition to the war was also on the rise. During the Korean conflict, twenty-two nations had contributed forces to help turn back the Communist aggressors; in Vietnam only four countries—South Korea, Thailand, Australia, and New Zealand—sent combat troops to fight alongside U.S. and South Vietnamese forces. (Thai troops did not arrive in South Vietnam until mid-1968 and were not much of a factor in the war.) South Korea's participation came with numerous strings attached, including the Korean government's insistence that the United States provide large financial subsidies and other incentives.[4] NATO, America's long-time partner, evinced not the slightest interest in helping. A few Alliance leaders, like West German Foreign Minister Gerhard Schroeder, discerned a clear link between the outcome in Southeast Asia and the fate of Europe. Schroeder feared that, if the United States failed to prevail in Vietnam, it would expose Europe to renewed Soviet pressure. But his was a minority view. Far more prevalent among Europeans was the notion that Vietnam was a distraction, a needless diversion of American attention and resources that would end up weakening the Alliance and increase Europe's share of the defense burden.[5]

At no point did the Johnson administration attempt to develop or implement a defense policy that brought the allocation of U.S. resources for Vietnam into line with commitments elsewhere. While the Joint Chiefs were well aware of this gap in planning, they could never persuade either President Johnson or Secretary McNamara to take the necessary steps to bridge it. The foreseeable result was a draw-down of personnel and equipment assigned to or earmarked for Europe and other contingencies. Calling up the Reserves, a course the JCS consistently favored, would have alleviated some of these problems. Yet any time they raised the issue, the President and the Secretary of Defense rejected it as politically infeasible. As a result, planning for Vietnam followed no coherent blueprint and became instead a series of ad hoc responses to an increasingly intractable situation that consumed more and more American lives and treasure.

In the autumn of 1966, Westmoreland launched a major offensive aimed at putting maximum military pressure on the Viet Cong and North Vietnamese. According to the available intelligence, infiltration of regular NVA units from the North had subsided and there were signs that the Viet Cong was having trouble replacing its losses.[6] In light of these findings and a recent surge in U.S. troop levels, Westmoreland believed he had at his disposal sufficient strength to deal a crippling blow that would turn the war around. To augment the offensive in the south, the Joint Chiefs sought permission for Admiral U.S. Grant Sharp, commander of U.S. forces in the Pacific (CINCPAC), to step up Operation *Rolling Thunder* attacks against North Vietnam. By then, with Westmoreland in firm control of the ground war in the South, about the only place where the JCS could make a difference was in the air war against the North. Following a lengthy debate with the White House, they finally persuaded the President in June 1966 to relax some of the restrictions on bombing petroleum facilities near Hanoi and Haiphong.[7] Though the ensuing attacks had limited effect, they set the stage for the submission in August of a more ambitious *Rolling Thunder* program package that included industrial and transportation targets in North Vietnam's Red River Delta. President Johnson approved the new bombing scheme in November, just as the ground campaign was getting under way, but at the State Department's urging he deferred its full implementation pending the outcome of a British initiative exploring the possibility of negotiations. As it turned out, it was not until February 1967 that the President allowed the approved program to proceed in toto.[8]

The uncoordinated execution of these measures, and delays in carrying them out, virtually assured that they would have a limited impact on the course of the war. While Westmoreland's ground offensive scored some notable successes at the outset, it proved more difficult to sustain than expected with the forces available. Taking territory held by the Viet Cong was easier than holding it and making it secure. By the early spring of 1967, the Viet Cong and North Vietnamese had begun a counterattack that reclaimed lost ground as the American offensive became overextended and bogged down. Meanwhile, stepped-up enemy activity in the northern I Corps region along the Demilitarized Zone (DMZ) and against the heavily fortified American base at Khe Sanh suggested that the North Vietnamese were massing for a conventional invasion of the South, causing COMUSMACV to divert troops and airpower from other operations. Seeing no other choice, Westmoreland (at Wheeler's urging) served notice in mid-March 1967 that he would need a minimum of 100,000 more troops within the coming year just to hold his existing positions in I Corps, and probably double that number to maintain the momentum of operations elsewhere. If approved, the additional buildup would bring the American presence in South Vietnam to over 670,000 troops.[9]

In Washington, Westmoreland's request for more troops touched off a heated internal debate that lasted well into the summer. One reason the review dragged on was that it had to compete with a sudden emergence of other critical problems—the ABM deployment issue, growing tensions between Greece and Turkey over Cyprus, the escalating militancy of the antiwar movement at home, and the outbreak of war in early June 1967 between Israel and its Arab neighbors resulting in a series of Israeli victories that recast the balance of power in the Middle East. Finding time to address these issues challenged the Joint Chiefs no less than it did McNamara and others in the Johnson administration and made it difficult to pursue an orderly and systematic assessment of the situation in Vietnam.

Once they got down to business, the Joint Chiefs rallied in support of Westmoreland, feeling that now was not the time to cut and run. All the same, there were continuing differences among them over basic strategy, with the Army and Marines favoring a greater effort on the ground in the south and the Air Force and the Navy urging stronger air and naval action against the North. To get around their disagreements, the chiefs linked a further buildup in the south such as Westmoreland proposed with an expansion and intensification of the *Rolling Thunder* air campaign against the North.[10] As far as Secretary McNamara was concerned, however, a renewed intensification of the war held no appeal. Since the previous autumn, he had shown growing frustration over the lack of military progress and could not help eyeing the rising financial costs of the war, which had grown steadily to more than a third of the defense budget.[11] Though he had once offered Westmoreland practically a blank check, he regretted having done so, and was inclined to level off U.S. military action in hopes of enticing the enemy into negotiations.[12] Still, he wanted Westmoreland's request to receive a fair hearing and called him back to Washington to explain his position directly to President Johnson. At one point in their meeting, the President turned to Westmoreland and asked testily: "When we add divisions, can't the enemy add divisions? If so, where does it all end?"[13]

As the debate progressed, it focused more and more on the air war against North Vietnam. Confident that the results would show up sooner or later, General Wheeler characterized the air campaign was one of two "blue chips" the United States possessed (the other being the capacity to mount an aggressive ground campaign) that could directly influence the outcome of the war.[14] In practice, however, the United States had never pursued the air war with the same degree of commitment it had shown on the ground. Under the allocation approved during the early days of the war, COMUSMACV had first call on air assets, with the result that about two-thirds of the sorties flown by the Air Force and the Navy had been either in support of combat operations in the South or for interdiction purposes against the Ho Chi Minh Trail; only about one-third of the sorties had been against the North. Moreover, under the "graduated response" rules that governed *Rolling Thunder*, many lucrative bombing targets in the Hanoi-Haiphong area and Red River Delta remained untouched. Arguing that the next step was obvious, Air Force Chief of Staff General John P. McConnell persuaded his JCS colleagues that with or without a buildup in the South, they should press McNamara and the President to lift restrictions on the air campaign and pursue the rapid and methodical destruction of North Vietnam's war-supporting infrastructure.[15]

While the chiefs' position on the air campaign had strong support in military circles, it met with unmitigated disdain from McNamara and the OSD "whiz kids." Labeling the air war as counterproductive, they considered JCS proposals for expanding it dangerous and risky. McNamara had never put much stock in the bombing to begin with, so it was no surprise to him as the war dragged on that study after study reaching his desk showed it as having limited success in curbing infiltration into the South or on North Vietnam's capacity to wage war. Citing the administration-imposed restrictions on targets and bombing under which the Air Force and the Navy operated, the JCS responded that such results were practically preordained. But under the cost-effectiveness criteria he applied to practically everything, McNamara concluded that the air war was becoming too expensive in terms of pilots and planes lost and other factors and ought to be sharply curtailed rather than expanded.[16]

The showdown between McNamara and the Joint Chiefs came in August 1967 during open hearings before the Senate Armed Services Preparedness Investigating Subcommittee, chaired by John C. Stennis of Mississippi. The instigator of the hearings was Senator W. Stuart Symington of Missouri, first Secretary of the Air Force in the Truman administration and an outspoken advocate of more vigorous use of airpower against North Vietnam. Like other conservative Democrats, Stennis and Symington had become impatient and thought that more could be done with airpower to win the war and to avoid the need for additional ground troops. Not wanting to give the committee any more opportunities than it already had to second-guess his conduct of the war, President Johnson sent McNamara and Wheeler to Saigon in July to work out a new statement of troop requirements and to review the air campaign. Following a busy round of briefings, McNamara and Wheeler returned to Washington bearing a revised request from Westmoreland for an additional 50,000 troops, the most the United States could muster without calling up the Reserves or vastly curtailing draft deferments. But despite heavy pressure from CINCPAC and the theater air staff, McNamara refused to endorse an expansion of the bombing operations.[17] Knowing how Stennis and his colleagues would react, Johnson took matters into his own hands and on July 20, 1967, he approved a modification to the *Rolling Thunder* campaign that included about a dozen new targets, some in the Hanoi-Haiphong sanctuary area.[18]

Already severely strained, relations between McNamara and the Joint Chiefs became even worse once the Stennis committee's hearings began. Testifying in executive session, the JCS, Admiral Sharp, and Lieutenant General William W. Momyer, USAF, commander of the Seventh Air Force in Vietnam, all insisted that the administration's "doctrine of gradualism" toward bombing had proven ineffective and that the air campaign they were allowed to carry out was too little too late. In rebuttal, McNamara defended the current concept of operations as carefully thought out and "directed toward reasonable and realizable goals." Indicating that JCS proposals to ramp up the bombing were exactly the opposite, McNamara left the clear impression that he considered his military judgment superior to that of the professionals, while his choice of words challenged their soundness of mind.[19] The chiefs were dismayed and in the aftermath of the hearings, relations between the JCS and the Secretary of Defense sank to a new low. "Leaks" to the press of growing dissension within the Pentagon inevitably followed. Attempting to repair the damage, President Johnson held a news conference at which he insisted that there was "no deep division" within the administration over the prosecution of the war.[20] A lame defense, it convinced no one and only added to the administration's widening credibility gap. Like the conflict in Vietnam, the policy process in Washington had come practically to a standstill, unable to cope or to find new ideas.

TET AND ITS AFTERMATH

The impasse over Vietnam was short-lived, broken by the tightening NVA siege of Khe Sanh and the massive Viet Cong offensive launched in late January 1968 during the Tet holidays. Though not the "bolt out of the blue" that the Korean invasion of 1950 was, the Tet uprising still caught American and South Vietnamese forces off guard by its nationwide scale and scope and by the Viet Cong's determination to take and hold urban areas. Fighting in the ancient capital city of Hue was especially intense and required nearly a month of bloody house-to-house combat to dislodge the enemy. In all, 2,100 American and 4,000 South Vietnamese soldiers died in combat during the uprising. Viet Cong losses were put at 50,000 or more.[21]

The enemy's dramatic Tet offensive almost obscured the ongoing struggle for Khe Sanh, a strategic outpost in the northwest corner of South Vietnam's I Corps region. Defended by a combined force of U.S. Marines and South Vietnam (SVN)

Rangers, Khe Sanh straddled Route 9, a key east-west highway, and was an ideal launching point for search-and-destroy operations against the Ho Chi Minh Trail. In January 1968, the North Vietnamese started massing three divisions around Khe Sanh, laying a siege that evoked memories of the 1954 contest for Dien Bien Phu. While there were strong arguments for abandoning the base, the consensus among the Joint Chiefs was that it should be held at all cost. Indeed, General Wheeler termed Khe Sanh "the anchor of our whole defense of the northern portion of South Vietnam," and argued that defending it would tie down many North Vietnamese who otherwise would be free to attack elsewhere.[22]

Though confident that the outpost would hold, Westmoreland wanted to minimize the risk and ordered what became the most intense air bombardment of the war against enemy positions around Khe Sanh. Toward the end of January, taking matters a step further, he notified the Joint Chiefs that he was exploring a plan, code-named *Fracture Jaw*, to use nuclear or chemical weapons to relieve the enemy pressure. Referred to the Joint Staff for review, *Fracture Jaw* remained a topic of discussion between Washington and Saigon for several weeks. But as Khe Sanh's prospects improved, Westmoreland lost interest in any further nuclear planning. Eventually, the plan reached President Johnson, who wanted nothing to do with it and ordered it summarily withdrawn, thus bringing to a close the first and only episode in which the Joint Chiefs contemplated the specific use of nuclear weapons in Vietnam.[23]

A failure militarily, the enemy's Tet offensive was a stunning political success that broke the back of support for the war in the United States. Almost overnight, opinion in Washington and across the country changed, leaving the Joint Chiefs practically alone in clinging to the administration's original objectives. Instead of insuring the survival of an independent, non-Communist South Vietnam, President Johnson now declared that bringing the war to a peaceful resolution was his top concern. Earlier, McNamara had made known his decision to leave office and in late February 1968, disillusioned and demoralized, he finally stepped down.[24] His successor, Clark M. Clifford, promptly initiated a top-to-bottom review of the war. Unsure of what to expect, the Service chiefs turned to Wheeler, who did his best to bolster their morale and keep the momentum of the war going even while the President was renewing his call for negotiations and ordering cutbacks in air operations against the North, all with an eye toward eventual withdrawal. The result was a continuation of the conflict, but at a reduced tempo that left the outcome more in doubt than ever.[25]

With McNamara's departure, the JCS were cautiously optimistic that in reassessing its options, the administration would not stray too far from its original course. By then, Wheeler was both the dominant figure in JCS deliberations and an accepted member of President Johnson's inner circle. Shortly after the Stennis committee hearings in late summer 1967, he had suffered a mild heart attack. Despite a swift recovery, he indicated he might have to retire. Johnson refused to let him go. "I can't afford to lose you," the President told him. "You have never given me a bad piece of advice."[26] Starting in October 1967, Wheeler was a regular participant in the Tuesday lunch, attended by the President and his senior advisors. On March 22, 1968, Johnson announced that Wheeler would serve an unprecedented fifth year as Chairman.[27] Yet proximity to power did not equate with influence and, as was often the case, Wheeler returned to the Pentagon from his meetings with the President appearing to his staff tired and discouraged.[28]

Day in and day out, Wheeler and the other chiefs waged an uphill battle to be heard. In fact, intelligence reports affirmed that the Tet offensive had decimated the Viet Cong, resulting in an improved military situation across Vietnam. It was the opportunity the Joint Chiefs had been waiting for and, wasting no time, Wheeler urged Westmoreland to exploit the enemy's weakness through a series of new operations. Accordingly, Westmoreland revived his earlier request for another 200,000 troops to finish the job. Wheeler knew that an increase of that size was bound to be controversial and that the odds of approval were against it, but he felt the war was entering a new and more "critical phase" and couched his endorsement of Westmoreland's request in an ominous assessment of the alternative.[29] What the President wanted, however, was less conflict, not more, and with that end in mind he accepted the advice of his new Secretary of Defense and others whose political instincts he trusted, that the time had come to deescalate the war, turn it over to the South Vietnamese, and get American troops out in an orderly manner.[30]

Disappointed by the turn of events, the Joint Chiefs of Staff felt increasingly beleaguered and isolated. They regarded the President's decision of March 31, 1968, to stop bombing above the twentieth parallel and to expedite the search for a negotiated settlement as ill-advised and militarily unsound. As Wheeler characterized it, the bombing halt amounted to an "aerial Dien Bien Phu."[31] Yet neither he nor the Service chiefs had anything better to offer that the President, Congress, or the American public would have considered acceptable. As during bombing pauses in the past, the JCS expected the North Vietnamese to use the respite to build up their defenses and to resupply their troops, and were not disappointed. Yet even airpower enthusiasts acknowledged that there was not much they could do for the next month or so due to the onset of the monsoon season and poor flying weather. Everything, it seemed, was conspiring against JCS efforts to keep the war on track.[32]

Setting the stage for an American withdrawal became the de facto policy. On October 31, 1968, President Johnson suspended the entire bombing campaign against the North, a gesture aimed at jump-starting the stalled Paris peace talks. Only armed reconnaissance flights continued. By now, the JCS realized that there was virtually nothing they could say or do that might convince the President to change his mind. Treating the bombing halt as inevitable, they minimized the risks, accepting them as "low and manageable," even though they remained uneasy over the ultimate consequences for South Vietnam. Slowly but surely, the United States was winnowing its participation in the war and shifting the burden to the South Vietnamese, a process that came to be known as "Vietnamization."[33]

Carrying out the draw-down fell to the new COMUSMACV, General Creighton W. Abrams, who succeeded Westmoreland in mid-1968 when the latter returned to Washington to become Army Chief of Staff. A leading expert in tank warfare, Abrams' combat experience dated from World War II when he commanded an armored task force. As Vice Chief of Staff of the Army from September 1964 to May 1967, he had been deeply involved in the massive deployments of Army units to Vietnam. Though aggressive by instinct, he could sense that the war was winding down and that he would soon be under strong political pressure to limit casualties with low-risk operations and a more defensively oriented deployment of his forces. The Joint Chiefs would have preferred a more proactive posture to keep the enemy off balance. But by the time the Johnson administration left office, the pursuit of a military outcome was no longer a credible option. The best the chiefs could hope for from that point on was a holding action to allow a graceful exit.[34]

NIXON, THE JCS, AND THE POLICY PROCESS

It fell to a new President, Richard M. Nixon, to create something positive out of the previous administration's fiasco in Vietnam. As a candidate for the White House in 1968, Nixon promised to bring American troops home and to end the war "with honor." Even so, he opposed a precipitous withdrawal because it might damage American prestige and trigger a chain reaction of Communist takeovers in Southeast Asia. Once in office, he ruled out seeking "a purely military solution," but affirmed his determination to use force as necessary to achieve his goals.[35] At the same time, he and his assistant for national security affairs, Henry A. Kissinger, sought to enhance the prospects for a negotiated settlement by pursuing "détente" with the Soviet Union and a rapprochement with Communist China (see chapter 11). Though more open to JCS advice than Kennedy and Johnson, he also had no qualms about second-guessing or even belittling the chiefs' advice. Indeed, he was fond of citing H. G. Wells' observation that military people had mediocre minds because intelligent people would never contemplate a military career.[36] But he had the good sense to realize that it was better to have the Joint Chiefs on his side than against him. The result was a somewhat smoother relationship than in the past between military and civilian authorities, even if at times Nixon followed a separate, secret agenda and seemed to have little use for professional military advice if it conflicted with his political objectives.

Those serving on the JCS during Nixon's first year in office were holdover appointments from the Johnson administration. As their terms of service expired, Nixon gradually brought in people of his own choosing. Like Kennedy and Johnson, Nixon found it easier and more convenient to deal with the Chairman. Once a year, he held a formal Oval Office meeting with JCS for picture-taking. Otherwise, he seldom met with them as a group. At Nixon's request, Wheeler stayed on as CJCS until July 1970, but his deteriorating health caused him to share his responsibilities with his heir-apparent, Admiral Thomas H. Moorer, Chief of Naval Operations. An aviator in World War II with a distinguished record of combat experience, Moorer had a reputation around the Pentagon for being blunt but affable, cantankerous yet effective. As Commander in Chief of the Pacific Fleet (CINCPACFLT) in 1964–1965, Moorer had a personal hand in planning and overseeing the early stages of the *Rolling Thunder* air campaign against North Vietnam. Known as a "hawk" on the war, he was definitely the right choice for carrying out the administration's strategy of stepping up military pressure on North Vietnam. Following in Taylor's footsteps, Moorer shunned the role of "team player" and viewed himself first and foremost as an agent and spokesman for the administration. According to one official account, Moorer's influence as Chairman was so thoroughly pervasive that he "was now the only JCS member who really counted."[37]

Moorer's JCS colleagues were a typically diverse group with diverse interests. General John D. Ryan, who succeeded McConnell in August 1969 as Air Force Chief of Staff, was a leading airpower strategist in the Curtis LeMay tradition. An outspoken advocate for his Service, he touted the efficacy of strategic bombing whenever he could. His Army counterpart, General William C. Westmoreland, was the former COMUSMACV, whose frustration and brooding over his recent experiences in Vietnam were all too apparent. Though not yet a full-fledged member of the Joint Chiefs, the Commandant of the Marine Corps, General Leonard F. Chapman, Jr., acted as if he were. Described as "quiet, articulate, and thoughtful," he was an active contributor during JCS deliberations.[38] But with ending the Vietnam War now a foregone conclusion, most of the chiefs showed less interest in joint matters than in protecting their respective Services against the inevitable effects of postwar cutbacks.

The exception was Moorer's successor as CNO, Admiral Elmo R. Zumwalt, Jr., who professed determination to demonstrate that Service and joint interests were not mutually exclusive, as some in uniform believed. The first surface commander to become CNO since Arleigh Burke, Zumwalt wanted to augment the Navy's fleet of expensive nuclear-powered aircraft carriers (CVANs) with smaller, conventionally-powered carriers and surface ships that could be built in greater numbers for less money. He also stressed the need for improved inter-Service cooperation and collaboration to maximize available resources. One of his suggestions was that Army helicopter pilots and Air Force fliers train to operate from Navy vessels. While the Army warmed to the idea, the Air Force wanted no part of it. Still, it did not stop Zumwalt from continuing to explore other joint ventures for sharing assets.[39]

While the policy process in which the Joint Chiefs operated remained outwardly similar to that of previous administrations, decisionmaking became more entrenched than ever in the White House, where Nixon and Kissinger, the national security advisor, played the key roles. A complex and controversial figure, Nixon was exceptionally well versed in world affairs. In Peter W. Rodman's estimation, he had "the deepest intuition and shrewdest strategic judgment of any modern president."[40] Kissinger was equally well informed. Like McGeorge Bundy and Walt Rostow, he came from an academic background, but was far more practical and better steeped in the history of great power politics. As a professor of government at Harvard University before joining the Nixon administration in 1969, Kissinger had published at length on balance-of-power politics and the concept of "limited" nuclear wars. He had built his reputation around studying the tactics and behavior of historic power brokers who excelled in the behind-the-scenes art of Realpolitik—like Otto von Bismarck, Germany's 19th-century "iron chancellor," and Prince Clemens von Metternich of Austria. His biographers generally agree that he saw himself in a similar light, operating as an Old World diplomatist when raison d'état and personal diplomacy reigned supreme.

Coordination between the White House and the JCS took two forms— through the resuscitated mechanisms of the National Security Council, and through backchannel communications. One of Nixon's declared goals was to restore the NSC to an approximation of the system that had existed under Eisenhower. Toward that end, he directed that the Council function as his "principal forum for the consideration of policy issues."[41] Initially, the Joint Chiefs welcomed this reaffirmation of the NSC's central role since it promised to restore more structured, reliable, and predictable procedures to the policy process. But according to Zumwalt, it was not long before the JCS began to question how much they could rely on Nixon and Kissinger to match words with deeds.[42] As time passed, Nixon relied less and less on the NSC and held fewer and fewer meetings.[43] Helmut Sonnenfeldt, a member of the NSC Staff at the time, recalled that Nixon studiously reserved the right of final decision and treated NSC deliberations as "purely advisory meetings."[44] Nor did Nixon bring back Eisenhower's practice of adopting detailed, all-encompassing basic policy papers to guide budgetary decisions, the development of programs, and the allocation of resources. Instead, he attacked problems piecemeal—an effective means of keeping others off balance and concealing his overall purpose—with a barrage of directives, known as national security decision memoranda (NSDM) and requests for reviews, called national security study memoranda (NSSM).[45]

Below the NSC, JCS access to policy guidance was through a battery of interagency committees, all closely overseen, if not personally chaired, by Kissinger. These included the NSC Review Group, headed by Kissinger, to screen matters for submission to the full NSC, and four specialized advisory bodies organized at the Deputy Secretary level for Vietnam, defense policy, arms control, and crisis management.[46] Outside this structure, Kissinger also established informal contacts with the Pentagon through the JCS liaison office. The proper channel of communication was from the White House through the Office of Secretary of Defense to the Chairman of the Joint Chiefs. Kissinger, however, often bypassed OSD by calling Moorer directly and by transmitting documents to him through the JCS liaison office, housed next door in the Old Executive Office Building. Melvin R. Laird, Secretary of Defense during Nixon's first term, deeply resented Kissinger's circumvention of his authority and after an unseemly episode in 1971 involving the mishandling of classified documents by a Navy yeoman assigned to the NSC as a stenographer, he closed the JCS liaison office. Whether the yeoman, Charles E. Radford, was "spying" for the JCS or acting on his own was never conclusively ascertained. But despite the closure of the office, backchannel contacts continued to be one of Kissinger's preferred methods of doing business, a habit he found impossible to break.[47]

WINDING DOWN THE WAR

Nixon's first order of business in Vietnam was to create a politico-military environment favorable to the withdrawal of U.S. forces. When he became President in 1969, the United States still had over half a million troops engaged there and no concrete plans for getting them out.[48] Modeling his policy on Eisenhower's strategy for ending the Korean War, he sought to apply a combination of diplomacy and "irresistible military pressure" to achieve a negotiated settlement with the North Vietnamese that would include the mutual withdrawal of U.S. and NVA forces.[49] Known as "linkage," his diplomatic strategy was to encourage détente with the Soviet Union and exploit signs of a Sino-Soviet ideological split to weaken Communist bloc support of Hanoi. Simultaneously, he extended the war through covert means into Cambodia and accelerated the Vietnamization and pacification programs to cover the phased withdrawal of U.S. ground forces and to provide the government of South Vietnam with increased capabilities for future self-defense. At the outset of his Presidency, Nixon announced to his Cabinet that he expected the war to be over in a year. Almost immediately, he was backtracking from his prediction.[50]

VIETNAM: RETREAT AND WITHDRAWAL

While the Joint Chiefs of Staff took close note of the negotiations, they were rarely directly involved. Even though they had representatives on the various interagency bodies dealing with the peace talks, the governing assumption within the JCS organization was that negotiating strategy did not lie "within the normal purview" of the Joint Chiefs of Staff.[51] Their more direct and immediate concern was to figure out ways of keeping military pressure on the enemy while the United States scaled back its participation in the war. With the loss of U.S. nuclear superiority in the 1960s, Nixon was in no position, as Eisenhower was in 1953, to threaten the use of atomic weapons. Casting about for options, he and Kissinger flirted with the idea of resuming the air war against the North and briefly considered a plan (Operation Duck Hook) to launch a series of quick, intense, and "brutal" strikes against key North Vietnamese targets. But they quickly dropped the idea owing to the lukewarm support it enjoyed among the Joint Chiefs, the political repercussions such actions could have at home, and the danger of derailing plans for détente with the Soviet Union.[52]

With the range of options limited, the preferred approach both at the White House and in the Pentagon became a concerted bombing campaign with B–52s against Viet Cong and NVA sanctuaries in neighboring Cambodia, targets previously off limits to U.S. air attack. The Joint Chiefs, COMUSMACV, and CINCPAC had long favored the destruction of these enemy bases, but had had no luck persuading the previous administration to accept the political and diplomatic risks such an operation might entail. With Nixon's advent, they found a more receptive audience and on March 15, 1969, they received a green light to proceed.[53]

Like the decision to intervene with ground troops in 1965, the "secret" bombing of Cambodia was one of the most controversial episodes of the war. Lasting into May 1970, the attacks concentrated on six enemy bases along the Cambodian-South Vietnamese border and involved the expenditure of over 180,000 tons of munitions.[54] To keep the operation quiet, the White House, the Joint Staff, and COMUSMACV resorted to elaborate deception measures that concealed flight plans and the expenditure of bombs. Privately, members of the Joint Chiefs grumbled at being party to Nixon's duplicity, some complaining that efforts to hide the bombing were "stupid" and bound to fail.[55] But in Nixon's view, preserving secrecy was essential in order to avoid antiwar protests.[56] Actually, there was not much secret about the whole affair. Cambodian leader Prince Norodom Sihanouk knew about the bombing from the outset and obligingly looked the other way. The North Vietnamese were well aware, as were the Soviets, the Chinese, and key figures on Capitol Hill. About the only group not privy to the secret was the American public.

While putting pressure on the enemy through the secret bombing campaign, Nixon sought to expedite the U.S. withdrawal under cover of the Vietnamization program, the incremental substitution of SVN troops for U.S. forces. As the Joint Chiefs repeatedly cautioned, however, the Vietnamization program devised under the Johnson administration and inherited by Nixon was intended solely to develop a security force and would not result in a SVN army that could tackle the North Vietnamese.[57] After taking a personal look at the program in operation, Secretary of Defense Laird came back from a trip to Southeast Asia in March 1969 with an alternative plan to increase the arming, training, and equipping of the South Vietnamese so they could take on not only the Viet Cong but also the NVA.[58] Though Nixon viewed Vietnamization as an integral part of his strategy, he had never envisioned developing and refining South Vietnam's military capabilities quite as fast or to the same degree. Initially skeptical of Laird's proposal, Nixon and Kissinger quickly changed their minds after the Secretary of Defense, without consulting the White House, publicly outlined his program on national television and "leaked" a story to the press, intimating that it was agreed administration policy. "It was largely on the basis of Laird's enthusiastic advocacy," Nixon recalled, "that we undertook the policy of Vietnamization."[59]

Whether the South Vietnamese were up to the task became a recurring issue in JCS deliberations over the next several years. On paper, the South Vietnamese military was a formidable force. With nearly a million men under arms, it ranked as one of the largest in the world. Except for a few elite units, however, it was a heavily conscripted army in which desertion rates were high and morale low. Barely a match for the Viet Cong, it was virtually untested against North Vietnamese regulars. Recognizing the ARVN's weaknesses, the Joint Chiefs urged a paced withdrawal of U.S. forces, coordinated with periodic assessments of the progress of Vietnamization, pacification, and the enemy situation.[60] Nixon agreed that the chiefs' "cut-and-try" approach made a lot of sense and should be followed as much as possible.[61] But for economic reasons he needed to curb defense spending and was under strong political pressure to bring U.S. troops home at an accelerated pace. As a consequence, in setting timetables for the redeployment of U.S. forces, the Joint Chiefs came to realize that "other considerations" than the progress of Vietnamization tended to be the decisive factors.[62]

An early test of Vietnamization occurred during the allied invasion of Cambodia in the spring of 1970. The results were inconclusive, however, owing to the heavy involvement of U.S. forces alongside the South Vietnamese, the extensive presence of U.S. advisors among SVN units, and because the NVA elected for the most part not to engage the invaders. The event precipitating the invasion was a political crisis in neighboring Cambodia, brought on by anti-Communist demonstrations culminating in March 1970 in a coup d'état that replaced the nominally neutralist regime of Prince Sihanouk with a pro-Western one headed by Premier Lon Nol. As one of his first acts, Lon Nol closed the port of Kampong Son (Sihanoukville) to

NVA transfers, thus denying the enemy a major entrepôt for weapons and supplies destined for South Vietnam. A wave of Communist counterattacks led by North Vietnamese regulars soon followed, prompting COMUSMACV, CINCPAC, and the Joint Chiefs to coordinate the development of contingency plans to shore up Lon Nol's regime and, at the same time, to complete the destruction of enemy sanctuaries along the border. The plan initially presented by the Joint Chiefs called for a cross-border operation into Cambodia with U.S. ground forces spearheading the effort.[63] At the time, there were still substantial numbers of U.S. combat troops in Vietnam and no clear picture of how well the ARVN would perform. Nixon and Kissinger, however, wanted the South Vietnamese to be in the vanguard, partly to deflect expected criticism at home and to underscore the lowering of the U.S. profile in accordance with recently announced troop reductions.[64]

In late April, a combined U.S.–SVN invasion force entered Cambodia. Though they captured large quantities of supplies, documents, and military hardware, the allies made little contact with the enemy after the first day. General Abrams wanted to exploit the situation with deeper probes into Cambodia to draw the enemy out. Back in the United States, the Cambodian invasion had aroused some of the largest and most strident protests to that point in the war, suggesting that political support was weak and continuing to decline. Feeling the pressure, President Nixon rejected Abrams' proposal to expand the operation and ordered U.S. troops back across the border by the end of June. While it was not much of a test for the Vietnamization program, Abrams praised the performance SVN forces and relayed word to Washington that he considered their planning and execution "very impressive."[65]

With growing confidence in South Vietnamese forces, Abrams (with encouragement from Nixon and Kissinger) began to envision even bigger operations. Thus, as the Cambodian incursion drew to a close, he received the go-ahead from Admiral Moorer for a new operation known as LAM SON 719, a "dry season" search and destroy foray into Laos to disrupt enemy movement along the Ho Chi Minh Trail. Initiated with the expectation of large-scale U.S. combat ground support inside Vietnam and heavy U.S. air support in Laos, LAM SON 719 was the product of planning done late in 1970 at MACV headquarters in Saigon and in Hawaii by Commander in Chief, Pacific, Admiral John S. McCain, Jr.[66] By then, Nixon and Kissinger had more or less given up trying to negotiate a mutual reduction of forces with the North Vietnamese and had decided to concentrate on a unilateral U.S. withdrawal. The function of LAM SON 719, as Kissinger envisioned it, was to cut enemy supply lines, curb infiltration into the south, and buy time to complete an orderly pull-out of U.S. forces.[67]

LAM SON 719 may have been doomed before it started. With advance warning from their spies in Saigon, the North Vietnamese had ample time to reinforce units and strengthen their defenses along the Ho Chi Minh Trail. By their own account, the NVA had amassed a force of 60,000 troops, against an ARVN invasion force of 17,000. In Washington, meanwhile, following a lengthy and contentious debate, Congress finally passed a foreign military sales bill early in 1971 incorporating the Cooper-Church amendment banning U.S. advisors from assisting in operations outside Vietnam. With U.S. advisory assistance thus curtailed, the South Vietnamese faced serious problems coordinating their air and artillery support. Still, from all the Joint Staff had seen and heard of the plan, there was nothing overtly objectionable about LAM SON 719 and, indeed, much to recommend it, including Abrams' budding confidence in the ARVN and a growing awareness that this might be the last time the South Vietnamese could conduct a dry-season offensive while U.S. forces were still present in Vietnam in substantial numbers to provide backup.[68]

As the operation began in early February 1971, however, confidence in it began to fade. Most skeptical of all was Army Chief of Staff General Westmoreland. Reluctant to second-guess the commander on the scene, Westmoreland had stifled his reservations, much as the JCS had muffled their misgivings about plans for the Bay of Pigs invasion a decade earlier. When pressed by Kissinger for his views, however, Westmoreland lashed out against LAM SON 719, declaring it to be "a very high risk" enterprise with a slim chance of success. Several times as COMUSMACV, Westmoreland had studied the possibility of mounting a similar attack into Laos. But he had never followed through due to the Johnson administration's concern that it would be too risky and would require an inordinate commitment of resources—probably no fewer than four U.S. divisions, or nearly half the U.S. in-country fighting force. In lieu of the invasion taking place under LAM SON 719, Westmoreland urged the White House to consider short raids, feints, and mobile operations to keep the North Vietnamese off balance and to interrupt traffic along the Ho Chi Minh Trail.[69] Bothered by Westmoreland's comments, Kissinger turned to Moorer, who downplayed the general's concerns and offered his assurances, based on Abrams' assessments, that the concept behind the plan was sound.[70]

Once underway, LAM SON 719 began running into one problem after another. Outnumbered and outgunned, the South Vietnamese found their search-and-destroy mission turned into a sustained conventional battle in which the enemy had the initiative. Determined not merely to repel the attackers and protect their lines of communication, the NVA sought to inflict a crushing defeat on the South Vietnamese army that would discredit the American policy of Vietnamization. At a meeting with the Secretary of Defense on March 15, Westmoreland criticized ARVN tactics and, in Moorer's words, "badmouthed the whole LAM SON 719 operation." The next day Moorer assured President Nixon that "things were going pretty well." Nixon wanted the ARVN to keep the operation going into April, when he intended to announce further U.S. troop withdrawals. But under heavy attack from the enemy, the ARVN began a precipitous withdrawal. The tide had turned and, as Kissinger put it, the South Vietnamese were "bugging out." What the administration tried to depict as an orderly tactical withdrawal, journalists on the scene described as a tragic and chaotic rout.[71]

BACK TO AIRPOWER

Though it was not the total catastrophe some observers depicted, LAM SON 719 was clearly a major setback for the United States and its Vietnamese allies. Most serious of all, it had exposed glaring shortcomings in the administration's Vietnamization program. Given enough time and training, perhaps, the ARVN might someday become a formidable fighting force; but for the foreseeable future, it was in no position to stop aggression from the North on its own. One of the few positive things to come out of the whole episode was Secretary of Defense Laird's increased interest in providing more effective measures to block enemy infiltration along the Ho Chi Minh Trail. Toward the end of 1971, with this in mind, he assigned a new Army Brigadier General, John W. Vessey, to the U.S. Embassy in Laos. Working with the Ambassador and CIA station chief, Vessey oversaw the allocation of funds for covert operations against North Vietnamese infiltration. In 1982, under the Reagan administration, Vessey would again attract high-level attention and become the President's choice to chair the Joint Chiefs of Staff.[72]

Despite ongoing efforts by the Nixon administration to shore up South Vietnam's security, the danger from the North continued to grow, while U.S. troop strength continued to drop. By the beginning of 1972, there were fewer than 150,000 American Servicemen left in Vietnam, and under approved troop withdrawal schedules half of those would be gone in a few months. Shrugging off the ARVN's disappointing performance in LAM SON 719, the Nixon White House repeatedly urged the Saigon regime to undertake new forays into Laos and Cambodia. At the same time, to offset the loss of U.S. ground strength, Admiral Moorer, often on his own initiative, pressed Secretary of Defense Laird to relax restrictions on air attacks against North Vietnam and to increase the use of "protective reaction strikes" against surface-to-air missile (SAM) and antiaircraft (AAA) sites that threatened U.S. planes conducting interdiction flights over South Vietnam and Laos. Laird had no objection to American pilots protecting themselves, but as for other attacks against the North, he turned them down more often than not, feeling that they would re-escalate the war and delay U.S. troop withdrawals. President Nixon, however, proved more flexible, and by the end of 1971 bombing against targets in North Vietnam below the 20th parallel was again on the rise.[73]

Convinced that even more was needed, General John D. Lavelle, USAF, Commander of the Seventh Air Force in South Vietnam, took matters into his own hands by stepping up air attacks against the North. Whether he had authority to do so was never fully clear. Adopting "a liberal interpretation" of the rules of engagement, Lavelle later estimated that he carried out "in the neighborhood" of 20 such raids (the real number was closer to thirty) between November 1971 and March 1972. He defended his actions, however, on the grounds that he had the tacit encouragement of his superiors in Washington, including both Admiral Moorer and Secretary Laird, who had urged him to "make maximum use" of existing authority to put pressure on the North.[74] Still, in mounting preplanned attacks Lavelle had gone overboard and risked reigniting the still smoldering bombing controversy between Congress and the administration. Upon learning of the general's interpretation of orders, Moorer and Laird quickly arranged with Air Force Chief of Staff General John D. Ryan to have Lavelle quietly relieved of his duties. But as rumors of the incident spread, they prompted several well-publicized, albeit inconclusive, congressional investigations.[75]

Meanwhile, across Vietnam, the threat of stepped-up combat continued to mount. The showdown came around Easter, on March 30, 1972, when the North Vietnamese launched a coordinated attack against the South, which they initiated with a full-scale conventional invasion across the DMZ, using tanks and self-propelled artillery. Allied intelligence had known for months that the North Vietnamese were preparing a large-scale operation but could not pinpoint either the date or place. Throughout the ensuing crisis, Nixon and Kissinger frequently ignored established lines of communication with the Pentagon and in the interest of expediency dealt directly with Admiral Moorer and the Joint Staff, whose views were more in harmony with those of the White House than Laird's. Seeing the invasion as a challenge to the credibility of his whole foreign policy, President Nixon believed that only a vigorous military response would convince Hanoi and its allies in Moscow and Beijing that he meant business. With battlefield success his uppermost concern, Nixon saw no choice but to remove all restrictions on the use of airpower, something he had been loath to do earlier. In view of the North's blatant aggression, American public and congressional opinion largely acquiesced. Moorer agreed that Hanoi's leaders respected nothing more than the unstinting application of military force, and to that end he helped arrange a swift buildup of airpower. Among the forces added for action were 189 F–4 fighter-bombers, 210 B–52s (half of SAC's bomber force), and four carrier task forces, bringing to six the number of carriers on station, the largest concentration of naval airpower yet seen in the war.[76]

With the increased availability of airpower came friction between Washington and the command in Saigon over how and where to apply it. Nixon, Kissinger, and

Moorer envisioned a fairly broad-brush campaign aimed not simply at curbing the current aggression, but at carrying out punitive raids against the north to break the enemy's morale and force the North Vietnamese back into serious negotiations. Abrams, supported by Laird, wanted the additional airpower available for operations in the South, on the assumption that that was where the war would be won or lost. After the LAM SON 719 debacle, however, Moorer grew increasingly frustrated with Abrams. At one point during the early days of the enemy's Easter offensive, with Kissinger present, Moorer related the substance of a rambling telephone call they had just had in which the COMUSMACV complained that he was "sick and tired" of civilians in Washington telling him what to do and would resign if he did not have his way. Eventually, Abrams calmed down. But the damage was done. Thenceforth, Moorer often bypassed the COMUSMACV and dealt with Abrams' subordinate and Lavelle's successor as Commander of Seventh Air Force, General John W. Vogt, USAF, who until recently had been Director of the Joint Staff. By transferring Vogt to Saigon, the Chairman had a trusted ally on the scene whose appraisals and advice he valued more than Abrams'.[77]

Like Lyndon Johnson, Richard Nixon took a strong personal interest in the air campaign and participated actively in planning and overseeing its execution. Yet there was none of the soul-searching or hemming and hawing that had gone on during the Johnson years. In deference to Abrams' expressed concerns, Nixon gave first priority to supporting the South Vietnamese and blunting the NVA invasion. According to Vogt, the intensity of these air strikes on the invaders resembled the effects of a "meat-grinder."[78] Operations against the North, code-named *Linebacker*, harkened to the "hard knock" bombing strategy advocated by the Joint Chiefs in the mid-1960s, and stressed repeat attacks on bridges, rail lines, fuel supplies, cement and power plants, airfields, and other high-profile military, industrial, and transportation targets. In giving his approval to launch *Linebacker*, Nixon admonished Moorer to mount an all-out effort and to avoid wasting bombs on "secondary targets."[79] Going further, he wanted to restrict North Vietnam's resupply from external sources, and on May 8, 1972, he announced the unprecedented step of mining Haiphong harbor, something the Joint Chiefs had urged since the early stages of the war.[80]

For a variety of reasons, *Linebacker* achieved results that were never feasible under the *Rolling Thunder* campaign of 1965–1968. By shifting from guerrilla tactics to conventional warfare and by incorporating tanks and other mechanized equipment into their battle plan, the North Vietnamese became dependent, like other modern armies, on long, readily identifiable supply lines that made ripe targets for air attack. Interdiction under the *Linebacker* campaign thus became more successful than during *Rolling Thunder*. A further difference between the two campaigns was the increased availability by 1972 of precision-guided munitions (PGMs or "smart bombs"), which allowed more accurate attacks against targets previously off limits in congested urban areas. While guided munitions had been around since the late stages of World War II, they had been difficult to use and not very effective. Improved models made their first appearance in Southeast Asia toward the conclusion of *Rolling Thunder* in 1968. Thereafter, technical problems limited their use to lightly defended targets in Laos and South Vietnam. But by 1972, more sophisticated electronics employing laser guidance systems opened the way for PGM raids against fixed targets in the heavily built-up Hanoi-Haiphong area.[81]

By early June, the North Vietnamese offensive was beginning to lose steam and there were indications from Hanoi of a renewed willingness to negotiate. In the United States, Nixon's decision to resume bombing had provoked predictable reactions from antiwar groups and liberals in Congress. But compared with the Cambodian invasion and earlier episodes, the protests and demonstrations were relatively mild, a sign that troop withdrawals and ending the draft were having the desired effect of diffusing the war as a political issue. Nixon's popularity at home was in fact at an all-time high, pointing toward an easy reelection in November. With his position thus fairly secure at home, Nixon kept up the bombing pressure on the North and did not call a halt until late October, when he was satisfied that the negotiations were on course toward an agreement.

THE CHRISTMAS BOMBING CAMPAIGN

While Nixon had used airpower to thwart an NVA military victory in the spring of 1972, he also hoped that it would pay diplomatic dividends by coercing the North Vietnamese back to the negotiating table and into a peace settlement. Once the bombing stopped in late October, however, unexpected problems arose in convincing not only leaders in Hanoi but also the regime in Saigon, headed by President Nguyen Van Thieu, to accept a ceasefire. One of Thieu's main objections to the deal, which Kissinger negotiated, was that it would leave huge numbers of Communist troops in place in South Vietnam. As many as 160,000 NVA regulars remained in the South and another 100,000 were in Laos and Cambodia.[82] Despite months of heavy air attacks, neither Kissinger nor the Joint Chiefs saw any way of dislodging them without the large-scale reintroduction of U.S. ground forces.

Frustrated by this turn of events, Nixon again resorted to bombing to put pressure on Hanoi to abide by the accords and to demonstrate to the Thieu government that the United States would stand behind it once the peace settlement took effect. A secret letter from Nixon to Thieu, pledging that the United States would "react strongly" if South Vietnam were threatened again sealed the bargain.[83] However,

Nixon informed no one of his promise, not even the Joint Chiefs. Yet even if he had, it probably would have made little difference. Congress, with antiwar liberals in the vanguard, felt bound by no such guarantees, and when the Communists resumed their offensive in 1975, it fell back on earlier legislation blocking U.S. forces from intervening.

The resumption of bombing in December 1972 thus helped to facilitate the signing of a peace agreement which, in the long run, was largely inconsequential. Its major accomplishment was to facilitate the return of U.S. prisoners of war.[84] Codenamed *Linebacker II*, the operation covered an 11-day period over the holidays and became known as the Christmas bombing campaign. Militarily, the main difference between *Linebacker II* and previous bombing operations was the concerted use of B–52s against targets in and around Hanoi and Haiphong. Ever since the secret bombing of Cambodia, Nixon had had a fascination with the use of B–52s and during the buildup for *Linebacker I*, increasing B–52 deployments to Guam and Thailand had been his top priority. The big bombers appeared for the first time over the North Vietnamese heartland in five raids in April 1972. Without much evidence, Nixon boasted to his staff that these attacks had been "exceptionally effective, the best ever in the war."[85] In fact, the results had not been particularly impressive, and the need for heavy fighter escort had diverted assets from other missions. Meantime, Abrams was clamoring for more B–52 support to help thwart the Communist offensive in the South. The net result was that, from early May on, the B–52s ceased operations against the North and concentrated on targets below the twentieth parallel.[86]

As he contemplated launching *Linebacker II*, Nixon resolved that B–52s would spearhead the effort. Underlying the operation was his determination to mount a show of force that would break enemy leaders' will to resist. Initially, both Moorer and Kissinger doubted whether using B–52s would produce better results than fighter-bombers. But as it became clear that Nixon was less interested in specific military objectives than in achieving a strong psychological impact, their reservations evaporated. Working in unison, the Joint Staff, the Strategic Air Command, the Air Staff, and the Pacific Air Forces quickly assembled a list of 55 key targets, aiming in each case for "mass shock effect in a psychological context." On December 7, Moorer met at Camp David with the President, who reviewed the target plan and "seemed to be pleased with it." A few days later, Moorer notified the Commander in Chief of Strategic Air Command, General John C. Meyer, USAF, that a major air offensive against the North was "definitely on the front burner" and that Hanoi and Haiphong would be the primary target areas. "I want the people of Hanoi to hear the bombs," Moorer told him, "but minimize damage to the civilian populace." Moorer also consulted by secure telephone with the CINCPAC, Admiral Noel Gayler, and confirmed the punitive purpose of the bombing.[87]

Attacks commenced on December 18, 1972, and lasted, with a brief pause over Christmas, until December 29. Though Air Force and Navy fighter-bombers also took part, SAC's B–52s dropped 75 percent of the total bomb tonnage during *Linebacker II*. In wave after wave, night after night, they pounded targets from Hanoi and Haiphong to the Chinese border. The most impressive display to date of American military power, these raids came closer than anything yet to threatening the survival of the North Vietnamese regime. Realizing what was at stake, the North Vietnamese put up a ferocious defense and during the first few nights they inflicted unexpectedly high losses on U.S. aircraft. The most serious losses came on the third night (December 20–21) when enemy surface-to-air missiles claimed six B–52s out of an attacking force of ninety. B–52 crews were used to flying over Laos and South Vietnam and were unaccustomed to a hostile environment, so the downing of planes during the early stages of *Linebacker II* came as a shock. Morale problems ensued, and there was a jump in the number of crewmen reporting for sick call. A change in bombing tactics and the compression of attacks into closer intervals, allowing the North Vietnamese defenders less time to reload their SAMs, helped overcome the problem. "It worked out beautifully," Moorer confided to his diary. "I don't think anybody in the world could have coordinated an operation as well as we did."[88]

For the Joint Chiefs of Staff, the success of *Linebacker II* was the high-water mark of the war. After years of frustration and setbacks, they had finally dealt the North Vietnamese a crippling blow. Meyer and Moorer believed that the North Vietnamese probably had to give up because they were running low on SAMs. With another week of raids, Meyer estimated, "we could fly anywhere we want over North Vietnam with impunity."[89] Nixon, however, had other plans. Feeling that he had made his point, he ordered the B–52s to stand down rather than risk the loss of more planes and crews or possibly jeopardize his budding détente with the Soviets and his rapprochement with the Chinese. The Joint Chiefs had long contended that an unrestricted air campaign would be decisive in Vietnam, and in December 1972 their advice appeared vindicated.

THE BALANCE SHEET

The ceasefire signed in January 1973 lasted barely 2 years. During this interval, the Joint Chiefs completed the withdrawal of the few U.S. troops still in Vietnam and progressively redeployed their other forces from the region. For a while, the United States continued to bomb NVA and Communist base camps in Cambodia, but in August 1973 Congress called a halt. Congressional pressure likewise led to the cessation of air reconnaissance flights over Laos a year later. Moorer suspected that the Communists would use the ceasefire to regroup and rearm, and they did. Launching a major offensive in April 1975, they quickly overwhelmed South Vietnamese defenders, who were practically helpless without American airpower. While Vietnamization had shielded the withdrawal of American ground troops, it had not done much to strengthen South Vietnam's security or to assure its continued independence. The Joint Chiefs had no plans to rush U.S. forces back into Southeast Asia or to intervene on the SVN government's behalf. Yet even if such plans had existed, political pressures at home doubtless would have blocked their implementation.

Despite the war's outcome, the Joint Chiefs never felt that the United States had erred by going into Vietnam. What they saw instead was a misguided effort, pursuing flawed goals and blunders in the way the war was planned, organized, and fought. Some of these blunders, they admitted, were of their own making; others were not. In World War II and initially in Korea, the attainment of military objectives had taken priority. But in Vietnam the Joint Chiefs had found themselves from the outset prosecuting a limited war heavy in diplomatic and political overtones. The initial objective was to apply military power to achieve a stalemate, an outcome which from the chiefs' point of view squandered their resources and ran counter to the American military ethos. Against an enemy bent on victory at any cost, such war aims were utterly unrealistic as well. Set within these parameters, the American effort in Vietnam was doomed to fail.

After the Vietnam War, the Joint Chiefs' role fell under close scrutiny. Calls for reform proliferated and were eventually instrumental in passage of the Goldwater-Nichols Department of Defense Reorganization Act of 1986, an attempt by Congress to improve future JCS effectiveness through institutional reorganization (see chapter 15). The most trenchant critique of the chiefs' performance in Vietnam was by an Army major (later brigadier general), H.R. McMaster. In his thoroughly researched and well-written book, *Dereliction of Duty*, published in 1997, McMaster took the chiefs to task for not being more forthright in offering advice to the Secretary of Defense and the President. More than a generation removed from Vietnam, McMaster found it hard to understand how the Joint Chiefs could disagree so strenuously with the Johnson administration's "graduated response" strategy, yet remain so compliant as their superiors blatantly ignored their advice. Relegated to what he describes as a "peripheral position in the policy-making process," the chiefs became, in McMaster's words, the "five silent men."[90]

What McMaster overlooks is that by the mid-1960s, when American intervention in Vietnam took place, the Joint Chiefs of Staff had passed their prime. Though they remained, as the National Security Act decreed, the President's top military advisors, their stature and institutional influence had diminished considerably since the 1940s when they came into being as a corporate body. During World War II, they met regularly with the President and accompanied him to meetings around the world. They knew every allied leader personally and were key figures at the high-level wartime conferences at which strategy and postwar planning took place. In terms of authoritative advice and influence, they had no rivals.

By the 1960s, the situation had changed. For one thing, the wartime grandees were long gone, succeeded by men who had been junior officers in World War II. Those who made up the Joint Chiefs of Staff during the Vietnam era were highly dedicated and decorated military officers. No one seriously questioned their professional credentials or competence. But they operated on a different plane from those who had served on the Joint Chiefs in World War II, the leaders who had shaped the allied victory over the Axis. McMaster's complaint that the JCS should have been more outspoken on Vietnam overrates their stature and influence. Had they been Marshall, King, and Arnold or their immediate successors, their advice would have been hard if not impossible for the President, Congress, and the American public to ignore. But the men who served on the JCS by the 1960s lacked the gravitas of their predecessors. Little wonder, then, that Army Chief of Staff Harold Johnson dismissed the suggestion that he and his colleagues ought to have resigned in protest as a hollow and pointless gesture.[91]

Moreover, a new policy- and decisionmaking system had replaced the one in effect when the JCS came into existence, resulting in a proliferation of overlapping agencies and organizations, some in direct competition with the Joint Chiefs. By the mid-1960s, the chiefs' most formidable competitor was the Office of the Secretary of Defense, which had grown steadily in influence and importance since its creation in 1947. Under McNamara, it had amassed a wealth of additional authority and capabilities for analyzing military strategy and for offering alternative advice to that rendered by the JCS. Given McNamara's forceful personality and the precarious relationship between the JCS and the White House under Presidents Kennedy and Johnson, it was hardly surprising that the chiefs' credibility and influence were on the wane.

Unable to bring their views to bear directly, the Joint Chiefs adopted an incremental approach to the war. They assumed that any steps toward greater military involvement would sooner or later develop into the course they advocated. In the process, they lent their support to a military strategy they considered fundamentally flawed and became complicit in the administration's folly. At the same time, as the decision to intervene in force was taking shape, inter-Service bickering over whether to stress ground operations in the South or a concerted air and naval campaign against the North denied them a clear voice and focus. Yet even if the Joint Chiefs had spoken as one, their limited influence within the wider sphere of the policy process effectively undercut their ability to sway key decisions on the conduct of the war.

VIETNAM: RETREAT AND WITHDRAWAL

With the advent of the Nixon administration, the strategy debate came full circle back to the chiefs' original premise that the most effective approach was to mount heavy military pressure directly against North Vietnam. Owing to the ongoing reduction in U.S. ground forces and limited South Vietnamese capabilities, however, recourse to a combination of air and sea power became the only viable option. Fearing Chinese intervention or a nuclear confrontation with the Soviets, President Johnson had consistently scorned the chiefs' advice in that regard. But by Nixon's time, the emergence of détente and the opening with China allowed the President a degree of leverage and flexibility that had not previously existed. Given the decisive results achieved by the *Linebacker* operations, coupled with the mining of Haiphong, one is tempted to speculate that a bolder strategy earlier might well have avoided a long, drawn-out war. Yet without the diplomatic groundwork painstakingly laid by Nixon and Kissinger, the more aggressive strategy advocated by the JCS in 1964–1965 could just as well have backfired.

As disappointing to the Joint Chiefs as the outcome in Vietnam may have been, it was not the serious setback to American global interests that many had feared a Communist victory might be when the United States went into Vietnam. All the same, the nature and pervasive impact of the war had a devastating effect. Not only did the war shatter the national consensus that had supported and sustained faith in the containment concept for nearly two decades; it also left American conventional forces in a state of near-disarray, weaker and less sure of themselves than at any time since the 1930s. Especially hard-hit was the Army, which emerged from the conflict a shambles. Recovering from the trauma of Vietnam became the Joint Chiefs' first order of business, and for the next decade and a half, through the end of the Cold War, it would overshadow practically all other aspects of their deliberations.

NOTES

1. "President's Address to the Nation Announcing Steps to Limit the War in Vietnam," March 31, 1968, *Public Papers of the Presidents of the United States: Lyndon B. Johnson, 1968–69* (Washington, DC: GPO, 1970), Pt. I, 469–476; Lyndon Baines Johnson, *The Vantage Point: Perspectives of the Presidency, 1963–1969* (New York: Holt, Rinehart and Winston, 1971), 365.

2. Johnson, *The Vantage Point*, chap. 9.

3. Merle Pribbenow, trans., *Victory in Vietnam: The Official History of the People's Army of Vietnam, 1954–1975* (Lawrence: University Press of Kansas, 2002), 153–205 and passim.

4. Stanley Robert Larsen and James Lawton Collins, Jr., *Allied Participation in Vietnam* (Washington, DC: Department of the Army, 2005), 120–159.

5   Lawrence S. Kaplan, "McNamara, Vietnam, and the Defense of Europe," in Vojtech Mastny, Sven G. Holtsmark, and Andreas Wenger, eds., *War Plans and Alliances in the Cold War: Threat Perceptions in the East and West* (London: Routledge, 2006), 286–300.

6   George L. MacGarrigle, *Taking the Offensive: October 1966 to October 1967* (Washington, DC: Center of Military History, 1998), 25.

7   Jacob Van Staaveren, *Gradual Failure: The Air War Over North Vietnam, 1965–1966* (Washington, DC: Air Force History and Museums Program, 2002), 279–297.

8   Wayne Thompson, *To Hanoi and Back: The United States Air Force and North Vietnam, 1966–1973* (Washington, DC: Air Force History and Museums Program, 2000), 41–43.

9   Graham A. Cosmas, *The Joint Chiefs of Staff and The War in Vietnam, 1960–1968* (Washington, DC: Office of Joint History, 2009), Part 3, 43–45; MacGarrigle, 216–217.

10 JCSM-218-67 to SECDEF, April 20, 1967, "Force Requirements—Southeast Asia FY 1968," JCS 2339/255-3; Cosmas, *JCS and the War in Vietnam, 1960–68*, Pt. 3, 48–50.

11 Robert S. McNamara with Brian VanDeMark, *In Retrospect: The Tragedy and Lessons of Vietnam* (New York: Times Books, 1995), 264–265.

12 See Draft Memo, SECDEF to President, May 19, 1967, "Future Actions in Vietnam," *FRUS 1964–68*, V, 423–438.

13 Notes on Discussions with President Johnson, April 27, 1967, *FRUS 1964–68*, V, 350.

14 CJCS Statement, February 7, 1967, attached to Letter, JCSJ to SecState, February 8, 1967, JMF 9155 (February 18, 1965), NARA.

15 JCSM-288-67 to SECDEF, May 20, 1967, "Worldwide US Military Posture," JCS 2101/538-5.

16 Draft Memo, SECDEF to President, June 12, 1967, "Alternative Military Actions Against North Vietnam," *FRUS 1964–68*, V, 475–481; McNamara, *In Retrospect*, 265–271.

17 William W. Momyer, *Airpower in Three Wars* (Washington, DC: Department of the Air Force, 1978), 25–26.

18 Cosmas, *JCS and the War in Vietnam, 1960–68*, Pt. III, 62–67; Cosmas, *MACV: Years of Escalation, 1962–67*, 419; Mark Clodfelter, *The Limits of Airpower: The American Bombing of North Vietnam* (New York: Free Press, 1989), 106–108.

19 McNamara testimony, August 25, 1967, in U.S. Congress, Senate, Committee on Armed Services, Preparedness Investigating Subcommittee, *Hearings: Air War Against North Vietnam*, 90:1, Pt. 4, 281 (quote) and passim.

20 "President's News Conference," September 1, 1967, *Public Papers of the Presidents of the United States: Lyndon B. Johnson, 1967* (Washington, DC: GPO, 1968), Pt. I, 817.

21 Graham A. Cosmas, *MACV: The Joint Command in the Years of Withdrawal, 1968–1973* (Washington, DC: Center of Military History, 2007), 59.

22 CM-2922-68, January 19, 1968, quoted in Jacob Van Staaveren, "The Air Force in Southeast Asia: Toward a Bombing Halt, 1968" (MS, Office of Air Force History, September 1970), 8 (declassified).

23 Cosmas, *JCS and the War in Vietnam, 1960–68*, Pt. III, 142–143; Cosmas, *MACV: Years of Withdrawal*, 41; CM-2944-68 to SECDEF, February 3, 1968, "Khe Sanh," *FRUS 1964–68*, VI, 120.

24 Deborah Shapley, *Promise and Power: The Life and Times of Robert McNamara* (Boston: Little, Brown, 1993), 432.

25 Walter S. Poole et al., "Chairmen and Crises: The Vietnam War" (MS, Joint History Office, Joint Chiefs of Staff, n.d.), 129–136 (declassified/publication forthcoming) gives an excellent overview of JCS thinking during the immediate post-Tet period.

26 Interview by Dorothy P. McSweeny with Gen Earle G. Wheeler, May 7, 1970, 21–22, Oral History Collection, Johnson Library.

27 "President's News Conference," March 22, 1967, *Johnson Public Papers, 1968*, 430.

28 Interview with Lt. Gen. John B. McPherson, USAF (Ret.), April 3, 1990, JHO Collection. See also Cosmas, *MACV: Years of Withdrawal*, 94.

29 Cosmas, *JCS and the War in Vietnam, 1960–68*, Pt. III, 149–160; Cosmas, *MACV: Years of Withdrawal*, 88–97; Westmoreland, *A Soldier Reports*, 351 (quote); "Report of CJCS on Situation in Vietnam and MACV Force Requirements," February 27, 1968, JCS 2472/237.

30 Clark Clifford, with Richard Holbrooke, *Counsel to the President: A Memoir* (New York: Random House, 1991), 492–526; Johnson, *Vantage Point*, 365–424.

31 Quoted in Van Staaveren, "Toward a Bombing Halt," 39.

32 Momyer, 26–27.

33 "Notes on the President's Meeting," October 29, 1968, *FRUS, 1964–68*, VII, 399–401; Cosmas, *JCS and the War in Vietnam, 1960–68*, Pt. III, 230–231.

34 Cosmas, *MACV: Years of Withdrawal*, 244–245.

35 "Address to the Nation on Vietnam," May 14, 1969, *Nixon Public Papers, 1969*, 369–375. See also Walter Isaacson, *Kissinger: A Biography* (New York: Simon & Schuster, 1992), 159–160.

36 Entry, January 17, 1972, in H.R. Haldeman, *The Haldeman Diaries: Inside the Nixon White House* (New York: G.P. Putnam's Sons, 1994), 397.

37 Poole et al., "Chairmen and Crises," 224.

38 Palmer, *25-Year War*, 92.

39 Elmo R. Zumwalt, Jr., *On Watch: A Memoir* (New York: Quadrangle, 1976), 69–70.

40 Peter W. Rodman, *Presidential Command* (New York: Knopf, 2009), 69.

41 *Department of State Bulletin* 60, no. 1548 (February 24, 1969), 163. See also David J. Rothkopf, *Running the World: The Inside Story of the National Security Council and the Architects of American Power* (New York: Public Affairs, 2004), 114–115.

42 Zumwalt, 317–318.

43 During Nixon's first year as President, the NSC met 37 times, 21 times in 1970 and only 10 times during the first 9 months of 1971. See John P. Leacacos, "Kissinger's Apparat," *Foreign Policy* 5 (Winter 1971–1972), 5.

44 Helmut Sonnenfeldt, "Reconstructing the Nixon Foreign Policy," in *The Nixon Presidency*, ed. Kenneth W. Thompson (Lanham, Md.: University Press of America, 1987), 319.

45 NSDM 1, January 20, 1969, "Establishment of NSC Decision and Study Memoranda Series," *FRUS 1969–76*, II, 29–30.

46 NSDM 2, January 20, 1969, "Reorganization of the NSC System," *FRUS 1969–76*, II, 30–33. See also Chester A. Crocker, "The Nixon–Kissinger National Security Council System, 1969–1972: A Study in Foreign Policy Management," in *Report of the Commission on the Organization of the Government for the Conduct of Foreign Policy: Appendices* (Washington, D.C.: GPO, June 1975), vol. 6, 79–99.

47 Zumwalt, 369–376; Henry Kissinger, *White House Years* (Boston: Little, Brown, 1979), 722–723; Mark Perry, *Four Stars* (Boston: Houghton, Mifflin, 1989), 234–235.

48 Kissinger, *White House Years*, 226.

49 Richard M. Nixon, *No More Vietnams* (New York: Arbor House, 1985), 101–107.

50 Alexander M. Haig, Jr., with Charles McCarry, *Inner Circles: How America Changed the World: A Memoir* (New York: Time Warner, 1992), 224.

51 Willard J. Webb, *History of the Joint Chiefs of Staff: The Joint Chiefs of Staff and the War in Vietnam, 1969–1970* (Washington, DC: Office of Joint History, 2002), 283.

52 Memo, Lake and Morris to Robinson, September 29, 1969, "Draft Memorandum to the President on Contingency Study"; Memo, Kissinger to Nixon, October 2, 1969, "Contingency Military Operation Against North Vietnam"; "Conceptual Plan of Military Operations," undated, all declassified in box 89, NSC Files: Subject Files, Nixon Presidential Materials, NARA; and <http://www.gwu.edu/~nsarchiv/NSAEBB/NSAEBB195/index.htm#1>. See also William Burr and Jeffrey Kimball, "Nixon's Secret Nuclear Alert: Vietnam War Diplomacy and the Joint Chiefs of Staff Readiness Test, October 1969." *Cold War History* 3:2 (January 2003), 113–156.

53 Cosmas, *MACV: Years of Withdrawal*, 283–286.

54 Webb, *JCS and War in Vietnam, 1969–70*, 136–137.

55 Palmer, *25-Year War*, 96.

56 Richard Nixon, *RN: The Memoirs of Richard Nixon* (New York: Grosset & Dunlap, 1978), 382.

57 JCSM-6-69 to SECDEF, January 4, 1969, "Republic of Vietnam Armed Forces Improvement and Modernization," JCS 2472/272-28; JCSM–40–69 to SECDEF, January 21, 1969, "Republic of Vietnam Armed Forces Improvement and Modernization," JCS 2472/272-30.

58 Webb, *JCS and War in Vietnam, 1969–70*, 11–12, 64.

59 Dale Van Atta, *With Honor: Melvin Laird in War, Peace, and Politics* (Madison: University of Wisconsin Press, 2008), 182–183; Nixon, *RN*, 392 (quote).

60 JCSM-522-69 to SECDEF, August 25, 1969, "Vietnamizing the War," JCS 2472/467-4.

61 NSDM 24, September 17, 1969, "Vietnam," *FRUS 1969–76*, VI, 407–408.

62 Webb, *JCS and War in Vietnam, 1969–70*, 75–82.

63 JCSM-149-70 to SECDEF, April 3, 1970, "Ground Strikes Against Base Areas in Cambodia," excerpted in "JCS Recommendations and SECDEF Actions with Respect to Cambodia, 1 January 1969–15 February 1975" (MS, Historical Division, Joint Secretariat, JCS, February 26, 1975), 14.

64 Kissinger, *White House Years*, 483–505; Nixon, *RN*, 446–451.

65 Willard J. Webb and Walter S. Poole, *The Joint Chiefs of Staff and The War in Vietnam, 1971–1973* (Washington, DC: Office of Joint History, 2007), 2.

66 Ibid., 2–3; Cosmas, *MACV: Years of Withdrawal*, 320–327.

67 Kissinger, *White House Years*, 990–991; Jeffrey Kimball, *The Vietnam War Files: Uncovering the Secret History of Nixon-Era Strategy* (Lawrence: University Press of Kansas, 2004), 133–134.

68 Cosmas, *MACV: Years of Withdrawal*, 326–335; Webb and Poole, *JCS and the War in Vietnam, 1971–73*, 3–4.

69 Webb and Poole, *JCS and the War in Vietnam, 1971–73*, 2; Kissinger, *White House Years*, 995–996.

70 Westmoreland, *A Soldier Reports*, 271–273; Webb and Poole, *JCS and War in Vietnam, 1971–73*, 2–9.

71 Cosmas, *MACV: Years of Withdrawal*, 329–335; Webb and Poole, *JCS and War in Vietnam, 1971–73*, 12–16; Nixon, *RN*, 498–499.

72 Van Atta, 351–353.

73 Webb and Poole, *JCS and War in Vietnam, 1971–73*, 109–117.

74 Memo for the Record by Col Robert M. Lucy, USMC, Asst to CJCS, 21 Sep 72, "Responses for the Record by General Lavelle to Questions from Senator Smith," C, Material on Replacement of Gen Lavelle 1972 folder, JHO 18-0020.

75 Thompson, *To Hanoi and Back*, 199–210; Webb and Poole, *JCS and War in Vietnam, 1971–73*, 126–128.

76 Clodfelter, *Limits of Airpower*, 152–153.

77 Webb and Poole, *JCS and War in Vietnam, 1971–73*, 156. On Nixon's lack of confidence in Abrams, see Haldeman, *The Haldeman Diaries*, 436–437.

78 Entry, May 4, 1972, Moorer Diary, cited in Webb and Poole, *JCS and War in Vietnam, 1971–73*, 163.

79 Webb and Poole, *JCS and War in Vietnam, 1971–73*, 162.

80 "Address to the Nation on the Situation in Southeast Asia," May 8, 1972, *Nixon Public Papers, 1972*, 585.

81 Wayne Thompson, *To Hanoi and Back: The United States Air Force and North Vietnam, 1966–1973* (Washington, DC: Air Force History and Museums Program, 2000), 230–231. Only Air Force F–4s had the necessary modifications to carry PGMs.

82 Palmer, *25-Year War*, 131.

83 Letter, Nixon to Thieu, January 16, 1973, quoted in Nixon, *RN*, 749–750.

84 See Vernon E. Davis, *The Long Road Home: U.S. Prisoner of War Policy and Planning in Southeast Asia* (Washington, DC: Historical Office, Office of the Secretary of Defense, 2000), 453–490.

85 Entry, April 16, 1972, Haldeman, *The Haldeman Diaries*, 441.

86 Thompson, *To Hanoi and Back*, 224–227.

87 Clodfelter, *Limits of Airpower*, 184; Webb and Poole, *JCS and War in Vietnam, 1971–73*, 291–294.

88 Moorer quoted in Webb and Poole, *JCS and War in Vietnam, 1971–73*, 298; Clodfelter, *Limits of Airpower*, 192–193, discusses morale problems; Thompson, *To Hanoi and Back*, 264–265 summarizes strategy and changes in tactics.

89 Quoted in Webb and Poole, *JCS and War in Vietnam, 1971–73*, 298.

90 H.R. McMaster, *Dereliction of Duty: Lyndon Johnson, Robert McNamara, the Joint Chiefs of Staff, and the Lies that Led to Vietnam* (New York: HarperPerennial, 1997), 300–329.

91 Lewis Sorley, *Honorable Warrior: General Harold K. Johnson and the Ethics of Command* (Lawrence: University Press of Kansas, 1998), 223–224.

Admiral Thomas H. Moorer, USN, Chairman, Joint Chiefs of Staff, 1970–1974

Chapter 11

DÉTENTE

As the war in Southeast Asia wound down, the Joint Chiefs of Staff began a slow and sometimes uncomfortable reassessment of their military plans and policies. Similar reassessments had followed previous wars and invariably had given rise to passionate inter-Service rivalries and intense competition for resources. Some of these elements, to be sure, were present in the aftermath of Vietnam. But compared to the build-downs that followed World War II and Korea, the transition following Vietnam was relatively smooth and easy. Indeed, the most serious problems that arose were in developing military policies and a force posture compatible with a rapidly changing international environment dominated by the prospect of a new era in Soviet-American relations known as "détente."

An evolving process, détente was the outgrowth of a series of Soviet-American initiatives, some dating from the 1950s, to establish what political scientist Stanley Hoffmann termed "a stable structure of peace."[1] Coming to fruition in the early 1970s, détente lasted roughly from the signing of the SALT I accords in 1972 until the Soviet invasion of Afghanistan in 1979. Historians generally agree that, while the two sides shared certain common interests, they approached them from different perspectives and expected different outcomes. Hence the friction and disagreements that sometimes accompanied détente and ultimately brought its demise. For President Nixon and his national security advisor, Henry A. Kissinger, détente was integral to the post-Vietnam restructuring of American foreign relations and related defense policies. Persuaded that the two previous administrations had concentrated too much on the Third World, Nixon and Kissinger set about redefining the country's vital interests. Shifting the focus from Asia to Europe, they wanted to strengthen relationships with traditional allies and revitalize NATO, which had gone into decline during the American preoccupation with Vietnam. At the same time, acknowledging that the United States could never regain the strategic superiority it had enjoyed into the early 1960s, they accepted parity in strategic nuclear power with the Soviet Union as a fact of life and sought agreements with Moscow that would curtail growth in both sides' strategic arsenals. Overall, they envisioned a new "era of negotiations" that would ease East-West tensions, facilitate the resolution of long-standing Cold War issues (e.g., Vietnam and Berlin), break new ground in arms control, and improve avenues of communication with the two Communist behemoths, the Soviet Union and the People's Republic of China. As for obtaining lasting results, Nixon was cautiously optimistic. "All we can hope from détente," he later wrote, "is that it will minimize confrontation in marginal areas and provide, at least, alternative possibilities in major ones."[2]

In assessing the military requirements of détente, the Joint Chiefs found themselves under more pressure than usual to exercise restraint and to hold down requests for new programs, despite a continuing buildup in Soviet military forces. Looking beyond Vietnam, the JCS contemplated a list of requirements that included not only the replacement of weapons and equipment worn out or lost in the war, but also the modernization of the force structure to stay current with emerging technologies and recent increases and improvements in Soviet capabilities. Strategic retaliatory forces, they believed, were in especially urgent need of attention. Yet with détente the watchword, a buildup on the scale and scope the JCS believed necessary became increasingly unlikely. The Services might receive some of the modernization and improvements they wanted, and the Armed Forces would continue to be an important instrument in American foreign policy. But after Vietnam, the emphasis for nearly a decade would be increasingly on nonmilitary solutions to Cold War problems.

SALT I

The linchpin in the Nixon-Kissinger strategy of détente was the arms control process, organized around the Strategic Arms Limitation Talks (SALT). Conceived under the Johnson administration, SALT was supposed to have started in the fall of 1968 but was called off at the last minute by the United States to protest the Soviet-led Warsaw Pact invasion of Czechoslovakia that snuffed out the reformist government of Alexander Dubček. Revived under Nixon, SALT finally got underway in November 1969. Once a distant adjunct of defense policy, arms control by the late 1960s was becoming a critical element in shaping the size and capabilities of the country's strategic arsenal. After years of heavy military spending and bloodshed in Vietnam, SALT seemed a welcome respite and soon acquired a high degree of popular and congressional support. For many it also became a fairly accurate barometer of U.S.-Soviet relations in general. Indeed, by the time SALT I was underway, the idea had taken hold, both in the executive branch and in Congress, that progress in controlling nuclear weapons would give impetus to progress in resolving other thorny Cold War issues as well.

The Joint Chiefs of Staff welcomed progress in arms control that led to improved U.S.-Soviet relations, but not if it meant crippling the country's strategic deterrent or postponing its modernization. Still chafing from the constraints imposed by McNamara, the JCS felt increasingly hard-pressed to maintain credible strategic nuclear deterrence in the face of a Soviet missile buildup of unprecedented proportions. By 1969, while still inferior in the overall number of intercontinental delivery vehicles, the Soviets had surpassed the United States in operational ICBM launchers.[3] To cope with this threat, even if arms control talks proved productive, the Joint Chiefs wanted a new manned strategic bomber (the B–1) and a new fleet of ballistic missile submarines (the *Trident* class) and were awaiting the outcome of further developmental studies by the Air Force concerning an advanced ICBM.[4]

As for the specifics of an arms control accord, the Joint Chiefs insisted that, above all, it should be fully verifiable, a view shared by key members of Congress who would be passing judgment on whatever agreements the administration might reach with the Soviets.[5] For years, the Joint Chiefs had argued that on-site inspections were the only ironclad way of determining whether the Soviets were in compliance. But in March 1967, they amended their position and agreed to accept the results of unilateral verification derived from space-based satellites, known in arms control parlance as "national technical means." Under these rules, it would be up to each side to determine whether the other was in compliance. This requirement virtually assured that any agreement reached between the United States and the Soviet Union would deal, in the first instance, with numerical limitations on launchers and only secondarily with payload, deployment mode, and performance characteristics.[6]

The preparatory round of SALT I opened in Helsinki on November 17, 1969, and lasted about 4 weeks. For the next 2½ years, negotiations alternated between Helsinki and Vienna, averaging a round of talks every 3 months.[7] A major difference between these negotiations and earlier arms control efforts like the negotiation of the Test Ban Treaty under the Kennedy administration, was the presence throughout SALT of JCS representation on the U.S. delegation owing to the persistence of General Earle G. Wheeler, USA, the JCS Chairman. During Senate deliberations over the Test Ban Treaty, General Wheeler heard grumblings from Congress over the exclusion of the JCS from the negotiations. Using these signs of discontent as his opening wedge, he arranged for the JCS, in the summer of 1968, to be part of an ad hoc arms control study group in the Office of the Secretary of Defense, chaired by Morton H. Halperin, that was starting to draft a negotiating position. To represent the JCS, Wheeler brought in Major General (later Lieutenant General) Royal B. Allison, USAF, who had headed strategic planning at CINCPAC. Authorized a small staff, Allison acquired the title of Assistant to the Chairman for Strategic Arms Negotiations (ACSAN), but reported to the Joint Chiefs collectively through the Director, Joint Staff. According to John Newhouse's generally reliable behind-the-scenes account of SALT I, *Cold Dawn*, Wheeler bypassed the Joint Staff in selecting Allison because he lacked confidence in the arms control component in J-5 to provide reliable advice. When the Nixon administration took office, Allison continued to represent the Joint Chiefs in the interagency arena and became their member on the U.S. delegation to SALT I.[8]

SALT's ostensible goal, from the American standpoint, was to put a cap on the further buildup of strategic arms. U.S. intelligence estimates routinely confirmed that the Soviets were continuing to add to their arsenal of ICBMs, but shed little light on the intentions behind the buildup. As a rule, the CIA and the State Department downplayed the danger of a fundamental shift in the strategic balance, whereas the Joint Chiefs, OSD, and the Defense Intelligence Agency (DIA) refused to rule out such a possibility. Speaking publicly, Secretary of Defense Melvin Laird declared that the Soviets were seeking nothing less than a disarming "first-strike" strategic capability.[9] President Nixon, however, refused to be quite so specific. According to Kissinger, Nixon disdained the technicalities of arms control (the details bored him) and regarded SALT mainly as a vehicle for improving relations with Moscow.[10] Going into the talks, the President approved a highly generalized set of instructions that glossed over disagreements among his advisors on Soviet intentions. For negotiating purposes, the President left the door open to a wide range of limitations as long as they were verifiable and did not hinder efforts by the United States to preserve "strategic sufficiency," a rather vague concept that the White House defined as rough parity in strategic nuclear power with the Soviet Union.[11] Adopting a wait-and-see attitude, the Joint Chiefs declined to recommend their own specific proposals, arguing that as advisors to the President and the Secretary of Defense, it was not their place.[12]

The Soviets, on the other hand, had a fairly firm SALT agenda that included protecting the gains they had made in offensive strategic missiles in the 1960s and curbing U.S. progress in ballistic missile defense (BMD). At the outset of the talks, the Soviets also sought a broad definition of strategic systems that embraced any nuclear weapon capable of hitting the other side's homeland. This definition would have encompassed all American aircraft carriers and nearly every theater system in Europe and Asia, but not similar forward-based systems deployed by the Soviet Union. Not surprisingly, the American side found it unacceptable.[13] Eventually, the talks concentrated on only two sets of offensive systems—ICBMs and submarine-launched ballistic missiles (SLBMs).

In addition to the formal talks held in Helsinki and Vienna, Kissinger and Soviet Ambassador Anatoly Dobrynin engaged in substantive "backchannel" negotiations in Washington. Carefully concealed from practically everyone, including the

Joint Chiefs, these backchannel talks gave Nixon and Kissinger a direct link to the Kremlin and quickly became the true forum of the SALT I negotiations. Out of these exchanges, it soon became clear that the best result SALT I could hope to produce on offensive strategic arms was a temporary moratorium or "freeze" on "new starts."[14] On May 20, 1971, Washington and Moscow jointly issued a brief statement dampening the immediate prospects for a permanent offensive arms accord and instructing negotiators to devote their energies for the next year to a treaty limiting antiballistic missiles (ABMs).[15] This "breakthrough," as the White House characterized it, completely surprised the Joint Chiefs and left them somewhat confused. Indeed, Admiral Moorer, the JCS Chairman, initially misunderstood the deal and thought it continued to link an agreement on offensive weapons with an agreement on defensive ones.[16] Broadly worded and open to several interpretations, the freeze imposed loose restrictions and left both sides more or less free to complete additions and improvements to their arsenals where construction was already underway.

In the absence of progress on controlling offensive weapons, missile defense became the only area of U.S.-Soviet competition to be subjected to permanent constraints as a result of SALT I. While both sides had ABM programs, the consensus within the American intelligence and scientific communities was that the United States had a definite advantage owing to its work on phased-array radars. Even so, the systems under consideration were exceedingly expensive and far from foolproof. Citing high costs and continuing technical difficulties, the Johnson administration had rejected JCS arguments in favor of a nationwide system and had endorsed only a "point defense" ABM, known as Sentinel, to protect Minuteman missile fields. But it had left the decision on actual deployment up to the next administration.[17] Though ambivalent about the military value of BMD, Nixon recognized its potential as a bargaining chip with the Soviets and in March 1969 announced that the United States would proceed with deployment of a limited ABM system, now called Safeguard. Nixon's decision kept the program alive, but it also touched off a sharp debate in Congress that came down to a narrow Senate victory for the administration's authorization bill in August 1969.[18]

Like the agreement to "freeze" offensive forces, negotiations on the ABM issue took place to a considerable extent outside the official SALT framework. The accord finally reached was largely the product of an informal exchange of views between Paul H. Nitze, the OSD representative to SALT, and his Soviet counterpart, Aleksandr Shchukin, an expert in radio wave electronics. A Deputy Secretary of Defense in the Johnson administration and most well known as the "author" of NSC 68, Nitze had been in the forefront of the lobbying effort as a private citizen to preserve ABM during the congressional debate in the summer of 1969. Now, as a member of the SALT delegation, he took a leading role in negotiating ABM away. The shift in Nitze's thinking came from his realization, based on that experience, that for political reasons the current U.S. ABM effort faced an uncertain future. "If the negotiations failed," he believed, "we still were not going to have an ABM program because the Senate wasn't going to give it to us." Out of his talks with Shchukin between late 1971 and early 1972 emerged an agreement on radars and associated technical matters that set the stage for the ABM Treaty. According to Gerard Smith, who headed the U.S. SALT negotiating team, Nitze's persistence resulted in far more precise constraints on ABM radars than delegation members expected to achieve or than agencies in Washington, including the Joint Chiefs, would have preferred.[19]

At their Moscow summit in May 1972, Nixon and Soviet General Secretary Leonid I. Brezhnev unveiled the results of SALT I: an "interim" agreement imposing a 5-year freeze on both sides' offensive strategic missile launchers as of the date of the agreement; a permanent treaty sharply limiting ABMs; and a set of statements explaining and interpreting the agreements. For verification purposes, each side was on its own. Disagreements would be referred to a U.S.-Soviet Standing Consultative Commission, which would assist with implementation.[20] Reveling in the accomplishments of SALT I, the Soviets clearly saw it as confirmation of their superpower status on a par with the United States. That U.S. warplanes were at the time engaged in a heavy bombardment of Moscow's ally, North Vietnam, and Communist positions in South Vietnam in retaliation for Hanoi's "Easter Offensive," seemed outwardly of little consequence to Brezhnev and his colleagues. To them, all that mattered was that détente had officially arrived.

Back home, critics assailed SALT I as a limited success. In defense of the accords, the Nixon administration insisted that the interim agreement and the ABM Treaty were mutually reinforcing and that a permanent, more restrictive offensive arms accord would follow shortly. Even so, there were murmurings of dissatisfaction with the deal, especially among the Joint Chiefs. As far as the JCS were concerned, the "frozen" numbers spoke for themselves: an American arsenal of 1,054 ICBMs, 41 missile submarines, and 656 SLBM launchers, versus a Soviet force of more than 1,600 ICBMs, 43 missile submarines, and 740 SLBMs.[21] The JCS were incredulous that between the announcement of May 20, 1971, that had supposedly suspended the negotiation of an offensive arms treaty and the signing of the SALT I accords a year later, Nixon and Kissinger had allowed the Soviets to add 91 ICBM launchers to their arsenal (silos under construction at the time of the announcement) without a word of protest. At the same time, the White House had dawdled on nailing down an SLBM agreement, and in the end, much to the Joint Chiefs' consternation, had given the Soviets virtually free rein to upgrade their fleet ballistic missile submarine force.²² The administration's defenders took the position that the United States still had a two-and-a-half to one lead in long-range bombers (unaffected by SALT I) and a substantial advantage in targetable warheads through the ongoing retrofitting of many U.S. missiles with MIRVed reentry vehicles (RVs). But as the chiefs and others were quick to point out, land-based bombers were the most vulnerable part of the strategic triad, and the American lead in MIRVed RVs was temporary since the Soviets were now well along on their own MIRV program.²³

Despite misgivings, the Joint Chiefs supported the SALT I accords, provided the administration and Congress took the necessary steps to monitor Soviet compliance, modernize the U.S. strategic deterrent, and support "vigorous" research and development.²⁴ During a briefing for congressional leaders just prior to the signing of the SALT I agreements, Admiral Moorer acknowledged that the Soviets "were outstripping U.S. in every category with the exception of bombers." To prevent the United States from slipping farther behind, Moorer stressed the need for continuing modernization of the U.S. strategic arsenal and mentioned specifically the B–1 bomber and the Trident missile submarine. Without these improvements, he insisted, "we could not live with this proposed agreement."²⁵

While sympathetic to the chiefs' concerns, most in Congress shared the President's view that the SALT I agreements marked a major turning point in U.S.-Soviet relations and that their political and diplomatic benefits outweighed their military drawbacks. With major restrictions on the further deployment of ABMs and emerging parity in strategic offensive power, some theorists contended that a new era, based on deterrence through "mutual assured destruction," or MAD, had arrived. The Senate approved the ABM Treaty on August 3, 1972, and the interim agreement on September 14. Acceptance of the latter, however, carried an amendment, sponsored by Democratic Senator Henry M. Jackson of Washington, stipulating that there should be equality in the number of launchers in any future treaty on ICBMs.²⁶

Several years later, columnist Marquis Childs asserted that Senator Jackson had harassed witnesses who had helped to negotiate SALT I when they appeared before the Senate Armed Services Committee. Childs said that the Chairman's arms control assistant, Lieutenant General Allison, had received the "heaviest Jackson fire" because he had publicly gone along with the agreement even though privately he believed it would leave the United States vulnerable to a Soviet first strike. Convinced that Allison had not been completely candid with the committee, Jackson sent word to the JCS, Childs said, that he would "blackball" any promotion for Lieutenant General Allison in the Air Force or his nomination to any future government post. The upshot was that the Joint Chiefs relieved Allison of his duties in February 1973 and he took early retirement.[27]

Tape recordings made by Nixon of Oval Office conversations confirm that Jackson did indeed put pressure on the White House to "purge" the American SALT negotiating team and that Lieutenant General Allison was one of those he singled out.[28] That the senator's views could have had such an impact suggests not only the influential role he played in arms control and related issues, but also the highly charged politics that surrounded the SALT process. In fact, within a year of signing the SALT I accords, U.S. intelligence detected Soviet tests of four new ICBMs, three of them—the SS–17, the SS–18, and the SS–19—with a demonstrated MIRV capability.[29] SALT I had not provided much respite from the competition in strategic arms. Time would tell if SALT II would do a better job.

SHORING UP THE ATLANTIC ALLIANCE

At the same time as the Joint Chiefs were wrestling with SALT, they faced the equally challenging problem of revitalizing the Atlantic Alliance. The war in Vietnam had shifted American attention from Europe to the Far East and in the process had raised serious questions about whether the United States remained committed to Europe's security and welfare. Lacking the consistent American interest and leadership it had known in the past, NATO had begun to drift. To be sure, MC 14/3, the 1967 NATO strategy blueprint endorsing the "flexible response" doctrine, and the Harmel Report, approved around the same time and calling for stepped-up negotiations with the Soviets, had helped to paper over some of the emerging differences and disagreements. But for the longer term, the repairs needed to go deeper, perhaps as far as forging a new transatlantic partnership.

On paper, the American commitment to NATO at the end of the 1960s appeared nearly as sound and robust as ever—41/3 divisions, 2 armored cavalry regiments, 32 air squadrons totaling 640 planes, and 25 combatant ships of the Sixth Fleet, all at NATO's disposal in the event of emergency. Under the "swing strategy" adopted in the 1950s, the Joint Chiefs also earmarked certain air and naval units for emergency transfer from the Pacific to Europe. Although Vietnam had depleted the strategic reserve available from the United States, plans initiated under the Johnson administration to preposition equipment in Europe promised to help surmount these problems, save money, and over time improve NATO's conventional capabilities. But as the Joint Chiefs were acutely aware, these plans were still in the early stages of implementation. Moreover, many of the units stationed in Europe were in "hollow" condition, stripped of experienced personnel and lacking up-to-date equipment. Overall, U.S. troop strength was about 28 percent below what it had been toward the beginning of the decade. Despite the increased emphasis in NATO planning on forward defense and flexible response, the Alliance's true capacity to deter continued to rest on a combination of U.S. strategic power and NATO's tactical nuclear arsenal.[30]

Even though the Nixon administration wanted to demonstrate a renewed interest in European security, it had no plans for deploying additional forces or going beyond routine modernization of those that were there. President Nixon wanted to appear tough and strong to the Europeans and restore their confidence in the United States, but he also wanted to avoid precipitous action that might jeopardize détente or drive up defense costs at home. Relying on diplomacy to achieve their objectives, Nixon and Kissinger embarked on a series of initiatives in keeping with the spirit of détente and the Harmel Report to lessen tensions by opening a broad dialogue with the East. Among the results were a quadripartite modus vivendi on Berlin, the normalization of relations between East and West Germany, the creation of an East-West confidence-building forum (the Conference on Security and Cooperation in Europe, or CSCE), and the launching of talks, parallel to SALT, on mutual and balanced force reductions (MBFR) in conventional capabilities between NATO and the Warsaw Pact. Under the "Guam Doctrine," a concept casually disclosed during a trip to Asia in the summer of 1969, President Nixon acknowledged that there were limits to American power and that thenceforth, apart from its existing treaty commitments, the United States would avoid anything other than financial or military aid to the Third World. Europe, by implication, had moved back to the top of the U.S. agenda.[31]

With diplomacy in the forefront, NATO's military problems practically slipped from general view. Almost unnoticed was a progressive erosion of its capabilities that left the Alliance effectively incapable of fighting as a single entity by the late 1960s. Despite its unified command and elaborate mechanisms for consultation and collaboration, NATO remained a hodgepodge of armies, having made little progress since the 1950s toward standardizing equipment or integrating communications. Practically no one, least of all the West Germans, seriously entertained the idea of fighting a war in Europe, conventional or otherwise. To save money, the European allies had cut back on stockpiling to the point that the FRG had only enough artillery shells for a week of fighting, rather than the 30-day combat period prescribed in NATO planning documents. Worst of all, these deficiencies appeared to be fully known to Warsaw Pact commanders, who claimed to have ready access to such information from well-placed spies inside NATO headquarters.[32]

Meantime, NATO faced an increasingly imposing Warsaw Pact threat. While paying lip-service to détente, the Soviets pursued a steady modernization of Warsaw Pact forces, with the apparent purpose of enabling them to operate effectively in either a nuclear-chemical or conventional environment. Dating from the mid-1960s, the Warsaw Pact's modernization plan stressed the introduction, at almost double the normal replacement rate, of new and improved tanks, artillery, armored personnel carriers, and tactical fighter aircraft.[33] In the Joint Chiefs' estimation, however, the most dramatic and unsettling new development was the emergence of a significant Warsaw Pact tactical nuclear capability, organized around a new generation of more accurate and more usable short-range surface-to-surface missiles. Comparing nuclear capabilities, the JCS rated the Warsaw Pact's as "militarily superior to NATO's." Whereas NATO's tactical nuclear weapons were mainly aging show pieces for deterrence, the Warsaw Pact's were more tailored-effect weapons for waging war. The Joint Chiefs further found that NATO's conventional forces alone could not survive a concerted tactical nuclear attack by the Warsaw Pact. For years, the Joint Chiefs and other Western military planners had taken it for granted that, despite NATO's conventional inferiority vis-à-vis the Warsaw Pact, its superiority in tactical nuclear weapons gave it a definite edge in a showdown. Now the tables were turned.[34]

Efforts by the Joint Chiefs to draw attention to the Warsaw Pact buildup and to elicit support for offsetting measures met with limited success. On Capitol Hill, American involvement in Vietnam, the strategic arms race, and worries about the mounting expense of keeping U.S. troops abroad continually overshadowed European security concerns. The issue of costs had been a constant refrain in congressional debates since the early 1950s, when the United States first assigned large numbers of troops to the Alliance. By the early 1970s, with inflation on the rise and the dollar weakened by heavy expenditures on the Vietnam War, it was a cause célèbre in some circles. Especially active in drawing attention to the problem was Senate Majority Leader Mike Mansfield of Montana, whose quasi-isolationist views dovetailed neatly with the antiwar, antimilitary sentiments of his liberal Democratic colleagues. Convinced that the Europeans could—and should—contribute more, Mansfield thought the United States could halve its presence "without adversely affecting either our resolve or ability to meet our commitment under the North Atlantic Treaty."[35]

Though the Nixon administration successfully fought off Mansfield's attacks, it had its hands full and in the process became all the more cautious in considering measures to bolster the Alliance.[36] According to Secretary of State William P. Rogers, the United States would be doing well to keep forces at "essentially present levels."[37] At the same time, disagreements within the Intelligence Community over how to assess the Warsaw Pact buildup—whether it constituted an attempt by the Soviets to achieve outright military superiority, as DIA, J-2, and the military intelligence staffs believed, or whether, in the CIA's view, such dangers were overblown—further complicated the administration's efforts to develop a response.[38] In November 1970, following a lengthy interagency debate, President Nixon finally approved policy guidance (NSDM 95) that leaned toward the JCS on the need for preserving a strong U.S. posture in Europe, with near-term emphasis on improving conventional deterrence. Whether tactical nuclear capabilities should be addressed as well was held over for further study.[39] Ostensibly a victory for the Joint Chiefs, the triumph was short-lived when, in implementing the President's decision, Secretary of Defense Laird gave the lead to his Systems Analysis organization, which took a more flexible view of NATO requirements than did the JCS. There ensued 7 more months of bickering in the Pentagon between the Joint Staff and OSD, culminating in yet another Presidential decision (NSDM 133) that relaxed overall improvement goals.[40]

Whether NATO could ever achieve a level of conventional capabilities on a par with those of the Warsaw Pact remained a matter of debate and conjecture, both in Washington and in Europe, throughout the remainder of the Nixon administration and on into the Presidencies of Gerald Ford and Jimmy Carter. Although practically everyone agreed that there was room for improvement, there was no consensus on what to do or how to go about it. As a general objective, Secretary Laird suggested the Allies aim for a 4 percent real increase in their annual military spending, a goal the Europeans summarily rejected as beyond their means. More to their liking was the European Defense Improvement Program (EDIP), a low-budget approach to upgrading communications and infrastructure put forth by Britain, West Germany, and eight other European nations in December 1970. A broader initiative, drafted at NATO headquarters and known as AD–70, appeared at the same time. Projecting across-the-board improvements, AD–70 was the brainchild of General Andrew J. Goodpaster, USA, who had become Supreme Allied Commander the year before. A former aide to President Eisenhower and once Director of the Joint Staff, Goodpaster was a highly respected figure on both sides of the Atlantic. An inventory of deficiencies and anomalies rather than a plan of action, AD–70 elicited mixed pledges of support from the Allies. But because it bore Goodpaster's imprimatur, it probably received a more favorable reception than would otherwise have been the case.[41]

By 1973—the "Year of Europe" as the Nixon administration proclaimed it—NATO realized that it faced major problems and was taking steps to upgrade its equipment and improve interallied coordination and integration of functions.

Slowly but surely, the EDIP and AD–70 were bearing fruit.[42] Just how far the Alliance had progressed toward strengthening itself became the subject of yet another Nixon administration internal review (NSSM 168), launched early in 1973, with the Army and OSD (Systems Analysis) leading the effort.[43] Additional inputs came from the new Secretary of Defense, James R. Schlesinger, who followed in McNamara's footsteps in believing that NATO could indeed mount a credible conventional defense. All things considered, Schlesinger found NATO to be better prepared and equipped to deal with a conventional threat at the outset of a war than it had been only 3 or 4 years earlier. Because of limitations on naval forces, however, he was less sanguine about NATO's prospects in the event of a prolonged conflict requiring U.S. reinforcements who might not arrive in time to stave off an escalation of the conflict.[44]

Though obviously more committed than they had been for some time, European NATO leaders continued to shy away from elaborate and expensive modernization plans. As a rule, they preferred the less costly piecemeal approach that involved improvements in selected areas such as anti-armor, aircraft shelters, and stockpiling. Moreover, just as NATO was beginning to take a closer look at its deficiencies and do something about them, the Yom Kippur War of October 1973 erupted in the Middle East, causing the United States to divert equipment and munitions to Israel, much of it drawn from stockpiles allocated to NATO. Meanwhile, the Watergate scandal continued to engulf Washington. Increasingly preoccupied with its domestic difficulties, the Nixon administration had significantly less time for NATO and saw its influence and authority within the Alliance steadily recede. Others, most notably the West Germans, stepped up to take America's place, so that by the mid-1970s the initiative in nuclear modernization and other key areas had passed from Washington to Bonn.

NATO was making strides to improve itself, but it was still an Alliance with serious problems. Raw numbers purporting to show enhancements to NATO capabilities covered up the underlying malaise. According to General Alexander M. Haig, Jr., who succeeded Goodpaster as SACEUR in 1974, NATO forces faced pervasive morale and discipline issues. "Alcoholism and drug abuse were serious and widespread," Haig found. "Our state of readiness was way below acceptable standards.... There was little sense of organized purpose imposed from above, little communication among subordinate commands."[45] In assessing NATO's prospects, the Joint Chiefs remained confident that the Alliance would survive and even prosper as the bulwark of Western security. But despite the end of the Vietnam War and the redeployment of U.S. forces, NATO seemed to be achieving limited headway toward making a difference and redressing the strategic balance in Europe.

DÉTENTE

CHINA: THE QUASI-ALLIANCE

Nixon and Kissinger realized that NATO's chronic difficulties in raising and maintaining forces could not be solved in isolation. Thus, instead of trying to meet the Soviet threat to Europe head on, they sought to offset Soviet power via other means—by attempting to curb the buildup of arms, encouraging détente, and last but not least, exploiting the Soviet Union's deteriorating relationship with the People's Republic of China (PRC).[46] Evidence of worsening relations and ideological conflict between the two Communist giants had been accumulating for years, steadily undermining the concept theretofore accepted in the West of a Communist monolith.[47] By the late 1960s, there were reports of a buildup of opposing forces and armed clashes along the Sino-Soviet border. Sensing a golden opportunity, Nixon had indicated that forging a rapprochement between the United States and the PRC would be part of his agenda if he was elected.[48] Once in office, he and Kissinger made a determined effort not only to mend differences with Beijing, but also to convince skeptics—the Joint Chiefs of Staff among them—that a rapprochement with China would in the long run pay handsome dividends for the United States. The resulting improvement in Sino-American relations, as Kissinger later described it, amounted to nothing less than a "quasi-alliance."[49]

The opening gambit in the White House's effort to bring the Joint Chiefs around to its point of view on China was a military posture review (NSSM 3) ordered by President Nixon the day after taking office.[50] Characterized by Kissinger as "a highly esoteric discussion of military strategy," the review's unstated purpose was to reexamine the Johnson administration's practice of developing military plans in the expectation of waging two-and-a-half wars—one in Europe, another in Korea or Southeast Asia, and a smaller third contingency like the 1965 intervention in the Dominican Republic. Though there had never been sufficient forces to execute such a strategy with any confidence of success, it remained an integral part of the Joint Strategic Objectives Plan (JSOP), the Joint Chiefs' annual assignment of assets to meet theater and strategic requirements.

From the review they had ordered, Kissinger and Nixon envisioned a wholesale reordering of strategic priorities. At issue was whether it was still realistic and feasible to allocate resources on the basis of a two-and-a-half war scenario, or whether a more limited definition of risks, assuming minimal chances of a major conflict involving the PRC, would serve American interests just as well, if not better.[51] To be sure, a major incentive for downgrading the prospects of a war with China was budgetary, since a key finding of the posture review was that a fully-funded two-and-a-half war strategy would cost at least twenty percent more annually than adoption of a one-and-a-half war strategy.⁵² But there were also important political and diplomatic considerations involved. "The reorientation of our strategy signaled to the People's Republic of China," Kissinger said, "that we saw its purpose as separable from the Soviet Union's, that our military policy did not see China as a principal threat."⁵³

The Joint Chiefs of Staff initially took a different view, arguing that the changes Nixon and Kissinger were proposing would invite aggression, complicate the allocation of resources, and invite the early use of nuclear weapons in certain circumstances.⁵⁴ No one doubted that one key underlying purpose was simply to save money. Yet throughout the defense establishment, the implications were nothing short of ominous. Indeed, for those in uniform, Communist China remained a hostile power whose interests and worldview were sharply at variance with those of the United States. Less than 2 decades earlier, U.S. and Communist Chinese forces had fought pitched battles on the Korean Peninsula. Long-range appraisals done since then by the Joint Chiefs and by the Intelligence Community had routinely stressed China's commitment to achieving political dominance in Asia, its support for Communist insurgencies, and its close identification with leftist revolutionary causes around the globe.⁵⁵ A nuclear power since 1964, China had also acquired a thermonuclear capability in 1967, ostensibly the motivating factor in Secretary of Defense McNamara's decision to propose the deployment of a limited ABM system. Against this background of conflict and antagonism, a rapprochement with China was, in the Joint Chiefs' eyes, both hard to imagine and ill advised.

Brushing aside JCS objections, President Nixon formally embraced the one-and-a-half war strategy in his first annual report on U.S. foreign policy issued in February 1970. In explaining the change, the President insisted that he was only trying to harmonize strategy with capabilities.⁵⁶ By then, however, Nixon had firmly made up his mind to improve relations with Beijing and was heavily engaged in exploratory talks using the American and Chinese Ambassadors to Poland. The change in American military strategy was meant as an inducement to the Chinese. Later, not getting the cooperation they wanted from the State Department, Nixon and Kissinger turned to sensitive backchannel contacts established through Pakistan to finalize a deal with the Chinese. The net effect was an extraordinarily high degree of secrecy that sealed off the talks from practically anyone outside the White House (including the Joint Chiefs) who had an interest in the matter and to present them when the time came with a fait accompli.

Meanwhile, Nixon and Kissinger kept the State and Defense Departments occupied by commissioning a succession of studies through the NSC examining various aspects of the China issue. The most serious impediment to a Sino-American rapprochement to be identified was the U.S. relationship with the Republic of China (ROC), the rival government on Taiwan headed by the venerable Generalissimo Chiang Kai-shek, which the United States recognized as the de jure regime. A staunch anti-Communist and long-time U.S. ally, Chiang once had a loyal following in the United States, which had made sure over the years that the ROC received unstinting American assistance. On at least two occasions—in 1954 and in 1958—the United States had almost gone to war with Communist China in support of the ROC's continuing occupation of several offshore island groups in the Taiwan Strait.

Since the late 1950s, however, things had changed. Tensions over the offshore islands had eased, the China Lobby that had been so active on Chiang's behalf had lost its clout in Washington, and more and more countries were recognizing the PRC as the legitimate government of China. In October 1971, the UN General Assembly expelled the ROC, forcing it to cede its seat on the Security Council to the People's Republic. Despite its declining fortunes, however, the ROC retained a corps of supporters in Congress and continued to play a key role in American defense policy for East Asia. Once described by General Douglas MacArthur as "an unsinkable aircraft carrier," Taiwan provided the United States with access to basing and staging areas from which to control the Taiwan Strait, to assist in maintaining lines of communication, and to bring military power to bear quickly against the mainland should the need arise.[57] All in all, the Joint Chiefs of Staff considered Taiwan to be an essential link in their Pacific defense perimeter and, as such, a crucial part of the "close-in" containment strategy applied against Communist China.[58]

Still, as the Vietnam War wound down, Taiwan's usefulness to American defense planners steadily diminished, resulting in the closure of numerous installations, the withdrawal of personnel, and reductions in U.S. subsidies and assistance to the ROC. One of the cutbacks was the elimination of the Taiwan Strait Patrol, a money-saving move instigated by the Nixon administration in mid-November 1969. Initiated by President Truman in 1950 to protect Nationalist China from Communist attack, the Taiwan Strait Patrol tied up the use of two U.S. destroyers. Recognizing that the patrol had become largely symbolic, the Joint Chiefs accepted its elimination as a sensible alternative to the reduction of naval forces elsewhere. Thenceforth, ships of the Seventh Fleet transiting the Taiwan Strait would do the job. A small "gesture to remove an irritant," as Kissinger described it, the elimination of the Taiwan Strait Patrol figured squarely, along with the adoption of the one-and-a-half war strategy concept, in the administration's ongoing effort to improve Sino-American relations.[59]

Continuing to pursue a conciliatory approach, Kissinger wanted to offer further concessions—a nonaggression pact and/or the withdrawal of U.S. forces from

Taiwan—to demonstrate U.S. readiness to extend détente to the mainland. But he met with stiff resistance from the JCS, who urged caution in dealing with Beijing and no change in security arrangements with Chiang's regime. Any new concessions, the Joint Chiefs insisted, should be on a quid pro quo basis.[60] Undaunted, Kissinger set off for a secret rendezvous with Chinese Premier Zhou Enlai (Chou En-lai) in the summer of 1971. Even though Kissinger tried to disguise the purpose of his trip, Admiral Moorer, the JCS Chairman, later confirmed that he was able to follow developments closely because Kissinger and Nixon used a special Navy communications system part of the time to stay in touch.[61] Directed mainly at improving the atmosphere of Sino-American relations, the principal accomplishment of Kissinger's meeting with the Chinese was the "announcement that shook the world" on July 15, 1971, that President Nixon would visit China during the early months of the new year.[62]

Despite the prospect of improved relations with mainland China, the Joint Chiefs continued to oppose any major concessions. Still unresolved by the time of the President's visit to Beijing in February 1972 was a firm administration position on the future of U.S.-ROC defense arrangements, a matter of key importance to the JCS.[63] But in attempting to raise the matter and make their views known, they encountered repeated rebuffs from the White House and were unsuccessful in securing an interagency review prior to the President's departure.[64] Kissinger alone handled the agenda and other details of the summit in one-on-one talks with Zhou Enlai during a return trip to Beijing in October 1971.[65]

A momentous event that attracted intensive news coverage, Nixon's trip to China seemed to herald a new era in Sino-American relations. Despite a large entourage, no members of the Joint Chiefs accompanied the President, an apparently intentional omission aimed at playing down military matters. Still, there were strong politico-military overtones throughout the visit, with the threat posed by the Soviet Union a subject of mutual interest and, from all appearances, the number-one Chinese security concern. Though there were no discussions of specific collaboration against that threat, President Nixon recalled that the Chinese took great pleasure in the discomfort his visit seemed to cause to leaders in Moscow.[66]

Taiwan also figured large in the discussions, though it was not the obstacle that many (including the JCS) expected it to be. During an earlier exchange of views, the Chinese had indicated that the withdrawal of U.S. forces from Taiwan and from the Taiwan Strait should be the "first question" addressed at any summit meeting.[67] During his talks with Zhou in October 1971, however, Kissinger had served notice that the United States was not prepared to take a definitive position on Taiwan's future. The Chinese had backed off and during their meetings in February 1972,

Nixon and his hosts—Zhou and Chinese Communist Party Chairman Mao Zedong—downplayed the Taiwan issue. While the summit's communiqué confirmed that the United States regarded the withdrawal of its forces from Taiwan as the "ultimate objective," it mentioned nothing about a timetable or other commitments.[68] "The overwhelming impression left by Chou, as by Mao," Kissinger recalled, "was that continuing differences over Taiwan were secondary to our primary mutual concern over the international equilibrium."[69]

The Joint Chiefs greeted the outcome of the President's trip with relief and reassurance. Major changes were clearly taking place in Sino-American relations. But for the time being, the American security posture in the Far East remained essentially unchanged. Even so, a Sino-American entente was beginning to take shape. About a month after the President's trip, the North Vietnamese launched their "Easter Offensive" against South Vietnam, to which Nixon retaliated with the mining of Haiphong harbor and two massive air campaigns (*Linebacker I* and *II*) that brought American planes perilously close to the Chinese border. A few years earlier, such actions by the United States might have provoked an overtly hostile Chinese response, perhaps even direct intervention in the war. But by 1972, in light of the recent Sino-American rapprochement and continuing tensions between Moscow and Beijing, the threat of Chinese intervention barely figured in Nixon's calculations. In the event, Chinese forbearance spoke for itself. Though there were the customary public denunciations of American behavior, the PRC veered toward neutrality and offered only token help to the North Vietnamese.[70] Most telling of all was Beijing's rejection of a plan, jointly put forth by the Kremlin and Hanoi, to bypass the American bombing and mining of Haiphong by off-loading cargos at Chinese ports and bringing supplies overland into North Vietnam.[71] Equally important was the Joint Chiefs' tacit appreciation of Chinese restraint. Indeed, from that point on, JCS objections to further improvements in relations with the PRC became less frequent and their tone in support of Taiwan less strident. The establishment of formal diplomatic relations between the United States and the PRC was still some years away. But increasingly, it seemed to the Joint Chiefs to be the next logical step.

DEEPENING INVOLVEMENT IN THE MIDDLE EAST

If the American rapprochement with China seemed to test the durability of détente with the Soviet Union, developments in the Middle East toward the end of Nixon's Presidency nearly brought it to a premature end. Here, more than anywhere else, Soviet-American relations threatened to come full circle back to the confrontational policies and behavior of the 1950s and early 1960s. The precipitating event was the Arab-Israeli Yom Kippur War of October 1973. In many respects a "proxy conflict," the Yom Kippur War tested Eastern Bloc weapons and tactics used by the Arabs against those of the West as adapted by the Israelis. At the outset, it seemed that cooperation between Washington and Moscow would succeed in containing the conflict. Intensifying instead, it brought the threat of Soviet intervention and prompted the Joint Chiefs to place U.S. nuclear forces on increased alert, making it the most serious East-West confrontation since the 1962 Cuban Missile Crisis. That détente survived the ordeal, at least for a while, suggests an underlying degree of mutual respect brought on not only by the general improvement in U.S.-Soviet relations, but also by the realities of nuclear parity and the resulting caution that both sides felt compelled to observe.

Behind the headlines of the October War was the larger issue of American involvement in Middle East security, a role that had been growing steadily since the Suez crisis of 1956. With heavy obligations in Europe and the Western Pacific, the Joint Chiefs had generally been averse to commitments in the Middle East and had been content to rely on diplomacy and/or intervention by the British or the French to hold matters in check. But after the Suez debacle, the 1958 coup in Iraq, and the collapse of the Baghdad Pact, the Joint Chiefs had found themselves taking a more direct hand in the management of the region's security. Three issues predominated—the containment of Soviet power and influence, the protection of Western access to Persian Gulf oil fields, and the security of Israel.

While all three issues were interrelated, the Israeli situation overshadowed all others. Offering arms and other assistance from the mid-1950s on, the Soviets played on Arab nationalism and hostility toward the Jewish state in order to make inroads across the region, notably in the confrontation states of Egypt, Syria, and Iraq. To check the growth of Moscow's influence, the United States cultivated closer ties with Saudi Arabia, Jordan, and other Arab moderates, and encouraged Iran (a Muslim but non-Arab country) to become an anti-Soviet bulwark protecting the Persian Gulf. For domestic political reasons, however, shoring up Israel's security became Washington's top regional priority, and led to a policy of occasional, selective sales of sophisticated weapons, including tanks and Hawk antiaircraft missiles. By the 1960s, U.S. arms transfers to Israel well outpaced American military assistance to the Arab world. Predicting Middle East "polarization" should this trend continue, the Joint Chiefs found the United States increasingly identified with Israeli interests and the Soviet Union with those of the Arabs.[72]

Tensions peaked during the Six Day War of June 1967, in which Israeli forces seized the Sinai Peninsula from Egypt, the Golan Heights from Syria, and East Jerusalem and the West Bank from Jordan. Never close to begin with, relations between the Joint Chiefs and Israel's high command grew even farther apart during the conflict when Israeli warplanes and torpedo boats attacked USS *Liberty*, an American electronic intelligence ship operating in international waters off the Sinai coast. Owing to an almost complete breakdown of inter-Service cooperation in transmitting communications, the *Liberty* was actually operating closer to shore than the Joint Chiefs had intended; orders for it to pull back were in transit at the time of the attack.[73] Later, the Israelis insisted that they had mistaken the *Liberty* for an Egyptian ship known to be in the area. The attack inflicted heavy casualties on the U.S. crew and elicited deep regrets from the Israeli government. Insisting that the United States was as much to blame as they were, however, the Israelis refused to acknowledge any negligence and characterized the incident as an unfortunate "chain of errors."[74] The Joint Chiefs did not belabor the point, but at both the Pentagon and the White House suspicions lingered that the attack had not been accidental.[75]

In the aftermath of the Six Day War, the Joint Chiefs found the Middle East becoming more polarized than ever, a ripe environment for further strife.[76] Alarmed by the rapidity with which Moscow replenished Egypt's depleted arsenal, the Johnson administration responded in kind, by stepping up deliveries of tanks, fighter aircraft (including Navy A–4 "Skyhawks," then in critically short supply in Vietnam), and other weapons to bolster Israel's defenses.[77] A "war of attrition" ensued, during which Israeli and Egyptian gunners routinely exchanged fire across the Suez Canal, Israeli commandos launched attacks across the Gulf of Suez, and the Israeli Air Force, flying freshly acquired U.S.-made F–4 "Phantoms," carried out deep-penetration raids into Egypt. Worried that the United States might find itself isolated, the JCS urged the Nixon administration to curtail arms sales to Israel and to use its leverage to expedite a regional peace settlement through the United Nations. As part of an overall agreement, the Chairman, General Wheeler, suggested a protocol, backed by the "Big Four" (the United States, the Soviet Union, Britain, and France), guaranteeing enforcement of any settlement.[78]

While the Nixon administration's declared intention was a more balanced policy in the Middle East, popular and congressional pressure preserved the tilt toward Israel. As a result, there were few significant curbs on arms deliveries and no significant pressure applied on the Israelis to make concessions toward a peace settlement.[79] To maintain a "military balance" in the Middle East, the Joint Chiefs advocated selling sufficient weapons and equipment to the Israeli armed forces to defend Israel against an Arab attack "without destabilizing losses."[80] But as a practical matter, until the October War exposed serious shortfalls and weaknesses in Israeli defenses, it was hard to gauge how this principle applied. At the same time, despite progress elsewhere on détente, U.S.-Soviet competition in the Middle East reached a new level of intensity. By the early 1970s, the Soviets were augmenting their naval forces in the eastern Mediterranean and had markedly increased their personnel strength in Egypt. Soviet pilots flew patrols in MiG–21s with Egyptian markings and Soviet technicians operated a network of SA–3 surface-to-air point defense missiles to prevent the Israelis from conducting further deep-penetration raids.[81] A U.S.-brokered ceasefire ended the war of attrition along the canal in August 1970, but beneath the superficial calm that settled over the region, Arab-Israeli tensions remained high.

A key turning point was the decision by Egyptian President Anwar Sadat to expel his Soviet advisors in July 1972, barely more than a year after signing a Treaty of Friendship and Cooperation with Moscow. Exactly how many Soviets were involved is unclear, though an Egyptian source states that as many as 21,000 went home.[82] A career army officer, Sadat had come to power shortly after the death of Egypt's charismatic Gamal Abdul Nasser in September 1970. While he vowed to follow in Nasser's footsteps, Sadat found the current situation of "no war, no peace" an intolerable obstacle to his first priority—reviving Egypt's economy. Seeking Western investment, he knew he would have to create an economic and political environment more hospitable to capitalism, which meant moving away from socialism (manifest most clearly by the Soviet presence) and making peace with Israel.

While relations between Cairo and Moscow remained cool for some months after the expulsion of July 1972, Egypt continued to need Soviet military and economic support. For the moment, Sadat only wanted to change the basis of his relationship with the Soviet Union, not end it. At the same time, seeking to improve his contacts with the West, he reopened a backchannel, originally established in April 1972, with the Nixon White House to discuss ending the Israeli occupation of the Sinai.[83] Sadat would have preferred a negotiated settlement, but he knew he had little bargaining power and resolved to improve his position through the only means available—military action against Israel. Expelling the Soviets was the first phase of his plan, since he suspected, not without cause, that Moscow would never risk jeopardizing détente by overtly cooperating in launching a war. Sadat did not expect to achieve a clear-cut military victory, but if Egypt could demonstrate a credible limited war capability, he thought he stood a good chance of restoring his country's self esteem and prestige and of forcing the Israelis into negotiations.[84]

On October 6, 1973 (Yom Kippur in Israel, Ramadan in Arab countries), Egyptian forces mounted a successful surprise assault across the Suez Canal, timed to coincide with a Syrian attack against Israeli positions on the Golan Heights. According to the "leaked" findings of a congressional investigation, NSA intercepts routinely available to the Joint Chiefs would have confirmed that the Egyptians were planning an attack; however, the sheer volume of the message traffic and the inability of the NSA and DIA to process all the data efficiently gave rise to an "intelligence failure." The Israelis were similarly caught off guard.[85] Named Operation "Badr" after the first victory of the Prophet Mohammad in 630 AD, the Egyptian assault quickly breached Israeli defenses (the Bar-Lev Line) but carried only a few kilometers into the Sinai. Following an initial period of indecision and confusion, the Israelis regrouped and on October 8 launched a counterattack that thwarted Egyptian efforts to extend their bridgehead.

A see-saw battle ensued over the next few days, during which time the Soviet Union and the United States made half-hearted attempts to arrange a ceasefire through the UN. Meanwhile, Washington and Moscow both expedited the airlift of weapons and supplies to their clients. By October 15, Israeli forces had gained the offensive. As a UN-brokered ceasefire was about to take effect, they broke through Egyptian lines, crossed the Suez Canal with makeshift pontoon bridges, and proceeded to envelop the Egyptian Third Army—45,000 troops in all—trapping it on the eastern side of the canal. With some of his best forces facing imminent annihilation or surrender, Sadat appealed to Brezhnev for help. On October 24, the Soviet leader responded. Declaring Israel to be in violation of the ceasefire, he served President Nixon with an "ultimatum," as Kissinger characterized it, warning that the Soviet Union was prepared to take "appropriate steps unilaterally" to bring the conflict to an end.[86]

Until Brezhnev's ultimatum, the Joint Chiefs played a low profile in the crisis. For coordination they relied on the Washington Special Action Group (WSAG), an interagency crisis-management subcommittee of the NSC that included Admiral Moorer among its members and Kissinger as chairman.[87] Preoccupied with the escalating Watergate affair and a separate scandal involving allegations of financial wrongdoing by Vice President Spiro Agnew (culminating in Agnew's resignation on October 10, 1973), Nixon deferred increasingly to Kissinger and the WSAG to guide American policy. The Yom Kippur war, Nixon later observed, "could not have come at a more complicated domestic juncture."[88] The Joint Chiefs' job during those hectic days was to monitor events on the battlefield, expedite the transfer of supplies to the Israelis, and take precautionary steps by reviewing contingency plans for the evacuation of Americans and the deployment as necessary of U.S. forces.

Once the fighting began, the WSAG sought to establish and maintain a position of quasi-neutrality insofar as the pro-Israeli bent of the United States would allow. Despite a surge of Soviet naval power into the eastern Mediterranean, the United States confined its presence there to a single naval task group. Organized around the carrier *Independence*, the task group took up station southwest of Crete on October 7 where it remained for the duration of the crisis. Sixth Fleet's urgent function was surveillance of the Soviet naval presence. To carry out his mission, Sixth Fleet Commander Vice Admiral Daniel J. Murphy, Sr., proposed to move his ships closer to the conflict and augment them with a second carrier, the *Franklin D. Roosevelt*, then operating off Sicily. Moorer and the WSAG, however, refused Murphy's request. Citing policy constraints, they reminded him that he was to distance himself from possible involvement in keeping with the administration's "low-key, even-handed approach toward the hostilities."[89]

A similar policy of restraint initially governed American assistance to Israel. For the first few days of the conflict, the Israelis could have whatever they reasonably required as long as they transported it themselves. While the Joint Chiefs never had occasion to adopt a corporate position, most of them—Moorer especially—believed the United States was playing a dangerous game by giving the Israelis even limited help. Only the CNO, Admiral Zumwalt, a self-described "strong proponent of resupplying Israel," felt the United States should be more forthcoming.[90] But as the fighting intensified and Israeli losses climbed, the voices of caution at the Pentagon became drowned out by those in Congress and the public who demanded that restrictions on aid to Israel be relaxed, if not lifted altogether. By October 10, Israel's situation had become precarious. Even if Israel did not lose the war, it would emerge from the conflict severely battered and crippled. Rumors spread that in a last-ditch effort to save the country, the Israeli cabinet had authorized the deployment of nuclear-armed Jericho missiles.[91] Adding further to the tension were indications that the Soviets had mobilized several elite airborne divisions for possible deployment to Egypt and the "ominous news," as Kissinger called it, that Moscow had launched an airlift to Syria.[92]

Meanwhile, a standoff had developed between Kissinger and Secretary of Defense James R. Schlesinger over the processing of Israeli assistance requests. The results were a slow-down of deliveries and a rising level of irritation on Capitol Hill that threatened President Nixon's chances of surviving the Watergate scandal. On October 12, demonstrating that he was still in charge, the President flung open American arsenals to the Israelis. Authorizing the use of jumbo C–5A transport planes to expedite deliveries, he brushed aside objections from the Joint Staff and the Office of the Secretary of Defense that his actions "might blight our relations with the Arabs" and dangerously deplete U.S. war reserves. In the end, the American airlift allowed the Israelis to prevail. But as the Joint Chiefs had feared, it raised other problems in the form of Arab retaliation through an oil embargo against the West, and friction within NATO over the draw-down of supplies allocated to Alliance defense and the use of European bases for intelligence-gathering.[93]

Still, it was Brezhnev's ultimatum that captured the Joint Chiefs' attention more than anything. By itself, Brezhnev's threat to take unilateral action might have been dismissed as diplomatic bluster. In all likelihood, as British foreign policy expert Gordon S. Barrass has pointed out, it did not reflect his true views.[94] But coming on top of the Soviet naval buildup in the eastern Mediterranean, the mobilization of combat divisions trained in rapid deployment, and stepped-up Soviet air activity, there was every reason for the chiefs to be concerned. The ensuing decision to place U.S. nuclear forces on heightened alert (DEFCON 3) emerged from a late night WSAG meeting in the White House Situation Room on October 24.[95] Nixon took no part in the deliberations and remained well out of the way, attended by Alexander Haig, who was then Kissinger's deputy.[96] Immediately after the meeting, Moorer returned to the Pentagon and arranged that the alert be carried out in conspicuous fashion to attract the attention of Soviet intelligence. A few hours later, the fully assembled Joint Chiefs met with Secretary of Defense Schlesinger to discuss further moves, including the possibility of raising the alert to DEFCON 2, a level not used since the Cuban Missile Crisis. But by morning, a fresh message from Moscow couched in conciliatory language laid the matter to rest. By late the next day all U.S. commands had resumed their normal alert posture.[97]

While Israel prevailed in the October War, it was at a tremendous cost that approached a Pyrrhic victory: as many as 2,800 dead and another 9,000 wounded. Arab losses were substantially larger. JCS estimates of the outcome hesitated to proclaim a clear-cut winner. Most predicted that another war was only a matter of time and that in the long run the continuing identification of the United States with Israel would work to the detriment of U.S. interests in the Middle East. Indeed, the more closely the United States became aligned with Israel, the less influence and credibility it was apt to have in Arab countries and in the economically and strategically important Persian Gulf. It followed, in the JCS view, that the most important objectives in the aftermath of the October War were to reestablish stable relations with the moderate Arab states like Jordan and Saudi Arabia, and to shore up ties elsewhere in the Muslim world, especially with Iran and Pakistan. Yet given the political realities in the United States, it was altogether likely that Washington would continue to pursue a divided policy that supported Israel while trying to placate the Arabs and curb further Soviet inroads.

Whether the Middle East was ready for peace remained to be seen. Détente had helped to avoid a great power confrontation during the October War, but in the aftermath of the fighting it did little to promote a more hospitable environment for resolving the Arab-Israeli conflict. Celebrated in the Arab world as a great victory, the October War demonstrated that Israel was far from invincible and lifted Sadat's reputation and prestige to unprecedented heights. Yet in moving toward a peace settlement, he was practically alone. A multinational Geneva peace conference, co-organized by the United States and the Soviet Union in November 1973, attracted little participation from the Arab world and broke up inconclusively almost as soon as it began. Thenceforth, it would be up to the Egyptians and Israelis themselves, negotiating bilaterally and relying on the United States as intermediary, to reach a modus vivendi.

Meanwhile, the Cold War, like the Arab-Israeli conflict, refused to go away, détente notwithstanding. In his final posture statement to Congress, submitted shortly before the end of his term as Chairman in July 1974, Admiral Moorer cited the ABM Treaty and the SALT I interim agreement as "first steps . . . to establish some control over the deployment of significantly increased strategic forces by both the U.S. and the USSR." As encouraging as these agreements might have been, however, Moorer remained concerned by the Soviet Union's "aggressive modernization programs" in everything from strategic offensive weapons to general purpose forces for ground, sea, and air warfare. Drawing on the recent experience of the October War, he saw lessons to be learned. One was "that the military balance must be assessed on the capabilities of potential adversaries rather than on their announced or estimated intentions." Détente, he argued, had created an atmosphere of increased "good will" between the United States and the Soviet Union. But it had yet to slow the arms race or curb the potential for confrontation that the competition implied.[98]

NOTES

1. See Stanley Hoffmann, "Détente," in Joseph S. Nye, Jr., ed., *The Making of America's Soviet Policy* (New Haven: Yale University Press, 1984), 231–263.

2. Richard M. Nixon, *RN: The Memoirs of Richard Nixon* (New York: Grosset & Dunlap, 1978), 941.

3. NIE 11-8-69, September 9, 1969, "Soviet Strategic Attack Forces," in Donald Steury, ed., *Intentions and Capabilities: Estimates on Soviet Strategic Forces, 1950–1983* (Washington, DC: Central Intelligence Agency, 1996), 253–254.

4. See JSOP 74-81, Book II (Strategic Forces), Vol. II, circulated under SM-802-71, December 21, 1971, JCS 2143/397; Walter S. Poole, Lorna S. Jaffe, and Wayne M. Dzwonchyk, "History of the Joint Chiefs of Staff: The Joint Chiefs of Staff and National Policy, 1969–1972" (MS, Historical Division, Joint Secretariat, Joint Staff, March 1991), 70–71 (declassified/publication forthcoming).

5. JCSM-377-69 to SECDEF, June 17, 1969, "Preparation of U.S. Position for Possible Strategic Arms Limitation Talks, JCS 2482/28-13; Poole, Jaffe, and Dzwonchyk, 140–141.

6. JCSM-143-67 to SECDEF, March 14, 1967, "Proposal on Strategic Offensive and Defensive Missile Systems," JCS 1731/966-2.

7   Thomas W. Wolfe, *The SALT Experience* (Cambridge: Ballinger, 1979), 8–9.

8   John Newhouse, *Cold Dawn: The Story of SALT* (New York: Holt, Rinehart and Winston, 1973), 113–114; "JCS Organizational and Procedural Arrangements for Arms Control Matters" (MS, Historical Division, Joint Chiefs of Staff, Joint Secretariat, December 31, 1980), 5–6.

9   Lawrence Freedman, *U.S. Intelligence and the Soviet Strategic Threat*, 2d ed. (Princeton, NJ: Princeton University Press, 1986), 131–133.

10 Henry Kissinger, *White House Years* (Boston: Little, Brown, 1979), 148.

11 NSDM 33, "Preliminary Strategic Arms Limitation Talks," November 12, 1969, Nixon Papers.

12 SM-766-69, November 10, 1969, "SALT Preparation," JCS 2482/58; Poole, Jaffe, and Dzwonchyk, 145–146.

13 Kissinger, *White House Years*, 149.

14 Ibid., 820.

15 "Strategic Arms Limitation Talks," May 20, 1971, *Nixon Public Papers, 1971*,

16 MFR by CJCS (M-41-71), "Meeting with the President, 20 May 1971," entry May 20, 1971, Moorer Diary, cited in Poole, Jaffe, and Dzwonchyk, 165.

17 Ibid., chap. 8.

18 Donald R. Baucom, *The Origins of SDI, 1944–1983* (Lawrence: University Press of Kansas, 1992), 39–50.

19 Interview No. 2 with Nitze by James C. Hasdorff, May 20, 1981, U.S. Air Force Oral History Collection, Maxwell Air Force Base, Alabama, 480 (quote); Paul H. Nitze, *From Hiroshima to Glasnost: At the Center of Decision—A Memoir* (New York: Grove Weidenfeld, 1989), 314–320; Gerard Smith, *Doubletalk: The Story of SALT I* (New York: Doubleday, 1980), 41–42.

20 For the text of these agreements, see U.S. Arms Control and Disarmament Agency, *Arms Control and Disarmament Agreements: Texts and Histories of Negotiations* (Washington, DC: GPO, 1982), 132–157.

21 Roger Labrie, ed., *SALT Hand Book: Key Documents and Issues, 1972–1979* (Washington, DC; American Enterprise Institute, 1979), 13–14. Under the interim agreement, the United States could build up to 44 fleet ballistic submarines and 710 launchers; the Soviets could go as high as 950 SLBMs and 62 submarines but would have to make off-setting reductions in land-based ICBMs to stay within the freeze.

22 Freedman, *U.S. Intelligence and the Soviet Strategic Threat*, 165–166; Poole, Jaffe, and Dzwonchyk, 178–191.

23 Raymond L. Garthoff, *Détente and Confrontation: American-Soviet Relations from Nixon to Reagan* (Washington, DC: Brookings, 1994, rev. ed.), 213–223 and passim.

24 JCSM-258-72 to SECDEF, June 2, 1972, "National Security Assurances in a Strategic Arms Limitation Environment," JCS 2482/152; summarized in Poole, Jaffe, and Dzwonchyk, 189–91.

25 ADM Moorer, May 19, 1972, quoted in Poole, Jaffe, and Dzwonchyk, 185.

26 Anna Kasten Nelson, "Senator Henry Jackson and the Demise of Détente," in Anna Kasten Nelson, ed., *The Policy Makers: Shaping American Foreign Policy from 1947 to the Present* (Lanham, MD: Rowman & Littlefield, 2009), 92–93.

27 Marquis Childs, "Jackson and the Generals," *The Washington Post*, December 14, 1976, 19.

28 See Nelson, "Senator Henry Jackson and the Demise of Détente," 93–94.

29 Admiral Thomas H. Moorer, *United States Military Posture for FY 1975* (Report by the Chairman of the Joint Chiefs of Staff, February 26, 1974), 15–16.

30 Memo, SECDEF to President, February 20, 1969, "NATO Defense Issues," JCS 2450/695; "NSC Review—US Policy Toward NATO," undated, attach. to Memo, Davis to Pedersen et al., March 17, 1969, JCS2450/676-1; DJSM-1644-69 to SECDEF, October 23, 1969, "NSSM-65—Relationships Among Strategic and Theater Forces for NATO," JCS 2101/561-1; Poole, Jaffe, and Dzwonchyk, 193–195.

31 "Informal Remarks in Guam with Newsmen," July 25, 1969, *Nixon Public Papers, 1969*, 549.

32 Gordon S. Barrass, *The Great Cold War: A Journey Through the Hall of Mirrors* (Stanford, CA: Stanford University Press, 2009), 196.

33 Richard L. Kugler, *Commitment to Purpose: How Alliance Partnership Won the Cold War* (Santa Monica, CA: RAND, 1993), 250–251; Thomas W. Wolfe, *Soviet Power and Europe, 1945–1970* (Baltimore: Johns Hopkins Press, 1970), 471–472.

34 "ISA Summary of JCS Report and Supporting Study," enclosure to Memo, Ware (ISA) to DJS, December 3, 1969, "NSSM 65," JCS 2101/561-2; Poole, Jaffe, and Dzwonchyk, 195–196.

35 Quoted in Lawrence S. Kaplan, *NATO and the United States: The Enduring Alliance* (Boston: Twayne Publishers, 1988), 131.

36 See Henry Kissinger, *Years of Upheaval* (Boston: Little, Brown, 1982), 134.

37 *The New York Times*, December 7, 1969, 18; see also John S. Duffield, *Power Rules: The Evolution of NATO's Conventional Force Posture* (Stanford, CA: Stanford University Press, 1995), 196.

38 Kugler, 252.

39 NSDM 95, November 25, 1970, "US Strategy and Forces for NATO," box H-208, Nixon Presidential Material; and <http://nixon.archives.gov/virtuallibrary/documents/nsdm/nsdm_095.pdf>.

40 Poole, Jaffe, and Dzwonchyk, 207–210; NSDM 133, September 22, 1971, "US Strategy and Forces for NATO; Allied Force Improvements," box H-208, Nixon Presidential Materials, available at <http://nixon.archives.gov/virtuallibrary/documents/nsdm/nsdm_133.pdf>.

41 Lewis Sorley, "Goodpaster: Maintaining Deterrence during Détente," in Robert S. Jordan, ed., *Generals in International Politics: NATO's Supreme Allied Commander, Europe* (Lexington: University Press of Kentucky, 1987), 130–133; Kugler, 269–271; Duffield, 197–201. AD-70 was short for "Alliance Defense in the Seventies."

42 How fast and effectively NATO was moving on these issues remained a matter of debate. According to Goodpaster, "in true NATO style, you have to measure [progress] in millimeters." Quoted in Sorley, "Goodpaster," 131.

43 NSSM 168, February 13, 1973, "US NATO Policies and Programs," box H-207, Nixon Presidential Materials, available at <http://nixon.archives.gov/virtuallibrary/documents/nssm/nssm_168.pdf>.

44 Kugler, 302–304.

45 Alexander M. Haig, Jr., *Inner Circles: How America Changed the World—A Memoir* (New York: Warner Books, 1992), 521.

46 Kissinger vehemently denied that the administration's policy was to play off China against the Soviet Union. He used the curious explanation: "We could not 'exploit' that rivalry; it exploited itself." See Kissinger, *White House Years*, 763.

47 Donald S. Zagoria, *The Sino–Soviet Conflict, 1956–1961* (New York: Atheneum, 1964), was among the earliest to expose the "split" within the communist bloc.

48 Richard M. Nixon, "Asia After Vietnam," *Foreign Affairs* 56 (October 1967), 111–125.

49 Henry Kissinger, *On China* (New York: Penguin, 2011), 275 and passim.

50 NSSM 3, January 21, 1969, "US Military Posture and the Balance of Power," box H-207, Nixon Presidential Materials, available at <http://nixon.archives.gov/virtuallibrary/documents/nssm/nssm_003.pdf>.

51 Kissinger, *White House Years*, 220–221.

52 "US Military Posture and the Balance of Power: General Purpose Forces Section," September 5, 1969, 24, enclosure to Memo, DepSECDEF to NSC Members et al., September 5, 1969, "Final Report on U.S. Military Posture and the Balance of Power," JCS 2101/554-54, cited in Poole, Jaffe, and Dzwonchyk, 23–25.

53 Kissinger, *White House Years*, 221.

54 JCSM-743-69 to SECDEF, December 4, 1969, "Strategic Concept and Force Planning Guidance for Military Planning (Revised)—FY 1972–1979, JCS 2458/632-4; Poole, Jaffe, and Dzwonchyk, 28–30.

55 SNIE 13-69, March 6, 1969, "Communist China and Asia," in Allen, Carver, and Elmore, eds., *Tracking the Dragon*, 527–539.

56 "First Annual Report to the Congress on U.S. Foreign Policy for the 1970s," February 18, 1970, *Nixon Public Papers, 1970*, 176–177.

57 MacArthur quoted in D. Clayton James, *The Years of MacArthur*, vol. III, *Triumph and Disaster, 1945–1964* (Boston: Houghton, Mifflin, 1985), 460.

58 Poole, "JCS and National Policy, 1965–68" (MS), 717–728 (declassified/publication forthcoming).

59 CM-4764-69 to SECDEF, December 6, 1969, "Taiwan Strait Patrol," JCS 1966/172; *CINCPAC Command History, 1969*, I, 141–144; Kissinger, *White House Years*, 187; Poole, Jaffe, and Dzwonchyk, 445–446.

60 Talking Paper for SECDEF and CJCS for NSC Meeting, March 25, 1971, "United States China Policy (NSSM 106)," JCS 2118/262-8; Poole, Jaffe, and Dzwonchyk, 431–432.

61 See Poole, Jaffe, and Dzwonchyk, "JCS and National Policy, 1969–72," 434, fn. 46.

62 "Remarks Announcing Acceptance of an Invitation to Visit the PRC," July 15, 1971, *Nixon Public Papers, 1971*, 819–820; Kissinger, *White House Years*, 742–755.

63 JCSM-446-71 to SECDEF, October 6, 1971, "Military Considerations for the Pending Presidential Visit to the PRC," JCS 2270/44-1; Poole, Jaffe, and Dzwonchyk, 436–437.

64 CNOM 18-72 to JCS, February 3, 1972, "Military Considerations Concerning the President's Discussions with the PRC," JCS 2270/47; JCSM-58-72 to SECDEF, February 16, 1972, same subject, JCS 2270/47-1; Poole, Jaffe, and Dzwonchyk, 437–438.

65 Kissinger, *White House Years*, 774–784.

66 Nixon, *RN*, 567–568.

67 Message, Zhou Enlai to Nixon, May 29, 1971, *FRUS 1969–76*, vol. XVII, *China, 1969–72*, 332.

68 "Joint Statement Following Discussions with Leaders of the PRC," February 27, 1972, *Nixon Public Papers, 1972*, 378.

69 Kissinger, *White House Years*, 783–784, 1073–1074 (quote).

70 The most significant assistance provided by the PRC was the dispatch of 12 minesweepers to North Vietnam, which cleared 46 mines between July 1972 and August 1973. Official U.S. sources indicate that the Navy laid down "thousands" of mines during the operation. See Qiang Zhai, *China and the Vietnam Wars, 1950–1975* (Chapel Hill: University of North Carolina Press, 2000), 202–204; and Edward J. Marolda, ed., *Operation End Sweep: A History of Minesweeping Operations in North Vietnam* (Washington, DC: Naval Historical Center, 1993), xi.

71 Garthoff, 291.

72 JCSM-337-65 to SECDEF, May 6, 1965, "Impact on Area Arms Balance of Military Sales to Israel," JCS 2369/12-2; Poole, "JCS and National Policy, 1965–68," Pt. II, chap. X.

73 Thomas R. Johnson, *American Cryptology during the Cold War, 1945-1989*, Book II, *Centralization Wins, 1960-1972* (Washington, DC: National Security Agency, 1995), 432–39 (declassified), available at <http://www.nsa.gov/public_info/_files/cryptologic_histories/cold_war_ii.pdf>, details the *Liberty's* mission.

74 Israel Defence Forces, *Attack on the "Liberty" Incident, 8 June 1967* (Tel Aviv: History Department, Research and Instruction Branch, June 1982), 31.

75 Edward J. Drea, *McNamara, Clifford, and the Burdens of Vietnam, 1965–1969* (Washington, DC: Historical Office, Office of the Secretary of Defense, 2011), chap. 16; James Scott, *The Attack on the Liberty* (New York: Simon & Schuster, 2009), 138–141, 153–155.

76 JCSM-374-67 to SECDEF, June 29, 1967, "US Military Interests in the Near East," JCS 1887/720-.

77 Drea, *McNamara, Clifford, and Vietnam*, chap. 16.

78 Poole, Jaffe, and Dzwonchyk, 297–300.

79 See Kissinger, *White House Years*, 363–377.

80 JCSM-521-71 to SECDEF, November 30, 1971, "Combat Aircraft Sales to Israel," JCS 2369/46; also quoted and summarized in Poole, Jaffe, and Dzwonchyk, 328–329.

81 "The Military Balance in the Middle East," May 27, 1970, enclosure to JCS 2369/37-1; Poole, Jaffe, and Dzwonchyk, 300–301.

82 Mohamed Heikal, *The Road to Ramadan* (New York: Quadrangle, 1975), 175. Other sources place the size of the Soviet contingent in Egypt at between 10,000 and 15,000.

83 Kissinger, *White House Years*, 192–193; Kissinger, *Years of Upheaval*, 204–205.

84 See Anwar Sadat, *In Search of Identity: An Autobiography* (New York: Harper & Row, 1977), 215 and passim.

85 CIA, *The Pike Report* (Nottingham, UK: Spokesman Books, 1977), 26–94. See also Gerald K. Haines, "The Pike Committee Investigations and the CIA," *Studies in Intelligence* (Winter 1998–1999; unclassified edition), 81–92; and Christopher Andrew, *For the President's Eyes Only* (New York: HarperPerennial, 1995), 390–392.

86 Letter, Brezhnev to Nixon, October 24, 1973, Nixon Presidential Materials, Henry A. Kissinger Files, box 69, Dobrynin/Kissinger vol. 20 (October 12–November 27, 1973); Kissinger, *Years of Upheaval*, 583.

87 Kissinger at this time served in a dual capacity. Appointed Secretary of State in September 1973, he continued to serve also as the President's Assistant for National Security Affairs.

88 Nixon, *RN*, 922.

89 Zumwalt, *On Watch*, 434–436; Robert W. Love, Jr., *History of the U.S. Navy* (Harrisburg, PA: Stackpole Books, 1992), II, 654.

90 Zumwalt, *On Watch*, 434.

91 Walter Isaacson, *Kissinger: A Biography* (New York: Simon & Schuster, 1992), 517. According to U.S. intelligence, the Jericho had a range of 260 miles and was unsuitable for conventional munitions. See SNIE 4-1-74, "Prospects for Further Proliferation of Nuclear Weapons," August 23, 1974 (declassified with deletions), 2, 22, National Security Archive, GWU, and <http://www.gwu.edu/~nsarchiv/NSAEBB/NSAEBB240/snie.pdf>.

92 Kissinger, *Years of Upheaval*, 497.

93 Nixon, *RN*, 924; Kissinger, *Years of Upheaval*, 493.

94 Barrass, *Great Cold War*, 185, recounts Brezhnev's growing frustration with the Arabs.

95 Message, CJCS 2733 to CINCPAC et al., October 25, 1973, National Security Archive, GWU.

96 Haig, *Inner Circles*, 415–416.

97 Kissinger, *Years of Upheaval*, 586–589; Entry, October 26, 1973, Moorer Diary; Strategic Air Command, "Chronology: Middle East Crisis" (SAC History Study #139, December 12, 1973), 13 (declassified with deletions); Message, CJCS 5694 to CINCPAC et al. October 28, 1973, National Security Archive, GWU.

98 *United States Military Posture for FY 1975*, 1–6.

James R. Schlesinger, Secretary of Defense, 1973–1975

Chapter 12

THE SEARCH FOR STRATEGIC STABILITY

Détente lasted for roughly 7 years, from the signing of the SALT I agreements in 1972 until the Soviet invasion of Afghanistan in 1979. During that time, with the exception of the 1973 October War in the Middle East, there were no repetitions of the tense encounters that had been so commonplace in the 1950s and 1960s. From all outward appearances, détente was a huge success. Barely below the surface, however, the situation was different. The Soviet military buildup in both conventional and strategic nuclear forces continued, and with it came increased Soviet activity in Southeast Asia, the Middle East, and Africa. Often employing Cuban "proxies," the Soviets seemed more intent than ever on extending their power and influence into new areas where conditions were ripe for Communist penetration and U.S. interests were most vulnerable.

For the Joint Chiefs, these were exceedingly trying times. With the military's reputation and credibility in tatters after Vietnam, they were hard put to mobilize support for what they considered essential requirements to bolster the country's defense posture. Concentrating on disparities in strategic forces, they saw an especially urgent need for modernization but faced budgetary and political constraints that allowed only parts of their program to go forward as planned. Basically, the country was in no mood for a postwar military buildup. Instead, the approach most people preferred was a lowered profile abroad in line with the Nixon administration's projections under the Guam Doctrine, and further pursuit of reduced tensions with the Soviets through SALT and détente.

THE PEACETIME "TOTAL FORCE"

As they gradually shifted from a wartime to a peacetime footing in the early 1970s, the Joint Chiefs expected demobilization and cutbacks in military spending to take a heavy toll. What they failed to anticipate was a public and congressional backlash brought on by Vietnam which, when coupled with competition for funds from domestic social programs, would depress military spending for nearly a decade. The result was virtually no real growth in the U.S. military budget, compared to a net annual increase of 3 percent in Soviet military spending.[1] Once the Vietnam "bulge" was gone by the early 1970s, the Defense Department's annual budget authority, as measured in constant dollars, almost steadily declined. By FY80, it was about 1 percent less than what it had been a decade earlier in FY71. During that time, U.S. defense spending dropped from 7.2 percent of the country's gross national product to 5.2 percent. Since the 1970s were a decade of high inflation, the impact on the Services' buying power and their ability to modernize weapons and equipment was more than an inconvenience—it was nearly crippling.[2]

Faced with no-growth and negative-growth budgets, the Joint Chiefs strained to meet obligations abroad which until the end of the 1960s had revolved around a two-and-a-half war planning scenario. Though that was reduced by the Nixon White House to a one-and-a-half war requirement in 1970, the JCS still found themselves facing the possibility of simultaneous conflicts on two separate fronts—a major conflict, most likely in Europe, and a lesser one in Korea or the Middle East. Politically, this change had much to recommend it. Not only did it accord with the administration's desire to improve relations with China, but also it limited overseas commitments, as enunciated under the Guam Doctrine. A further advantage was that it simplified the work of JCS and Service planners (the Army's especially) by allowing them to focus their research and development (R&D) and acquisition policies more closely on supporting NATO.[3] At the same time, the one-and-a-half war strategy allowed air and naval assets deployed in the Far East to be redeployed to Europe or the Mediterranean more readily than in years past. But in the Joint Chiefs' eyes, the new concept still left U.S. forces spread exceedingly thin around the globe and took little or no account of the ever-present danger of unforeseen contingencies.

In keeping with its limited view of U.S. obligations abroad, the Nixon administration also endorsed a peacetime "total force" that was smaller than any the JCS had seen since the 1950s. Two key innovations were an all-volunteer Army (more expensive to maintain than a conscripted force but less politically troublesome) and increased reliance on Reserve capabilities. Once the Vietnam War was over, the administration projected a peacetime defense establishment organized around an Army of 13 active divisions (down from 18 at the height of the Vietnam conflict) and 8 divisions in the National Guard, a Navy of approximately 400 surface ships, 93 submarines, and 16 carriers, a Marine Corps of 3 Active divisions and 1 Reserve, and an Air Force of 21 Active and 11 Reserve wings.[4]

While the Joint Chiefs would have preferred a larger active peacetime force, inter-Service skirmishing over the allocation of resources prevented them from coming up with firm, prioritized recommendations. Unable to agree among themselves, the Joint Chiefs effectively ceded the determination of force levels to OSD, the White House, and the Office of Management and Budget (OMB). In these circumstances, fiscal considerations invariably triumphed over military ones. Most impacted of all was the Army, which faced a 20 percent cut in strength, compared with 10 percent cuts in the Air Force and Navy. As Admiral Moorer described the scene at one JCS meeting in February 1970, Army Chief of Staff General Westmoreland, was "running scared," disparaging the contributions of the other Services, and "grasping in every direction" for ways to stave off troop reductions.[5]

In fact, the force reductions after Vietnam were no more severe than those the Services experienced after Korea and far less debilitating than the massive post-World War II demobilization. The retention of air and naval power rather than large ground forces also followed earlier patterns and reflected the continuing practice of turning to technology to shore up the country's security in peacetime. Meanwhile, ending the draft allowed the Army to be more selective in the recruitment of personnel, a major step toward creating a more elite, cohesive institution. The net result was a smaller, more professional defense establishment with a lowered overall public profile, which was a distinct advantage at a time of strong skepticism toward the military in Congress and lingering anti-war sentiment in the country at large.

MODERNIZING THE STRATEGIC DETERRENT

The most urgent task facing the Joint Chiefs as the Vietnam War drew to a close was to reequip and modernize the Armed Forces. Hardware worn out in Vietnam had to be replaced, while advances in technology offered the possibility of a refurbished arsenal of more sophisticated and versatile weapons. Much of the attention focused on improving conventional forces: a new main battle tank (the M–1) and a new armored personnel carrier for the Army, new fighter aircraft for the Air Force and Navy, and new ships for the fleet. But as important as these acquisition programs may have been, they paled in comparison to what loomed in the strategic arena—arresting the ongoing decline in U.S. nuclear power through a concerted modernization of the strategic deterrent.

By the early 1970s, the Joint Chiefs of Staff agreed that bolstering strategic forces could no longer wait. Decisions taken in the mid-1960s at McNamara's instigation to freeze the number of launchers in the U.S. nuclear arsenal and President Nixon's acceptance of "strategic equivalence" with the Soviet Union in strategic forces, all the while negotiating arms control accords, had unsettling effects on JCS assessments of the military balance. Worried that the Soviets were on the verge of achieving a decisive advantage, the Joint Chiefs continued to look at a broad range of improvements to bolster the U.S. strategic posture. They realized that these improvements were unlikely to restore the strategic superiority the United States had previously enjoyed. But without them, the chiefs were skeptical of their ability to preserve effective deterrence or stability in future crises.

A tenuous consensus had emerged among the JCS in support of three new strategic systems by the early 1970s—the B–1 strategic bomber, the Trident fleet ballistic missile submarine, and the MX, a third generation ICBM. The oldest of the three, the B–1, dated unofficially from 1961 when the Air Force began exploring alternatives to the cancelled B–70. By the mid–1960s, the project had evolved into a formal request for a supersonic (Mach 2) low–level penetration bomber which Air Force Chief of Staff General John P. McConnell labeled "the top priority program within the Air Force" at the time.[6] Designated to replace older B–52 models, the proposed new plane (then known as the advanced manned strategic aircraft, or AMSA) encountered stiff resistance from McNamara and his civilian advisors, who considered manned strategic aircraft obsolete and less cost-effective than missiles.[7] In place of the AMSA, McNamara insisted that the Air Force make do with the F–111, a medium-range fighter-bomber with limited capabilities. Try as he might, however, McNamara was never able to kill the AMSA, which remained alive as a drawing board concept owing to the combined support of the Air Force and key members of Congress. Weighing the pros and cons, neither the Army, Navy, nor Marines saw an urgent need for the AMSA. All wanted closer study before going into production. But like the ABM issue, they endorsed the AMSA program seemingly in defiance of McNamara, as much as anything, and out of frustration over his persistent refusal to pay attention to military advice and to authorize new systems.[8]

With the advent of the Nixon administration, the AMSA became the B–1 and the Air Force received authorization to develop several prototypes for testing. If all went well, the JCS expected an initial operational capability (IOC) in FY78. Under the division of labor in effect at the time, Secretary of Defense Melvin Laird concentrated on Congress and Vietnam, while his deputy, David Packard (cofounder of the computer giant Hewlett-Packard), looked after procurement and administration. In an effort to control costs, Packard adopted a "fly-before-you-buy" acquisition policy which required hardware demonstrations of new weapons at predetermined intervals before the Defense Department would commit to full-scale production and procurement. For planning purposes, the Air Force estimated an eventual force of 241 planes, but could not guarantee the prime contractor, Rockwell International, that the government would purchase that many aircraft owing to the fly-before-you-buy requirement. A complex plane with state-of-the-art electronics and avionics, the B–1 was an expensive undertaking to begin with and became even more so as the project gathered momentum.⁹ With the Vietnam War winding down and money again becoming tight, pressure was growing for the JCS to take a more critical look at the B–1 and other new weapons.

Like the B–1, the Trident program faced chronic criticism and money troubles. Originally known as the undersea long-range missile system (ULMS), Trident was an outgrowth of the Strat-X study, an effort organized by Secretary of Defense McNamara in the mid-1960s through the Institute for Defense Analyses (IDA) to explore alternative strategic systems of the future. Treated as a follow-on to the Polaris and Poseidon programs, the original ULMS design was for a slow-moving underwater platform carrying up to 24 long-range missiles. To stay within McNamara's cost-effectiveness criteria, the Navy's Special Projects Office proposed using extended range Poseidon missiles and an existing nuclear power plant, but ran afoul of Vice Admiral Hyman G. Rickover, head of the Navy's nuclear propulsion program, who insisted on a new reactor system. By 1970, costs had escalated dramatically as requirements became more sophisticated and as the size of the boat grew to more than twice that of a Polaris submarine. A source of controversy within the Navy, the ULMS project (renamed Trident in May 1972) soon attracted widespread congressional attention as well and became a favored object of attack by Capitol Hill liberals, who considered it a wasteful and redundant drain on resources that could be better spent on other projects.

To distinguish Trident from other submarines and to increase its appeal, the Navy proposed to equip it with two new missiles. Initially, Trident boats would carry the C4 missile (also known as Trident I), virtually identical in size to the Poseidon missile but with up to twice the range. For boats going to sea in the mid- to late 1970s, the Navy proposed to deploy the D5 (Trident II) which would have the range, payload, and accuracy approximating a land-based ICBM, giving Trident a counterforce potential to threaten the highest priority enemy targets. Until then, to avoid charges of duplicating Air Force functions, the Navy had eschewed the development of sea-based missiles that could effectively attack military facilities other than Soviet submarine pens and similar "soft" targets. With Trident, the Navy would be moving into a new realm of military strategy by acquiring a true counterforce capability for the first time, one less vulnerable than the Air Force's ICBMs but no less effective.¹⁰

Even though the JCS agreed that Trident had unique potential, opinions differed on taking the next step and putting it into production. A majority of the Joint Chiefs—the CNO, the CMC, and the CJCS, Admiral Moorer—saw no reason to hesitate and wanted boats in the water by the mid to late 1970s. In contrast, the CSA and the CSAF, citing the uncertainties of the program and the Nixon administration's determination to negotiate arms control accords, adopted a wait-and-see attitude and urged that Trident be limited to the R&D phase for the time being.[11] Secretary of Defense Laird initially sided with the Army and Air Force, and in September 1971 he issued a formal public statement indicating that design studies and other work on a new missile submarine would proceed at a measured pace, with a production decision held in abeyance. But under pressure from the White House, he reversed course almost immediately and agreed to accelerate the Trident program, with a view toward strengthening the U.S. negotiating position in SALT and blunting possible conservative opposition in Congress to an arms control agreement.[12]

If Trident thus seemed headed for production and deployment, the same could not be said for the MX, the Air Force's proposed new state-of-the-art ICBM, which ran into one niggling problem after another. Like the B–1, the MX reflected the Air Force's annoyance with McNamara for blocking new programs and for refusing to countenance a strategic posture with predominantly counterforce capabilities. Emerging from design studies done in the mid-1960s, the MX (known at that time as the Advanced ICBM, or AICBM) grew directly out of the Air Force's desire for a weapon that would be larger, more powerful, and more accurate than the Minuteman, with an initial operational capability by the early to mid-1970s. Design specifications stipulated that it should be able to lift a payload of 7,000 pounds and have a range of 6,500 nautical miles and a circular error probable (CEP) of .2 nautical miles. A formidable undertaking in and of itself, the development of such a missile proved to be less of an obstacle than finding a survivable, politically plausible basing mode, an issue that would dog the MX throughout its checkered history and delay its deployment for more than a decade.[13]

During the Nixon administration, the MX had joined the B–1 and Trident as a staple in the Joint Chiefs' inventory of future weapons systems in the JSOP.[14] Even so, assessments of the missile's importance and ultimate role in the strategic arsenal varied from Service to Service. Least enthusiastic of all was the Navy, which saw the MX competing directly with Trident for funds and mission. At issue was whether the United States needed, and could afford, two new strategic systems performing roughly the same functions.[15] To observers with long memories, the situation was analogous to the competition between the Air Force and the Navy during the carrier-B–29 controversy in the late 1940s. In this instance, however, the Navy had the edge with a more versatile weapons system. Perhaps with Louis Johnson's untoward experience in mind, Secretary of Defense Laird and his immediate successors made no attempt to adjudicate the dispute and instead adopted the course of least resistance by allowing both programs to go forward simultaneously, reserving judgment on their relative merits for later. A temporizing approach, this solution avoided what could have been an ugly inter-Service battle. Yet it also left important decisions dangling with steadily diminishing prospect of ever finding a clear resolution acceptable to all involved.

THE SEARCH FOR STRATEGIC STABILITY

TARGETING DOCTRINE REVISED

As the competition between Trident and the MX heated up, it boiled over into two other areas—arms control and strategic targeting. A moderate-to-low priority since the Kennedy administration tried with limited success to introduce greater flexibility in the early 1960s, targeting doctrine emerged during the Nixon years to become the source of renewed interest and controversy. Shortly after taking office, Nixon and Kissinger visited the Pentagon and received their first formal briefing on the Single Integrated Operational Plan (SIOP) then in effect detailing programmed attacks against the Sino-Soviet bloc in the event of a general war. According to published accounts, Nixon was "appalled" by the high levels of death and destruction that a nuclear exchange would cause and by the corresponding lack of flexibility in the SIOP to limit and control attacks. Seeking a remedy, Kissinger secured the President's approval in the summer of 1969 for a reexamination of targeting practices "to meet contingencies other than all-out nuclear challenge."[16]

Several factors reinforced Kissinger's concern that targeting policy needed reform. One was the inexorable increase during the 1960s in Soviet strategic nuclear power, which had gone beyond what most intelligence analysts had predicted. Once the Soviets reached strategic parity with the United States, Kissinger believed, the concept of assured destruction was less likely to deter and the Soviets might be tempted to launch a less than full-scale nuclear attack against the West. The results might not be incapacitating, but without the ability to respond in kind, the President's only practical choice under the existing SIOP would be a suicidal act of all-out destruction—something Kissinger felt no sane individual would seriously countenance. Ever since the revisions introduced under Kennedy and McNamara's subsequent institutionalization of the assured destruction concept, the Joint Chiefs had held the line on all but piecemeal changes to the SIOP.[17] Now, Kissinger argued, the time had come to think in more flexible and creative terms, where nuclear war "is more likely to be limited" and "smaller packages will be used to avoid going to larger one[s]."[18]

The outcome of the ensuing inquiry—NSDM 242—was nearly 5 years in the making. Part of the explanation for why the project took so long was the continuing lack of urgency associated with targeting policy, compared with the immediate demands of other issues such as SALT and Vietnam. Also, there was a widely shared reluctance on the part of JCS planners to grant civilians (other than the President and the Secretary of Defense) access to the inner workings of strategic nuclear war plans and the process by which they were formulated. Highly classified, these plans were rarely discussed outside a restricted circle of uniformed strategic planners who scoffed at the notion that all they had to do was push a button to alter a plan. Initiating even limited changes in the SIOP was a time-consuming and complex process. To be sure, with the pending introduction of more sophisticated weapons systems like the B-1, the MX, and Trident, and ongoing improvements to command and control capabilities, the amount of time and effort needed to amend a plan and reprogram forces was shortening. But it was still an onerous, difficult, and sensitive technical process that JCS planners guarded with utmost care.

The Joint Chiefs' uneasiness over the whole question of strategic nuclear targeting was further exacerbated by difficulties in determining what Kissinger and the President hoped to accomplish. Even if the United States exercised restraint in launching nuclear attacks, there was no assurance the Soviets would respond in a similar fashion. On the contrary, JCS targeting planners operated on the assumption that any use by the West of strategic nuclear weapons, even in a limited capacity, was almost certain to elicit a wholesale nuclear response from the Soviet Union.[19] At various points during the deliberations surrounding NSDM 242, Kissinger asked the Joint Chiefs for examples of how limited strategic nuclear power might be applied. But according to David Aaron, who served on the NSC Staff, Kissinger rejected every JCS response. Either the proposed uses were excessive, in Kissinger's opinion, or too limited to convey a clear message and serve a constructive purpose.[20]

NSDM 242 had its origins in an intradepartmental study initiated at the Pentagon under the supervision of John S. Foster, Jr., the long-time Director of Defense Research and Engineering (DDR&E) and a highly respected figure among military planners. Secretary of Defense Laird had become worried that unless the Defense Department took a firm hand in the matter, Kissinger might unilaterally produce a new targeting directive. Accordingly, in January 1972, Laird gave Foster practically carte blanche to review targeting practices and to explore the feasibility of a more "flexible range of strategic options." While the Chairman of the Joint Chiefs, Admiral Moorer, was a designated member of Foster's study panel, the Director of the Joint Staff usually served in his stead. Until then the Joint Chiefs had done their best to discourage a reworking of targeting doctrine. But with an array of new strategic weapons awaiting the nod for production, they were hard pressed not to cooperate without acknowledging that the new arsenal they wanted would be no better or more versatile than the old.[21]

Throughout the review process, the Joint Chiefs and Foster's task force carried on a brisk exchange of opinions and ideas. Not since the preparation of the first SIOP in 1960 had the JCS played such an active role in shaping targeting doctrine. Drawing on advice from the JCS, the Director of the Joint Strategic Target Planning Staff, and others, Foster and his colleagues came up with an extensive, but not fundamental, reworking of targeting guidance, which it submitted to the Secretary of Defense in tentative form in May 1972. In late July, Foster briefed Kissinger and members of the NSC on the panel's findings, which one NSC Staffer characterized as a "radical departure from the current policy."²² A more accurate description would have been the reaffirmation of assured destruction under conditions of controlled escalation. A final report, reflecting further inputs from the Office of the Secretary of Defense and the Joint Chiefs of Staff followed, and Secretary of Defense Laird forwarded it to President Nixon in December. As Laird described it to the President, the purpose behind the proposed changes in targeting doctrine was to satisfy "your expressed desire for useable nuclear options other than mass destruction, and the needs of our basic strategy of realistic deterrence."²³

On the basis of the Defense Department's report, Kissinger moved the targeting review up a notch to the interagency level in February 1973. Again, Foster took charge of the effort.²⁴ Though it made minor alterations and additions, the interagency panel essentially concurred in the Pentagon's findings, and by the summer of 1973 a draft Presidential directive had emerged. Approved by President Nixon the following January, NSDM 242 reaffirmed that the assured destruction concept remained basic U.S. strategic doctrine, but with modifications in targeting practices that interjected a greater degree of flexibility into attack plans. The principal innovation was the requirement for "limited employment options" that would enable the United States "to conduct selected nuclear operations, in concert with conventional forces, which protect vital U.S. interests and limit enemy capabilities to continue aggression." Should these limited attacks fail to deter the Soviets from further military action, the United States might then launch large-scale attacks against the Soviet Union that would limit damage to the United States and its allies and cripple enemy recovery for years to come, a concept known as "counter-recovery" targeting.²⁵

Translating this guidance into a working doctrine fell mainly to the new Secretary of Defense, James R. Schlesinger, who, as an analyst at the RAND Corporation in the 1960s, had been involved in critiquing the old strategy. Since then, having served as director for national security affairs at the Bureau of the Budget, as Chairman of the Atomic Energy Commission, and as Director of Central Intelligence, Schlesinger had come to certain conclusions on his own about what constituted effective deterrence. Sworn in as Secretary of Defense in July 1973, he took charge at the Pentagon too late to have an impact on the content of NSDM 242, but just in time to interpret how the directive ought to be applied. To Kissinger's chagrin, it was Schlesinger's name, not his, that came to be associated with the new strategy.

The public unveiling of the "Schlesinger doctrine" occurred on January 10, 1974, during a question-and-answer period before the Overseas Writers Association in Washington, DC. Though it had been an open secret for months that the administration was conducting a targeting review, Schlesinger's comments were the first official confirmation. The United States, he said, had decided to amend the assured destruction concept and embrace, on a selective basis, attacks against "certain classes" of Soviet military installations. Missile silos and airfields were among those he specifically mentioned. Realizing that this was an exceedingly sensitive issue, he added that he was speaking "hypothetically" and repeatedly stated that the United States had no intention of using such attacks to attempt a disarming first strike. Rather, the intention would be to convince the other side that the United States was bent on protecting its interests without necessarily resorting to all-out nuclear war. While outwardly similar to the counterforce/no-cities doctrine that McNamara had unsuccessfully pushed 12 years earlier, Schlesinger's approach was more discriminating and restricted, keeping counterforce targeting within reach of current and projected JCS capabilities. Insisting that this was not a fundamental departure from current targeting practices, Schlesinger also affirmed that sufficient forces would be held in reserve to achieve assured destruction goals, should the conflict escalate. But if the United States could achieve its aims without going that far, so much the better.[26]

Reactions to the Schlesinger doctrine were mixed. While some strategic theorists proclaimed it potentially destabilizing to the new era of "mutual" assured destruction, or MAD, that the SALT I agreements had ushered in, others reserved judgment.[27] A key consideration that contributed to muting criticism was Schlesinger's caution and obvious reluctance to use the new strategy as justification for expensive new weapons or other requirements. The Foster Panel had looked into that question but had refrained from making detailed recommendations because it did not believe that weapon systems acquisition policy could be formulated solely or even primarily on the basis of employment policy. Secretary Schlesinger drew a similar distinction. In assessing requirements, he acknowledged the eventual need for the B–1 and the MX, but saw no urgency in proceeding with the acquisition of either pending the resolution of technical problems. Until then, he favored keeping both programs in an advanced state of testing and development. Instead of rushing to deploy new land-based delivery systems, he stressed modest improvements in existing Air Force capabilities—a higher yield and more accurate MIRVed reentry vehicle (the Mark 12A) for the Minuteman III, and two more powerful and sophisticated thermonuclear bombs (the B–61 and the B–77) carried aboard B–52s. At the same time, part of the Poseidon fleet would be fitted with C4 (Trident I) missiles to improve their range and effectiveness. The only new system he envisioned playing a key role under the recently adopted strategy was Trident—first, because it was farther along than either the MX or B–1, and second, because it combined a potential counterforce capability with relative invulnerability.[28]

All in all, the targeting review leading to adoption of the Schlesinger doctrine probably came out better for the Joint Chiefs than they initially expected. While laying down new targeting priorities, it generally reinforced their preferences, especially in the counterforce category, and provided a strong rationale for completing the strategic modernization program. What it failed to do was establish a specific link between the need for the B–1 and the MX, on the one hand, and on the other, the execution of tasks delineated in NSDM 242, including the additional functions entailed in carrying out limited options. Only Trident emerged with a definite mandate to proceed under the new targeting scheme. But with the foundations thus laid, the chiefs could be reasonably confident that if they continued to press their case, sooner or later resources would catch up with the changes in employment policy.

SALT II BEGINS

Like the targeting review leading to adoption of the Schlesinger doctrine, arms control negotiations figured prominently in the post-Vietnam debate over U.S. strategic modernization. The JCS position was that with or without arms control, modernization should go forward to stay abreast of increases and improvements in Soviet capabilities. But in the wake of SALT I, there was considerable caution, both at the White House and on Capitol Hill, about pressing ahead with new strategic weapons that might poison the atmosphere of future negotiations and provoke, in Kissinger's words, "an explosion of technology and an explosion of numbers" in delivery vehicles.[29] Not everyone agreed that slowing down or postponing modernization was a wise move, certainly not the Joint Chiefs of Staff and certainly not Democratic Senator Henry M. Jackson of Washington, who had done as much as anyone to draw attention to the imperfections of the SALT I accords. But from the momentum generated by the earlier talks, there was growing optimism for the prospects of SALT II and a corresponding reluctance to jeopardize those negotiations with hasty spending on new weapons.

The Soviets were less reticent about their programs. Though eager for SALT II, they were not about to let it get in the way of efforts to bolster their strategic forces, an ongoing process since the mid-1960s. While SALT I had "frozen" long-range offensive launchers (ICBMs and SLBMs) at existing levels, it had left both sides more or less free to replace those weapons with newer models and to conduct research and development as needed. During 1973, with the ink on the SALT I accords barely dry, the Soviets began testing four new ICBMs, three with MIRV capability. All had new guidance and reentry systems, making them more accurate and lethal than the missiles they were slated to supersede. According to intelligence sources, the impetus behind developing these new weapons was "almost certainly . . . a desire for improved ability to strike at U.S. strategic forces—a factor long stressed in Soviet strategic doctrine."[30] The disclosure that the United States might be moving in the same direction under the Schlesinger doctrine—toward an enhanced counterforce capability—met with typically sharp criticism and stern warnings from the Kremlin, which accused the United States of jeopardizing the strategic balance and endangering arms control. What the Soviets conveniently overlooked was that the United States was taking its time in upgrading its capabilities and had categorically ruled out trying to regain strategic superiority or to acquire a disarming first-strike capability.[31]

Begun under Nixon's Presidency in December 1972, SALT II stretched over two subsequent administrations and was supposed to provide a permanent replacement for the temporary SALT I interim agreement on offensive arms. Instead, it yielded only a limited-duration treaty that the United States never ratified. Shortly after the negotiations began (now conducted on a permanent basis from Geneva), Senator Jackson insisted that the Joint Chiefs replace Lieutenant General Royal B. Allison, USAF, as their representative to SALT. His successor, appointed in March 1973, was Lieutenant General Edward L. Rowny, USA. Insisting that Allison had been ineffectual, Jackson wanted someone with tougher negotiating instincts and "dragooned" Rowny, a personal friend, into the job. A West Point graduate with additional degrees from the Johns Hopkins University, Yale, and American University, Rowny had commanded troops in World War II, the Korean War, and Vietnam and had served as a nuclear planner at NATO. His friendship with Senator Jackson dated from the 1950s, when Rowny was assigned to the Infantry School at Fort Benning, Georgia, and Jackson, then a Congressman, was doing his 2-week obligated tour of duty as an Army Reservist.[32]

At the time of his appointment to SALT, Rowny was deputy chairman of the NATO Military Committee, in charge of organizing the Mutual and Balanced Force Reduction (MBFR) Talks. Rowny was personally skeptical whether SALT would ever accomplish much and would have preferred to remain with the MBFR negotiations where he saw more opportunities, both for an agreement and for career advancement. He distrusted Kissinger, who returned the sentiment by lumping Rowny in the category of the "undisputed hawks."[33] Leery of the Soviets as well, Rowny became even more so the longer he was associated with SALT and the more contact he had with them at the negotiating table.

Rowny's appointment was only one of several key personnel changes that affected the JCS role in SALT II. Though not directly engaged in the negotiations, the Joint Chiefs were part of a large and complex arms control "community" in Washington that had grown up over time to develop and assess proposals, evaluate verification measures, and monitor the progress of the talks.[34] In keeping with the pattern of JCS involvement in other areas of national policy, the Service chiefs looked to the Chairman to handle the day-to-day chores connected with SALT, arrange interagency representation, and convey their views to the appropriate authorities. In other words, arms control work was increasingly concentrated around the Chairman.

With the departure of Admiral Moorer in July 1974, the Chairmanship fell for the first time in nearly a decade and a half to an Air Force officer, General George S. Brown. A bomber pilot in Europe in World War II, Brown's career had been a succession of high-profile command and staff jobs that led him steadily up the ladder to become Chief of Staff of the Air Force in 1973. Though he stayed in that job only a year before Nixon appointed him CJCS, he established himself as a strong proponent of the B–1 and other Air Force interests. As Chairman, he continued to champion the plane, terming it "a virtually indispensable element of our deterrent force."[35] At the same time, he adopted a cautious outlook on arms control and relied heavily on Rowny (a friend from their days at West Point) to help shape JCS positions on SALT.

The Joint Staff acquired a fresh look under Brown. Responding to budget cuts and criticism growing out of the Vietnam War that the JCS organization was inefficient and ineffective, Brown decided to streamline the Joint Staff by abolishing two directorates, Personnel (J-1) and Communications-Electronics (J-6).[36] As part of a Defense-wide effort to reduce costs, he also cut extraneous Joint Staff billets in line with a targeted 25 percent personnel reduction in the OSD-JCS headquarters staff, and supported the consolidation of analytical functions, a process that included the dissolution of the Weapons Systems Evaluation Group (WSEG). Created in 1949 to provide analytical support for the Joint Chiefs, WSEG had grown increasingly independent of and less useful to the JCS. By the mid-1970s, about three-quarters of its work was for non-JCS interests. Ordered abolished by the Secretary of Defense in March 1976, most of WSEG's ongoing projects for the Joint Chiefs transferred directly to the Studies, Analysis, and Gaming Agency (SAGA), a JCS in-house analytical body that operated in conjunction with but separately from the Joint Staff.[37]

Around the same time that General Brown became Chairman, the Joint Chiefs acquired three other new members, making it the most extensive turnover in JCS membership since the end of World War II. Brown's successor as Chief of Staff of the Air Force was General David C. Jones, a B–29 bomber pilot during the Korean War and former aide to Curtis E. LeMay. With Zumwalt entering retirement, Admiral James L. Holloway III, a highly decorated aviator, became Chief of Naval Operations. Finally, in October 1974, General Fred C. Weyand became Army Chief of Staff, succeeding General Creighton W. Abrams, who had died in office the month before. The only holdover was the Commandant of the Marine Corps, General Robert E. Cushman, Jr., a veteran of three wars and one time deputy director at the CIA.

The most dramatic personnel change was at the White House. On August 9, 1974, barely a month after Brown's appointment as Chairman, Nixon finally succumbed to the pressures of the growing Watergate scandal and relinquished the Presidency to Gerald R. Ford, a former Republican Congressman from Michigan. Appointed Vice President the previous October following Spiro Agnew's ignominious resignation, Ford had little experience in defense and foreign affairs. To maintain continuity, he turned to Kissinger, who was then serving as both Secretary of State and National Security Advisor. "Henry," he said, "I need you.... I'll do everything I can to work with you."[38] As a result, the NSC, with its elaborate structure of committees and support groups, all either chaired or overseen by Kissinger to afford the President and his national security assistant maximum control, remained the focal point of interdepartmental deliberations and decisionmaking. Normally, the Joint Chiefs would have welcomed the retention and reaffirmation of what was outwardly a carefully structured and predictable policy environment. But after the discovery of Kissinger's backchannel negotiations with Dobrynin during SALT I, there was a growing awareness at the Pentagon that formal policy mechanisms might not count for much since Kissinger seemed inclined to circumvent them whenever it suited his purpose.

If the Joint Chiefs were by then deeply suspicious of Kissinger, their immediate boss, Secretary of Defense Schlesinger, was even more so. Indeed, not since the days of Louis Johnson and Dean Acheson had a Secretary of Defense and a Secretary of State been more at odds. Following a custom adopted during Laird's tenure, Schlesinger and Kissinger met regularly for breakfast to discuss common problems and to try to narrow their differences. Rarely were they totally successful. As Kissinger described it, the two became locked in a "personal rivalry" that amounted to "an old-fashioned struggle for turf."[39] According to Zumwalt, their differences went deeper and amounted to an intellectual tug-of-war. "In Jim Schlesinger," he claimed, "Henry Kissinger met his superior as a strategic theorist. But since Henry is a superior bureaucrat, he was able to impose his policy positions on Jim most of the time."[40]

VLADIVOSTOK

It was against this background of rivalries, feuds, intrigue, and turf wars that the new Ford administration attempted to carry forward the work begun by its predecessor in shaping a SALT II treaty. Realizing that they had been overly reticent in expressing their views at the outset of SALT I, the Joint Chiefs resolved that in SALT II they would take a more active and prominent role in shaping U.S. policy. All the same, they were in no rush to conclude an agreement and generally worked closely with Schlesinger and his staff to develop common OSD-JCS positions that would give the Pentagon more unity and better leverage in dealing with Kissinger and the White House. According to Admiral Zumwalt, JCS members further sought to strengthen their position by establishing "backchannel" contacts with Senator Jackson and others in Congress who were sympathetic to military views.[41]

The most critical stumbling block in SALT II was the limitation of multiple independently targetable reentry vehicles (MIRVs), a subject that SALT I had ignored. As SALT II began, Kissinger wanted to constrain MIRV deployment by limiting ICBM throw-weight, but could not convince the Joint Chiefs that such arrangements were sound or workable. Arguing that Kissinger's approach would be too hard to verify, the Joint Chiefs favored equality ("equal aggregates") in numbers of delivery vehicles—missiles and heavy bombers—with each side free to MIRV its missiles to the extent it saw fit. To keep MIRV deployment contained, the Joint Chiefs suggested a maximum of around two thousand strategic delivery vehicles on each side. Actually, the JCS position came closer to that proposed by the Soviets than Kissinger's, but would have required cuts in the number of Soviet launchers to bring them into compliance with the U.S. ceiling, something Moscow was initially loath to accept. In an attempt to bridge differences at home and make the American position more palatable to the Soviets, President Nixon in February 1974 approved a new negotiating offer (NSDM 245) calling for equal overall aggregates (2,350 ICBMs, SLBMs, and bombers) and equal ICBM MIRV throw-weight.[42]

Despite the new offer, the talks remained deadlocked, needing something imaginative or dramatic to break the impasse. By the spring of 1974, with the Watergate affair bearing down on Nixon more heavily than ever, the Soviets lost confidence in the President's capacity to lead and for all practical purposes suspended serious negotiations.[43] Efforts by Kissinger to jump-start the talks during a visit to Moscow in March 1974 came up empty.[44] Desperate for a SALT II deal to help resuscitate his reputation and to stave off impeachment, Nixon began exploring further concessions. At the Pentagon there were growing suspicions that the President's judgment had become clouded and that his behavior was suspect. Attempting to make Schlesinger and the Joint Chiefs his scapegoats, Nixon accused them of intentionally sabotaging détente by taking "an unyielding hard line against any SALT II agreement that did not ensure an overwhelming American advantage" in offensive strategic power.[45] The charge was patently untrue and unfair. But it put Schlesinger and the Joint Chiefs on the defensive. They had to justify themselves anew when the Ford administration took over.

Under Ford, Kissinger quickly solidified his position as the President's closest advisor, while Schlesinger and the JCS suffered repeated setbacks that reduced them to marginal roles. Ford had the utmost respect for military power and was inclined to grant the Defense Department modest increases in its budget, the first in several years. But he struggled to mobilize support for the idea after the JCS Chairman, General Brown, delivered a tirade against "Jewish bankers" during a seminar at Duke University in October 1974. A gross indiscretion, Brown's remarks came at an especially inopportune time when the United States was trying to engage Israel and the Arab states in peace talks and as the new administration sought to establish a working relationship with Congress. Furious condemnations of the Chairman's behavior followed promptly from Capitol Hill. Brown apologized for the gaffe and insisted to friends that he was in no way anti-Semitic, as critics claimed. But his comments remained an embarrassment that reflected poorly on the JCS and the military in general.[46]

The most visible evidence of the Joint Chiefs' limited influence was their exclusion from the Vladivostok mini-summit between Ford and Brezhnev in late November 1974. Hurriedly arranged by Kissinger, the summit's purpose was to breathe new life into the practically moribund SALT II negotiations. Since the agenda at Vladivostok was heavily weighted toward military issues, it would have made sense for the White House to include JCS representation in its party. But apparently there was no room on the plane, even though 140 other people accompanied the President.[47] In preparation for the meeting, Schlesinger and the Joint Chiefs urged Ford not to be hasty but to hold out for equal aggregates. Rather than risk the talks breaking down, Kissinger made a pre-summit trip to Moscow, where he and Soviet Foreign Minister Gromyko worked a deal.[48] What emerged at Vladivostok was a numerical-parity formula that imposed an overall ceiling of 2,400 on strategic launchers, giving the appearance of strategic equality (as mandated by Congress), and a sub-limit of 1,320 on the number of MIRVed vehicles. The net effect was to reconfirm the status quo by allowing the Soviets to retain their lead in ICBMs and the United States to keep its relative advantage in SLBMs and bombers. But since the Joint Chiefs had no plans to build up to the allowed numbers under the Vladivostok formula, the only side that stood to gain was the Soviet Union.[49]

While there was probably not much that the chiefs' presence at Vladivostok could have done to change the overall outcome, it might have helped avoid later controversy over two issues—cruise missiles and the Soviet "Backfire" bomber. Experiencing a revival, the U.S. cruise missiles under development in the 1970s were updated versions of a technology dating from the German V-1 "buzz bomb" of World War II. Equipped with exceedingly precise guidance systems, the new cruise missiles could fly at low altitudes, carry either a conventional or nuclear warhead, and penetrate existing radar nets virtually at will. While the precise mission of these weapons had yet to be defined, the operating assumption in R&D circles was that they could have both tactical and strategic uses. The Soviets also had cruise missiles, but had not as yet shown any interest beyond tactical applications.[50]

The Soviets knew that one of the variants being developed by the U.S. Air Force was an air-launched cruise missile (ALCM) for deployment aboard B–52s, and at Vladivostok they sought to curb the program indirectly by proposing range limitations on air-to-surface missiles. The Joint Chiefs of Staff opposed range constraints on cruise missiles, but with no representative present during the talks, they were unable to advise on how to address the issue. Later, while briefing Congress, Kissinger insisted that there had been no agreement to limit the range of ALCMs and that only *ballistic* missiles were affected. The Soviets, however, disagreed, setting off a dispute that lasted for years.[51]

The most serious faux pas committed at Vladivostok that the chiefs' presence might have avoided was the decision to treat the new Soviet Backfire bomber as an intermediate range weapon and not as a strategic one. While there were few details known about the plane in the West, the Joint Chiefs expected it to be deployed in significant numbers within a few years and were convinced from its general design and performance characteristics that it was fully capable of intercontinental missions.[52] The Soviets, however, wanted the Backfire to be accorded the status of an intermediate range bomber, a designation Kissinger saw no reason not to accept.[53] In exchange, Brezhnev offered at Vladivostok to drop previous Soviet demands to bring French and British nuclear forces and U.S. forward-based systems in Europe and the Far East under SALT counting rules. At Kissinger's urging, Ford accepted the tradeoff Brezhnev proposed, only to discover upon his return to Washington that the Joint Chiefs and others thought the Backfire decision had been ill-advised.[54]

Despite imperfections, the Vladivostok accords received a generally favorable reception in the United States. Among those offering their endorsements, albeit somewhat grudgingly, were Secretary of Defense Schlesinger and the Joint Chiefs of Staff. Others, like Senator Henry Jackson, would have preferred lower numerical ceilings. But by and large public and congressional opinion welcomed the agreements as a major step toward curbing the arms race. In January and February 1975, both houses of Congress passed resolutions endorsing the Vladivostok accords. SALT II was back in business.

MARKING TIME

Based on the outcome at Vladivostok, the Ford administration estimated that it would be only a few months before a SALT II treaty materialized. In fact, negotiations dragged on for 4 more years. Part of the problem was the lack of formal or authoritative minutes of the decisions taken at Vladivostok. The "official record" comprised a broadly worded joint press release handed out at the end of the conference, a subsequent aide-mémoire, and the conflicting recollections of the participants.[55] Trying to sort out what had been decided at Vladivostok proved beyond the capacity of the negotiators in Geneva. By the summer of 1975, it was clear that the talks were for all intents and purposes again at an impasse and that key provisions of the Vladivostok agreement needed to be renegotiated.[56]

At the same time, the Soviets showed no sign of being in a hurry to conclude a treaty and seemed content to mark time. Many in Moscow, including some of Brezhnev's top military advisors, thought the General Secretary had been too accommodating at Vladivostok by making needless concessions to the Americans. Seeing the United States as a spent force with its power in decline, they argued that Brezhnev should have held out for better terms. According to one account, Brezhnev had to force his defense minister, Marshal Andrei Grechko, to "eat the Vladivostok agreement." Even though Brezhnev's views prevailed, he remained under intense personal and political pressure, and on the trip home from the Far East he suffered a stroke. Brezhnev recovered and resumed his duties in a short while, but his health deteriorated from that point on, and he was less and less able to keep the hard liners in check.[57]

If waning American military power was apparent to the Soviets, it was even more visible to the Joint Chiefs of Staff. In his annual posture statement summarizing the situation at the outset of 1975, General Brown characterized the U.S.-Soviet military balance as being in a state of "unstable equilibrium." Decisions made earlier by Moscow and programs already in progress, he warned, "display massive momentum toward significant force increase and modernization." In contrast, the United States, with its "modest programs," was barely keeping up. Mindful of the President's injunction against openly criticizing the Vladivostok accords, Brown acknowledged the agreement as a stabilizing influence, but pointed out that arms control by itself was no guarantee of security. "Arms control is a means, not an objective," he argued. "The objective is peace."[58]

Without a stronger defense commitment from Congress and the White House, however, the Joint Chiefs saw little chance of turning the situation around. Certainly the most stunning evidence of U.S. decline was the collapse of South Vietnam in the spring of 1975. In early April, with the North Vietnamese offensive in full swing, President Ford sent General Fred C. Weyand, former COMUSMACV and now Army Chief of Staff, to Saigon on a fact-finding mission. Based on what he saw, Weyand returned to Washington convinced that the South Vietnamese were "on the brink of total military defeat," a view shared by Schlesinger, Brown, and senior members of the Intelligence Community.[59] Refusing to give up, however, Weyand recommended immediate emergency assistance to the South Vietnamese totaling over $700 million in military aid. A face-saving gesture at best, Weyand's proposal received grudging approval from the White House but fell on deaf ears when it reached Congress.[60] South Vietnamese resistance collapsed shortly thereafter, and within a few years bases like the sprawling facility at Cam Ranh Bay that had once played host to American forces were being used by the Soviets to project their air and naval power into the Western Pacific and Indian Oceans.

South Vietnam's demise ushered in a progressive erosion of U.S. power and influence across the Third World. Seizing on American weakness, Moscow launched vigorous efforts to restore its position in the Middle East and the Arab world by shoring up ties with Syria, Iraq, and Yemen, establishing close relations with Libya, and stepping up covert assistance to Palestinian terrorist groups.[61] In Somalia and south of the Sahara, the Soviets made further inroads. Almost as soon as the Portuguese empire collapsed, Soviet advisors and thousands of Cuban military "volunteers" began arriving in Angola and Mozambique to help prop up Marxist regimes. A decade and a half earlier, in 1960–1962, when Communist influence threatened to overtake the Congo, the Joint Chiefs had favored strong countermeasures, including military intervention if necessary. But by the mid-1970s, with the experience of Vietnam behind them, they were far more cautious and reserved and generally urged diplomacy and covert operations to counter Soviet moves rather than direct military action.

An exception to this pattern was the *Mayaguez* affair in May 1975 involving the seizure of a U.S. cargo ship by the Khmer Rouge, who had taken control of Cambodia about the same time South Vietnam collapsed. As news of the capture of the *Mayaguez* reached Washington, it brought back memories of the 1968 *Pueblo* incident when the United States had done nothing more than vent its "outrage" at North Korea's seizure of one of its spy ships and, later, offer an abject apology to secure the crew's release. Resolving not to be put in a similar position, President Ford took a tough line from the beginning and wound up authorizing military action to take back the ship and its crew.

As the debate over what to do unfolded, it became a test of wills between Kissinger and Schlesinger, with the Joint Chiefs caught in the middle. Frustrated by the recent setback in Vietnam, Kissinger encouraged Ford to believe that only a strong show of force would suffice, while Schlesinger adopted a wait-and-see attitude. Schlesinger knew that the Khmer Rouge had detained ships sailing near the Cambodian coast on previous occasions and usually released them without incident within a day or so. So it stood to reason that sooner or later they would let the *Mayaguez* go free. Kissinger, however, disagreed and in making his case convinced Ford that this was too serious a provocation to go unpunished.[62]

Despite the *Pueblo* incident, the Joint Chiefs had no contingency plans for such situations and had to improvise by relying on a hastily assembled operational concept prepared under the supervision of Admiral Noel Gayler, commander in chief of the Pacific theater. Among the options on the table for putting pressure on the Cambodians were air attacks from Navy carriers, punitive raids using B–52s, and the massing of a surface naval force off the Cambodian coast. Eventually, drawing on Gayler's inputs, the JCS recommended, and President Ford approved, a more limited operation involving a rescue party of several hundred Marines backed by tactical air. While one party of Marines boarded and secured the ship, the others would land on a small island, Koh Tang, just off the Cambodian coast, where the crew was thought to be held. Securing the ship, which the Cambodians had abandoned, went without incident. The landing at Koh Tang, however, was a different matter. Operating from sketchy intelligence, the Marines encountered stronger resistance than expected and suffered heavy casualties. Soon withdrawn under fire, they discovered that the *Mayaguez* crew had been released unharmed 4 hours before they landed on Koh Tang. Small wonder that some historians rate the *Mayaguez* operation as a prominent "military failure."[63]

Still, the *Mayaguez* episode was not without useful lessons. By revealing gaps in JCS planning and organization, the operation stimulated interest in the theretofore neglected field of "special operations" and by extension helped generate support for JCS organizational reform resulting in the Goldwater-Nichols Act of 1986 and the subsequent Nunn-Cohen amendment. Given the heavy emphasis on preparing for large-scale conventional conflict since adoption of the "flexible response" concept in the mid-1960s, the Joint Chiefs and the Services had not paid much attention to developing the necessary doctrine, arms, and forces for rescue missions and other specialized tasks. Nor had the political and budgetary climate at the time been conducive for it. But as a result of the *Mayaguez* affair and the rising tempo of international terrorism during the 1970s, interest in special operations began to grow to the point that by the end of the decade each Service was taking a closer look at its requirements.[64]

A further consequence of the *Mayaguez* incident was to set the stage for a high-level "purge" within the Ford administration, with Schlesinger the primary target. Never comfortable with Schlesinger to begin with, Ford considered him aloof, patronizing, and arrogant; after *Mayaguez*, he lost confidence in Schlesinger altogether.[65] The precipitating event leading to the Secretary's dismissal was Schlesinger's decision to call off a final air strike against the Cambodians once news reached the Pentagon that the crew was free and the Marines had withdrawn from Koh Tang. Secretaries of Defense going back to Forrestal had routinely taken it upon themselves to cancel Presidential orders when they judged them to be "OBE" (overtaken by events). Kissinger, however, seems to have gone out of his way to put it in Ford's mind that Schlesinger had been willfully insubordinate.[66]

With relations between Ford and Schlesinger continuing to deteriorate, the President finally decided in late October 1975 that the time had come to find a new Secretary of Defense. Named as Schlesinger's successor was Donald H. Rumsfeld, then White House director of operations. In what the press called the "Halloween Massacre," Ford also recalled George H.W. Bush from his post as envoy to China to replace

William E. Colby as Director of Central Intelligence and stripped Kissinger of his title as Assistant for National Security Affairs. The ouster of Kissinger from his national security job (his former deputy, retired Air Force Lieutenant General Brent Scowcroft, replaced him) was aimed at quieting criticism from Congress that Kissinger had grown too powerful through occupying two major positions. But with his former deputy now managing the NSC, Kissinger's power and influence were little diminished.

The Joint Chiefs, as they were prone to do, took these changes in stride. Like President Ford, they had found Schlesinger's detached manner off-putting at times, but they had the utmost respect for his intellectual ability and his commanding grasp of nuclear strategy. Rumsfeld, in contrast, came from a political background, and his experience in defense affairs was confined primarily to recently serving as Ambassador to NATO. According to the Washington rumor mill, he had his sights set on someday becoming President. Kissinger remembered him as "tough, capable, personally attractive, and knowledgeable."[67] Whatever else, he made a favorable impression on the chiefs and, being well connected at the White House, increased the military's profile where it counted.

Under Rumsfeld, the Joint Chiefs moved several steps closer to realizing the aims of their strategic modernization program. Echoing JCS concerns that the strategic balance was shifting in favor of the Soviet Union, Rumsfeld urged a go-slow approach to further arms control talks until the United States could reassess the full range of its strategic requirements. "The level of deterrence suitable for Brezhnev," he argued, "is not necessarily the level of deterrence suitable for us."[68] Meanwhile, he advocated a modest strategic buildup that included continuation of Trident, acceleration of both the B–1 and MX programs to get them ready for production, and deployment of the Mark 12A warhead (previously authorized but delayed for technical reasons) to enhance the effectiveness of the Minuteman III force. Abandoning the no-growth defense budgets of the past, he proposed modest increases to keep military spending slightly ahead of inflation. Not all of these decisions would survive the scrutiny of the incoming Carter administration in 1977, but at the time they were cause for cautious optimism among the JCS that senior policymakers were aware of U.S. weakness and prepared to do something about it.[69]

While détente survived the stresses and strains of this period, the reasons probably had less to do with the commonality of U.S. and Soviet interests than with the reluctance of either side to admit that this latest version of "peaceful coexistence" was not bound to last. "If détente unravels in America," Nixon warned Brezhnev shortly before he relinquished the Presidency, "the hawks will take over, not the doves."[70] Brezhnev could well have said the same thing about the situation in the Soviet Union. Neither leader liked to think of the Cold War as having become a winner-take-all or zero-sum game. But that in effect was what it had become—and how increasingly it seemed destined to play out.

NOTES

1 *United States Military Posture for FY 1983* (Washington, DC: Organization of the Joint Chiefs of Staff, n.d.), 15 (unclassified edition).

2 U.S. Department of Defense, Office of the Assistant Secretary of Defense (Comptroller), *National Defense Budget Estimates for FY 1985* (Washington, DC, March 1984), 108–109, 135.

3 Richard L. Kugler, *Commitment to Purpose: How Alliance Partnership Won the Cold War* (Santa Monica, CA: RAND, 1993), 264.

4 U.S. Department of Defense, *Fiscal Year 1972–76 Defense Program and the 1972 Defense Budget* (Washington, DC: GPO, 1971), 77–82, 181.

5 Admiral Moorer, February 21, 1970, quoted in Walter S. Poole, Lorna S. Jaffe, and Wayne M. Dzwonchyk, "History of the Joint Chiefs of Staff: The Joint Chiefs of Staff and National Policy, 1969–1972" (MS, Historical Division, Joint Secretariat, Joint Staff, March 1991), 96 (declassified/publication forthcoming).

6 Quoted in *History of Strategic Air Command, 1965,* Vol. I (History and Research Division, HQ Strategic Air Command, April 1967), 141.

7 Marcelle Size Knaack, *Post-World War II Bombers, 1945–1973* (Washington, DC: Office of Air Force History, 1988), 575–579.

8 Lawrence S. Kaplan, Ronald D. Landa, and Edward J. Drea, *The McNamara Ascendancy, 1961–1965* (Washington, DC: Historical Office, Office of the Secretary of Defense, 2006), 486, 490; "Discussion" section (Enclosure C) to JCSM-925-64 to SecDef, October 31, 1964, "The Strategic Aircraft Program," JCS 1800/900-1.

9 Knaack, 579–583.

10 D. Douglas Dalgleish and Larry Schweikart, *Trident* (Carbondale, IL: Southern Illinois University Press, 1984), 41–44; Graham Spinardi, *From Polaris to Trident: The Development of U.S. Fleet Ballistic Missile Technology* (Cambridge: Cambridge University Press, 1994), 113–123.

11 Poole, Jaffe, and Dzwonchyk, 53–68.

12 Spinardi, 119–120; Henry Kissinger, *White House Years* (Boston: Little, Brown, 1979), 1129.

13 Bernard C. Nalty, "USAF Ballistic Missile Programs, 1964–1966" (MS, USAF Historical Division Liaison Office, March 1967), 45–47, available at <http://www.gwu.edu/~nsarchiv/nukevault/ebb249/doc04.pdf> (accessed July 18, 2011); and Barnard C. Nalty, "USAF Ballistic Missile Programs, 1967–1968" (MS, Office of Air Force History, September 1968), 56–59, available at <http://www.gwu.edu/~nsarchiv/nukevault/ebb249/doc05.pdf> (accessed July 18, 2011).

14 Walter S. Poole, "The History of the Joint Chiefs of Staff and National Policy, 1965–1968," (MS, Historical Division, Joint Secretariat, Joint Chiefs of Staff, May 1985), part I, 45 (declassified/publication forthcoming).

15 See CNO views as presented in JSOP FY71–78, Book II, "Strategic Offensive and Defensive Forces," Vol. II, enclosure to Note by Secretaries to the JCS, January 27, 1969, JCS 2143/339 (Book II); summarized in Poole, Jaffe, and Dzwonchyk, 52–53.

16 William Burr, "The Nixon Administration, the 'Horror Strategy,' and the Search for Limited Nuclear Options, 1969–1972—Prelude to the Schlesinger Doctrine," *Journal of Cold War Studies* 7 (Summer 2005), 34, 40–46; Kissinger, *White House Years*, 216. See also NSSM 64, July 8, 1969, "US Strategic Capabilities," National Security Council Institutional File, box H-207, Nixon Presidential Materials; available at <http://nixon.archives.gov/virtuallibrary/documents/nssm/nssm_064.pdf>.

17 Poole, 24–31; Desmond Ball, "The Development of the SIOP, 1960–1983," in Desmond Ball and Jeffrey Richelson, eds., *Strategic Nuclear Targeting* (Ithaca, NY: Cornell University Press, 1986), 62–70.

18 Notes on NSC Meeting, February 14, 1969, quoted in Burr, 49.

19 Poole, 29–30.

20 Fred Kaplan, *The Wizards of Armageddon* (New York: Simon & Schuster, 1983), 370–371; Janne E. Nolan, *Guardians of the Arsenal* (New York: HarperCollins, 1989), 114–115.

21 Kissinger, *White House Years*, 217; Burr, 70; Terry Terriff, *The Nixon Administration and the Making of U.S. Nuclear Strategy* (Ithaca, NY: Cornell University Press, 1995), 61–62.

22 HAK Talking Points, DOD Strategic Target Study Briefing, July 27, 1972, NSC Files, Nixon Papers; and <http://www.gwu.edu/~nsarchiv/NSAEBB/NSAEBB173/SIOP-18.pdf> (accessed July 18, 2011).

23 Memo, Laird to Nixon, December 26, 1972, "Nuclear Weapons Planning" (declassified), National Security Archive collection, copy in JHO.

24 NSSM 169, February 13, 1973, "US Nuclear Policy," NSC Institutional Files, Box H-207, Nixon Presidential Materials, available at <http://nixon.archives.gov/virtuallibrary/documents/nssm/nssm_169.pdf> (accessed July 18, 2011).

25 NSDM 242, January 17, 1974, "Policy for Planning the Employment of Nuclear Weapons," NSC Institutional Files, Box H-208, Nixon Presidential Materials, NARA; and <http://nixon.archives.gov/virtuallibrary/documents/nsdm/nsdm_242.pdf> (accessed July 18, 2011). See also Scott D. Sagan, *Moving Targets: Nuclear Strategy and National Security* (Princeton: Princeton University Press, 1989), 44–45;

26 Remarks by Schlesinger to Overseas Writers Association, Washington, DC, January 10, 1974, in U.S. Department of Defense, *Public Papers of James R. Schlesinger, Secretary of Defense, 1974* (Washington, DC: Historical Office, Office of the Secretary of Defense, n.d.), I, 17–31.

27 Lawrence Freedman, *The Evolution of Nuclear Strategy* (New York: St. Martin's Press, 1983), 379–382, summarizes reactions pro and con.

28 U.S. Department of Defense, *Report of the Secretary of Defense James R. Schlesinger to the Congress on the FY 1975 Defense Budget and FY 1975–1979 Defense Program* (Washington, DC: GOP, 1974), 42, 51–57; U.S. Department of Defense, *Report of the Secretary of Defense James R. Schlesinger to the Congress on the FY 1976 and Transition Budgets* (Washington, DC: GPO, 1975), chaps. 2 and 3.

29 Henry Kissinger, *Years of Upheaval* (Boston: Little, Brown, 1982), 1175.

30 NIE 11-8-73, "Soviet Forces for Intercontinental Attack," January 25, 1974, in Donald Steury, ed., *Intentions and Capabilities: Estimates on Soviet Strategic Forces, 1950–1983* (Washington, DC: Central Intelligence Agency, 1996), 326.

31 Raymond L. Garthoff, *Détente and Confrontation: American–Soviet Relations from Nixon to Reagan* (Washington, DC: Brookings, 1994, rev. ed.), 466–467, summarizes Soviet responses to the Schlesinger doctrine.

32 Edward L. Rowny, *It Takes One to Tango* (Washington, DC: Brassey's, 1992), 1–20.

33 Henry Kissinger, *Years of Renewal* (New York: Simon & Schuster, 1999), 849.

34 Mark M. Lowenthal, "US Organization for Verification," in William C. Potter, *Verification and SALT: The Challenge of Strategic Deception* (Boulder, CO: Westview Press, 1980), 77–94, gives a comprehensive overview of the formal arms control structure in the 1970s.

35 George S. Brown, *United States Military Posture for FY 1976* (Washington, DC: Department of Defense, February 5, 1975), 30–31.

36 Historical Division, *Organizational Development of the Joint Chiefs of Staff, 1942–1989* (Washington, DC: Joint Chiefs of Staff, November 1989), 48–50.

37 Memo, Rumsfeld to CJCS et al., March 9, 1976, "Organizational Changes—Disestablishment of WSEG," U, JCS 1977-380; John Ponturo, *Analytical Support for the Joint Chiefs of Staff: The WSEG Experience, 1948–1976* (Washington, DC: Institute for Defense Analyses, 1979), 343–363.

38 Gerald R. Ford, *A Time to Heal* (New York: Harper & Row, 1979), 30.

39 Kissinger, *Years of Upheaval*, 1154, 1187.

40 Elmo R. Zumwalt, Jr., *On Watch: A Memoir* (New York: Quadrangle, 1976), 432.

41 Ibid., 429.

42 NSDM 245, February 19, 1974, "Instructions for the SALT Talks, Geneva," National Security Council Institutional Files, Box H-207, Nixon Presidential Materials, available at <http://nixon.archives.gov/virtuallibrary/documents/nsdm/nsdm_245.pdf> (accessed July 18, 2011).

43 See Garthoff, 473–485.

44 Kissinger, *Years of Upheaval*, 1020–1031.

45 Richard M. Nixon, *RN: The Memoirs of Richard Nixon* (New York: Grosset & Dunlap, 1978), 1024.

46 Edgar F. Puryear, Jr., *George S. Brown, General, U.S. Air Force: Destined for Stars* (Novato, CA: Presidio Press, 1983), 246–257; George M. Watson, Jr., *Secretaries and Chiefs of Staff of the United States Air Force: Biographical Sketches and Portraits* (Washington, DC: Air Force History and Museum Program, 2001), 150–151; *The Chairmanship of the Joint Chiefs of Staff, 1949–1999* (Washington, DC: Joint History Office, Joint Chiefs of Staff, 2000), 123.

47 Garthoff, 497. Unlike Ford, Brezhnev brought many of his senior military advisors with him to Vladivostok.

48 Kissinger, *Years of Renewal*, 277–279.

49 Thomas W. Wolfe, *The SALT Experience* (Cambridge, MA: Ballinger, 1979), 174–175.

50 Robert Emmet Moffit, "The Cruise Missile and SALT II," *International Security Review* 4 (Fall 1979), 271–293; Richard K. Betts, ed., *Cruise Missiles: Technology, Strategy, Politics* (Washington, DC: Brookings, 1981), 83–100, 339–358.

51 Garthoff, 497–499.

52 Joint Chiefs of Staff, *Report by the Chairman of the Joint Chiefs of Staff: United States Military Posture for FY 1975* (Washington, DC: Department of Defense, February 1974), 24 (unclassified edition).

53 Kissinger, *Years of Renewal*, 301; and Rowny, 73–74.

54 Kissinger, *Years of Renewal*, 301, 847.

55 Wolfe, 174–175.

56 Garthoff, 501–503.

57 Gordon S. Barrass, *The Great Cold War: A Journey Through the Hall of Mirrors* (Stanford, CA: Stanford University Press, 2009), 190–191.

58 Brown, 53–54, 200–201.

59 Minutes, NSC Meeting on Indochina, April 9, 1975 (declassified), National Security Advisers Files, Gerald R. Ford Library; available at <http://www.fordlibrarymuseum.gov/library/document/nscmin/750409.pdf> (accessed July 18, 2011).

60 Willard J. Webb and Walter S. Poole, *The Joint Chiefs of Staff and The War in Vietnam, 1971–1973* (Washington, DC: Office of Joint History, Joint Chiefs of Staff, 2007), 359.

61 SNIE 11/2-81, "Soviet Support for International Terrorism and Revolutionary Violence," May 1981, in Gerald K. Haines and Robert E. Leggett, eds., *CIA's Analysis of the Soviet Union, 1947–1991: A Documentary Collection* (Washington, DC: Center for the Study of Intelligence, Central Intelligence Agency, 2001), 106.

62 Ford, 275–281; Kissinger, *Years of Renewal*, 547–566.

63 See Richard A. Gabriel, *Military Incompetence: Why the American Military Doesn't Win* (New York: Hill and Wang, 1985), 61–83.

64 David W. Hogan, Jr., *Raiders or Elite Infantry? The Changing Role of the U.S. Army Rangers from Dieppe to Grenada* (Westport, CT: Greenwood Press, 1992), 205–210.

65 Ford, 320–324.

66 Kissinger, *Years of Renewal*, 570–572. Kissinger incorrectly asserted (ibid., 573) that "in our system" the Secretary of Defense was "not directly in the chain of command," and that Schlesinger therefore had no authority to issue the orders he did. Kissinger was apparently unaware of the 1958 amendments to the National Security Act, which established the chain of command as running from the President, through the Secretary of Defense, to the combatant commanders.

67 Kissinger, *Years of Renewal*, 177.

68 Minutes, NSC Meeting on SALT, January 19, 1976, 22 (declassified), National Security Advisers' Files, Ford Library; available at <http://www.fordlibrarymuseum.gov/library/document/nscmin/760119.pdf>. Library file copy reads "necessary," an apparent typo.

69 Department of Defense, *Annual Defense Department Report, FY 1977* (Washington, DC: Department. of Defense, January 27, 1976), i–vii; U.S. Department of Defense, *Report of Secretary of Defense Donald H. Rumsfeld on FY 1978 Budget, FY 1979 Authorization Request, and FY 1978–1982 Defense Programs* (Washington, DC: Department of Defense, January 17, 1977), 123–130.

70 Nixon, *RN*, 1031.

General David C. Jones, USAF, Chairman, Joint Chiefs of Staff, 1978–1982

Chapter 13

THE RETURN TO CONFRONTATION

By January 1977, when President Jimmy Carter took office, détente was beginning to show unmistakable signs of wear. In both Washington and Moscow, opposition to further accommodations with the other side was on the rise. While Brezhnev had managed to force the hard-liners to "eat" the Vladivostok accords, the prevailing mood within the Soviet elite was that the United States was losing the arms race and that the correlation of forces had turned in favor of the Kremlin.[1] Many in the West—including the Joint Chiefs—agreed that U.S. military credibility was at its lowest ebb since World War II and that the balance of power was in a precarious state. Never, it seemed, had America's prestige been lower or its status as a superpower so uncertain.

The new Carter administration was, if anything, even more committed to preserving détente than its two immediate predecessors. If he achieved nothing else during his Presidency, Jimmy Carter wanted to reduce the threat of nuclear war, cut the number of opposing strategic weapons, and lessen the drain that military expenditures placed on the world's resources. Ultimately, he hoped to shift attention from the Cold War to other issues—the global crisis in energy supplies, the protection of human rights, and especially the need to improve relations and the distribution of resources between the developed and developing worlds. Instead of military power, Carter proposed to rely more on diplomacy and moral suasion to achieve American security objectives.[2] While he did not dismiss the need for armed force in support of foreign policy, he thought it had been overused in the past. Thenceforth, he said in his inaugural address, the United States would "maintain strength so sufficient that it need not be proven in combat—a quiet strength based not merely on the size of an arsenal but on the nobility of ideas."[3]

CARTER AND THE JOINT CHIEFS

Almost from the moment the Carter administration arrived, the Joint Chiefs of Staff found themselves on the defensive, with their advice treated as suspect and their methods and procedures under close scrutiny. Despite a succession of austere budgets since Vietnam, defense spending remained at what many in the incoming administration deemed excessive, driven by outmoded force-sizing practices, lax management, and inefficient allocation and use of resources. As an immediate target upon taking office, Carter proposed to trim five to seven billion dollars from the military budget. Hoping eventually to reduce military spending even further, Carter never ceased to push and prod the Pentagon to save money, do more with less, and above all keep in mind the greater humanitarian good.

Carter was unlike any President the Joint Chiefs had known. An Annapolis graduate (Class of '46), he resigned from the Navy in 1953 to manage his family's Georgia peanut business. After turning the business around, he went into politics, became governor of Georgia, and acquired a national following. A populist, he identified himself with the center-left wing of the Democratic Party. Like John F. Kennedy, he appeared uneasy, almost awkward, around "the brass." If he dealt with the JCS at all, it was generally through his Secretary of Defense, Harold Brown, or the Chairman. Despite his celebrated penchant for mastering detail, he had little patience for lengthy JCS threat assessments and posture statements and preferred crisp summaries prepared by White House aides. In his memoirs, he insisted that he enjoyed "good relations" with the JCS during his 4 years in office.[4] Yet he rarely met with the chiefs as a corporate body. His trips to the Pentagon were few and usually for ceremonial functions rather than substantive discussions. On one of the few occasions when he did listen to JCS advice—in planning the failed Iran hostage rescue mission in 1980—the results were a disaster, confirming Carter's belief that the military was anything but infallible.

Under Carter, as under his immediate predecessors, the CJCS continued to be the pivotal link between civilian authority and the military. Though there was some speculation as the new administration took office that General George S. Brown, USAF, the serving Chairman, would be replaced, Carter brushed such talk aside and kept Brown on until he stepped down from active duty for health reasons in June 1978, 10 days before the expiration of his term. His successor, General David C. Jones, had previously been the Air Force Chief of Staff. A Curtis LeMay protégé, Jones had served on the Air Staff in Washington while McNamara was Secretary of Defense, when civil-military relations were at low ebb. As Chairman, he made it his goal to achieve a harmonious partnership between OSD and the JCS.[5]

Carter's choice of another Air Force officer as Chairman was the source of endless speculation. Some thought that it was a reward for Jones's acquiescence in Carter's decision a year earlier, fulfilling a campaign pledge, to cancel the B–1 bomber. Proponents of the B–1, feeling that Jones had accepted the cancellation order too easily, argued that he should have fought harder to keep the plane. Jones disagreed. "There were those who said I should have fallen on my sword," he recalled. But he doubted whether it would have served a useful purpose. "Carter had campaigned on cancellation of the B-1. Who am I to sit in judgment?"[6]

Jones's appointment as CJCS seemed to some observers to be consistent with an emerging pattern by the Carter administration of naming competent yet low-profile officers to sit on the Joint Chiefs of Staff. At the same time he nominated Jones, Carter also sent the names of two other new JCS members to the Senate: General Lew Allen, Jr., to become the Air Force Chief of Staff; and Admiral Thomas D. Hayward to succeed the popular and respected Admiral James L. Holloway III as Chief of Naval Operations. Allen, a Ph.D. in physics, was at heart a scientist, while Hayward's background was in naval aviation and program analysis. Both were able and dedicated officers. But they were virtually unknown outside their respective Services and came from technical backgrounds that did little to prepare them as high-level politico-military advisors. The net effect, wrote Bernard Weinraub of the *New York Times*, was "an awareness within the defense hierarchy that the influence of the Joint Chiefs is on the decline."[7]

Efforts by the Joint Chiefs to reestablish their influence and authority initially met with limited success. A case in point was their handling of reforms to the joint strategic planning system, which one administration after another had deplored. As initiated under the Carter administration, these reforms targeted the Joint Strategic Objectives Plan (JSOP), the Joint Chiefs' mid-range (7-year) estimate of military requirements which they updated annually as their major contribution to the budget process. Urged by the administration to modernize their planning methods, the Joint Chiefs introduced the Joint Strategic Planning Document (JSPD) in place of the JSOP over the course of 1978–1979. Like its predecessor, the function of the JSPD was to appraise the threats to U.S. interests and objectives and to recommend a level of programmed forces to address those dangers. Even so, the JSPD was little better than the JSOP in providing a strategic framework for the allocation of resources since it made no attempt to prioritize programs; instead, it treated each Service's needs as having more or less equal importance. Since allocating resources invariably posed the most difficult problems at budget time, the absence of a prioritized list rendered the JSPD almost useless. As a result, few outside the Joint Staff paid any more attention to the JSPD than they had to the JSOP.[8]

Carter was convinced that the only way to make the Joint Chiefs more efficient and effective was through a top-to-bottom reorganization, the subject of a Defense-wide review initiated in November 1977 by Secretary of Defense Harold Brown.[9] Richard C. Steadman, a former Deputy Assistant Secretary of Defense, chaired the panel that examined the role of the JCS. Reporting its findings in July 1978, the Steadman group recommended streamlining Joint Staff procedures and increasing the power and authority of the CJCS. Arguing that it was virtually impossible for the Service chiefs to render wholly objective advice, the panel looked to the Chairman as the only military officer with no current or prospective Service responsibilities to interfere with providing the necessary leadership and administrative authority to make the JCS organization more responsive and effective.[10]

The Joint Chiefs were notably unenthusiastic about the Steadman group's findings. At his first press conference as Chairman, General Jones downplayed their probable impact, indicating that there was as yet no consensus on how to proceed. "We have a long ways to go," he said, "before we can really figure out how to merge all of these conflicting views in the joint arena and come up with recommendations on some of these difficult issues."[11] Privately, Jones told Secretary of Defense Brown that while he saw "a number of things" that would improve JCS performance, he expected the changes, if any, to be minor. "I firmly believe," he added, "that the fundamental organizational structure is sound."[12] Commenting as a corporate body, the Joint Chiefs concurred that the Steadman report contained many "innovative, positive suggestions," but cautioned that implementation efforts should be "evolutionary in nature."[13]

Undaunted, President Carter continued to treat JCS reform as unfinished business. Had he been reelected in 1980, he undoubtedly would have proposed legislation along the lines the Steadman report recommended. But as a one-term President, he never had the time or opportunity to go beyond piecemeal changes. The only legislative reform enacted during Carter's Presidency was a law he signed on October 20, 1978, granting the Commandant of the Marine Corps coequal status with the Service chiefs, thereby recognizing in statute what had become commonplace in practice. While the movement for JCS reform was indeed beginning to take definite form, it would still be some time before it gathered sufficient momentum to produce more than superficial changes.

STRATEGIC FORCES AND PD-59

The most striking difference between Jimmy Carter and the Joint Chiefs was in their respective views of the world and the threat posed from Moscow. While Carter acknowledged the Soviet Union as a hostile power, animated by an ideology sharply at odds with Western values, he entered office brimming with optimism that he could do business with the Soviets and reach an early SALT agreement that would obviate both sides' need for new or additional strategic forces. He liked the idea of tailoring basic national security policy accordingly, and favored refinements in strategic-targeting and weapons-employment policy that would reduce the death and destruction from a possible military confrontation. Even though the Joint

Chiefs applauded the President's idealism, they also considered it somewhat naïve and could not help but question the practicality of some of his proposals. As they had for years, the chiefs continued to measure Soviet intentions in terms of Moscow's large and growing military arsenal. As time went on, to be sure, Carter's views on the Soviet Union changed, especially after the Soviet invasion of Afghanistan late in 1979. Yet he never accepted the chiefs' basic premise that the United States was falling dangerously behind the Soviet Union in effective military power and could only redress this situation through a major strengthening of U.S. capabilities.

For Carter, as for his two immediate predecessors, strategic modernization was often a source of intense friction between the Pentagon and the White House. Carter was determined to reduce military spending and saw no better place to begin than with the increasingly expensive B–1 bomber, which he summarily cancelled in June 1977, thus fulfilling a campaign pledge.[14] Carter and Secretary of Defense Brown both questioned the B–1's penetration capabilities and concluded that it had become superfluous with the advent of air-launched cruise missiles which could be delivered from existing B–52s and other platforms. Later, they argued that emerging "stealth" technology offered more promising possibilities than the B–1.[15] But according to General David Jones, the Air Force Chief of Staff at the time, stealth R&D then concentrated on developing smaller planes and cruise missiles and was not seriously involved in producing a bomber alternative to the B–1.[16]

Carter's cancellation of the B–1 proved far more contentious than the White House expected and came in the wake of another controversial decision, announced in May 1977, to withdraw U.S. troops from Korea. Like his handling of the B–1, Carter had been thinking about pulling troops out of Korea well before the election. According to published accounts, he was heavily influenced by analysts at the Brookings Institution who believed that the continuation of a large U.S. presence amounted to a dangerous "trip wire" that could easily ensnarl the United States in another unpopular Asian war.[17] That the pullout of U.S. forces would in the long run save money, free up assets for deployment elsewhere, and distance the United States from what President Carter considered the South Korean government's wobbly human rights record became in the final analysis the decisive factor in his thinking.[18]

Underlying Carter's foreign and defense policies was his faith in détente to move the United States and the Soviet Union permanently away from the confrontational politics of the past. While he acknowledged the contributions of military power to an effective foreign policy, he was satisfied with maintaining "essential equivalence" in strategic forces and a balance of power "at least as favorable as that that now exists."[19] But after the contretemps over cancellation of the B–1, he was under constant pressure to reassure the JCS and pro-defense members of Congress that he remained committed to preserving a credible deterrent posture. Thus, he showed continuing strong support for the Trident program in the face of allegations of shoddy management and enormous inflation-driven increases in construction costs, and let stand the Ford administration's decision to proceed with production of the Mark 12A warhead to increase Minuteman III's accuracy and effectiveness against hardened targets. Most significant of all was his determination to resolve the MX controversy, resulting in his approval in 1979 of a plan to deploy 200 MX missiles in a mobile-basing mode. Yet it was a decision he found personally repugnant, and in his diary he characterized the MX deployment as "a nauseating prospect to confront, with the gross waste of money going into nuclear weapons of all kinds."[20]

Perhaps because he disliked nuclear weapons so much, Carter was determined to exercise the closest possible control over them. Not since Harry S. Truman had a President been so personally involved in the management of the country's nuclear arsenal, its configuration, and how it would be used. Most far-reaching of all were the changes President Carter made in the targeting and employment policies governing U.S. nuclear forces. The Joint Chiefs regarded these matters as basically closed after adoption of the Schlesinger doctrine (NSDM 242) in 1975. But to Carter and his national security advisor, Zbigniew Brzezinski, NSDM 242 was merely the first step. Convinced that targeting doctrine should have specific political as well as military objectives, Brzezinski wanted it to include refinements that amounted to the "ethnic cleansing" of the Soviet Union by threatening the heaviest casualties among the Great Russian population, as opposed to the Latvians, Ukrainians, and other nationalities that had been more or less coerced into joining the Soviet state.[21]

The upshot was the appearance in November 1978 of the "countervailing strategy," the product of an interagency review headed by Leon Sloss, a respected strategic analyst and consultant to Secretary of Defense Brown.[22] Presented to the Joint Chiefs as more or less a fait accompli, the countervailing strategy was in many respects a logical extension of the Schlesinger doctrine. As Secretary Brown described it, its function was the maintenance of "military (including nuclear) forces, contingency plans, and command-and-control capabilities to convince Soviet leaders that they cannot secure victory, however they may define it, at any stage of a potential war."[23] But in carrying out these tasks, it imposed a far more sophisticated and rigorous set of targeting requirements. A formidable assignment, Chairman Jones promised to give it his utmost attention but was somewhat skeptical of achieving quick results. In fact, Jones believed strategic nuclear planning and targeting had become so exceedingly complex that he foresaw few significant changes resulting anytime soon, no matter what the declared targeting policy might be.[24]

The most strenuous objections to the countervailing strategy came from the State Department. According to Brzezinski, Secretary of State Cyrus R. Vance found the whole inquiry into nuclear targeting emotionally disturbing and gave it limited cooperation.[25] As a direct result, approval of a Presidential directive (PD–59) sanctioning the new strategy was held up until July 1980.[26] By then, however, the Joint Chiefs were well along toward putting the countervailing strategy into operation since many of its provisions could be implemented on orders of the Secretary of Defense. The main function of PD–59 was to pave the way for issuance of a new Nuclear Weapons Employment Policy (NUWEP-80), which the Joint Chiefs received in October 1980.[27]

President Carter and Secretary of Defense Brown both insisted that it was never their intention under the countervailing strategy to make sweeping changes in U.S. policy or doctrine. According to Brown, the countervailing strategy amounted to nothing more than a "modest refinement in U.S. nuclear strategy as a response to charges that the USSR had achieved strategic nuclear superiority." Its aim, he insisted, was to strengthen deterrence and not to boost war-fighting capabilities.[28] Carter's view was essentially the same. As much as he abhorred nuclear weapons, he accepted the necessity of their role in U.S. defense policy, but sought to narrow their use for strategic purposes in carefully pre-planned ways, avoiding wholesale destruction. Hence the emphasis on options that would theoretically allow the President to choose from an almost endless array of measured responses to almost any level of Soviet provocation.[29]

By and large, the Joint Chiefs agreed that the more options they and the President might have, the better. As during previous strategic reviews, however, their main concern was one of feasibility. The most complex and demanding targeting policy to that point, the countervailing strategy required them to prepare for almost any contingency, from a limited nuclear exchange to a fully generated nuclear war. Most military professionals involved in this process shared the view of the Chairman, General Jones, that implementing the new doctrine would be a slow and laborious process, testing the patience and resourcefulness of all involved. That it would require significant improvements in technology, from weapons in the field to command, control, and communications, was practically a given. In other words, implementing the countervailing strategy was a long-term process that JCS planners approached with mixed feelings about achieving ultimate success.

SALT II

As intent as President Carter was on exercising closer command and control over the targeting and use of nuclear weapons, he was even more determined to reach agreement with the Soviets on reducing nuclear arms. Not satisfied with the tentative ceilings set at Vladivostok in 1974, he wanted "deep cuts" and speculated at one point that he saw no need for either side to keep more than 200 ICBMs.[30] Indicative of his thinking was the sweeping statement in his inaugural address that his "ultimate goal" was nothing less than "the elimination of all nuclear weapons" from the face of the earth.[31] "I want the level of our capability as low as possible," Carter told his senior advisors, "but I'm not naive. Possibly 1,000 ICBMs, each with one warhead, with some limitations on the size of the warhead." In any case, Carter added, "we should work for dramatic reductions, carefully monitored and not unfavorable to either side."[32]

Carter's deep cuts plan left the Joint Chiefs stunned. Having never envisioned reductions on the scale Carter proposed, they found it hard to imagine how they could effectively deter the Soviets with such a small strategic arsenal. While the Chairman dutifully pledged his support in helping the President realize his goal, he was uncertain how much cooperation to expect from his JCS colleagues, whom he described as "staunch proponents of reductions, but with caution." "Trying to lead the [Service] Chiefs on this issue," he warned, "is like putting three wild dogs through a keyhole."[33]

With his authority as President, Carter could finesse any objections raised by the Joint Chiefs. The Soviets, however, were another matter. To the leadership in Moscow, as one Soviet foreign minister later estimated, arms control constituted "95 percent of the total relationship, more or less," with the United States.[34] Complaints from the hardliners notwithstanding, SALT more often than not had yielded handsome dividends for the Soviets (most notably confirmation of strategic parity with the United States) which many now saw Carter trying to wrest away. Irritated also by the President's human rights campaign and its strident support for the celebrated dissident physicist Andrei Sakharov, the Soviets viewed the deep cuts proposal not merely with suspicion but with utter dismay. Thus, when Secretary of State Vance arrived in Moscow in late March 1977 to discuss the matter, he received both a chilly reception and a flat rejection of the offer.[35]

Despite the setback in Moscow, President Carter continued to believe that a SALT II treaty with significant reductions was attainable. But from that point on, he was more cautious and never substantially departed from the Vladivostok formula. Nevertheless, the negotiations proved more difficult than expected and moved forward slowly, requiring high-level intervention from time to time to revive the momentum and to overcome deadlocks on key details. Finally, in June 1979, Carter and Brezhnev met in Vienna to sign the SALT II treaty modeled on the Vladivostok accords. A complicated agreement, SALT II was to run for 5 years and imposed a series of ceilings and subceilings on strategic weapons, including not only missiles but also manned bombers, to create a complex web of quantitative and qualitative constraints. Already a source of growing controversy in the United States, the SALT II treaty faced an uncertain fate in the Senate, where sentiment was almost evenly divided for and against.

Throughout the negotiation of the SALT II treaty, the JCS played a limited role in shaping U.S. policy. A common complaint among officers on the Joint Staff and from the JCS representative to SALT II, Lieutenant General Edward L. Rowny, USA, was that they were often excluded from high-level deliberations and denied access to sensitive exchanges of information between the delegations. While Rowny respected Carter's idealism and enthusiasm, he considered the President inexperienced, closed minded, and ill-served by advisors like Secretary of State Vance and Paul Warnke, the administration's chief arms control negotiator, who seemed to Rowny overly eager to cut a deal with the Soviets. Frustrated and disappointed, Rowny retired from the Army shortly after the Vienna summit to devote his energies to defeating the SALT II treaty in the Senate.[36]

Before stepping down, Rowny tried to persuade the JCS to come out in opposition to the treaty. Broadly speaking, the charges that he and others lodged against it were four-fold: 1) it did nothing to reduce the threat to U.S. land-based forces (missiles and bombers) and risked weakening deterrence by preserving the Soviet Union's overwhelming superiority in "heavy" ICBMs; 2) it failed to impose effective constraints on the Soviet Backfire bomber; 3) it mandated undue curbs on the U.S. cruise missile program; and 4) as a limited duration agreement, it would require immediate renegotiation. The net effect, opponents argued, was an unequal agreement slanted toward the Soviets. Some critics also argued that the treaty would be hard to verify, but most opponents dismissed verification concerns as inconsequential since there were so many concessions to the Soviets that it would be pointless for them to cheat.[37]

The Joint Chiefs agreed that the SALT II treaty was flawed. But they rejected Rowny's basic contention that the United States would be better off without the treaty than with it, and were prepared to accept it provided there were no further delays in deploying the MX and in completing the other remaining elements of the strategic modernization program. Thus, during the debate in Congress, the Joint Chiefs steered clear of evaluating the treaty's merits and concentrated on giving their assessment of its strategic implications. They adopted the position that SALT II was "a modest but useful contribution to our national interests" and could produce effective results only in conjunction with improvements in the overall U.S. defense posture. "Our priority must go to strategic nuclear force modernization," General

Jones told Congress, "but increases are needed across the board for nuclear and non-nuclear forces." A tepid endorsement, it still satisfied the White House and avoided the embarrassment the administration would have suffered had the JCS followed Rowny's advice and opposed the treaty.[38]

Just as support for ratification seemed to be building in the Senate, there came disclosures in August-September 1979 that U.S. intelligence had confirmed the existence of a Soviet combat brigade in Cuba. At issue was whether the presence of these forces violated the precedents barring the reintroduction of Soviet military power set by the 1962 Kennedy-Khrushchev agreements ending the Cuban Missile Crisis. In fact, rumors and reports of Soviet military activity in Cuba circulated almost constantly and normally caused little stir. But with the SALT II debate ongoing, the Soviet brigade became a *cause célèbre* that played into the hands of the treaty's opponents, dimming its chances of approval. The fatal blow to SALT II's prospects was the Soviet invasion of Afghanistan in December 1979, which Carter himself later admitted doomed any chance the administration might have had of gaining the two-thirds vote needed for approval.[39] To demonstrate U.S. displeasure with Soviet behavior, President Carter withdrew the treaty from Senate consideration and made no recommendation that it be rescheduled for a vote in the foreseeable future. But having come this far with the treaty, he refused to repudiate it outright and in May 1980 announced that the United States would abide by its terms as long as the Soviet Union did the same.[40]

To the Joint Chiefs, Carter's decision to withdraw the SALT II treaty while abiding by its terms seemed a reasonable if not altogether satisfying outcome. Even though the JCS disliked the treaty, they were more concerned by what could happen should there be no treaty at all, a situation that could arguably open the way to a further buildup of Soviet strategic forces and an expensive escalation of the arms race. Flawed as it might be, the JCS were prepared to accept SALT II and work within its terms until something better came along.

NATO AND THE INF CONTROVERSY

The same concerns that prompted uneasiness in the Senate over the SALT II treaty were also reshaping attitudes toward the security of Europe. While NATO leaders had initially welcomed the improved atmosphere of détente, many were increasingly apprehensive as the 1970s wore on lest Moscow exploit this situation to extend its power and influence, undermine support within the Alliance for strong defense policies, and ultimately drive NATO apart. By mid-decade, with the ongoing buildup of Warsaw Pact capabilities showing no evidence of abating, alarm bells began sounding throughout NATO capitals. While the evidence was by no means conclusive, signs indicated that the Soviets were building up for a large-scale confrontation and posturing their forces for a Blitzkrieg-style attack against the West should war erupt. In assessing the probable outcome, Joint Staff and intelligence analysts in Washington reached the uncomfortable conclusion that the chance of stopping a Warsaw Pact attack with minimal loss of territory "appears remote at the present time."[41]

The most ominous development was the appearance of the SS–20, a land-based triple–warhead mobile missile that the Soviets began deploying in March 1976, apparently as a replacement for their aging SS–4s and SS–5s. Derived from an experimental ICBM (the SS–X–16), the SS–20 had a range of 5,000 kilometers and thus fell just outside the SALT I limits, making it an intermediate-range ballistic missile. Some observers described it as the "pocket battleship" of its time. Like the German and Japanese heavy cruisers built in the 1930s, it eluded arms control constraints but still had almost the same range and payload as a fully functional strategic weapon.[42] "Our new SS–20 missile," boasted one Soviet general, "was a breakthrough unlike anything the Americans had. We were immediately able to hold all of Europe hostage."[43] With the SS–20, the Soviets could target not only every major capital in Europe, but also much of North Africa and practically the entire the Middle East.

As President Carter took office, JCS and NATO planners were still in the preliminary stages of assessing the SS–20's military and strategic impact. Still unknown were how many launchers the Soviets might eventually deploy or how they intended to use them. Operating in this thin air of uncertainty, the new administration downplayed the need for an immediate response and decided to concentrate on improving NATO's conventional capabilities under a new initiative known as the Long-Term Defense Program (LTDP). Calling for a 3 percent real increase in NATO funding (at the same time the United States was preparing cuts in its overall defense budget), the LTDP stressed the increased prepositioning of U.S. supplies and equipment in Europe, better management of resources, and across-the-board upgrades in the Alliance's conventional forces. Even though previous administrations had espoused similar goals, the enthusiasm shown by Washington for the LTDP, coupled with President Carter's well-known antipathy for nuclear weapons, brought to the fore what many Europeans (the West Germans especially) had feared since the adoption of the flexible response strategy in the 1960s—that Washington would try to move NATO away from nuclear deterrence toward almost exclusive (and more costly) reliance on conventional forces.[44]

Shortly after taking office, with a view to allaying these concerns, the administration dispatched State-Defense briefing teams to update European leaders on the status of U.S. nuclear planning and to reassure them that the United States remained committed to maintaining robust nuclear capabilities by developing the neutron bomb and ground- and sea-launched cruise missiles. Apparently, however, these briefing teams were exceedingly frank in discussing the technical difficulties associated with cruise missiles and conveyed the impression that these weapons would have a limited bearing on the strategic balance if and when they became operational.[45] Shortly thereafter, in June 1977, came the inadvertent and premature public disclosure of the neutron bomb—a tactical nuclear warhead capable of generating high levels of lethal radiation with a small explosion—that left the Carter administration mired in a public relations debacle that eventually sidelined the program. The net result was that many European leaders remained uneasy about American promises and wanted to see more in the way of concrete programs to strengthen their security, lest they take matters into their own hands.[46]

The catalyst for what became the most far-ranging reassessment of NATO's nuclear requirements since the 1950s was a speech by West German Chancellor Helmut Schmidt before the International Institute for Strategic Studies (IISS) in London in late October 1977. Based on the progress made thus far in SALT II, Schmidt was convinced that the United States and the Soviet Union were moving toward an agreement that would suit the superpowers but bargain away capabilities like cruise missiles that could be crucial to European security. Schmidt believed that once the Soviet SS–20 force became fully operational, Europe would be increasingly at the mercy of Soviet military and political pressure, no matter how strong its conventional forces might be, unless it had its own comparable, offsetting nuclear forces. Accordingly, in his speech to the IISS, he called for preserving "the full range of deterrence strategy," and implied that the United States was not doing enough either to curb the SS–20 threat through arms control or, failing that, to provide NATO with more credible theater nuclear forces.[47]

The White House's answer to Schmidt's challenge was to turn the question over to the High Level Group (HLG), a new advisory body to NATO's Nuclear Planning Group. Averse as ever to nuclear weapons, President Carter favored exercising restraint and looked to the HLG to explore policy options that would avoid or lessen the need for additional new deployments. While the President refused to countenance a one-for-one deployment with the Soviets, he knew he had to do something to show his support for NATO or risk irrevocably weakening Alliance solidarity. After much personal agonizing, he finally yielded and endorsed a compromise, formally adopted by NATO in December 1979, that called for the limited modernization of the Alliance's intermediate-range nuclear forces (INF). Simultaneously, NATO announced unilateral plans to reduce its nuclear arsenal by 1,000 warheads (weapons scheduled for decommissioning anyway) and extended an offer to scale back or cancel its INF modernization program if the Soviets would do likewise with their SS–20s.[48]

Slated to begin around the end of 1983, NATO's modernization measures consisted of deploying 572 mobile launchers, broken down into 464 U.S. ground-launched cruise missiles (GLCMs) and 108 U.S. Pershing II (P–II) ballistic missiles to replace an identical number of obsolescent Pershing IAs based in West Germany. The decision to include ballistic missiles in the mix was largely at the instigation of the Joint Chiefs and aimed at placating the West Germans, who wanted the reassurance of an up-to-date, fast-reaction weapon. Though the U.S. programs were still in the developmental stage, each GLCM and P–II would have sufficient range to threaten targets along the western edge of the Soviet Union, a capability that land-based NATO forces had previously lacked. The cruise missiles would be dispersed around Western Europe, while the P–IIs would be based entirely in the Federal Republic. All would be subject to NATO authority under the operational command and control of the U.S. Army.

Though not as extensive as the Joint Chiefs had hoped, NATO's nuclear modernization program satisfied their basic requirements and seemed to point the Alliance in what the JCS considered the right direction. For Carter, however, it was a dreadful setback—the acceptance of more weapons he loathed and an acknowledgement that, at bottom, the security of the NATO area continued to rest directly on the threat to use them. Worse still, from Carter's standpoint, the pending deployment was a further tacit admission that détente in Europe was on the wane. Yet it was probably the only sound decision he could have made without risking a permanent rupture within the Alliance. Whether arms control negotiations would obviate the need for NATO to follow through on its INF deployment plans remained to be seen. But by the time the Carter administration left office, there were few signs that NATO and the Soviet Union would soon reach a deal, if ever. Not until the Reagan administration would talks begin in earnest.

THE ARC OF CRISIS

While they were instrumental in shaping the Carter administration's policy toward nuclear modernization in Europe, the Joint Chiefs were less successful in persuading the White House to adopt a tougher stand against Soviet encroachment on the Third World. In some ways, the chiefs had only themselves to blame. Insisting that they lacked sufficient resources, they had consistently downplayed U.S. military involvement in Third World conflicts in the aftermath of Vietnam and had instead urged the use of diplomacy and covert operations to block the Soviets from making further inroads. This remained the basic JCS position throughout the Carter administration and on into Ronald Reagan's Presidency. But as the 1970s drew to a close, it was increasingly apparent to the Joint Chiefs that Third World problems were more intractable than they had assumed and that a larger military role for the United States was becoming unavoidable.

Though President Carter eventually came to a similar conclusion, he remained apprehensive about the application of military force to solve problems in Asia, Africa, and Latin America and believed the key to countering Communism in those parts of the world lay in promoting democratic values, economic improvements, and better living conditions. His preference, as always, was for diplomatic initiatives that would ease the threat of future conflicts and improve the North-South dialogue. Toward those ends, he managed to broker two significant breakthroughs: a new treaty with Panama, approved by the Senate in April 1978, ending both the American colonial presence and American control of the Panama Canal; and the Camp David peace accords reached later that year between Egypt and Israel. Cautiously optimistic about both, the Joint Chiefs welcomed the peace deal between Israel and Egypt in hopes that it would strengthen the U.S. strategic posture in the Middle East, but were decidedly cool toward giving up the Panama Canal, which they continued to regard as a vital American interest. Eventually, they gave the treaty a tepid endorsement.[49]

Meanwhile, avoiding U.S. involvement in Third World conflicts was proving increasingly difficult. At the outset of the Carter administration, perhaps the most volatile situation likely to engage the United States was the simmering dispute between Ethiopia and Somalia for control of the barren Ogaden plateau in the Horn of Africa. Overshadowing all was the apparent determination of Moscow to extend its influence throughout the region. Once strong allies of the Somalis, the Soviets had changed sides and thrown their support to the self-proclaimed Marxist regime in Ethiopia that had overthrown the decrepit monarchy of Haile Selassi in 1974. Toward the end of November 1977, on the heels of a series of secret aid agreements, the Soviets launched a massive airlift—larger than anything they had undertaken in Angola or elsewhere in Africa—to fortify Ethiopia with an estimated $1 billion in new arms and supplies and 17,000 elite Cuban combat troops. During the ensuing conflict, Ethiopia and its Soviet bloc allies easily overwhelmed the Somalis, setting off alarm bells in Washington that would reverberate for years to come.[50]

The leading advocate for a more forceful policy to counter Soviet encroachments in the Third World was the President's national security advisor, Zbigniew Brzezinski. During the deliberations surrounding the Ogaden crisis, it became clear that U.S. options were limited. About the most the United States could do to influence the situation directly was to deploy a naval task force off the coast of Somalia. Long before the crisis erupted, Brzezinski had foreseen an urgent need for a broader range of capabilities and had persuaded President Carter, as part of the administration's review of basic policy in the summer of 1977, to include a requirement for a "force of light divisions with strategic mobility," backed by adequate air and naval support, that could respond quickly to emergencies. Out of the bureaucratic process thus set in motion eventually emerged the Rapid Deployment Joint Task Force and its successor, the U.S. Central Command.[51]

Despite high-level endorsement, the creation of a rapid reaction force languished "on the back burner" for the next several years.[52] A reluctant supporter to begin with, President Carter gradually lost interest in the idea and seems to have forgotten it altogether once the Ogaden crisis eased early in 1978. Previous efforts to create such a force, starting in 1962 with the establishment of U.S. Strike Command (USSTRICOM) at McNamara's instigation, had little success owing to the initial reluctance of the Navy and Marine Corps to dedicate forces. During the Vietnam War, Strike Command's role further declined as available units for rapid reaction missions virtually disappeared. In 1971, acknowledging that USSTRICOM had outlived its usefulness, the Joint Chiefs replaced it with a new organization they called U.S. Readiness Command (USREDCOM). Based at MacDill Air Force Base, Florida, USREDCOM operated without assigned geographical responsibilities and mainly performed training, doctrinal, and advisory functions connected with joint deployments.[53]

Meanwhile, the "arc of crisis," as Brzezinski called it, was moving steadily eastward from the Horn of Africa into the Indian Ocean, the Persian Gulf, and Southwest Asia. Across the Middle East and on into Central Asia, conflict and political turmoil were the order of the day. Although the origins of many of these problems had more to do with local feuds and rivalries than with the Cold War, the perception in Washington was that conditions were ripe for Soviet penetration. In light of the West's heavy dependence on Persian Gulf oil, the Carter administration had all the more reason to be alarmed.

The most dangerous threat to U.S. interests was the declining power and authority of Shah Mohammed Reza Pahlavi of Iran, a longtime ally of the United States. Awash in oil revenues, the Shah aspired to modernize his country and turn it into a major partner of the West. Eager to cooperate, the Nixon administration had supplied Iran with an arsenal of sophisticated weapons and advanced technologies, including help for a nascent atomic energy program. The policy that emerged was to develop a "twin pillar" system of security relying on Iran and Saudi Arabia to police the region. Of the two, however, Iran was clearly the preferred partner. Henry Kissinger remembered the Shah as "an unconditional ally . . . whose understanding of the world situation enhanced our own."[54] The Joint Chiefs, after some initial hesitation, became similarly impressed with the Shah's leadership. By the early 1970s they regarded Iran's role as an anti-Soviet bastion as practically indispensable. In an area where American friends were few and far between, the Joint Chiefs of Staff pointed out that Iran was "a stabilizing influence" and as "strong and trusted [an] ally" as the United States was likely to find.[55]

The Carter White House had a somewhat different image of Iran. Brzezinski dismissed the Shah as a megalomaniac whose overly ambitious policies sowed the seeds of his destruction. Although President Carter was more charitable, finding much about the Shah and his regime to admire, he also saw much that left him uneasy. A founding member of the Organization of the Petroleum Exporting Countries (OPEC), the international oil cartel, Iran had played a key part in setting the high energy prices that were a major contributing factor to the soaring inflation of the 1970s. At the same time, as a direct result of his efforts to liberalize Iranian society, the Shah had alienated a number of powerful interest groups, including conservative Muslim religious leaders. As opposition to his policies mounted, the Shah turned increasingly to his secret police (SAVAK) to quell the dissent, a practice replete with alleged human rights abuses that President Carter found especially repugnant. But despite challenges to the Shah's regime, intelligence estimates soft-peddled the severity of the disturbances and in so doing contributed to a false sense among the Joint Chiefs and others in Washington that Iran was a safe and stable ally.[56]

The collapse of the Shah's power was as sudden as it was unexpected. In late November 1978, with unrest, strikes, and antigovernment demonstrations escalating, it became clear that the level and intensity of the demonstrations were sufficiently serious to threaten the survival of the monarchy itself. Amid the turmoil, the Joint Chiefs endorsed precautionary measures that included the evacuation of American citizens from Iran and stepped-up naval deployments in the Indian Ocean.[57] As the crisis deepened, talk turned to the possibility of a military solution, a discussion cut short by the realization that about anything the United States did would be too little too late.[58] The dénouement began on December 27, 1978, which one press account described as "a day of wild lawlessness and shooting in the capital and a strike that effectively shut down the oil industry."[59] By then, many middle-class Iranian moderates had joined the religious radicals in calling for the Shah to step down. Hoping that the Iranian generals might intervene and restore order, Brzezinski persuaded President Carter to send General Robert E. Huyser, USAF, the Deputy Commander of U.S. forces in Europe, to Tehran on a fact-finding mission. What Huyser found was an Iranian military in utter disarray, thoroughly demoralized and too poorly organized to make a difference. In mid-January 1979, the Shah fled the country, leaving it in the hands of a weak civilian government with little experience and even less popular support. For all practical purposes, the real head of state in Iran was Ayatollah Ruhollah Khomeini, a charismatic cleric recently returned from exile in France who was determined to rub out Western influence and establish a way of life based on fundamentalist Muslim principles. As bad as the Shah's downfall may have seemed for U.S. interests in the region, worse things were yet to come.

RISE OF THE SANDINISTAS

Half a world away in the Central American country of Nicaragua, a similar drama was playing out, though on a far smaller scale than the crisis in Iran. Relatively stable and prosperous by Latin American standards, Nicaragua was the virtual fiefdom of a right-wing dictator, Anastasio Somoza, whose family had ruled the country with the help of the U.S.-trained and equipped *Guardia Nacional* since the 1930s. Vehemently anticommunist, the Somoza regime had earned a reputation in U.S. military circles as being a strong and dependable ally against the threat of Cuban-instigated Communist expansion. But by the mid-1970s, Somoza's support both at home and in Washington was beginning to erode. Though known more for political corruption than brutality, Somoza had come under fire from the Carter administration for alleged human rights abuses and soon became the target of U.S. sanctions that included a cut-off of military aid. Starting with the Panama Canal treaty, President Carter hoped to change the U.S. image in Latin America. Withdrawing support for Somoza was part of that process.[60]

In September 1978, with opposition mounting and his back against the wall, Somoza authorized the National Guard to launch an all-out offensive against the most immediate threat to his regime—a leftist insurgency led by the Sandinista National Liberation Front (FSLN). Even though the Guard dealt the rebels a severe military setback, it also destroyed many towns and villages and inflicted heavy civilian casualties. Significant segments of the population became alienated and went over to the FSLN. While the Carter administration had no use for Somoza, it was also leery of the FSLN, whose Marxist rhetoric and Cuban connections seemed certain to place it on a collision course with the United States should it ever come to power. Seeking middle ground, President Carter approved a diplomatic initiative early in 1979 aimed at persuading Somoza to step aside voluntarily to make way for a more representative regime. Negotiations broke down, however, and by late May Somoza's forces and the Sandinistas were again engaged in pitched battle.

As with the Shah of Iran, the consensus in Washington was that Somoza would survive the Sandinista challenge. By early June, however, it was clear that U.S. intelligence had misjudged the situation and that the Sandinistas were gaining the upper hand, in part as a result of weapons covertly supplied by Cuba. Operating without fresh supplies, the National Guard steadily disintegrated, paving the way for a Sandinista victory. When at last on July 17, 1979, Somoza finally stepped down and fled to Miami, he left a country in physical ruin and political disarray. In the view of the Joint Chiefs, Nicaragua was now a ripe target for Communist penetration and a potential launch pad for Cuban adventurism elsewhere in Central America.

During its remaining time in office, the Carter administration wrestled with limiting the consequences of the Sandinista victory. The stated goals in Central America were "the development of democratic societies, the observance of human rights, the ending or diminution of violence and terrorism, and the denial of the region to forces hostile to the U.S."[61] Persuaded that the United States had overreacted to Castro's takeover of Cuba in 1959, many on the NSC Staff and at the State Department believed that the United States should work with the Sandinistas to establish good relations, promote political pluralism in Nicaragua, and steer the country away from becoming "another Cuba." The Joint Chiefs were skeptical of this approach, but among the President's senior advisors, only Brzezinski seemed to share their concerns. Meanwhile, the security situation in Central America continued to deteriorate as intelligence reports confirmed an influx of Cuban-supplied Eastern Bloc arms. But in trying to persuade the White House or the State Department to act, the JCS found little interest in anything that smacked of a military solution. In consequence, development of a comprehensive policy toward Central American became practically impossible, leaving decisions to emerge in a fragmented, reactive manner.[62]

Toward the end of his Presidency, Carter evinced signs of having second thoughts about trying to work with the Sandinistas. Slowly but surely he began to adopt a position more akin to that advocated by the Joint Chiefs and Brzezinski. Hoping to get a better picture of the situation, he authorized the National Security Agency to step up its monitoring of developments in Nicaragua and to expand its coverage of Sandinista communications.[63] Even so, Carter continued to view military action as a last resort and refused to abandon his belief that a political settlement, acceptable to all involved, was ultimately feasible. But in Central America, as elsewhere, his faith in détente and nonviolent solutions had been badly shaken and, as he left the White House, it was with a clear awareness that the next administration would be less reticent and adopt a more forceful course of action to prevent the spread of Sandinista and Cuban influence.

CREATION OF THE RAPID DEPLOYMENT FORCE

The fall of the Shah and the ensuing collapse of the Somoza regime had distinctly unsettling effects on JCS thinking. Alarmed by the sudden escalation of threats to

U.S. interests, the Joint Chiefs acknowledged a pressing requirement for a more responsive force posture capable of interjecting a measure of stability into troubled parts of the world. Although the JCS had been moving steadily in that direction ever since Brzezinski raised the issue in the summer of 1977, it was not until 2 years later that they felt their studies had progressed far enough to begin seriously discussing a mission statement, the assignment of forces, and command arrangements. Assuming that the focus of such a force would be the Middle East, JCS planners generally agreed that the most practical solution would be a joint task force or perhaps a new joint command and that either way, because it was almost certain to have a high political profile, it would be "*the* force except Europe and Korea."[64]

The central figure throughout the subsequent planning and preparations culminating in activation of the Rapid Deployment Force (RDF) was Secretary of Defense Harold Brown. Having previously served as Secretary of the Air Force and in other high-level Pentagon positions, Brown was well aware of the potential for inter-Service competition and rivalry that new programs presented. He repeatedly cautioned the Joint Chiefs and Service planners against using the RDF as leverage for more money or resources. What Brown and the President envisioned was a rather small fast-reaction force drawn mainly from available assets. Nonetheless, the opportunities were too inviting for the Services to ignore. Competition became especially acute between the Army and the Marine Corps as each jostled for a larger role on the assumption that the new organization would be first and foremost a ground-based intervention force, with supplemental air and naval support. Advised by Brzezinski that the President was growing impatient, Brown nudged the Services along as best he could and achieved a tentative agreement breaking the impasse by November 1979.[65]

Meanwhile, the seizure of the U.S. Embassy in Tehran on November 4, 1979, and the Soviet military intervention in Afghanistan the following month sent a shudder through Washington resulting in a wholesale reassessment of U.S. defense and security requirements for the Middle East and Southwest Asia. Both events produced an escalation of tensions and posed serious challenges to the protection of U.S. interests. Of the two, it was the Soviet invasion of Afghanistan on December 24–25, 1979, that most alarmed the Joint Chiefs and their superiors. Aimed at assuring a pro-Moscow regime in Kabul, the Soviet invasion drew sharp and swift international condemnation. Remote as Afghanistan seemed, Carter and his advisors saw its fate tied directly to that of the United States. "A Soviet-occupied Afghanistan," the President told the country, "threatens both Iran and Pakistan and is a steppingstone to possible control over much of the world's oil supplies."[66] To be sure, many of the responses that followed had been set in motion earlier. But with Afghanistan providing the catalyst, they came to fruition sooner rather than later and helped expedite the transformation of the RDF from a drawing board concept into a functioning organization.

At the heart of this transformation was the Carter Doctrine, announced in the President's State of the Union Message on January 23, 1980. In effect, Carter confirmed publicly what he and his subordinates had been saying privately to one another and in off-the-record talks with reporters for some time—that Washington had major interests at stake in the Persian Gulf and that the necessary response was a military buildup. Under that policy, President Carter served notice that the United States would not allow the Gulf to fall into hostile hands, that it would pursue a "cooperative security framework" in the area, and that it would back up those initiatives with requisite military force.[67] As evidence of his resolve, the President pointed to the pending creation of the Rapid Deployment Force, which he said would "range in size from a few ships or air squadrons to formations as large as 100,000 men." Among the specific initiatives being taken to support the RDF, the President mentioned the development and production of a new fleet of large cargo aircraft with intercontinental range and the design and procurement of a force of pre-positioned ships to carry heavy equipment and supplies for three Marine brigades.[68]

Announcement of the Carter Doctrine caught the Joint Chiefs largely off guard. Learning of the decision only a few days before the President's speech, they saw the administration acting hastily and without adequate preparations. Nonetheless, from that point on, JCS planning accelerated quickly, culminating in the activation of the Rapid Deployment Joint Task Force (RDJTF) on March 1, 1980. Headquartered at MacDill Air Force Base, Florida, RDJTF was technically a subordinate element of USREDCOM. But because of its prominent political profile, RDJTF reported directly to the Joint Chiefs of Staff. At Secretary of Defense Brown's request, it also maintained a liaison staff at the Pentagon for politico-military interface with the Joint Staff, OSD, and other agencies.[69] The first commander of RDJTF, Lieutenant General P.X. Kelley, USMC, publicly described the new organization as "an exceptionally flexible force" that would eventually pull together "the capabilities of all four services into one harmonized fighting machine with a permanent command and control headquarters."[70] For the time being, however, RDJTF had no assigned forces and functioned mainly as a headquarters, planning, and advisory organization, much like USREDCOM. For the next several years, RDJTF's primary functions were to organize and supervise exercises acquainting U.S. forces with the peculiarities of operating in the Middle East, and to make plans and preparations for eventually establishing a permanent forward headquarters there.

With the creation of the RDJTF, the Joint Chiefs expected the United States to emerge as the predominant outside power in the Middle East. In years past it had been the British and, to a lesser extent, the French who had carried out that function. Now it was the turn of the United States. Looking ahead, the chiefs could not help but be uneasy. Apart from the political complications involved (most notably the American relationship with Israel), they saw Washington moving into unfamiliar territory where a continuous U.S. military presence could become unavoidable and require a far larger allocation of resources than the current administration was willing to make. The Carter administration had hoped to avoid such commitments. But as it departed in 1981, it passed along a growing list of obligations that left the United States more deeply embroiled in the Middle East and Southwest Asia than ever before.

THE IRAN HOSTAGE RESCUE MISSION

The creation of the RDJTF was a major step toward coping with the volatility that increasingly plagued the Middle East and Southwest Asia in the late 1970s. By the same token, it signaled a partial revival and resurgence of JCS influence within the policy process in Washington. While the chiefs' role had been growing steadily following the Shah's downfall and the ensuing acceleration of contingency planning for Southwest Asia, it came even more to the fore during the subsequent seizure of the American Embassy in Tehran and efforts by the U.S. military in the spring of 1980 to liberate those held hostage. Even though the mission ended in failure, it confirmed that the United States was far from averse to the use of force and that in keeping with the decisions that had given rise to the RDJTF, it would not hesitate to intervene militarily if its interests became threatened.

The event precipitating the hostage crisis was President Carter's decision in late October 1979 to allow the Shah, then in exile in Mexico, to enter the United States for emergency medical treatment. Outraged by what they considered continuing U.S. support of the Pahlavi regime, a mob of Iranian militants stormed the American Embassy in Tehran on November 4 and seized between fifty and sixty Foreign Service officers and Marine guards. Two days later, with the militants still controlling the Embassy and showing no sign of leaving, President Carter authorized Brzezinski to begin exploring options other than diplomacy for securing the hostages' release. One possibility mentioned by General Jones was a rescue effort using helicopters launched from aircraft carriers in or near the Persian Gulf. Yet even though a rescue attempt appeared feasible in theory, the consensus among the Joint Chiefs was that because of the uncertainties involved it stood a "very high risk of failure" and did not appear viable. Brzezinski disagreed and with President Carter's concurrence he ordered the JCS to proceed immediately with preparation of a contingency plan along the lines the CJCS had proposed.[71]

By November 12, 1979, the JCS had established the nucleus of a joint task force within the Joint Staff (J-3) under the command of Major General James E. Vaught, USA, with advisory support from Major General Philip C. Gast, USAF, the former chief of the U.S. Military Assistance Advisory Group in Iran. From that point on, Joint Staff planners ceased to be regularly involved in the rescue. To preserve secrecy, Vaught and his staff worked in isolation and reported directly to the Joint Chiefs through the Chairman. A high-level ad hoc committee chaired by Brzezinski provided overall coordination and interagency liaison from the White House. According to General Jones's retrospective account, he and his JCS colleagues "went through many, many different options." He recalled that, "In the initial stages, we did not see any option that had a reasonable chance of success."[72] But by late November 1979, he and Vaught agreed that the use of helicopters offered the most practical and effective means of conducting the rescue. From this decision evolved plans for Operation *Eagle Claw*. While Jones later denied any explicit deal-cutting to give each Service a share of the action, his assistant, Lieutenant General John S. Pustay, USAF, remembered things differently. According to Pustay, there was a feeling "that it would be nice if everyone had a piece of the pie." Pustay hastened to add, however, that in his view the multi-Service nature of the operation was dictated by its complex requirements and in no way interfered with its execution.[73]

Even though planning was continuous and intense from mid-November 1979 on, it was not until early March 1980 that Jones recalled feeling "a growing confidence" that the rescue mission was coming together in terms of a feasible plan, trained personnel, suitable equipment, and reliable intelligence.[74] To get the hostages out, the Joint Chiefs proposed launching helicopters from carriers in the Arabian Sea, which would then rendezvous with a Delta Force assault team at a remote location in Iran (code-named *Desert One*) and proceed to Tehran. There, they would liberate the hostages, secure the airport, and fly out. A complicated and risky plan, it rested heavily on exploiting the element of surprise and achieving effective inter-Service cooperation and coordination every step of the way.[75]

Whether President Carter would sanction such a hazardous and complex operation remained to be seen. Toward the end of an all-day meeting at Camp David on March 22, Jones presented what Brzezinski described as the "first comprehensive and full briefing on the rescue mission" the President had yet received. Disappointed over the latest failure of diplomacy to free the hostages, Carter was more ready than ever to contemplate military action. But he thought the plan that Jones presented "still needed more work." To help determine its feasibility, he authorized a reconnoitering mission deep inside Iranian territory, the first step toward establishing the *Desert One* base camp for the planned operation.[76]

With JCS preparations nearing the "go-or-no-go" point of the mission, pressure was growing for President Carter to make a decision. While he continued to favor a diplomatic settlement, he thought time was running out and concluded that forceful action was now his most viable—perhaps only option. Accordingly, on April 11, he assembled the National Security Council for a final look at the rescue plan. The meeting lasted nearly 2 hours. Using a pointer and visual aids to illustrate the logistics involved, General Jones insisted that the rescue option had been well rehearsed and was on schedule to commence in late April. Armed with a list of prepared questions, the President found Jones's answers to be much more satisfactory than at previous meetings. The only dissenting view came from Deputy Secretary of State Warren M. Christopher, sitting in for Cyrus Vance, who was on vacation in Florida. Christopher had attended earlier NSC meetings on the rescue mission but had taken no active part in the discussion. Opposed to military interventions in general, he urged caution and thought there were still important diplomatic and economic avenues to be explored. Carter, however, said he had already discussed the matter privately with First Lady Rosalynn, Presidential advisor Hamilton Jordan, Vice President Walter Mondale, Secretary of State Vance, and Jody Powell, the White House Press Secretary, and made up his mind. Shutting off further debate, he announced: "We ought to go ahead without delay."[77]

Despite over 5 months of intensive training and preparation, *Eagle Claw* remained a perilous undertaking in which much could—and did—go wrong. Launched on April 24, 1980, the operation experienced equipment breakdowns almost from the beginning. By the time of the rendezvous at *Desert One*, there were too few helicopters still operational to complete the mission. While preparing to turn around and go home, one of the helicopters collided with a C–130 transport, causing both aircraft to explode. Eight U.S. Servicemen and an Iranian translator died. The ignominious withdrawal that followed (leaving behind most of the dead) effectively doomed President Carter's hopes of ending the hostage standoff and represented a humiliating blow to the power and prestige of the United States.

In the aftermath of the *Desert One* disaster, the Joint Chiefs sought to piece together what happened and why and to learn how similar failures might be avoided in the future. By far the most detailed and thorough examination of the hostage rescue mission was that undertaken at the Chiefs' request by the Special Operations Review Group (SORG), chaired by retired Admiral James L. Holloway III, a former Chief of Naval Operations. The review group drew two general conclusions—that there had been undue emphasis on untested ad hoc arrangements throughout the operation, and that an overriding concern for operational secrecy (e.g., the exclusion of the National Security Agency) had crippled the planning process. Anticipating future missions of the kind, the SORG recommended, and the Joint Chiefs concurred, that there should be a permanent Counterterrorist Joint Task Force (CTJTF), with assigned staff and forces, backed by a special operations advisory panel comprised of high-ranking officers with backgrounds in special operations and joint planning.[78]

While the Joint Chiefs sought to draw constructive lessons, critics leapt on the failure of the hostage rescue mission as further evidence, along with Vietnam and the Mayaguez affair, that the JCS had become an ineffectual organization in urgent need of institutional reform. Many of the legislative changes later incorporated into the 1986 Goldwater-Nichols Act drew their immediate inspiration and impetus from the *Desert One* disaster. To be sure, *Eagle Claw* was a flawed operation. Yet its failure stemmed not from any one cause but from a variety of factors. As much as anything, it revealed the Joint Chiefs' lack of familiarity with the Middle East and the unforeseen difficulties of projecting military power into that part of the world. With tactics, weapons, and training oriented since the onset of the Cold War toward conflicts in Europe or East Asia, the Joint Chiefs of Staff were largely unacquainted with the unique problems of the Middle East and lacked a well-established infrastructure there to support military operations. The creation of the Rapid Deployment Force was supposed to help overcome these problems. But until it became a tested, working reality, the JCS had no choice but to rely on makeshift arrangements and learn as they went along.

The hostage crisis was a desperate, almost unprecedented situation, and it seemed to cry out for desperate, unprecedented measures. Carter knew that the rescue mission was a long shot and never blamed anyone other than himself for its failure. Still, like John F. Kennedy after the Bay of Pigs, he was clearly disappointed with the performance of his military advisors. Never strong to begin with, Carter's confidence in the Joint Chiefs sank even further in the aftermath of *Desert One*. By all accounts, the JCS had done the best they could, but with resources stretched thin in the aftermath of the Vietnam War, their ability to respond effectively in emergencies was severely constrained. In *Desert One* as elsewhere, the effects of the "hollow force" were all too apparent. Carter may have felt that if given a second term, he could have turned the situation around. However, he never had that opportunity. As it happened, that task fell instead to a new administration, operating from a different worldview and a different set of assumptions about national security.

NOTES

1. See Gordon S. Barrass, *The Great Cold War* (Stanford, CA: Stanford University Press, 2009), 190–191.
2. Jimmy Carter, *Why Not the Best?* (Nashville: Broadman Press, 1975), laid out the President's foreign policy agenda. See also Gaddis Smith, *Morality, Reason, and Power: American*

Diplomacy in the Carter Years (New York: Hill and Wang, 1986); and Burton I. Kaufman, *The Presidency of James Earl Carter, Jr.* (Lawrence: University Press of Kansas, 1993).

3 "Inaugural Address of President Jimmy Carter," January 20, 1977, in *Public Papers of the Presidents of the United States: Jimmy Carter, 1977* (Washington, DC: GPO, 1977), 3.

4 Jimmy Carter, *Keeping Faith* (Toronto: Bantam Books, 1982), 222.

5 Erik B. Riker-Coleman, "Political Pressures on the Joint Chiefs of Staff: The Case of General David C. Jones" (Paper presented before the Society of Military History Annual Meeting, Calgary, Alberta, May 2001), 3–5 and passim, available at <www.unc.edu/~chaos1/jones.pdf> (accessed July 18, 2011).

6 Quotations from Gen David C. Jones, USAF (Ret.), former CJCS, interviewed by Steven L. Rearden and Walter S. Poole, February 4, 1998, Arlington, VA, transcript, JHO. See also Mark Perry, *Four Stars* (Boston: Houghton, Mifflin, 1989), 268–269.

7 Bernard Weinraub, "Joint Chiefs Losing Sway Under Carter," *The New York Times*, July 6, 1978: A11.

8 Walter S. Poole, *The Evolution of the Joint Strategic Planning System, 1947–1989* (Washington, DC: Historical Division, Joint Secretariat, Joint Staff, September 1989), 15–18; letters, Lt Gen Richard L. Lawson, J-5, to Brzezinski and Aaron, January 14, 1980, U; Memo, Shoemaker, Utgoff, and Welch to Brzezinski, February 21, 1980, U; and Draft Memo to Dir, OMB, U, all in National Security Adviser Collection, Agency File, box 10, JCS 1/79-2/80 folder, Carter Library.

9 Memo, Carter to Brown, September 20, 1977, "Defense Reorganization," U, JCS 1977/392; OASD(PA) news release no. 529-77, November 17, 1977, cited in JCS 1977/409-5.

10 "Report to the Secretary of Defense on the National Military Command Structure (The Steadman Report)," July 1978, in U.S. Congress, House, Committee on Armed Services, *Hearings: Reorganization Proposals for the Joint Chiefs of Staff*, 97:2 (Washington, DC: GPO, 1982), 912–924.

11 Transcript of News Conference by Gen David C. Jones, CJCS, July 25, 1978, 2–3, National Security Adviser Collection, Agency File, box 10, JCS 3/77-12/78 folder, Carter Library.

12 CM-79-78 to SECDEF, September 1, 1978, "NMC Structure and Departmental Headquarters Studies," U, JCS 1977/409-5.

13 JCSM-290-78 to SECDEF, September 1, 1978, "Comments on NMC Structure and Departmental Headquarters Studies," U, Enclosure A JCS 1977/409-5.

14 "President's News Conference, June 30, 1977," *Carter Public Papers, 1977*, 1197–1200.

15 Carter, *Keeping Faith*, 80–83; Harold Brown, *Thinking About National Security* (Boulder, CO: Westview Press, 1983), 72–74.

16 Jones Interview, February 4, 1998.

17 See Larry A. Niksch, "U.S. Troop Withdrawal from South Korea: Past Shortcomings and Future Prospects," *Asian Survey* 21 (March 1981), 326–328; and Don Oberdorfer, "Carter's Decision on Korea Traced to Early 1975," *The Washington Post*, June 12, 1977: A15.

18 Ernest W. Lefever, "Withdrawal from Korea: A Perplexing Decision," *Strategic Review* 6 (Winter 1978), 28–35; Franklin B. Weinstein and Fuji Kamiya, eds., *The Security of Korea: U.S. and Japanese Perspectives on the 1980s* (Boulder, CO: Westview Press, 1980), 69–106.

Cyrus Vance, *Hard Choices* (New York: Simon and Schuster, 1983), 127–128, and Zbigniew Brzezinski, *Power and Principle* (New York: Farrar, Straus, Giroux, 1983), 127, both mention human rights abuses as a factor shaping the administration's attitude toward South Korea.

19. PD/NSC-18, August 24, 1977, "US National Policy," U, NSC Collection, Carter Library; available at <http://www.jimmycarterlibrary.gov/documents/pddirectives/pd18.pdf> (accessed July 18, 2011).

20. Diary entry, June 4, 1979, in Carter, *Keeping Faith*, 241.

21. Raymond L. Garthoff, *Détente and Confrontation: American-Soviet Relations from Nixon to Reagan* (Washington, DC: Brookings, 1994, rev. ed.), 868; and David M. Walsh, *The Military Balance in the Cold War: U.S. Perceptions and Policy, 1976–85* (London: Routledge, 2008), 26.

22. Leon Sloss and Marc Dean Millot, "U.S. Nuclear Strategy in Evolution," *Strategic Review* 12 (Winter 1984), 19–28; and Walter Slocombe, "The Countervailing Strategy," *International Security* 5 (Spring 1981), 18–27.

23. Brown, *Thinking About National Security*, 81.

24. Jones Interview with author, February 4, 1998.

25. Brzezinski, *Power and Principle*, 458–459.

26. PD/NSC-59, July 25, 1980, "Nuclear Weapons Employment Policy," National Security Advisor's Papers, Carter Library (declassified/sanitized), available at <http://www.jimmycarterlibrary.gov/documents/pddirectives/pd59.pdf> (accessed July 18, 2011).

27. Desmond Ball, "The Development of the SIOP, 1960–1983," in Desmond Ball and Jeffrey Richelson, eds., *Strategic Nuclear Targeting* (Ithaca, NY: Cornell University Press, 1986), 79.

28. Harold Brown, "Domestic Consensus and Nuclear Deterrence," in *Defence and Consensus: The Domestic Aspects of Western Security*, Adelphi Paper No. 183 (London: IISS, 1983), Part II, 21. See also Brown's testimony explaining the countervailing strategy in U.S. Congress, Senate, Committee on Foreign Relations, *Hearing: Nuclear War Strategy*, 96:2 (Washington, DC: GPO, 1981), 6–28.

29. See Carter's interview in Michael Charlton, *From Deterrence to Defense: The Inside Story of Strategic Policy* (Cambridge, MA.: Harvard University Press, 1987), 88.

30. Brzezinski, *Power and Principle*, 157.

31. "Inaugural Address of President Jimmy Carter," January 20, 1977, *Carter Public Papers, 1977*, 3.

32. Minutes, SCC Meeting, February 3, 1977 (declassified), sub: SALT, National Security Adviser Collection, Staff Offices, box 3, SCC Meeting No. 2 folder, Carter Library.

33. Ibid.

34. Former Soviet Foreign Minister Alexander Bessmertnykh quoted in Barrass, 206.

35. Vance, 53–55; Garthoff, 629–634, 883–892.

36. Edward L. Rowny, *It Takes One to Tango* (Washington, DC: Brassey's, 1992), 94–103.

37. Patrick Glynn, *Closing Pandora's Box: Arms Races, Arms Control, and the History of the Cold War* (New York: Basic Books, 1992), 301–304, summarizes arguments against the treaty. See also Thomas W. Wolfe, *The SALT Experience* (Cambridge, MA: Ballinger,

1979), 236–239, which inventories the pros and cons; and Dan Caldwell, *The Dynamics of Domestic Politics and Arms Control: The SALT II Treaty Ratification Debate* (Columbia: University of South Carolina Press, 1991), offers a balanced account of the ensuing debate in Congress.

38 Jones's testimony, July 24, 1979, in U.S. Congress, Senate, Committee on Armed Services, *Hearings: Military Implications of the Treaty on the Limitation of Strategic Offensive Arms and Protocol Thereto (SALT II Treaty)*, 96:1 (Washington, DC: GPO, 1979), Pt. 1, 151–160.

39 Carter, *Keeping Faith*, 264–265.

40 "Address Before the World Affairs Council," May 9, 1980, *Carter Public Papers, 1980–81*, 872.

41 PRM/NSC-10 "Final Report: Military Strategy and Force Posture Review," undated, 9, enclosure to Memo, SECDEF to SecState et al., June 6, 1977, "PRM-10 Force Posture Study" (declassified/sanitized), available at <http://www.jimmycarterlibrary.gov/documents/prmemorandums/prm10.pdf> (accessed July 18, 2011).

42 Glynn, 312.

43 Gen Andrian Danilevich quoted in Barrass, 212–213.

44 James A. Thomson, "The LRTNF decision: evolution of U.S. theatre nuclear policy, 1975–9," *International Affairs* 60 (Autumn 1984), 603–604.

45 Robert J. Art and Stephen E. Ockenden, "The Domestic Politics of Cruise Missile Development, 1970–1980," in Richard K. Betts, ed., *Cruise Missiles: Technology, Strategy, Politics* (Washington, DC: Brookings, 1981), 400–401.

46 For a colorful account of the problems surrounding the neutron bomb, see Sam Cohen, *The Truth About the Neutron Bomb* (New York: William Morrow, 1983). For more analytical treatments, see Sherri L. Wasserman, *The Neutron Bomb Controversy* (New York: Praeger, 1983), 21–36; and Milton Leitenberg, "The Neutron Bomb—Enhanced Radiation Warheads," *Journal of Strategic Studies* 5 (September 1982), 341–369.

47 Helmut Schmidt, "The 1977 Alastair Buchan Memorial Lecture," October 28, 1977, *Survival* 20 (January-Februay 1978), 3–4.

48 Special Meeting of Foreign and Defence Ministers, Brussels, Communiqué of December 12, 1979, *NATO Final Communiqués, 1975–80* (Brussels: NATO Information Service, n.d.), 121–123.

49 JCS views on the Panama Canal Treaty in U.S. Congress, Senate, Committee on Foreign Relations, *Hearings: Panama Canal Treaties*, 95:1 (Washington, DC: GPO, 1977), Pt. 1.

50 Jeffrey A. Lefebvre, *Arms for the Horn: U.S. Security Policy in Ethiopia and Somalia, 1953–1991* (Pittsburgh: University of Pittsburgh Press, 1991), 175–196.

51 PD/NSC-18, "US National Strategy," August 24, 1977, NSC Collection, Carter Library; available at <http://www.fas.org/irp/offdocs/pd/pd18.pdf>. See also Brzezinski, *Power and Principl*, 177–178, 455–456; and Robert P. Haffa, Jr., *The Half War: Planning U.S. Rapid Deployment Forces to Meet a Limited Contingency, 1960–1983* (Boulder, CO.: Westview Press, 1984), 50–52.

52 P.X. Kelley, "Rapid Deployment: A Vital Trump," *Parameters* 11 (March 1981), 51.

53 Ronald H. Cole et al., *The History of the Unified Command Plan, 1946–1999* (Washington, DC: Joint History Office, 2003), 30–31, 36–38.

54 Kissinger, *White House Years*, 1261.

55 JCSM-525-70 to SECDEF, November 10, 1970, "US Military Mission with Iran and U.S. Military Advisory Group to Iran," JCS 2315/498-2.

56 Brzezinski, *Power and Principle*, 354; Carter, *Keeping Faith*, 437–439.

57 Brzezinski, *Power and Principle*, 390–391; Vance, 335–338; Daniel L. Haulman, *The United States Air Force and Humanitarian Airlift Operations, 1947–1994* (Washington, DC: Air Force History and Museums Program, 1998), 360–361.

58 See Robert E. Huyser, *Mission to Tehran* (New York: Harper & Row, 1986), 283–284.

59 *The New York Times*, December 28, 1978: 1.

60 Robert A. Pastor, *Condemned to Repetition: The United States and Nicaragua* (Princeton, NJ: Princeton University Press, 1987), 50–56.

61 PRM/NSC-46, May 4, 1979, "Review of U.S. Policies Toward Central America," NSC Files, Carter Papers; available at <http://www.jimmycarterlibrary.gov/documents/prmemorandums/prm46.pdf> (accessed July 18, 2011).

62 Pastor, 192–194.

63 Matthew M. Aid, *The Secret Sentry: The Untold History of the National Security Agency* (New York: Bloomsbury Press, 2009), 166–167.

64 Handwritten notes labeled "Meeting w/CJCS and MG Dyke," August 14, 1979, with emphasis in original, U, RG 218, 898/320 (January 11, 1979).

65 Historical Division, "The Rapid Deployment Mission" (MS, Joint Secretariat, Joint Chiefs of Staf, February 4, 1981), U, 6; Amitav Acharya, *U.S. Military Strategy in the Gulf* (London: Routledge, 1989), 63–67; Henrik Bliddal, *Reforming Military Command Arrangements: The Case of the Rapid Deployment Joint Task Force*, Letort Papers Series (Carlisle, PA: U.S. Army War College, Strategic Studies Institute, March 2011), 25–34, and <http://www.strategic-studiesinstitute.army.mil/pdffiles/PUB1048.pdf> (accessed July 18, 2011).

66 "Soviet Invasion of Afghanistan: Address to the Nation," January 4, 1980, *Carter Public Papers*, 22.

67 "State of the Union Address," January 23, 1980; ibid., 194–199.

68 "State of the Union: Annual Message to the Congress," January 21, 1980; ibid., 166.

69 Cole et al., *History of the UCP*, 58–59.

70 *A Discussion of the Rapid Deployment Force with Lieutenant General P.X. Kelley* (Washington, DC: American Enterprise Institute, 1980), 3–4.

71 Carter, *Keeping Faith*, 459; Gary Sick, *All Fall Down: America's Tragic Encounter with Iran* (New York: Random House, 1985), 213–216.

72 "News Conference by SECDEF Brown and CJCS Jones," April 29, 1980, *Public Statements of Harold Brown, Secretary of Defense, 1980–1981* (Washington: Historical Office, Office of the Secretary of Defense, n.d.), IV, 1446–1447.

73 John E. Valliere, "Disaster at Desert One: Catalyst for Change," *Parameters* 22 (Autumn 1992), 78.

74 "News Conference by SECDEF and CJCS," April 29, 1980, *Brown Public Statements, 1980–81*, IV, 1450.

75 Charlie A. Beckwith and Donald Knox, *Delta Force* (San Diego: Harcourt Brace Jovanovich, 1983), 253–256; Carter, *Keeping Faith*, 509–510; Zbigniew Brzezinski, "The

Failed Mission: An Inside Account of the Attempt to Free the Hostages in Iran," *The New York Times Magazine*, April 18, 1982: 30–31; Sick, 285–287; Paul B. Ryan, *The Iranian Rescue Mission: Why It Failed* (Annapolis: Naval Institute Press, 1985), 1–2.

76 Carter, *Keeping Faith*, 501; Brzezinski, *Power and Principle*, 487.

77 Carter, *Keeping Faith*, 506–507; Vance, 409; and Brzezinski, *Power and Principle*, 492–493.

78 Iran Hostage Rescue Mission Report, August 1980, 57–62, Naval Historical Center files, available at <http://www.history.navy.mil/library/online/hollowayrpt.htm#execsum> (accessed July 18, 2011).

General John W. Vessey, Jr., Chairman, Joint Chiefs of Staff, 1982–1985

Chapter 14

THE REAGAN BUILDUP

By 1981, détente was dead, the victim of overoptimism by its proponents in Washington and presumptive behavior by Moscow. That it waxed and waned came as no surprise to the Joint Chiefs of Staff, who were skeptical all along of whether détente would last, let alone fundamentally alter East-West relations. Toward the end of his Presidency, Jimmy Carter reluctantly agreed and initiated upward adjustments in the military budget. The "Carter buildup" was a limited affair, however, and did not go much beyond bolstering capabilities for the Rapid Deployment Force. As useful as these increases may have been, they were not enough, in the opinion of the Chairman, General David C. Jones, USAF, to offset the gains made by the Soviets in nuclear and conventional arms over the past decade or to reverse the "long term decline in our defense spending."[1]

To the incoming Reagan administration, strengthening the country's defense posture was top priority. During the 1980 Presidential campaign, Ronald Reagan had mounted a relentless attack on the Carter administration for neglecting national defense and for accepting the expansion of Communist power and influence around the globe. Rekindling memories of the "liberation doctrine" advocated by John Foster Dulles 3 decades earlier, Reagan swept into office promising to "roll back" Communist influence and seeking a stronger military to back him up. The ensuing buildup, soon to become the touchstone of Reagan's Presidency, dwarfed any the country had seen since World War II.

REAGAN AND THE MILITARY

Underlying President Reagan's commitment to strengthening the Nation's military posture was his belief that for it to succeed he would need to change the country's image of the Armed Forces, which was little improved since Vietnam. If he accomplished nothing else during his Presidency, Reagan wanted to lay the ghosts of that war to rest and revive respect for those serving in uniform. The credibility of his whole approach to foreign and defense policy depended on it. An old-fashioned patriot who was proud of having been an Army officer in World War II (even though he remained in Hollywood making training movies for the War Department), Reagan regarded service in the Armed Forces as character-building and the military itself as an integral part of the country's historic greatness. "I told the Joint Chiefs of Staff," he recalled, "that I wanted to do whatever it took to make our men and women proud to wear their uniforms."[2]

The President's high esteem for the military notwithstanding, the leading architects of the Reagan buildup were predominantly civilians. Some, such as Secretary of Defense Caspar W. Weinberger and Director of Central Intelligence William J. Casey, were long-time political associates and personal friends of the President's. Others, including Under Secretary of Defense for Policy Fred C. Iklé, Assistant Secretary of Defense Richard N. Perle, arms control specialist Paul H. Nitze, and Russian expert Richard E. Pipes who came down from Harvard to work on the NSC, were people Reagan had met in the 1970s through his participation in the Committee on the Present Danger.[3] Only two high-level advisors—Alexander M. Haig, Jr., a former SACEUR, and retired Army Lieutenant General Edward L. Rowny—came from career military backgrounds. Haig served as Reagan's first Secretary of State and lasted barely a year and a half before policy disputes with Weinberger forced him to step down. His successor at State, George P. Shultz, was a corporate executive in private life and another of the President's personal friends. Rowny, the former JCS representative to SALT II, served as Reagan's senior advisor and chief negotiator on strategic arms control but made only limited inputs into shaping the buildup.

Reagan was business-like and initially cautious in dealing with the Joint Chiefs. Being holdovers from the Carter administration, the chiefs served under a cloud from that association. According to former Secretary of the Air Force Thomas C. Reed, they must have "cringed as Reagan and Weinberger talked about the 'decade of neglect' over which these officers had presided."[4] The most suspect was the Chairman, General David Jones, who had come under repeated attack from the President's conservative Republican supporters for being "too political" and too closely linked to the policies of the previous administration. Some wanted him sacked immediately. Seeing himself as "a nonpolitical moderate," Jones vowed to fight any attempt at dismissal. Weinberger, weighing the pros and cons, eventually became convinced that trying to fire him would be more trouble than it was worth.[5]

While Jones finished his term as Chairman, he was never part of "the family," as Reagan's inner circle of advisors was called. Nor did his advice carry much weight at the White House or with Weinberger. "Jones was an able man," Weinberger recalled, "but I never felt that he was quite as comfortable with me as his successors were."[6] Jones, for his part, observed tactfully that he and the Secretary of Defense labored under a "cool, never close relationship."[7] Personalities obviously entered in, but there were also serious substantive differences covering a wide range of issues, from the size and allocation of resources under the buildup, to strategic weapons policy, arms control, and the reenergized debate over JCS organizational reform. Though Jones and Weinberger tried to keep their differences behind closed doors, it was not long before their disagreements spilled over into high-level interagency deliberations and eventually into the public arena.

In March 1982, President Reagan announced that General John W. Vessey, Jr., USA, Vice Chief of Staff of the Army, would become Chairman when Jones retired that summer. At the same time, Reagan announced that Admiral James D. Watkins, a former nuclear submarine commander, would succeed Thomas Hayward as Chief of Naval Operations, and that General Charles A. Gabriel, a fighter pilot, would replace Lew Allen, Jr., as Air Force Chief of Staff.[8] A year later, Reagan completed the transformation by naming General John A. Wickham, Jr., to be the next Army Chief of Staff, succeeding General Edward C. Meyer, and General Paul X. Kelley, the organizer and former commander of the Rapid Deployment Joint Task Force, to be the next Commandant of the Marine Corps.[9]

As a practical matter, the "new" Joint Chiefs of Staff were not much different in outlook from the "old." But as Reagan appointees, they bore a greater degree of personal responsibility for helping to develop and implement the administration's policies and programs. The member who had the most dealings with the White House and who was by extension the most closely identified with the administration's policies was the Chairman, General Vessey. Lauded by President Reagan as a "soldier's soldier" in the tradition of General of the Army Omar N. Bradley, Vessey was the last World War II combat veteran to serve as Chairman. He was also the only one who had never been a Service chief or the head of a unified or specified command. His tour of duty in the 1970s as the U.S.-UN commander in Korea was, however, easily comparable to that of a unified command and in some ways far more demanding. Like Jones, his training and education had been on-the-job and at public universities, not at one of the Service academies. Popular and respected among his peers, he had fought in three wars and was well known and highly regarded by key foreign leaders. As Chairman, he spent a good deal of time traveling in troubled locales such as East Asia and the Middle East, where he had numerous friends and contacts.

Vessey and Weinberger formed a highly effective and productive partnership almost immediately. "I have rarely worked with anyone," Weinberger recalled, "for whom I had greater respect and admiration." In contrast to his intermittent contacts with Jones, Weinberger met with Vessey practically every morning for half an hour or more to go over business. Each Tuesday, Weinberger would meet collectively with Vessey and the Service chiefs in the "Tank," the JCS conference room in the Pentagon's National Military Command Center. Though Weinberger and the chiefs did not always see eye-to-eye, each one knew where the others stood on key issues.[10]

Vessey's advent sparked an immediate improvement in White House–JCS relations. Reagan and Vessey got on well (both were avid storytellers and enjoyed trading jokes) and usually saw one another two or three times a week at NSC meetings and other high-level functions. Operating in the "team player" tradition, Vessey saw himself as a bridge between the JCS, on the one hand, and the Secretary of Defense and the President, on the other. As one of the conditions for accepting the Chairmanship, Vessey insisted that the President meet regularly with the JCS as a group for an informal review of major issues and to become better acquainted. The first such meeting took place in early July 1982, shortly after Vessey became Chairman. Afterwards, Vessey sent Judge William P. Clark, the President's Assistant for National Security Affairs, a handwritten note recommending that similar meetings be held about every 6 months. The next meeting took place in December 1982 and proved so successful that Reagan wanted such sessions to be a regular part of his agenda. Thereafter, the meetings were held quarterly, continuing on through the end of Reagan's Presidency. In 1987, the chiefs broadened the format to include presentations by the combatant commanders at the President's invitation. Though the specific impact of these meetings is hard to assess, the overall impression is that they resulted in closer ties between the President and his military advisors and a more informed decisionmaking process all around.[11]

Improved cooperation at the top did not automatically translate into more useful and effective military policies. Indeed, the Reagan years were notorious for lapses and mismanagement of foreign and defense affairs that left the Joint Chiefs at times bewildered over what the President was trying to accomplish. In keeping with their military culture, the Joint Chiefs preferred clearly defined organizational roles and lines of authority. What they often got during the Reagan years were vague directives, lax assignments of authority, and contradictory behavior from the President and his subordinates. Whenever Reagan was personally interested or involved in a problem, things were apt to get done. Otherwise, the looseness of the overall structure led to a day-to-day system that often broke down and repeatedly failed to assure consistent policies or effective execution and follow-up. The most celebrated example was the 1986 Iran-Contra affair, in which rogue elements of the NSC Staff, including a Marine lieutenant colonel, Oliver L. North, used the proceeds from clandestine arms sales to Iran to finance unauthorized assistance to anti-Communist guerrillas in Central America. Though not involved directly in Iran-Contra, the Joint Chiefs were affected nonetheless. Iran was under a U.S.-imposed arms embargo that the JCS were responsible for helping to administer and enforce. But, with key elements of the President's own staff undercutting that policy, its credibility became immediately suspect and all the harder for the JCS to oversee and implement.

In addition to coordination breakdowns with the White House, the Joint Chiefs faced recurring difficulties with the Office of the Secretary of Defense. The explanation for these persistent confrontations lies in part in the basic friction-prone nature of civil-military relations and in the overlapping functions that OSD and the JCS had come to perform. But there was also a strong element of resentment on the part of the JCS and serving officers on the Joint Staff toward what many regarded as the unnecessary and unwarranted intrusion by the Secretary of Defense and members of his staff into the military planning process. In the critical areas of nuclear and conventional strategy, the allocation of resources, and defense budget planning, the oversight and direction exercised by upper- and mid-level OSD officials during Weinberger's tenure easily matched or exceeded that of previous administrations. While the Service chiefs welcomed and appreciated the budget increases the Reagan administration provided, they would have preferred less direct and heavy-handed supervision in managing the buildup, something more in line with the easygoing working partnership and sense of trust and understanding Vessey enjoyed with Weinberger and Reagan.

FORCES AND BUDGETS

The Reagan buildup was anything but orderly and systematic. According to David A. Stockman, Director of the Office of Management and Budget, the Reagan administration entered office with few specific plans and only a generalized estimate of military requirements.[12] Promises and pronouncements made during the campaign looked ahead to a 600-ship Navy containing "more aircraft carriers, submarines, and amphibious ships," early deployment of the MX intercontinental ballistic missile in "a prudent survivable configuration," revival of the B–1 bomber program, and an all-around increase in the readiness and industrial preparedness of the Armed Forces. Left unclear were the priority of programs and the strategic concept that would guide the allocation of resources in achieving these goals.[13]

The absence of a detailed blueprint notwithstanding, the new administration moved promptly with requests for line-item amendments to the already enacted FY81 defense budget and to the FY82 estimates President Carter submitted to Congress before he left office. Known as the "get well" budget, these amendments proposed immediate net increases of $32.6 billion in budget authority (i.e., cash and unfunded contracts), mostly in the form of add-ons to existing programs.[14] At the time, defense accounted for just over 5 percent of the country's gross national product (GNP). Assuming the President and his advisors were serious about making a difference, the Joint Chiefs recommended aiming for a target of between 6 and 7 percent of GNP, a fairly close approximation of what the administration achieved.[15]

Despite the Joint Chiefs' eagerness to be included in shaping the buildup, they found themselves more or less relegated to the sidelines while Weinberger and Stockman and their aides thrashed out the details of a future military spending program.[16]

Under the agreed formula, the target would be an annual 7 percent real increase in defense budget authority over the next 5 years, starting in FY83 and using FY82 as the base. The net increase, when inflation, the get-well additions, and other supplements were figured in, would boost defense spending by about 14 percent above projected levels in President Carter's last 5-year projection. Later, claiming that Weinberger had hoodwinked him, Stockman declared the proposed 7 percent increase "flagrantly excessive as a matter of pure fiscal affordability" and predicted it would cause crippling budget deficits.[17] But despite his fiscal conservatism, Reagan accepted the risk. "He said frequently to me," Weinberger remembered, "that if it ever came down to a choice between balancing the budget and spending enough to regain and keep our military strength, he would always come down on the side of the latter."[18]

These decisions produced the largest and most sustained expansion of military spending since World War II. In current dollars, defense budget authority rose from $178 billion in FY81 to $291 billion by the end of the decade, an increase of over 60 percent. The buildup would have been bigger had Congress not trimmed the administration's requests almost every year, starting with the FY82 budget. In constant (FY92) dollars, the picture was somewhat different and showed the administration reaping most of its gains during its first 5 years in office, when defense spending increased in real terms by roughly a third. At its peak in fiscal years 1986 and 1987, the buildup consumed 6.6 percent of the country's GNP. Thereafter, defense spending tapered off, so that by the end of the decade the military budget was again experiencing negative growth and had fallen to 5.8 percent of GNP (see figure 14-1). All the same, the Reagan buildup was an impressive "peacetime" accomplishment, with only the rearmament program initiated during the Korean War offering anything comparable in scale and scope.[19]

While the Joint Chiefs of Staff welcomed the Reagan buildup as long overdue, they were also concerned that it might not be enough to do the job the administration had set for itself. Even with the additions the President proposed, the JCS saw a yawning gap between available capabilities and the administration's perceived objectives. Along with the proffered increases in defense spending came heavier demands on the Armed Forces and a succession of unplanned tasks (e.g., naval deployments against Libya, peacekeeping operations in Lebanon, and the intervention in Grenada) that put unexpected stresses and strains on the military (see below). Also, the initial absence of a coordinated, high-level statement of basic national policy encouraged inter-Service competition for funds and stymied the development of JCS strategic plans that might have helped to clarify the allocation of functions and resources, especially in the Middle East and Southwest Asia. The only guidance the Joint Chiefs had before them came from the Office of the Secretary of Defense in the course of the normal budget process. In October 1981, in place of the one-and-a-half war planning scenario used since the Nixon administration,

Figure 14-1.

Department of Defense Budget Authority FYs 1981–1989 (in billions)									
	FY81	FY82	FY83	FY84	FY85	FY86	FY87	FY88	FY89
Current Dollars	$178	$214	$239	$258	$287	$281	$279	$284	$291
Constant Dollars (FY 92)	$272	$304	$328	$343	$366	$350	$337	$329	$324
Percentage Real Growth	13.1	11.5	8.1	4.6	6.7	-4.4	-2.1	-1.5	-2.4
Percentage Distribution of GNP	5.4	6.0	6.3	6.2	6.4	6.6	6.6	6.2	5.8

Source: U.S. Department of Defense, Office of the Comptroller, *National Defense Budget Estimates for FY92* (Washington, DC: Department of Defense, March 1991), 98, 147.

OSD substituted a more demanding requirement that the JCS prepare for up to three near-simultaneous conflicts—a major NATO–Warsaw Pact war in Europe, and lesser conflicts in Korea and the Persian Gulf. The Joint Chiefs duly complied but thought that OSD was setting an unrealistic agenda. In General Jones' opinion, the administration seemed bent on "trying to do everything."[20]

Even though the buildup was an across-the-board affair, its first order of business was to redress "the deteriorated strategic balance with the Soviet Union."[21] Reagan abhorred nuclear weapons as much as his predecessor and routinely called for their complete abolition. But he was even more averse to the risks entailed in falling farther behind the Soviet Union in effective strategic power and sided with the Joint Chiefs, believing the time had come for the United States to regain and maintain at least essential strategic nuclear equivalence with the Soviet Union. Strategic forces had been declining almost steadily as a share of the defense budget since the mid-1960s. Under Reagan, they rose from 7 percent of military spending in FY81 to a peak of 9.5 percent in FY85. Shortly after taking office, the administration set the stage for a major expansion of strategic capabilities to close what the President and his advisors often characterized as the "window of vulnerability." Among the strategic program objectives the President approved were the temporary deployment of up to one hundred MX ICBMs in existing Minuteman III or Titan silos, creation of a more survivable command and control system for nuclear war, modernization of the strategic bomber force with the introduction of two new types of bombers (a revived B–1 program and the recently inaugurated B–2 "Stealth"), an increase in the accuracy and payload of submarine-launched ballistic missiles, and stepped-up research and development of ballistic missile defense.[22]

The emphasis on improving strategic systems notwithstanding, spending on conventional capabilities accounted for the largest part of the buildup. During the 1970s, general purpose forces had averaged 35 percent of the military budget; under Reagan they rose to more than 40 percent. Much of the conventional buildup focused on improving power-projection capabilities, which involved nearly tripling the budget for air- and sealift support and building three new carrier battle groups for the Navy. Although the administration paid close attention to strengthening ground and tactical air capabilities, Weinberger acknowledged that naval expansion received preferential treatment because of its unique capacity to mount "offensive missions."[23] Aiming eventually for a fleet of 600 combatant vessels, Navy planners justified their shipbuilding and modernization goals under a "maritime strategy" involving forward deployment and vigorous offensive operations against the Soviet fleet in the North Atlantic.[24] Skeptics, pointing to the losses inflicted by Argentine stand-off EXOCET cruise missile attacks on the Royal Navy in the 1982 Falklands conflict, questioned the wisdom of heavy new investment in surface vessels. Nonetheless, the Navy's maritime strategy fascinated President Reagan and enjoyed strong backing among key members of Congress.[25] The Joint Chiefs never explicitly endorsed the maritime strategy but did support the expansion of naval forces in conjunction with the overall buildup. As a practical matter, they found the most urgent demands for naval power during the Reagan years arising from increased U.S. involvement in the Mediterranean and the Persian Gulf.

In May 1982, the President finally approved a statement of basic national security policy (NSDD 32) to help guide the buildup. The most detailed treatment of its kind in years, NSDD 32 rested on an alarming depiction of Soviet military power and sanctioned across-the-board preparations for possible conflict, from low-intensity encounters with Soviet "client" states like Libya, Cuba, and Nicaragua, to regional conventional wars and even nuclear exchanges between the United States and the Soviet Union. Adopting the long-haul philosophy that had guided the Eisenhower administration's defense policy, Reagan's strategic concept accepted the threat of tensions and confrontations with the Soviet Union as a continuous condition and suggested that maintaining a high level of military preparedness would have to go on indefinitely.[26]

As bleak as the outlook seemed, there was also cause for guarded optimism. Shortly after approving NSDD 32, President Reagan received a British intelligence report via the CIA pointing to a progressive disintegration of the Soviet system. Entitled *The Malaise of Soviet Society*, the British assessment cited extensive evidence of crime, corruption, and economic deterioration throughout the Soviet Union. For all its military power, the report suggested, the Soviet Union was a giant with feet of clay that was crumbling from within.[27] This was not the first such depiction of the Soviet Union that President Reagan saw, nor would it be the last. Yet it was a stark confirmation of what the President himself had been saying for years—that communism was a failed concept and that with time and patience, Western-style democracy and capitalism would triumph. With the military buildup, Reagan saw the United States not only protecting its interests but adding further to the economic and political pressure that would sooner or later end the Cold War and help bring down the Soviet Union.

MILITARY POWER AND FOREIGN POLICY

Even before the buildup's full effects could be felt, the prospect of a stronger defense establishment encouraged the President and his senior advisors to adopt more forceful foreign policies to push back the frontiers of Communist power and influence. While cautious in challenging Moscow directly, Reagan was less restrained when opportunities arose to undermine the Soviet Union elsewhere, by encouraging the Polish democratic trade union movement "Solidarity," for example, or by putting politico-military pressure on Soviet "puppet" regimes in the Middle East, Southern Africa, and Latin America. Basically, the administration's strategy combined overt and covert assistance to "those who are risking their lives . . . to defy Soviet-supported aggression," with the selective application of U.S. military power. The result, sometimes known as the Reagan Doctrine, was a more proactive anti-Communist foreign policy than anything seen since the Vietnam War.[28]

Early instances of this policy in operation included the U.S. naval exercises in the Gulf of Sidra in August 1981, held in an effort to destabilize Libyan strongman and Soviet collaborator Muammar Qaddafi; military intervention in Beirut in 1982–1984; the invasion of Grenada in October 1983; and the steady expansion of U.S. economic and military assistance to Central America from late 1981 on. Strictly speaking, the Beirut intervention was part of a multinational peacekeeping operation to promote stability in the aftermath of Israel's thrust into Lebanon to destroy the Palestine Liberation Organization, while the invasion of Grenada was a spur-of-the-moment rescue mission to protect U.S. citizens caught in the middle of a leftist putsch. Both, however, had similar underlying objectives that aimed at thwarting Soviet proxies—the Syrians who aspired to dominate Lebanon and the Cubans who were entrenching themselves in Grenada. Meanwhile, the administration dropped nearly all the Carter-era prohibitions on foreign arms sales, rescinded the previous administration's "leprosy letter" that had proscribed embassy assistance to American weapons dealers operating overseas, and declared arms transfers to friendly governments to be an "essential element" of American foreign policy.[29]

The stepped-up pace of American involvement abroad imposed unexpected requirements on the Joint Chiefs at a time when the buildup was still in its infancy and available resources were as yet little improved from those on hand in the 1970s. Exhibiting customary caution, the JCS urged restraint in dealing with the unpredictable Qaddafi and initially argued against military intervention in both Lebanon and Grenada until diplomatic and other options had been thoroughly explored and exhausted.[30] Invariably overruled, the JCS became the targets of sharp criticism when things went wrong, as exemplified by the flawed inter-Service coordination during the awkwardly executed Grenada operation and command and control problems preceding the Islamist terrorist attack on the Marine barracks in Beirut in October 1983 that left 241 U.S. Servicemen dead. In both instances, the JCS acknowledged that they could—and should—have done a better job. As a result, they found their image more tarnished than ever, with their methods and procedures increasingly under scrutiny by a skeptical Congress that was eyeing the possibility of wholesale JCS reorganization and reform.

Despite the problems the administration's proactive foreign policy posed, the Joint Chiefs welcomed the change. In reviewing President Carter's record, they found it lacking in long-range vision and replete with inconsistency in fulfilling U.S. commitments. Expecting the Reagan White House to do better, they were encouraged that the United States no longer appeared to be in retreat from problems abroad and soon found that they had Oval Office leadership that would back them up. During the planning for the Gulf of Sidra exercises, for example, the question arose of how the Navy should respond if threatened by Libyan aircraft. Without hesitating, Reagan assured the chiefs that they should not be afraid to let U.S. pilots chase opposing Libyan planes "right into the hangar." Critics denigrated this kind of guidance as "cowboy antics," but it was clear enough to the Joint Chiefs to give them rules of engagement they could readily understand and apply in difficult situations.[31]

Where the Joint Chiefs were least comfortable with the administration's foreign policy was in its lack of explicit sanction, other than the President's authority as Commander in Chief, to use U.S. military power. This issue was a recurring theme in administration debates and reflected the lingering effects of Vietnam on military thinking. It was especially troublesome for the JCS in developing responses to the Sandinistas and Soviet- and Cuban-sponsored insurgencies in Central America. The Joint Chiefs had no doubt that these movements posed a serious threat to U.S. security interests and required prompt and decisive action. However, they balked at the prospect of a protracted struggle undertaken without clear support from the public and Congress. President Reagan was no more inclined than his military advisors to see U.S. combat troops introduced into Central America. But he was equally determined to block any further Communist takeovers. The upshot was a quasi-covert war orchestrated by the CIA and organized around support of the Nicaraguan Contras, a coalition of anti-Sandinista insurgents formed after Somoza's downfall.³² Repeatedly attacked by liberal Democrats in Congress, the administration's Central America policy became intensely controversial and hard to manage amid frequently changing legislative mandates that restricted the types and amounts of aid the administration could provide. Though the Joint Chiefs arranged occasional air drops and other logistical support to the Contras, they studiously avoided direct U.S. military contact and involvement. Meanwhile, U.S. Southern Command, headquartered in Panama, simply looked the other way whenever the Contras mounted operations against the Sandinistas and their Cuban allies.³³

A similar pattern emerged with respect to administration policy toward Afghanistan where, again, the CIA had the lead. Here, however, given the recent creation of the Rapid Deployment Force and a growing security mission across Southwest Asia, the JCS would have preferred a more active role and a corresponding regional strategy with a clear-cut allocation of resources and responsibilities. Instead, as Secretary of State Shultz described it, the policy that evolved placed the CIA in charge of running the war and assisting the anti-Soviet "freedom fighters" (the *mujahideen*). Shoring up U.S. defense and security interests fell to the Joint Chiefs of Staff. A rather ambiguous governing directive (NSDD 99), adopted in the summer of 1983, acknowledged that the United States had vested interests across the region that would require involvement at various levels, the prepositioning of military supplies and equipment, and a U.S. presence for the indefinite future.³⁴

To handle their growing responsibilities in the Middle East and Southwest Asia, the Joint Chiefs upgraded the Rapid Deployment Force to a separate unified command on January 1, 1983. Known as U.S. Central Command (USCENTCOM), the new organization was supposed to be less susceptible to the pressures of inter-Service rivalry and friction than the RDJTF, in part because it operated with a more defined charter that placed it in charge of protecting U.S. interests within a specific area of responsibility (AOR) stretching from Pakistan to the Horn of Africa. Carefully excluded, however, was any USCENTCOM involvement with Israel, a move taken at JCS insistence to protect the new command's credibility and operational flexibility in the Arab world. As a practical matter, guaranteeing the security of the Persian Gulf and its oil supplies was USCENTCOM's uppermost concern.³⁵

The Joint Chiefs initially identified a Soviet invasion of Iran launched through neighboring Afghanistan as the primary threat to the region.³⁶ But as they became more familiar with the Persian Gulf and its problems, they recognized that the dangers were more complex than previously assumed. The turning point in JCS thinking was the 1984 "tanker war" between Iran and Iraq during which the United States, in cooperation with other Western powers, accepted responsibility for escorting neutral shipping through the Persian Gulf. As a rule, however, the Joint Chiefs shied away from major force commitments to the Middle East and embraced instead a "current force strategy" that relied on periodic show-of-force deployments, the expansion of support facilities on Diego Garcia, and combined exercises with friendly governments to underscore the U.S. commitment. Prior to the first Gulf War of 1990–1991, the only forces permanently attached to USCENTCOM were a small flotilla, the Middle East Force, which had routinely patrolled the Persian Gulf since the late 1940s. Meanwhile, under a deployment instigated by the Carter administration, two carrier task forces—normally one from U.S. Pacific Command and the other from U.S. Atlantic Command—operated periodically in the adjacent Arabian Sea. According to General Vessey, the Joint Chiefs wanted Central Command "to be very visible in the region" as a deterrent, but to carry limited capabilities. Vessey characterized this strategy as "deception of a grand order."[37]

Like the military buildup, the Reagan administration's foreign policy was part of the resurgence of American power. It served notice that the United States refused to accept Moscow's hegemonic ambitions and would take whatever steps it deemed necessary to block further Soviet inroads and, where opportunities presented themselves, to roll back Communist influence. Critics, arguing that the administration exaggerated the Soviet threat, viewed such behavior as unduly provocative and indifferent to the aspirations of struggling Third World countries, a throwback to the controversial practices of the 1950s and early 1960s that some felt had brought on Vietnam. The Joint Chiefs of Staff concurred that the administration's foreign policy had its limitations—that its authority to apply force was questionable and that it placed demands on the military without taking full account of available resources and commitments elsewhere. But as the Reagan buildup progressed and the chiefs became more familiar with what was expected of them, JCS objections became fewer and fewer. The net effect by the time Reagan left office was to increase the role of military power in foreign policy to a point where it was stronger and more pronounced than at any time since World War II.

THE PROMISE OF TECHNOLOGY: SDI

The Reagan buildup involved not merely expanding the capabilities of the Armed Forces, but doing it in the time-honored American tradition with the most up-to-date weapons and equipment. The 1970s had witnessed a stunning array of breakthroughs and improvements in technologies with military applications, from the expanded use of computers and space-based satellites for communications and battle management on the ground and in the air, to more sophisticated "smart bombs" and high-energy lasers. The potential seemed limitless. Some military analysts even suggested the possibility that in the not-too-distant future increasingly accurate and lethal conventional munitions could replace strategic nuclear weapons as the mainstay of the country's offensive deterrent force.[38] But until the Reagan buildup, limited funding had prevented the Services from fully exploring the opportunities these new technologies presented.

The Reagan administration's most ambitious effort to exploit this situation was the Strategic Defense Initiative (SDI), a program conceived and instigated in large part on the advice of the Joint Chiefs of Staff. The pivotal decision was the President's announcement in a nationally televised speech on March 23, 1983, that the time had come to draw a halt to the arms race between the United States and the Soviet Union and to move away from deterrence based on offensive arms and the threat of "mutual assured destruction," or MAD. In the absence of a negotiated bilateral agreement, Reagan was taking unilateral action. Summoning the scientific community to help, he called for an aggressive R&D program aimed at determining the feasibility of using new technologies to render ballistic missiles "impotent and obsolete." For more than 10 years, since the signing of the 1972 ABM Treaty, ballistic missile defense (BMD) had languished as a low priority. Now it was suddenly back atop the national agenda.[39]

While the Joint Chiefs played no part in drafting the President's speech, their role in shaping his decision to launch SDI was fundamental. As a rule, the Joint Chiefs had never taken a close corporate interest in research and development. Leaving R&D choices largely in the hands of the military Services, the JCS had concentrated on establishing general guidelines consistent with overall strategic plans. But by the early 1980s, they faced a unique situation that saw the rapid rise of new technological possibilities in BMD converge with a growing political controversy in the United States over the future of the ICBM force. The upshot was the emergence of a consensus among the chiefs that more needed to be done to coordinate and promote missile defense, a recommendation that appealed to the President's antinuclear prejudices without diminishing the administration's commitment a strong military posture. The resulting Strategic Defense Initiative (ridiculed as "Star Wars" by critics) undoubtedly went farther than anything the JCS had in mind. Yet once Reagan endorsed the program, it seemed to acquire a momentum of its own, which its proponents saw as having the potential of revolutionizing warfare, much like the Manhattan Project of World War II.

The revival of ballistic missile defense was initially a haphazard affair, drawn from scattered research carried out by the Army's ballistic missile defense organization at Huntsville, Alabama, the Defense Advanced Research Projects Agency (DARPA) in Washington, DC, and the two key government-run research laboratories—Los Alamos in New Mexico and Lawrence Livermore in California. Out of a dozen or more such projects dating from the late 1960s and 1970s came encouraging progress in such fields as kinetic kill vehicles, high energy lasers, particle beams, and other directed-energy systems operating from land-, sea-, and space-based platforms. With the Armed Forces increasingly dependent on space-based communications, the Joint Chiefs were mainly interested in these systems as a hedge against the possibility that the Soviet Union might develop an antisatellite (ASAT) capability to cripple U.S. communications systems.[40] But as the impact of these breakthroughs became more apparent, the JCS began to recognize that they offered a possible new counter to increasingly capable ground, air, and space-borne military threats.

Popular interest in these new technologies was also catching on. Among the more enthusiastic supporters of an increased BMD effort was Republican Senator Malcolm Wallop of Wyoming. A member of the Senate Select Committee on Intelligence, Wallop favored stepped-up work on U.S. space-based chemical lasers to counter the growing Soviet missile threat. Instead of a strategy of deterrence resting on mutual assured destruction, Wallop and a handful of others in Congress foresaw the coming of a new era they termed "mutual assured survival" built around defensive rather than offensive technologies.[41] A private citizens' group known as High Frontier envisioned a similar future. Headed by Lieutenant General Daniel O. Graham, USA (Ret.), a former director of the Defense Intelligence Agency and a defense adviser to Ronald Reagan's 1976 and 1980 campaigns, High Frontier gained widespread public notice by promoting a Buck Rogers-style, space-based missile defense system utilizing futuristic and existing technologies. But because of questionable technical data and dubious cost estimates, High Frontier's proposals received a tepid reception at the Pentagon.[42]

Even though the Joint Chiefs took no position on which specific BMD technologies to pursue, they made it plain that they were far from satisfied with the constrained level of R&D since the signing of the ABM Treaty. While it prohibited large-scale deployments, the treaty had not banned either the United States or the Soviet Union from conducting research and laboratory testing. With the advent of the Reagan administration, the JCS saw an opportunity to accelerate the pace and repeatedly urged the White House to do so. The chiefs' call for increased attention to ballistic missile defense took place against the backdrop of a growing popular movement at home and abroad for a "freeze" on further nuclear deployments. Adopting a "liberal-pacifist" orientation, the freeze movement used sit-ins and large-scale demonstrations to convey its message. Its members ranged from prominent liberal members of Congress to social activists and middle class professionals who had participated in antiwar causes during Vietnam. An international phenomenon, the freeze was especially active in the United Kingdom and Western Europe, where it became a prime candidate for penetration by the KGB, the Soviet espionage service. But as Christopher Andrew revealed in his authorized history of MI5, Britain's counterintelligence organization, the KGB backed off on discovering that competing Western intelligence agencies had already heavily penetrated the freeze movement.[43]

A favorite target of freeze advocates in the United States was the MX, the Air Force's new ICBM, which continued to enjoy a precarious fate. Shortly after taking office, President Reagan had jettisoned the Carter administration's "race track" deployment plan for the MX as too expensive and had ordered that one hundred of the missiles (half the planned force) be deployed temporarily in Minuteman silos.[44] In the spring of 1982, the Office of the Secretary of Defense advanced a longer-term solution it called the Closely Spaced Basing (CSB) plan, known commonly as "dense pack." Under the plan, the Air Force would deploy the remaining one hundred MX missiles in closely spaced, super-hardened silos near Cheyenne, Wyoming. The principle behind dense pack was that incoming Soviet warheads would destroy or divert themselves upon detonating, a phenomenon known as fratricide. But without the protection of a layered ballistic missile defense system, some analysts warned, dense pack was likely to become increasingly vulnerable to expected improvements in Soviet capabilities.[45]

The Joint Chiefs split over how to proceed. While the Air Force member, General Gabriel, praised the dense pack concept, the other chiefs doubted whether it offered sufficient survivability and recommended against it. Suggesting that long-range land-based missiles had outlived their usefulness, they thought the future lay in phasing out ICBMs and shifting primary reliance to a sea-based deterrent force organized around Trident submarines, backed by a heavy BMD overlay to prevent the Soviets from holding U.S. cities "hostage" in a crisis. Unwilling to give up on the MX so easily, General Vessey proposed a middle course that involved accepting dense pack and stepping up ballistic missile defense R&D, while holding an ABM deployment in abeyance pending a clearer picture of Soviet intentions. Warming to Vessey's proposal, President Reagan in late November 1982 gave it the green light to proceed.[46]

Despite the President's endorsement of the dense pack plan, Congress deemed it too risky and in December 1982 cut off MX funding while directing the administration to review the program. By then the Joint Chiefs were more divided than ever over the MX (recently named the Peacekeeper by the Air Force) and could only agree that, if it went forward, it should be in conjunction with a vastly enhanced ballistic missile defense R&D effort. The leading opponent of the MX was Admiral James D. Watkins, the Chief of Naval Operations, who considered further investment in ICBMs a waste of money. Deeply religious, Watkins saw the administration losing moral ground to the nuclear freeze movement and sought to shift the focus of the debate away from offensive weapons by stressing the deterrent potential of strategic defenses. While not as averse as Watkins to the MX, Vessey shared the

CNO's worry that it constituted a huge political liability and could interfere with completion of the rest of the JCS strategic modernization program. Matters came to a head during an executive session of the Joint Chiefs in the Chairman's office on February 5, 1983, held to prepare for a meeting less than a week later with the President. Citing the recent progress that BMD-related research had made, Watkins reiterated his support for an intensified ABM program and persuaded his colleagues that they should put the matter before the President for his consideration.[47]

Although not part of the formal agenda, strategic defense emerged as the principal topic at the chiefs' meeting with the President on February 11, 1983. Going into the meeting, Weinberger wanted to confine the discussion to issues relating to Peacekeeper, dense pack, and alternative basing modes. But at Vessey's request, he agreed to give the chiefs leeway. "I have asked the JCS to present their views to you today," Weinberger reportedly told the President, "because they differ from mine. . . . I don't agree with their recommendation, but you should hear it."[48] Vessey then devoted half an hour to presenting a detailed critique, with visual aids, of Soviet war aims, U.S. weapons employment policy, targeting concepts, and the capabilities of U.S. strategic and intermediate-range systems. The thrust of Vessey's talk was that the United States faced growing problems of maintaining effective deterrence with its existing and foreseeable arsenal of offensive weapons and that the time had come to take a fresh look at defensive alternatives. On the conclusion of Vessey's remarks, Reagan polled the Service chiefs to see if they agreed with the Chairman's analysis. Finding that they did, he opened the floor to discussion, whereupon Watkins jumped in with a strong endorsement of an enlarged BMD program. "Would it not be better," he asked the President, "if we could develop a system that would protect, rather than avenge, our people?" Deeply moved, Reagan seized on the idea and indicated that he wanted to pursue it. "Don't lose those words," he said.[49]

Exactly what the chiefs expected to achieve remains a matter of conjecture. At no point during the meeting did they indicate whether they were thinking about a comprehensive air and missile defense against nuclear weapons or against ICBMs only. Nor was it clear whether they were proposing a nationwide system or merely the protection of missile silo fields, or whether they favored a treaty-compliant, land-based BMD system or a more sophisticated space-based system that might require revision or abrogation of the ABM Treaty. Instead, their apparent aim was the more general wish to alert the President to the problems of the current system of deterrence, make him more aware of the possible alternatives, and stimulate greater interest in and funding for research and development. "It was the idea that defense might enter the equation more than in the past," Vessey recalled. "It was the idea that new technologies were more promising than they had been in the past."[50]

While the Joint Chiefs were fairly certain as they left the White House that they had an impact on the President's thinking, it was not until his speech to the Nation on March 23 that they finally learned the extent of it. By calling for increased efforts to render ballistic missiles "impotent and obsolete," Reagan was not merely setting an objective; he was charting his vision of the future, when dependence on deterrence through MAD would end and a new era of security resting on defense-based systems would begin. The avowed intent, in other words, was nothing less than a new strategic posture that would rid the world of having to live under the constant threat of nuclear annihilation. Critics were soon insisting that what Reagan had in mind was a leakproof, impenetrable shield—costly, risky, and doubtless unattainable. But others, including the Joint Chiefs, thought it was at least worth exploring the possibility while reaping whatever benefits the program might yield.

Reagan's speech of March 23, 1983, was, as it turned out, the high point of JCS influence on the President's Strategic Defense Initiative, as it soon became known. Indeed, from that point on, their role in the program steadily diminished as Weinberger, sensing enormous opportunities, gathered its components as closely as he could under his immediate control. Expecting the Services to carry out most of the R&D, the Joint Chiefs in May 1983 offered a plan known as "Project Defender" (the same name used for an experimental missile defense system in the 1960s) that would have created a joint JCS-OSD oversight body to provide the military departments with policy guidance and coordination for the program. Weinberger, however, had other ideas. Instead of two or three loosely linked Service-run programs, he wanted a centralized effort structured along functional lines. The upshot was the creation in April 1984 of the Strategic Defense Initiative Organization (SDIO), an OSD staff agency with its own director who reported to the Secretary of Defense.[51]

Although denied a direct hand in managing SDI, the Joint Chiefs remained strong advocates of the program and were confident that with the necessary resources and political support, it would show dramatic results in due time. They believed it only prudent to maintain a vigorous missile defense R&D program to stay abreast of what the Soviets might be doing and to take advantage of the spin-offs in such areas as antisatellite systems and space-based surveillance, reconnaissance, and communications. Adopting a long-term view, they projected a transition period of 30 years or more—time enough, if judiciously managed, to move away from reliance on offensive forces and MAD without the attendant risks that many strategic analysts predicted. In sum, the chiefs were upbeat about the program. But unlike many who found the President's vision enticing, they had no illusions about the problems involved and the chances of SDI succeeding.

Meanwhile, the Peacekeeper program experienced a slow but inexorable demise, ushered along by the findings of a bipartisan Presidential commission chaired by former national security advisor Lieutenant General Brent Scowcroft, USAF (Ret.).[52] The commission's report, released to the public in early April 1983, essentially signaled that, like the dreadnought, the era of the large, heavy ICBM was drawing to a close. Convinced that the dense pack deployment was flawed, the commission recommended against it and favored scaling back the Peacekeeper program to fifty or so units, all housed in Minuteman silos. Had it not considered the Peacekeeper to have some residual value as a "bargaining chip" in arms control, the Scowcroft Commission might well have advised terminating the program entirely. Looking ahead, the panel saw the Nation's security better served by a mobile fleet of small, single-warhead ICBMs dubbed "Midgetmen," which would be harder for the Soviets to target in a surprise attack.[53]

Scowcroft reportedly observed that after the President's SDI speech, the Peacekeeper was doomed no matter what the commission recommended.[54] Not everyone, to be sure, agreed, least of all the Joint Chiefs of Staff. Even though some members had doubts, they continued to endorse the MX as part of the country's mix of strategic forces for the foreseeable future. In assessing the impact of the Scowcroft Commission report, the JCS worried that its recommendation to downsize the Peacekeeper program was apt to delay other strategic modernization measures and generate pressures to shift money from strategic to conventional weapons. Consequently, even though they were divided over Peacekeeper's ultimate contribution, they remained unified in their support of full funding for all Service-recommended strategic programs in the interest of preserving a proper balance between strategic and conventional capabilities. But in light of the commission's report and the renewed interest in strategic defenses, political support for the Peacekeeper continued to wane and in 1985 Congress limited further deployment to the Scowcroft-recommended force of 50 launchers.[55]

ARMS CONTROL: A NEW AGENDA

With its focus on restoring U.S. military power, the Reagan administration seemed at times to have little patience for or interest in arms control. Like his predecessor, Jimmy Carter, however, Ronald Reagan wanted nothing more than to do away with nuclear weapons and end the threat of nuclear war. But his means of doing so often diverged sharply from those of previous administrations and could just as easily stress unilateral actions like SDI over negotiated agreements like SALT. Many within the administration, from the President on down, shared Secretary of Defense Weinberger's view that the 1970s had been "a melancholy chapter" in the history of arms control, resulting in agreements that had done more to increase U.S. vulnerability than to lessen it.[56] Learning from these presumed mistakes, the Reagan administration adopted the position that it was almost pointless to take new initiatives in the arms control field until the country had rebuilt its defenses and could negotiate from a stronger posture.

The Joint Chiefs concurred that the more the United States did to bolster its defenses, the stronger its negotiating position would be. Nevertheless, they considered it impractical not to include arms control as a factor in shaping the overall content and thrust of the buildup. During the 1950s and 1960s, the JCS had been among the staunchest skeptics of arms control, mainly because they had little faith in available verification measures. But as the arms control process gathered momentum during the 1970s, they began making adjustments in their thinking to accommodate the new reality. Though they may have preferred other outcomes, they found themselves operating under negotiated accords that impinged directly on the development, size, and configuration of military programs. Whether the JCS liked it or not, arms control had become an integral part of the military planning process.

During the debates over arms control policy in the Reagan years, the Joint Chiefs found themselves often advocating positions that, only a generation earlier, their predecessors would have dismissed out of hand. Given the limitations on available verification measures, they seriously questioned whether some arms control measures were feasible to carry out with confidence. Yet overall, they believed that the United States should continue to adhere to existing agreements and negotiate suitable replacement accords consistent with allowing improvements in the country's strategic posture. Since they considered the United States to have fallen behind the Soviet Union in effective military power, they looked to arms control as a means of buying time to protect programs that had yet to come to fruition and to preserve the tenuous military balance from a possible Soviet "breakout."

Under Reagan, arms control initially comprised three separate but related sets of negotiations—strategic nuclear forces, intermediate-range nuclear forces (INF), and conventional forces. In the last category were the Mutual and Balanced Force Reduction (MBFR) talks launched in 1973 in the euphoric early days of détente. Part of the "confidence-building" agenda at the time, the MBFR negotiations were supposed to help ease tensions in Europe by reducing conventional forces in the area surrounding East and West Germany where NATO and the Warsaw Pact had the greatest concentration of troops and equipment. The fundamental difficulty was that by the West's calculations the Warsaw Pact had significantly more troops, tanks, and other hardware deployed there than did NATO. Initially, the Western powers sought phased reductions to reach parity of forces, while the Soviet side wanted equal reductions in soldiers and equipment, a formula the West rejected because it would perpetuate and effectively institutionalize Soviet supremacy in conventional arms. Finding the talks essentially deadlocked as it entered office, the Reagan administration adopted a relaxed attitude and appeared in no rush to seek a breakthrough.[57]

Far more urgent—and unavoidable—were the problems growing out of NATO's 1979 "dual track" decision to deploy new American intermediate-range missiles in Europe to counter the Soviet SS–20 threat. A legacy of the Carter years, it fell to the Reagan administration to determine whether a negotiated settlement could be reached before NATO deployment began. The administration's publicly stated goal was the "zero-zero option" under which both sides would forego their INF deployments, dismantle their weapons, and restore the status quo ante. But with the Soviets so heavily invested in SS–20s, there was little optimism in the West that Moscow could be easily talked into doing away with its missiles.

Like the MBFR talks, the INF negotiations hit one snag after another. Convinced by the summer of 1982 that a zero-zero outcome was unattainable, the senior U.S. representative, Paul H. Nitze, on his own initiative, persuaded his Soviet counterpart, Yuli Kvitsinskiy, to entertain the possibility of an alternative solution known as the "walk in the woods" formula.[58] Under this the United States would deploy a reduced number of ground-launched cruise missiles, forego deployment of the Pershing II, and accept curbs on intermediate-range fighter bombers in exchange for scaled-back deployment of the Soviet SS–20s.[59] Over the years, officers in the military had come to have the highest regard for Nitze, who had a reputation as a hard-nosed negotiator and ardent proponent of a strong defense posture. But by offering concessions that the JCS had specifically cautioned against, Nitze tarnished his credibility with the chiefs. As it turned out, neither Reagan nor the leadership in Moscow thought very highly of the walk in the woods offer and the deal fell through. Lest there be similar episodes in the future, the JCS began to monitor the negotiations more closely.

In November 1983, the Joint Chiefs commenced the deployment of INF systems to Western Europe. When completed, NATO would have a refurbished arsenal of 572 up-to-date mobile INF launchers—464 ground-launched cruise missiles (GLCMs) and 108 Pershing II ballistic missiles, all armed with single warheads. Almost immediately, in protest over the deployment, Kvitsinskiy and his delegation walked out of the INF talks in Geneva and served notice that they were "discontinuing" further negotiations.[60] By coincidence, NATO at this time was wrapping up its annual *Autumn Forge* series of exercises with a command post exercise called *Able Archer 83*, a test of release options for nuclear and chemical weapons. More extensive and realistic than previous such exercises, *Able Archer 83* included the simulated use of Pershing II missiles. The KGB suspected that by coinciding with the INF deployments, *Able Archer* might be the cover for a surprise nuclear and chemical attack against the Warsaw Pact. The Soviets were so worried that Marshal Nikolai Ogarkov, Chief of the Soviet General Staff, reportedly moved his headquarters into a reinforced bunker buried deep below Moscow. From all indications, however, the Joint Chiefs were unaware that they were on the verge of a serious crisis. Fritz W. Ermarth, a senior intelligence analyst, recalled that none of the steps taken by the Soviets "crossed the thresholds that would have made our warning lights begin to flash." The incident passed without serious consequence and the INF deployments proceeded as planned, but in an atmosphere ripe for accidental war.[61]

At the same time the Soviets withdrew from the INF talks, they also suspended their participation in parallel negotiations on strategic arms. Now known as the Strategic Arms Reduction Talks (START), these negotiations had been underway since June 1982 and were supposed to find a replacement accord for the stillborn SALT II Treaty, which was due to expire toward the end of 1985. The chief U.S. negotiator at these talks was Edward Rowny, previously the JCS representative to SALT II. As a general objective, President Reagan wanted nothing less than "substantial reductions" in the strategic arsenals of both sides.[62] Adopting a different negotiating strategy from his predecessors, the President declined to engage in back-channel discussions of the sort Kissinger had conducted between 1971 and 1977, and discontinued the practice of separate high-level talks that had accompanied SALT II negotiations under President Carter. Thus, when INF and START both collapsed at the end of 1983, the United States and Soviet Union were for the first time in 14 years without a forum of any kind for discussing limitations and controls on nuclear arms.

Ignoring the dire predictions of the news media, the administration took the collapse of the talks in stride. Personally, President Reagan had a low opinion of the whole arms control process and privately characterized the unratified SALT II Treaty as a "lousy" agreement.[63] The Joint Chiefs basically agreed, but they were also uneasy that the longer the talks were in recess, the greater the temptation for the Soviets to take matters into their own hands. Without a SALT II replacement accord, the JCS were afraid that Moscow might forge ahead with new deployments that could negate the effects of the Reagan buildup. Rather than risk a renewed Soviet buildup, the JCS had been among those warning the President, almost from the moment he took office, to avoid any actions prior to the conclusion of a START accord that would be inconsistent with existing U.S.-Soviet strategic arms control agreements as long as the Soviet Union exercised similar restraint. The President had accepted this advice and in the jargon of the day had agreed not to "undercut" earlier SALT agreements. Later, he amended that position by refusing to abide by those accords if they came into conflict with "the survivability of our ICBM force" as the U.S. buildup progressed.[64]

To reach his goal of "substantial reductions," President Reagan in May 1982 publicly unveiled a complex formula that called for equal numbers of strategic warheads at levels one-third below current inventories, and further limitations leaving no more than half the remaining warheads on land-based ICBMs.[65] A departure from the previous philosophy toward arms control, which had emphasized numerical restraints on launchers rather than curbs on their destructive power, the President's proposal was the product of lengthy bureaucratic bargaining and compromise to address the differences among his advisors. The Joint Chiefs had doubts about the plan but recognized that, if implemented, it would practically eviscerate the Soviet ICBM force, reducing it by two-thirds, while leaving major elements of the U.S. buildup (the MX, the B-1, and Trident) virtually untouched. Not surprisingly, once the negotiations began, the Soviets countered with proposals, based largely on an extension of the SALT II accords, for curbing U.S. programs while leaving theirs basically intact. Some give and take ensued, yet by the time the Soviets walked out, there was nothing to suggest that a breakthrough was imminent.[66]

While the abrupt cessation of the START negotiations took many people by surprise, no one in the Reagan administration regarded it as a fatal setback. Those in Washington familiar with Moscow's negotiating techniques, including the Joint Chiefs, scoffed at the notion that the talks were dead and expected the Soviets to return to START in the spring or summer of 1984.[67] From everything the JCS could glean, the Soviets appeared interested in resuming a dialogue with the United States that would end the spiraling deterioration in relations between Washington and Moscow. But with the U.S. Presidential election coming up, Moscow was probably reluctant to do anything to enhance the current administration's prospects by being able to claim a major success in the areas of arms control or U.S.-Soviet relations. Hence, there was the likelihood of a continuing impasse even if the START negotiations resumed.

Increasingly, the issue that bothered the Soviets most was President Reagan's determination to press ahead with his Strategic Defense Initiative. While critics in the West dismissed SDI as a fanciful notion, the Soviets took it very seriously. Having invested enormous effort and resources into bolstering their strategic offensive forces, they now found themselves confronted with a revolutionary strategic paradigm that could seriously cripple, if not negate, everything they had accomplished. Condemning SDI as "irresponsible" and "insane," Soviet leaders saw it as nothing less than a "bid to disarm" their country.[68] Whether the decrepit and inefficient Soviet economy could rise to the occasion and compete with the United States in the new arena remained to be seen. Reagan suspected that no matter how Moscow responded, whether by trying to develop a competing missile shield or by embarking on a further offensive buildup in the hope of overwhelming U.S. defenses, the

Soviets would end up bankrupting themselves. One way or another, Reagan said, he was determined to "lean on the Soviets until they go broke."[69]

The Joint Chiefs agreed that bit by bit under the Reagan buildup the United States was regaining the initiative. But they were less optimistic than the President that the Soviets would come around to the West's way of thinking anytime soon, if ever. To them, the Soviet Union remained first and foremost a military colossus—a nuclear superpower whose military capabilities, if unleashed upon the West, could inflict enormous death and destruction. Like others who had wrestled with the problem over the years, the Joint Chiefs of Staff had grown so accustomed to the Cold War that they assumed it would go on indefinitely and paced themselves accordingly, with military programs designed for the long haul. Under the Reagan buildup, they were finally making strides toward redressing the strategic balance in ways they believed would carry the country's security into the next century. Little did they suspect that with the advent of new leadership in Moscow, the entire security environment was about to undergo a fundamental change and make way for the Cold War to end sooner than anyone expected.

NOTES

1. Jones's Statement, "Perspectives on Security and Strategy in the 1980s," in U.S. Joint Chiefs of Staff, *United States Military Posture for FY82* (Washington, DC: Organization of the Joint Chiefs of Staff, n.d.), vii (unclassified edition).

2. Ronald Reagan, *An American Life* (New York: Simon and Schuster, 1990), 235.

3. A public information organization, the Committee on the Present Danger had been a leading critic of the SALT II Treaty and had been in the forefront of advocating a stronger defense posture.

4. Thomas C. Reed, *At the Abyss: An Insider's History of the Cold War* (New York: Ballantine Books, 2004), 247.

5. Erik B. Riker-Coleman, "Political Pressures on the Joint Chiefs of Staff: The Case of General David C. Jones" (Paper Presented before the Society of Military History Annual Meeting, Calgary, Alberta, May 2001), 16–18; copy in JHO Files.

6. Caspar W. Weinberger, *In the Arena: A Memoir of the 20th Century* (Washington, DC: Regnery, 2001), 293.

7. Jones quoted in James R. Locher III, *Victory on the Potomac* (College Station: Texas A&M Press, 2002), 34.

8. "Nomination of Gen. John W. Vessey, Jr., To Be CJCS," March 4, 1982, and "Remarks Announcing the Nominees for CSAF and CNO," March 18, 1982, *Public Papers of the Presidents of the United States: Ronald Reagan, 1982* (Washington, DC: GPO, 1982), 265, 324–325. Hereafter cited as *Reagan Public Papers* with year.

9. "Nomination of General John A. Wickham, Jr., To Be CSA," March 15, 1983, and "Nomination of General Paul X. Kelley To Be Commandant of the Marine Corps," March 24, 1983, *Reagan Public Papers, 1983*, 396, 444.

10 Weinberger, *In the Arena*, 293.

11 Letter, Vessey to Clark, July 2, 1982, U, NSC Executive Secretariat Collection, Agency Files, box 11376, DOD Vol. VI folder, Reagan Library; Deborah Hart Strober and Gerald S. Strober, eds., *Reagan: The Man and His Presidency* (Boston: Houghton, Mifflin, 1998), 78. Records of JCS meetings with the President are filed randomly in the NSC Executive Secretariat Collection, System Files, Reagan Library.

12 David A. Stockman, *The Triumph of Politics: How the Reagan Revolution Failed* (New York: Harper & Row, 1986), 105; Caspar Weinberger, *Fighting for Peace: Seven Critical Years in the Pentagon* (New York: Warner Books, 1990), 47.

13 "Republican Party Platform of 1980," Adopted July 15, 1980 at Detroit, Michigan, available at <http://www.Presidency.ucsb.edu/ws/index.php?pid=25844> (accessed July 18, 2011).

14 Daniel Wirls, *Buildup: The Politics of Defense in the Reagan Era* (Ithaca, NY: Cornell University Press, 1992), 35–37.

15 Jones et al. testimony, February 4, 1981, U.S. Congress, House, Committee on Armed Services, *Hearings: Military Posture* 97:1 Pt. 1 (Washington, DC: GPO, 1981), 207; Memo, Weinberger to Reagan, March 9, 1981, "Joint Strategic Planning Document," U, White House Situation Room, Agency Files, box 91376, DOD Vol. VI folder, Reagan Library; *United States Military Posture for FY84* (Washington, DC: Organization of the Joint Chiefs of Staff, n.d.), 3 (unclassified edition).

16 Stockman, 106–108; Weinberger, *Fighting for Peace*, 49.

17 Stockman, 277. Carter budget estimates from *Congressional Quarterly Almanac, 1982*, 73.

18 Weinberger, *In the Arena*, 275.

19 U.S. Office of the Comptroller, *National Defense Budget Estimates for FY92* (Washington, DC: Department of Defense, March 1991), 98. Even though the Reagan administration set new records for long-term military spending, the largest single-year Cold War military budget (in constant dollars) remained that enacted under the Truman administration for FY52.

20 Jones quoted in Richard Halloran, "Military Forces Stretched Thin," *The New York Times*, August 10, 1983: 1.

21 NSDD 12, "Strategic Forces Modernization Program," October 1, 1981 (sanitized), NSC Executive Secretariat Collection, Reagan Library, available at <http://www.fas.org/irp/offdocs/nsdd/nsdd-12.pdf> (accessed July 18, 2011).

22 Budget percentages derived from *National Defense Budget Estimates for FY92*, 74–75. For strategic objectives under the buildup, see NSDD 12, loc. cit.

23 Weinberger testimony, February 2, 1982, in U.S. Congress, Senate, Committee on Armed Services, *Hearings: Department of Defense Authorization for Appropriations for Fiscal Year 1983* (Washington, DC: GPO, 1982), 18.

24 See John Allen Williams, "The U.S. Navy Under the Reagan Administration and Global Forward Strategy," in William Snyder and James Brown, eds., *Defense Policy in the Reagan Administration* (Washington, DC: National Defense University Press, 1988), 273–303.

25 See Entry, May 18, 1982, in Douglas Brinkley, ed., *The Reagan Diaries* (New York: HarperCollins, 2007), 85.

26 NSDD 32, "US National Security Strategy," May 20, 1982 (declassified), NSC Executive Secretariat, box 91311, NSDD 32 folder, Reagan Library, available at <http://www.fas.org/irp/offdocs/nsdd/nsdd-032.htm> (accessed July 18, 2011).

27 Gordon S. Barrass, *The Great Cold War: A Journey Through the Hall of Mirrors* (Stanford, CA: Stanford University Press, 2009), 284.

28 Quote from "Address Before a Joint Session of the Congress on the State of the Union," February 6, 1985, *Reagan Public Papers, 1985*, 135.

29 NSDD 5, "Conventional Arms Transfer Policy," July 8, 1981, in *Reagan Public Papers, 1981*, 615–617. See also Roy A. Werner, "The Burden of Global Defense: Security Assistance Policies of the Reagan Administration," in Snyder and Brown, eds., *Defense Policy in the Reagan Administration*, 143–165.

30 See Weinberger, *Fighting for Peace*, 143–144; Shultz, *Turmoil and Triumph*, 106–107; Edgar F. Raines, Jr., "The Interagency Process and the Decision to Intervene in Grenada," in Kendall D. Gott and Michael G. Brooks, eds., *The U.S. Army and the Interagency Process: Historical Perspective* (Fort Leavenworth, KS: Combat Studies Institute Press, 2009), 33–64.

31 Minutes, NSC Meeting on July 31, 1981, 4, (declassified), NSC Executive Secretariat Collection, NSC 00018, Reagan Library, available at <www.thereaganfiles.com> (accessed July 18, 2011); and Reagan, *An American Life*, 289.

32 Shultz, *Turmoil and Triumph*, 289; NSDD 17, "Cuba and Central America," January 4, 1982 (redacted), available at <http://www.fas.org/irp/offdocs/nsdd/nsdd–017.htm> (accessed July 19, 2011); NSDD 37, "Cuba and Central America," May 28, 1982 (declassified), available at <http://www.fas.org/irp/offdocs/nsdd/nsdd–037.htm> (accessed July 19, 2011); and NSDD 37A, "Cuba and Central America," May 28, 1982 (redacted), available at <http://www.fas.org/irp/offdocs/nsdd/nsdd–037a.htm> (accessed July 18, 2011).

33 Author's interview with former USCINCSOUTH Gen Paul F. Gorman, USA (Ret.), July 27, 2000.

34 George Shultz, *Turmoil and Triumph: My Years As Secretary of State* (New York: Charles Scribner's Sons, 1993), 570, 692; Robert M. Gates, *From the Shadows* (New York: Simon and Schuster, 1996), 251–252; NSDD 99, "United States Security Strategy for the Near East and South Asia," July 12, 1983 (sanitized), available at <http://www.fas.org/irp/offdocs/nsdd/nsdd–099.htm> (accessed July 18, 2011); entry, July 12, 1983, in Brinkley, ed., *Reagan Diaries*, 165.

35 Ronald H. Cole et al., *The History of the Unified Command Plan, 1946–1999* (Washington, DC: Joint History Office, 2003), 63–67.

36 *United States Military Posture for FY83* (Washington, DC: Organization of the Joint Chiefs of Staff, n.d.), 39 (unclassified edition).

37 Letter, Vessey to BG David A. Armstrong, USA (Ret.), September 23, 1995, U, JHO Files.

38 See Gen Bennie L. Davis, USAF, "Indivisible Airpower," *Air Force Magazine* (March 1984), 46–50; and Carl H. Builder, *The Prospects and Implications of Non–Nuclear Means of Strategic Conflict*, Adelphi Paper 200 (London: International Institute for Strategic Studies, 1985).

39 "Address to the Nation on Defense and National Strategy," March 23, 1983, *Reagan Public Papers, 1983*, 442–443.

40 Paul B. Stares, *The Militarization of Space: U.S. Policy, 1945–1984* (Ithaca, NY: Cornell University Press, 1985), 201–215.

41 Malcolm Wallop, "Opportunities and Imperatives of Ballistic Missile Defense," *Strategic Review* 7 (Fall 1979), 13–21. See also Angelo Codevilla, *While Others Build* (New York: Free Press, 1988), which sets forth Wallop's strategic defense rationale and blueprint.

Codevilla was Wallop's staff assistant. The two worked extremely closely together to promote legislation and interest in BMD.

42 Daniel O. Graham, *Confessions of a Cold Warrior* (Fairfax, VA: Preview Press, 1995), 117–146; Sanford Lakoff and Herbert F. York, *A Shield in Space? Technology, Politics, and the Strategic Defense Initiative* (Berkeley: University of California Press, 1989), 10–11.

43 Christopher Andrew, *Defend the Realm: The Authorized History of MI5* (New York: Knopf, 2009), 673–676.

44 John Edwards, *Superweapon: The Making of MX* (New York: W.W. Norton, 1982), 210; Patrick Glynn, *Closing Pandora's Box: Arms Races, Arms Control, and the History of the Cold War* (New York: Basic Books, 1992), 324–325.

45 Christian Brahmstedt, *Defense's Nuclear Agency, 1947–1997* (Washington, DC: Defense Threat Reduction Agency, 2002), 262; David H. Dunn, *The Politics of Threat: Minuteman Vulnerability in American National Security Policy* (New York: St. Martin's Press, 1997), 153–157.

46 Donald R. Baucom, *Origins of SDI, 1944–1983* (Lawrence: University Press of Kansas, 1992), 185; Minutes, NSC Meeting, November 18, 1982, "M–X Basing Decision" (declassified), NSC Executive Secretariat Collection, NSC 00066 folder, Reagan Library, and <http://jasonebin.com/nsc66.html> (accessed July 18, 2011); NSDD 69 (sanitized), November 22, 1982, "The M–X Program," NSC Executive Secretariat Collection, box 91286, NSDD 69 folder, Reagan Library, and <http://www.fas.org/irp/offdocs/nsdd/nsdd-069.htm> (accessed July 18, 2011).

47 Baucom, *Origins of SDI*, 184–190; Interview by Donald R. Baucom with ADM James D. Watkins, September 29, 1987, Revised by Watkins, October 22, 1989, U, 6, Missile Defense Agency Historian's Files.

48 Baucom Interview with Watkins, 6–7.

49 Weinberger, *Fighting for Peace*, 304–305. See also Baucom, *Origins of SDI*, 191–192; William J. Broad, *Teller's War: The Top-Secret Story Behind the Star Wars Deception* (New York: Simon & Schuster, 1992), 124–125; and Don Oberdorfer, *The Turn* (New York: Poseidon Press, 1991), 26–27.

50 Vessey quoted in Hedrick Smith, *The Power Game: How Washington Works* (New York: Random House, 1988), 607–608.

51 Memo, Weinberger to DepSECDEF et al., April 24, 1984, "Management of the Strategic Defense Initiative," U, JCS 1977/449.

52 NSDD 73 (declassified/sanitized), "Peacekeeper Program Assessment," January 3, 1983, NSC Executive Secretariat Collection, NSDD 73 folder, Reagan Library, available at <http://www.fas.org/irp/offdocs/nsdd/nsdd-073.htm> (accessed July 18, 2011).

53 President's Commission on Strategic Forces, *Report of the President's Commission on Strategic Forces* (Washington, DC: The President's Commission on Strategic Forces, 1983), 16–19. See also NSDD 91 (declassified), "Strategic Forces Modernization Program Changes," April 19, 1983, NSC Staff and Office Files: Office of the Secretariat, NSDD 91 folder, Reagan Library.

54 See Janne E. Nolan, *Guardians of the Arsenal: The Politics of Nuclear Strategy* (New York: New Republic/BasicBooks, 1989), 15–16.

55 Joseph Kruzel, ed., *American Defense Annual, 1986-1987* (Lexington, MA: Lexington Books and the Mershon Center, 1986), 68-71.

56 U.S. Department of Defense, *Annual Report of the Secretary of Defense to Congress: FY83* (Washington, DC: GPO, 1982), chap. I, 19.

57 Raymond L. Garthoff, *Détente and Confrontation: American-Soviet Relations from Nixon to Reagan* (Washington, DC: Brookings, 1994, rev. ed.), 533–537; Terrence Hopmann, "From MBFR to CFE: Negotiating Conventional Arms Control in Europe," in Richard Dean Burns, ed., *Encyclopedia of Arms Control and Disarmament* (New York: Charles Scribner's Sons, 1993), II, 970–974.

58 The name "walk in the woods" came from the manner in which Nitze and Kvitsinskiy negotiated the deal. To preserve secrecy, they met privately in the forested hill country outside Geneva.

59 Paul H. Nitze with Ann M. Smith and Steven L. Rearden, *From Hiroshima to Glasnost: At the Center of Decision—A Memoir* (New York: Grove Weidenfeld, 1989), 376–389.

60 Ibid., 397.

61 Ermarth quoted in Barrass, 300. See also Ben B. Fischer, *A Cold War Conundrum: The 1983 Soviet War Scare* (Washington, DC: Central Intelligence Agency, Center for the Study of Intelligence, 1997), 24–26; and Peter Vincent Pry, *War Scare: Russia and America on the Nuclear Brink* (Westport, CT: Praeger, 1999), 33–44.

62 "Remarks to the National Press Club on Arms Reduction and Nuclear Weapons," November 18, 1981, *Reagan Public Papers, 1981*, 1066.

63 Reagan's comments in Minutes, NSC Meeting, April 21, 1982, 6 (declassified), NSC Executive Secretariat Collection, NSC 00046 folder, Reagan Library; available at <http://jasonebin.com/nsc46.html> (accessed July 18, 2011).

64 NSDD 36 (declassified), "US Approach to START Negotiations—II," May 25, 1982, available at <www.fas.org/irp/offdocs/nsdd/nsdd-036.htm> (accessed July 18, 2011). President Reagan approved this modification to his no-undercut policy in the expectation of deploying the MX/PEACEKEEPER in the "dense-pack" mode which would violate the SALT I-imposed ban on the construction of new ICBM launchers.

65 "Address at Commencement Exercises at Eureka College in Illinois," May 9, 1982, *Reagan Public Papers, 1982*, 584–585.

66 Garthoff, *Great Transition*, 511–512.

67 Talks resumed, but not until March 1985. See below, chap. 15.

68 Soviet General Secretary Yuri Andropov quoted in *The New York Times*, March 27, 1983. See also Paul Lettow, *Ronald Reagan and His Quest to Abolish Nuclear Weapons* (New York: Random House, 2005), 114–115.

69 Quoted in Reed, *At the Abyss*, 227.

Admiral William J. Crowe, Jr., USN, Chairman of the Joint Chiefs of Staff, 1985–1989

Chapter 15

A NEW RAPPROCHEMENT

By the mid-1980s, as Ronald Reagan embarked on his second term, the military buildup launched at the outset of the decade was beginning to show results. Increasingly reassured, the Joint Chiefs believed that they had turned the corner and were now better poised to compete effectively in military power with the Soviet Union than at any time since the Vietnam War. Despite the re-imposition of congressionally mandated funding constraints, starting with the FY86 budget, they saw the balance of forces shifting back in their favor. As always, the JCS wanted more to be done than available money allowed and urged the President and Congress to be, if nothing else, consistent in their level of support for military programs. Yet, all things considered, the buildup seemed to be having the desired effect of restoring both a stronger defense posture and a renewed respect for the country's Armed Forces. Not since the early 1950s had the Nation's Military Establishment felt so assured.

Though more confident in the future than they had been for some years, the JCS were hardly complacent. As the President's second term began, changes in the Soviet Union, highlighted by the emergence of new leadership under the reform-minded Mikhail S. Gorbachev, created uncertainties in assessing the future direction of Soviet policy. At the same time, the ongoing modernization of Moscow's strategic forces, the heavy concentration of Soviet troops in Europe backed by SS–20 missiles, the continuing intervention in Afghanistan, and a surge of Cuban and Eastern Bloc "advisors" into Nicaragua suggested that the Communist threat remained as real and dangerous as ever. Against this backdrop, the Joint Chiefs saw no choice but to continue the defense policies and programs already in effect and to maintain a high level of military preparedness for the indefinite future.

DEBATING JCS REORGANIZATION

Of the challenges facing the Joint Chiefs at the outset of President Reagan's second term, none took up more of their time or was more frustrating than the growing movement in Congress for JCS reform. While dissatisfaction with the JCS system had existed ever since passage of the National Security Act of 1947, it had grown appreciably in the aftermath of Vietnam, the hurried execution of the 1975 *Mayaguez* rescue operation, and the failed *Desert One* mission in 1980 to free the Tehran hostages. Over the years it had become virtually an article of faith in some academic and congressional circles that the Joint Chiefs were little more than a committee of bickering military bureaucrats, wholly incapable of detaching themselves from parochial interests and rendering objective advice on such cross-Service matters as the allocation of resources and the impartial assignment of military functions.[1]

At the outset of the Reagan administration, some of the most severe critics of the JCS system were, in fact, its own members, including the serving Chairman, General David C. Jones, USAF. During his early days as CJCS, Jones had dismissed talk of restructuring the JCS as unwarranted and had taken the position "that the fundamental organizational structure is sound."[2] But he had changed his mind by the early 1980s. Having served on the JCS as Air Force Chief of Staff and as CJCS for a combined total of 8 years by the time he retired—longer than any other officer—he found himself increasingly frustrated with what he saw as a lengthening list of JCS lapses, failures, and "lowest common denominator" solutions. "The tough issues," he recalled, "got pushed under the rug."[3]

Jones' discontent first surfaced outside the Pentagon in early February 1982 when he and Secretary of Defense Weinberger appeared at a closed-door session of the House Armed Services Committee. During an exchange with committee members, Jones acknowledged his dissatisfaction with the current system and confirmed his support for measures to augment the powers of the Chairman, curb the heavy personnel turnover on the Joint Staff, and create a more efficient and responsive JCS organization.[4] A few weeks later, he went public with interviews to the news media and an article (cleared in advance with Secretary Weinberger), "Why the Joint Chiefs of Staff Must Change," in the February 1982 issue of *Directors & Boards*, which was reprinted a month later in *Armed Forces Journal International*, with a somewhat larger readership. Characterizing current arrangements as a "cumbersome committee process," Jones described the system as rife with inter-Service rivalry and competition. "We need to spend more time on our war fighting capabilities," Jones insisted, "and less on an intramural scramble for resources." Toward that end, Jones endorsed reforms to strengthen the authority of the Chairman over the combatant commanders, limit Service staff involvement in JCS actions, and broaden the training, experience, and rewards for joint duty. To facilitate attainment of these goals, Jones also favored providing the Chairman with a deputy.[5]

Among the Service heads at the time, only Army Chief of Staff General Edward C. Meyer showed any interest in Jones' proposals. Arguing that times had changed since World War II when the JCS came into existence, Meyer considered the existing system obsolete. Going well beyond Jones' proposals, Meyer wanted to abolish the Joint Chiefs and vest full authority over military planning and direction of the Joint Staff in the CJCS.[6] But after General Vessey's appointment as Chairman in the summer of 1982, Meyer muted his criticism. Vessey and Weinberger agreed that while the JCS system could be improved, its corporate structure and organization were sound and whatever reforms were needed could be achieved through administrative means. Indeed, for Vessey, the very essence of the JCS system was its corporate character, which he was loath to tamper with in the name of progress and reform.

After discussing the matter at length with the Service chiefs, Vessey notified the Secretary of Defense on November 22, 1982, that he could find no consensus among his colleagues in support of "sweeping changes." While conceding that their operations were not without "flaws," there was agreement among the Joint Chiefs that the problem stemmed largely from tensions that had developed over time between OSD and the JCS because of overlapping responsibilities. Vessey declined to assign blame for this situation but did acknowledge that the JCS needed to be more professional and objective in providing military advice. Still, he and his colleagues saw little they could do directly and felt that it was up to the Secretary of Defense to take corrective action by according them larger staffing and a more substantive role "on major decisions of strategy, policy, and force requirements."[7]

Meanwhile, inspired by Jones, Meyer, and a lengthening list of think-tank studies, key members of Congress began taking a closer look at alleged JCS shortcomings. Many on Capitol Hill initially agreed that Vessey's advent had improved the overall efficiency, effectiveness, and image of JCS operations. But after the bombing of the Beirut barracks and reports of breakdowns in coordination during the Grenada operation in October 1983, sentiment in Congress began to coalesce around the need for legislative action to strengthen the JCS system and make it more responsive. Stung by the untoward publicity, Vessey rushed through a series of administrative reforms aimed at improving JCS performance in the areas of resource allocation, the evaluation of cross-Service needs, and participation by the combatant commanders in the programming and budgeting process.[8] But it was too little too late, and in October 1984 Congress added a provision to the Defense authorization (P.L. 98-525) broadening the powers of the Chairman over the Joint Staff and simultaneously serving notice that it intended to revisit the entire question of JCS organization in the next session.[9]

Vessey now found himself unexpectedly at the center of a looming battle royal with Congress. While acknowledging that he faced "considerable outside pressure to reorganize," he continued to believe that through the stringent application of administrative reforms he could fend off the imposition of congressionally-mandated changes. If he could improve the effectiveness of the Joint Staff, he thought he could demonstrate that "we're doing our job as laid out in the law."[10] But despite Vessey's best efforts to find in-house solutions, support in Congress for legislative action continued to grow and by the summer of 1985 both the House and the Senate were actively considering bills to reform the JCS. In June 1985, hoping to head off a wholesale reorganization, President Reagan created a Blue Ribbon Commission on Defense Management, chaired by former Deputy Secretary of Defense David Packard, to review the overall status of defense organization and suggest appropriate remedies.[11] Undeterred, reformers in Congress refused to await the Packard Commission's findings and pressed ahead along a course of their own that would culminate in the 1986 Goldwater-Nichols Department of Defense Reorganization Act.

Feeling that he had done as much as he could, Vessey stepped down as Chairman of the Joint Chiefs on September 30, 1985, more than 6 months before the end of his term. His successor, Admiral William J. Crowe, Jr., USN, came with a lengthy résumé of staff and joint command jobs. Like Vessey, Crowe saw room for improvement in the quality and effectiveness of the Joint Staff.[12] But he was far less averse than his predecessor to accepting legislatively-mandated changes and had once testified before Congress in support of increased statutory powers for the Chairman and a stronger joint system.[13] Realizing that his views were at variance with the prevailing sentiments of his fellow Navy officers, he explained that his position was the result of experience. "I happened to be one of the people [in the Navy] who agreed that some reorganization was appropriate," Crowe recalled. "For three years, from 1977 to 1980, I had served as the Navy's JCS deputy, and during that time I had done a lot of thinking about the subject."[14] As Chairman, Crowe tempered his views somewhat to bring them more into line with Weinberger's. Yet overall, Crowe's advent was highly instrumental in tipping the balance in favor of the reform movement.

Soon after becoming Chairman, Crowe established informal staff-level contacts with the congressional committees considering the new legislation and sounded out the Service chiefs about a possible compromise. Crowe acknowledged that some degree of legislatively-imposed reorganization was unavoidable, but he shared his colleagues' concern that Congress, in its zeal to reform, had "overdramatized" the problem of inter-Service rivalry and its impact on JCS effectiveness.[15] While favoring measures to streamline the system, Crowe and the chiefs unanimously condemned any effort by Congress to abolish the JCS organization and replace it with a joint military advisory council. "While this proposal may have some theoretical appeal to some," they told the Secretary of Defense, "it has no 'real world' merit and, if adopted, would dramatically compromise the quality of advice to you and to the President."[16] Incorporating these views with his own, Weinberger notified the Senate Armed Services Committee on December 2, 1985, that while he was prepared to entertain modest changes, including a stronger advisory role for the Chairman and creation of a Vice Chairman to help expedite JCS business, he saw no need for the sweeping reorganization some in Congress insisted was needed.[17]

By now, differences had become so pronounced that an easy and amicable reconciliation of views between the congressional reformers and the administration was practically out of the question. The most contentious issues were those involving personnel policy centering on the creation of a joint officers corps, a proposal that had especially strong support in the House Armed Services Committee. Worried that a joint officer corps would deprive them of their best officers, the Service chiefs opposed the measure. In an effort at compromise, Crowe invited members of the committee, including Congressman Bill Nichols of Alabama, a key figure in shaping the emerging legislation in the House, to a breakfast meeting with the Joint Chiefs at the Pentagon on June 24, 1986. As the meeting progressed, the atmosphere became visibly strained. Finally, in an emotional outburst, the Chief of Naval Operations, Admiral Watkins, said: "You know, this piece of legislation is so bad it's, it's . . . in some respects it's just un-American." Nichols, who had lost a leg in combat in World War II, was personally offended and left the meeting indignant, less disposed than ever to listen to the chiefs or to accept Pentagon advice.[18]

After this regrettable incident, the Joint Chiefs played a diminishing role in the legislative process that culminated in passage of the Goldwater-Nichols Act. As often happens in the legislative process, the reorganization bills passed by the House and Senate required a conference to iron out differences. Working together, the co-chairs of the conference committee, Nichols and Barry Goldwater of Arizona, Chairman of the Senate Armed Services Committee, wrote the final law. As the conference was getting underway, Admiral Crowe made a last-minute appeal to delete all provisions relating to personnel policy.[19] But his request fell on deaf ears. The final legislation—approved in the Senate on September 16 and in the House the next day—reflected congressional preferences far more than anything the White House or the Pentagon wanted. Secretary of the Navy John F. Lehman, Jr., suggested that President Reagan ought to veto the legislation, but the President, facing other problems in Congress, signed it into law on October 1, 1986.[20]

THE GOLDWATER-NICHOLS ACT OF 1986

Culminating nearly 4 years of public debate and legislative maneuvering, the Goldwater-Nichols Department of Defense Reorganization Act of 1986 (P.L. 99-433) was the most extensive revision of the National Security Act since 1958. The most significant changes were those affecting the Joint Chiefs of Staff and the military command structure. Throughout the new law, the emphasis was on achieving a higher level of inter-Service cooperation and collaboration and a greater degree of integrated effort in practically every level and area of military activity, a concept increasingly referred to as "jointness." Though military leaders by and large agreed that it was a worthy objective, many if not most would have preferred a less detailed and less prescriptive law.

The most striking features of the law were those affecting the Chairman who now became "principal military advisor" to the President, the National Security Council, and the Secretary of Defense, superseding the JCS in that role. Functions and duties previously conferred collectively on the Joint Chiefs of Staff now passed to the Chairman, thus ending the days of corporate decisionmaking and consensus recommendations. In effect, the Service chiefs became a committee of senior military advisors to the Chairman. For assistance in discharging his expanded duties, the CJCS acquired a Vice Chairman and unfettered authority over the Joint Staff. Held to its current strength of 1,627 military and civilian personnel (a ceiling repealed in 1991), the Joint Staff remained barred from becoming "an overall Armed Forces General Staff," a prohibition first introduced in 1958. Still, with an added proviso in the law requiring officers to have joint duty for high-level promotion, the Joint Staff stood poised at last to gain primacy over the Service staffs.

In addition to increasing the Chairman's stature and authority, the new law gave him more specific responsibilities vis-à-vis the combatant commands and the military command structure. Although there had been talk of including the Chairman in the military chain of command, Goldwater-Nichols made only slight changes in the interests of protecting and preserving civilian control. Command lines, as laid out in 1958, continued to run from the President to the Secretary of Defense to the combatant commanders. However, the new law also authorized the Secretary to use the Chairman as his channel of communication with the combatant commanders, a practice already in effect. With the added authority of Goldwater-Nichols, the Chairman's role as the routine channel of communications between the National Command Authority (NCA) and the combatant commanders became fully institutionalized. In consequence, even though the Chairman had no statutory authority to exercise command, his responsibility for receiving political directives and translating them into operational orders gave him a de facto measure of command authority.[21]

The most controversial feature of the new law was its treatment of military personnel policy. Admiral Crowe and others had tried to persuade Congress not to include these provisions or, at least, to tone them down. But by the time the final legislation came to be written, relations between the Pentagon and Capitol Hill had become so strained that members of the conference committee were in no mood to listen. The result, bearing the designation of Title IV, was a highly prescriptive set of regulations for joint duty and promotion aimed at improving professionalism and eradicating alleged Service parochialism. Although the conferees dropped the idea of a joint officer corps, they agreed that officers should be encouraged to develop a "joint specialty" and affirmed a practice already in use requiring new flag officers to attend a "Capstone" course to prepare them for joint assignments with senior officers from other Services.

Implementing the Goldwater-Nichols Act fell largely to the Chairman, Admiral Crowe, who adopted an "evolution-not-revolution" philosophy modeled on Forrestal's approach to unification in the late 1940s. Crowe hoped to complete the process with "as little trauma and disruption as possible."[22] On November 6, 1986, he approved a directive restructuring the Joint Staff to meet expected Goldwater-Nichols needs. To augment the five existing directorates, Crowe revived the moribund Command, Control, and Communications Systems Directorate (J-6) and added two new ones—the Operational Plans and Interoperability Directorate (J-7), later renamed the Directorate for Operational Plans and Joint Force Development, and the Force Structure, Resource, and Assessment Directorate (J-8).[23] Crowe also put considerable personal effort into clarifying the role of the Vice Chairman (VCJCS), whose only assigned duty under the law was to preside at JCS meetings in the Chairman's absence. Secretaries of Defense had customarily regarded their deputies as their "alter ego" since Forrestal coined the phrase in 1948; Crowe believed the Vice Chairman should be prepared to function in a similar capacity.[24] The first Vice Chairman, General Robert T. Herres, USAF, took office on February 6, 1987, but did not receive a specific assignment of functions until April, when the Secretary of Defense, at Crowe's suggestion, directed that the VCJCS should concentrate on acquisition and resource management issues in order to free up time for the Chairman to deal with military policy and strategic matters.[25]

The toughest adjustments were those of redefining the Service chiefs' role under Goldwater-Nichols. Operating initially under a modified version of the old system, Crowe affirmed existing procedures that allowed his colleagues to present divergent views to the Secretary of Defense.[26] But since the JCS were no longer bound

Figure 15–1.

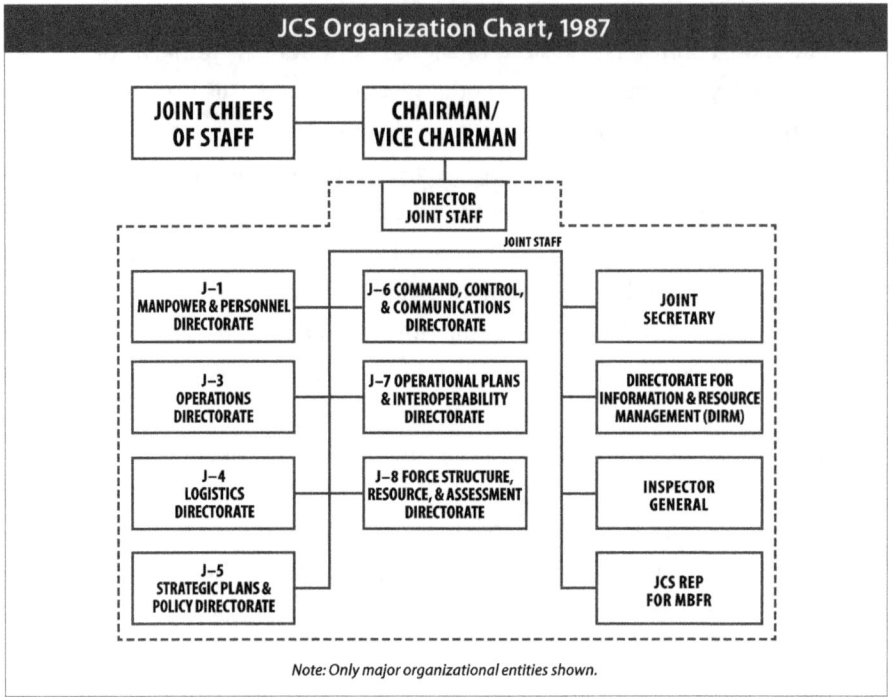

by the corporate unanimity rule, "split" recommendations became a thing of the past. As required by law, Crowe held "regular" (weekly) JCS meetings. In considering cross-Service matters such as arms control and the Strategic Defense Initiative (SDI), he routinely sought the collective advice of the Service chiefs and made it a practice to submit recommendations to the Secretary on a corporate basis. Crowe's caution and restraint disappointed those in Congress who expected the new law to have an immediate and dramatic impact on the way the JCS conducted business.[27] But it seemed to Crowe the right thing to do. "I started gently," he said, "but as time passed and the chiefs grew used to the idea of the new arrangements, I exerted my authority more and more."[28]

Like the original National Security Act passed in 1947, the Goldwater-Nichols amendments were a venture into uncharted territory. An intricate set of prescriptions, the law established many new responsibilities and created new relationships which only time and experience could sort out. It needed to be interpreted, applied, and tested. Within the military, it was a less than overwhelmingly popular piece of legislation, partly owing to some of its contents, but also because of the legislative process that brought it about. As the first Chairman to operate under Goldwater-Nichols, Crowe was understandably hesitant to make dramatic changes and sought to ease the Services into the new system. Subsequent Chairmen would be less patient and less reticent. But as far as Crowe was concerned, the implementation of the Goldwater-Nichols Act was an ongoing process and had barely begun by the time he left office.

NATO RESURGENT

While Congress and the Reagan administration were dueling over the future organization of the Joint Chiefs, a slow but steady transformation was taking place in Europe toward equalizing the military balance between East and West. For years, the Joint Chiefs had complained that NATO trailed the Warsaw Pact in effective military power and lacked the full spectrum of tactical nuclear and conventional capabilities to realize the goals set for itself under the flexible response doctrine and the forward defense strategy. But with the impetus of the Reagan buildup, the situation began to change. Determined to eliminate the deficiencies of the past, the Reagan administration lent its support to programs it saw as crucial to the restoration of NATO's power and credibility. Among them were the revival of the neutron bomb, which President Reagan announced in August 1981, and the decision to press ahead with deployment of a new generation of intermediate-range ballistic and cruise missiles. Both were controversial decisions that went forward despite public protests and sharp criticism. Yet as the process advanced, it became increasingly clear that the United States remained not only firmly committed to NATO but to reasserting its own influence and leadership within the Alliance as well.

The most difficult problems, as always, were those surrounding NATO's conventional capabilities, which routinely fell short of projected requirements. By the mid-1980s, having wrestled with this problem for decades to no avail, the Joint Chiefs and others in the Pentagon reached the sobering conclusion that the Europeans would probably never meet their agreed conventional force goals and that it was pointless to continue badgering them. Rather than seeking quantitative improvements in NATO's capabilities, U.S. defense planners looked to new and emerging technologies to provide qualitative multipliers to improve NATO's defenses. That approach had been tried numerous times, invariably with mixed results. But in light of the wide range of breakthroughs and improvements such as those driving the Strategic Defense Initiative, the chances of success seemed better than ever this time around. The upshot was the Conventional Defense Initiative (CDI), which the NATO defence ministers embraced at their May 1985 meeting. While many of the taskings were identical to those of the defunct Long-Term Defense Plan of the Carter years, the CDI was less ambitious than LTDP (thereby rendering it more attainable in theory) and relied squarely on advances in technology as a key means of improving NATO's conventional defense.[29]

Adoption of the CDI followed in lockstep with a related breakthrough in military thinking known as the Follow-On Forces Attack (FOFA) concept. Much of the impetus behind FOFA came from General Bernard W. Rogers, the NATO Supreme Commander from 1979 to 1987. As Army Chief of Staff immediately prior to becoming SACEUR, Rogers had encouraged the development of a new doctrinal concept known as AirLand Battle, which emphasized close coordination between land forces pursuing an aggressive maneuvering defense and air forces attacking the enemy's rear echelon units.[30] FOFA emerged from that broad operational concept. Meant as an enhancement to the flexible response strategy, FOFA envisioned the use of sophisticated surveillance aircraft (called JSTARS) to direct conventional attacks behind enemy lines against Warsaw Pact armored formations and other reinforcements. NATO would still need strong ready forces along the central front to meet the enemy's initial attack. But with FOFA, Rogers argued, NATO stood a better chance of reducing the number of Warsaw Pact reinforcements to "manageable proportions," thus lifting the nuclear threshold.[31]

While the Joint Chiefs applauded NATO's efforts, they cautioned against overoptimism and warned that the full impact of the CDI and FOFA initiatives was difficult to predict and, in any case, would not be felt for some time. Technically complex and expensive, FOFA relied on advanced computer systems and precision-guided munitions that were still experimental or in exceedingly limited supply. JSTARS, a joint Army-Air Force surveillance and tracking system around which the FOFA concept revolved, was barely more than a drawing-board concept. Initially, by speeding up the deployment of their reinforcements, the Soviets thought they could overcome whatever deep attacks NATO might launch.[32] But as they took a closer look at the situation and the possibility that not all would go according to plan, they came to the conclusion that they were steadily losing ground and that the initiative was passing to NATO. Publicly, the Soviets denounced FOFA as a veiled instrument of aggression, while privately Warsaw Pact military planners engaged in a frantic search for something to counter it. Increasingly they worried that the mainstay of their ground attack force—the heavy battle tank—might soon be obsolete. With the potential of President Reagan's Strategic Defense Initiative factored into the equation, Moscow's long-term military prospects had never seemed bleaker. NATO's, conversely, were looking up, though as those familiar with the Alliance's condition were well aware, a lot of work remained.[33] Still, according to British intelligence expert Gordon S. Barrass, "NATO leaders felt that they had finally gained the upper hand."[34]

A NEW RAPPROCHEMENT

GORBACHEV'S IMPACT

It was against this background of a resurgent NATO, the intensifying application of new technologies by the West, and signs of wavering confidence among Soviet defense planners that Mikhail S. Gorbachev ascended to power in Moscow as General Secretary of the Communist Party in March 1985. A dedicated Marxist, Gorbachev led a younger generation of reformers whose goal was to protect and preserve the Soviet system through the restructuring of the crumbling Soviet economy (*perestroika*), greater openness in public affairs (*glasnost*), and improved East-West relations. Curbing the drain caused by heavy defense expenditures was a top priority.[35]

While some in the West proclaimed Gorbachev's advent as the first step toward ending the Cold War, others—including the Joint Chiefs of Staff—adopted a more reserved outlook. Despite an improved atmosphere in East-West relations, JCS posture statements and threat assessments remained essentially unchanged throughout the 1980s. Outward improvements in U.S.-Soviet relations aside, the Joint Chiefs continued to view the Soviet Union as an implacable enemy with a "heavy dependence on military capabilities." Afraid of letting down their guard, the Joint Chiefs repeatedly recommended a high level of military preparedness across the entire spectrum of conflict contingencies, from sub-limited conventional conflicts to all-out nuclear war, until there was clear-cut evidence that the global force-to-force balance had shifted in favor of the United States and its allies.[36]

Still, the sincerity and seriousness of Gorbachev's overtures were hard to ignore. Wary at the outset, Reagan initially dismissed Gorbachev as "a confirmed ideologue," while Gorbachev looked on the President as "a product of the military-industrial complex" prone to "right-wing" extremism.[37] But as they became more familiar with one another, they reached a meeting of the minds and formed a close and productive partnership which, though far from perfect in solving problems, proved of fundamental importance in easing East-West tensions and eventually in ending the Cold War. Although the Joint Chiefs were slower to come around, their gradual acceptance of Gorbachev's initiatives as more than propaganda ploys effectively set the stage for a wholesale reconsideration of military requirements under the next administration.

Among the breakthroughs that Gorbachev's advent helped to facilitate, two in particular had a major impact on JCS thinking: the 1987 INF Treaty mandating the complete elimination of such weapons, and the withdrawal of Soviet troops from Afghanistan initiated a year later. Both involved significant concessions which in years past the Soviets had strenuously resisted and which the JCS had likewise been disinclined to contemplate without adequate assurances of Soviet compliance.

Resumed in the spring of 1985, the INF negotiations proceeded in tandem with talks on START and space-based defensive weapons (i.e., SDI). The ostensible goal was a comprehensive agreement. Unable to make headway on an overall accord, Gorbachev indicated in October 1985 that he would entertain dealing with INF separately from other systems, a change of procedure that allowed the INF talks to go forward at a faster pace.[38] The main concern raised by the Joint Chiefs was that as the elimination of nuclear weapons gathered momentum, the Soviets would be in an even stronger position than before because of their numerical superiority in conventional forces. President Reagan, however, was skeptical and sought to reassure the chiefs that their concerns would be addressed one way or another.[39] What finally emerged in the form of the INF Treaty, signed in December 1987, was practically unprecedented: a worldwide ban on all U.S. and Soviet ground-launched ballistic and cruise missiles with ranges of 500 to 5500 kilometers, backed by enforcement provisions allowing each side to conduct on-site inspections of the other's facilities.[40]

For Gorbachev, the INF Treaty was both a spectacular gesture of goodwill that cemented his reputation as a peacemaker in the West and the coup de grace to the Kremlin's hard-line defense planners who orchestrated the military buildup under Brezhnev. Soviet strategy as laid down from the mid-1970s on by Marshal Nikolai V. Ogarkov, chief of the General Staff, had relied on the SS–20 to spearhead a massive, surprise nuclear strike in conjunction with an immediate, high-speed conventional air and ground assault, to overwhelm NATO defenses.[41] What Ogarkov and other Soviet defense planners had failed to anticipate was that NATO would have the unity and resolve to respond with a theater missile modernization program resulting in the deployment of a new generation of more effective and usable weapons (the Pershing II especially) that could strike the Soviet homeland. Instead of an asset intimidating the West, the Soviet arsenal of SS–20s had become one of Moscow's most notorious liabilities.[42] All the same, the hard-liners gave way grudgingly. While Ogarkov's successor, Marshal Sergei F. Akhromeyev, dutifully endorsed the INF Treaty in public, he disparaged it in private as a "lopsided deal."[43] As yet, discontent within the Soviet military appeared manageable, but as a massive letter-writing campaign against the treaty by retired officers indicated, it was far from popular among the former rank and file.[44]

In the West, the most strenuous objections to the INF Treaty were raised by the former NATO Supreme Commander, General Bernard Rogers. Characterizing the treaty as the product of "short-term political expediency," Rogers believed that eliminating the Alliance's INF capability would cripple its capacity to offer the full range of effective deterrence.[45] Others, however, disagreed. While Crowe recalled some grumbling from Army and Marine Corps leaders, the consensus among the Joint Chiefs was that the INF Treaty marked a major breakthrough and was "too attractive a proposition to pass up."[46] As the most far-reaching arms control agreement thus far negotiated, President Reagan hailed it as "a realistic understanding" capable of providing a "framework" for a fundamentally improved relationship.[47] Likewise, it tended to confirm Reagan's philosophy that patience and persistence pay off in the long run and that the elimination of nuclear weapons, a goal his critics derided as a fanciful notion, was not so impractical after all. Buoyed by the positive outcome of the INF talks, the President indicated that he looked forward to signing a START treaty, incorporating a 50 percent reduction in heavy missiles, when he and Gorbachev met in Moscow in the summer of 1988. But as the date of the summit approached, continuing objections by the Soviets to SDI and a superabundance of unresolved details, many having to do with verification, prevented the two heads of state from consummating a deal. Not until 1991 did a START agreement materialize.[48]

No less significant than the INF Treaty in changing JCS thinking was the Soviet withdrawal from Afghanistan brought on by a combination of diplomatic pressure from the West and military pressure from the American-backed *mujahideen*. Dating from the waning days of the Carter administration, U.S. covert involvement in Afghanistan had remained a fairly low-key affair until President Reagan took steps in March 1985 to bolster the U.S. role.[49] As part of the effort, the Joint Chiefs waived their self-imposed prohibition on sharing high-technology weapons and released shoulder-fired Stinger antiaircraft missiles to the insurgents. A major turning point in the war, the advent of the Stingers severely restricted the Soviets' use of the air and compelled them to make significant changes in strategy and tactics. If not decisive, the introduction of the Stingers certainly helped to even the playing field and allowed the mujahideen to fight the Soviets and their allied Afghan forces to a virtual standstill.

Even before the Stingers were introduced, Gorbachev was convinced that the war in Afghanistan (increasingly costly and unpopular at home) could not be won, and in the autumn of 1985 he received approval from the Politburo to explore a strategy of withdrawal. Yet it was not until after the Stingers made their appearance on the battlefield that UN-brokered peace talks began to bear fruit. Eventually, under accords signed on April 14, 1988, the Soviet Union agreed to withdraw half its troops by August, and the rest by mid-February 1989.[50] Assuming Soviet compliance with the accord, the Joint Chiefs expected the logical result to be a steady decline in the power and authority of the Soviet-backed Islamic regime in Kabul. Whether it would be an inward-looking Islamic state, reserved in its dealings with the United States and the Soviet Union alike, or a "fundamentalist" regime comparable to neighboring Iran, remained to be seen.

The impending withdrawal of Soviet forces from Afghanistan was by any measure a triumph for the Reagan administration's hard-line foreign policy. Like the INF Treaty, it further validated the President's contention that steady pressure from all directions would elicit significant changes in Soviet behavior. A major defeat for Kremlin policy, analogous in many ways to the American setback in Vietnam, the withdrawal from Afghanistan was perhaps the clearest indication to that point that Soviet power and authority were in decline. Yet for the Joint Chiefs and others in Washington, recognition of the full implications of the Soviet withdrawal emerged slowly. All that seemed to matter at the time was that the Soviets had given up and, in so doing, had removed what the JCS had once considered a major menace to U.S. interests in Southwest Asia and the Middle East.

TERRORISM AND THE CONFRONTATION WITH LIBYA

With American military power on the rise and signs emerging that the Cold War might be winding down, the Reagan administration operated more freely in accepting risks. One of the areas where it stepped up U.S. involvement was against the growing threat of state-sponsored terrorism. Bolstered by assistance and coaching from Moscow, state-sponsored terrorist groups had become a favorite means among radical Third World regimes of putting pressure on the West. By the mid-1980s, one of the most notorious culprits in the eyes of President Reagan, the Joint Chiefs, and many others was Libyan strongman Muammar Qaddafi. Charismatic and unpredictable, Qaddafi pursued a unique brand of revolutionary ideology that combined militant Islam, popular democracy, and communal ownership of property to create something approximating an Islamic socialist state. In foreign policy, he aligned himself with the Soviet Union in return for military assistance and regarded Israel and the "bourgeois" countries of the West, led by the United States, as his enemies. He openly offered his support to international terrorist groups to bring them down. As one observer put it, "No country . . . not even Syria or Iran, matched the record of Libya under Qaddafi as an epitome of lawlessness and contempt for international norms."[51]

During his first term, President Reagan had authorized varying combinations of naval exercises, economic sanctions, and diplomatic pressure to try to persuade Qaddafi to moderate his policies and behavior, all to no avail. A major exporter of high-grade crude oil, Libya enjoyed close political and economic ties with many European countries, including Italy and France, despite its reputed links to terrorism. The net effect was lukewarm support for sanctions and other nonmilitary forms of pressure that Washington tried to apply. Then, in June 1985, Hizballah terrorists hijacked a U.S. airliner flying from Athens to Rome. During the episode the hijackers tortured and murdered an American passenger, Navy Petty Officer Robert Stethem. While there was no direct evidence connecting Libya to the hijacking, the assumption of the Joint Chiefs and others in Washington was that Qaddafi's role in terrorism overall was too pervasive to rule out the possibility and that curtailing that role would go far toward curtailing terrorism in general.[52]

Though committed to a strong stand against Qaddafi and terrorism, the Joint Chiefs wanted to avoid overreacting. Supported by Secretary of Defense Weinberger, they urged caution in responding and resisted efforts by Secretary of State George P. Shultz, National Security Advisor Robert C. McFarlane, and others who wanted to make greater use of military power. But during the waning months of 1985 came a rapid succession of bloody terrorist incidents—the seizure of the cruise liner *Achille Lauro*, the hijacking of an Egyptian airliner, and the machine gun attack on the passenger lounge of the Vienna, Austria, airport. As a result, the JCS found themselves under mounting pressure to conduct a major retaliatory campaign that would severely punish Qaddafi and weaken his power and prestige if not topple him. Finding the options limited, Chairman Crowe initially relied on a resumption of large-scale naval operations off the Libyan coast to convey the message to Qaddafi that the United States meant business. But after the April 5, 1986, terrorist bombing of a discotheque in West Berlin frequented by U.S. Service personnel, President Reagan ordered the JCS to prepare immediately for stronger measures. As the President characterized it, the intelligence was "pretty final" that the Libyans had helped plan the attack.[53]

The discotheque bombing set a planning process in motion culminating in the most deliberate and deadly military action yet taken by the United States against Qaddafi—the bombing raid on Libya carried out jointly by Air Force and Navy planes on April 14–15, 1986. Hurriedly assembled, the operational plan preferred in the Joint Staff drew on prior contingency planning and exercises conducted by the Air Force. It envisioned attacks carried out by F–111 medium-range fighter-bombers flying from bases in the United Kingdom. The President wanted to retaliate as soon as possible, and since the British had not as yet approved use of their facilities, the Joint Chiefs developed an alternative plan that relied on carrier-based planes already in the Mediterranean. A third option—to mount a raid with Tomahawk sea-launched cruise missiles—also received brief consideration but was soon dropped for lack of suitably armed and programmed missiles. Eventually, the British came around and gave the green light to use their bases. But by then the JCS, working in collaboration with the U.S. European Command, had settled on a composite operation (*Eldorado Canyon*), which incorporated attacks by land- and carrier-based air simultaneously.

The decision to use both land- and sea-based air was a practical move. Though derided by some naval aviation enthusiasts as a needless display of "jointness," it reflected the approved rules of engagement prescribing minimum collateral damage to civilians in urban areas. To obtain the accuracy the President wanted mandated the use of precision-guided munitions that Air Force F–111s were better equipped to deliver than Navy planes were at the time. Thus, while the F–111s spearheaded the raid with attacks on Tripoli, where the targets tended to be in built-up areas, carrier-based F–18s and A–6s hammered the more dispersed military targets across the Gulf of Sidra in Benghazi.[54]

Cleary punitive, *Eldorado Canyon* was never intended to inflict permanently crippling damage. Like the Doolittle raid on Tokyo in the early days of World War II, it was a demonstration of American resolve. Its objectives, as outlined by President Reagan prior to the attack, were to highlight Libya's vulnerability and to demonstrate that Qaddafi's continuing pursuit of terrorism would not go unpunished. "I have no illusion that these actions will eliminate entirely the terrorist threat," the President told his close friend, British Prime Minister Margaret Thatcher. "But it will show that officially sponsored terrorist actions by a government—such as Libya has repeatedly perpetrated—will not be without cost."[55]

Still, the raid on Libya moved the war on terrorism up a notch or two. A steadily growing menace, terrorism was destined in little more than a decade to succeed the Cold War as the number one security issue facing the United States and its allies. But in President Reagan's day, compared with the weighty issues of the Cold War, terrorism still seemed a problem of secondary importance and received ad hoc responses. Even so, it was beginning to loom larger and posed challenges that the JCS were as yet unsure how to handle. As the head of a country with close economic ties to the West through its oil sales, Qaddafi was in some respects a unique case. But he was also the same kind of leader, driven by fanatical religious zeal and messianic visions, that the Joint Chiefs were fated to come up against again and again. Inconclusive in its results, the clash with Libya during the Reagan years was a foretaste of the much more serious confrontations with terrorism and terrorist states yet to come.

SHOWDOWN IN CENTRAL AMERICA

Despite the new rapprochement in Europe and waning Soviet enthusiasm for the conflict in Afghanistan, the Cold War elsewhere continued almost unabated. Nowhere was that more true than in Central America, where the United States remained locked in an escalating struggle with the Soviet- and Cuban-backed Sandinista regime of Nicaragua. Throughout President Reagan's first term, the Joint Chiefs of Staff had consistently opposed direct military intervention in Central America and had encouraged the administration to rely on surrogates, known as counterrevolutionaries or "contras," to carry the fight to the Sandinistas. But as the President's second term was getting underway, there were growing signs that the contras were running out of steam, causing the JCS to reassess their position and to accept the possibility of a larger, more direct military role. Out of the ensuing give-and-take emerged a revised covert action program which President Reagan approved in January 1986, subject to the approval of legislative authority by Congress.[56]

The new program attempted both to revitalize the contra movement at the grass roots level in Nicaragua and to mobilize additional support in the United States. Controversial throughout their history, the contras resembled a rump version of the deposed Somoza regime and enjoyed barely lukewarm backing on Capitol Hill, where there was a general reluctance to provide much beyond humanitarian assistance. Under the new program, the administration proposed to expand its help to the contras with government-funded arms aid and professional training organized under the Joint Chiefs of Staff. In October 1986, after a lengthy and spirited debate, Congress finally approved the administration's request under its revised "covert" action program for $100 million to help the contras—$70 million in military aid and $30 million in humanitarian assistance.[57]

Almost immediately, however, implementation of the administration's program fell under the gathering cloud of the Iran-contra affair, a scandal that blew up over revelations of clandestine arms sales to Iran and the skimming of profits by members of the NSC Staff to subsidize the purchase of arms and ammunition for the contras. The precipitating event occurred on the morning of October 5, 1986, when a Soviet-made surface-to-air missile brought down a chartered C–123 cargo plane that was on a resupply mission to contras operating in northern Nicaragua. It turned out that the plane and its cargo were part of an off-the-books covert assistance program going back more than a year to circumvent aid prohibitions imposed by Congress in 1984. The Joint Chiefs knew of the contra resupply program, but they had no part in organizing it and assumed it to be part of a privately-financed and privately-run operation. If they had reason to think otherwise, they kept the information to themselves.

Reverberations from the Iran-contra affair extended far and wide, and by the summer of 1987 it was a full-blown scandal. Talk of impeaching the President was in the air. Ironically, at the same time the administration's Central America policy was falling under renewed attack in Washington, its revamped covert assistance program was beginning to show signs of turning the military situation to the contras' advantage. Better trained and indoctrinated, they were gradually becoming more effective fighters and more accepted by the local population. All the same, many Central American leaders, even those aligned with the United States in opposition to the Sandinistas, were uneasy about the contras' activities, and in August 1987 they joined in support of a new diplomatic initiative sponsored by Costa Rican President Oscar Arias to end the conflict through new, supervised elections.

With momentum building behind the Arias peace plan, Congress in February 1988 suspended further funding for the contras. Shortly thereafter, backed by Soviet attack helicopters and Cuban troops, the Sandinistas launched an all-out assault on the contras' base camps along the Nicaragua-Honduras border. Amid the escalating crisis, President Reagan met with his senior advisors and congressional leaders on the afternoon of March 16, but was unable to enlist the support of House Speaker Jim Wright and other key Democrats who were either noncommittal or opposed to any U.S. military action.[58] The next day, responding to a formal request from the Honduran government for U.S. assistance, President Reagan ordered a brigade-sized task force of the 82^د Airborne to conduct a 10-day "readiness exercise" in Honduras. Meanwhile, U.S.-piloted helicopters began ferrying Honduran troops into the battle zone.[59] Though the JCS rules of engagement governing these deployments made it highly unlikely that U.S. and Sandinista forces would ever confront one another, the implied threat of American military intervention appeared to have the desired effect, and within days the Nicaraguans curtailed their offensive. Yet even though the contras avoided annihilation, the fighting had taken a heavy toll on their numbers. On March 23, 1988, seeing no other choice, their leaders declared a unilateral ceasefire.

From that point on, the contras' fortunes entered a steep decline, a process hastened by political infighting within its leadership, dwindling resources, and the Reagan administration's grudging acceptance of the Arias peace plan. By May 1988, the contras were down to 400 front-line troops, too few to pose a serious threat to the Sandinista regime. Feeling that it had run out of options, the Reagan administration let matters drift until it left office in January 1989. By then, the incoming Bush administration, hoping to eliminate Central America as a source of continuing domestic political discord, had settled on a different course that abandoned further military pressure on Nicaragua in favor of negotiated solutions through multilateral diplomacy.[60]

Though disappointed by the turn of events in Central America, the Joint Chiefs took the outcome in stride. While it was a setback in certain respects, the emerging settlement was not the disaster that some within the Joint Staff had worried it might be. Indeed, as the dust settled, it became clear that the Sandinistas were far weaker politically than previously supposed. In agreeing to elections—finally held in 1990—the Sandinista regime virtually sealed its own demise. Even though the Joint Chiefs had not played a large or conspicuous role, their insistence that aid and training to the contras be placed on a more systematic and professional basis had gone far toward rescuing a faltering program and turning it around. All things considered, the chiefs' involvement helped to produce a more favorable outcome than would otherwise have been the case.

TENSIONS IN THE PERSIAN GULF

While the struggle for Central America tested the Joint Chiefs' capabilities and willingness to cope with low-intensity conflict, the resumption of tensions in the Persian Gulf challenged their resourcefulness in more traditional ways. Since taking office, using the prism of the Cold War, the Reagan administration had treated a Soviet invasion or attack against the Gulf oil fields as the primary danger in that part of the world and urged the JCS to plan accordingly.[61] Even so, the source of greatest volatility in the region was the ongoing conflict between Iran and Iraq. Precipitated by Iraqi President Saddam Hussein's attack on Iran in 1980 over a border dispute, the Iran-Iraq war had degenerated into a World War I-style conflict, complete with trench warfare, human-wave assaults, and chemical weapons. Deadlocked on the battlefield, the two antagonists took to crippling one another's economic base by attacking their respective capacities to produce and export petroleum products. So intense did the "tanker war" become that in the summer of 1984 the United States and other Western powers joined together to provide naval protection for nonaligned (primarily Kuwaiti) shipping. But by the end of the year, the attacks mostly stopped and the international protection effort relaxed.

The official policy of the United States toward the Gulf War was neutrality. Unofficially, the Reagan administration leaned in favor of Iraq. Characterizing Saddam as a "no good nut," President Reagan was fully aware that the Iraqi leader's regime was one of the most corrupt, ruthless, and repressive in the Middle East.[62] All the same, he was determined to block Iranian and radical Shia expansionism and worried that an Iranian victory over Iraq would destabilize the region. The policy in effect at the outset of President Reagan's second term was to do what was feasible and practicable, short of overt assistance or direct intervention, to avoid an Iraqi defeat or collapse. In practice, this meant seeking other governments' cooperation in enforcing an arms embargo against Iran (Operation *Staunch*), encouraging Saudi

Arabia, Kuwait, Egypt, France, and other countries friendly with Iraq to keep its war machine going, and from time to time providing the Iraqi armed forces with limited operational assistance and intelligence.

While tilting toward Iraq, the Reagan administration also pursued backchannel contacts with Iran that had the unintended side-effect of complicating JCS efforts to assure the safety of neutral shipping in the Gulf. The leading figures in this enterprise were former National Security Advisor Robert C. McFarlane and an assistant, Lieutenant Colonel Oliver L. North, USMC, who secretly helped arrange arms transfers to the Iranians in an effort to secure the release of Western hostages being held by Islamist militants in Lebanon. Limited initially to a handful of HAWK antiaircraft missiles purchased from Israeli stocks and a few hundred antitank TOW missiles, the arms-for-hostages deal was never large enough to tip the military balance in Iran's favor. But it carried immense weight as a symbolic gesture. Privately, as North and his associates expanded their contacts with the Iranians, President Reagan became concerned that they would send the wrong signal and lead Tehran to think that the United States was on its side.[63]

The initiation of U.S. covert aid to Iran late in 1985 roughly coincided with Tehran's decision to pursue a bolder, more aggressive strategy in its war with Iraq. Reeling from years of heavy casualties and mounting costs, Iran's leadership was desperate for a breakthrough, and in February 1986 it launched a two-pronged counterattack—a diversionary operation north of the Hawizeh Marshes followed by a major amphibious assault in the south that seized the strategically important Faw Peninsula. Eventually, the line stabilized, but only after heavy fighting that brought the Iranians to the outskirts of Basra, Iraq's second largest city. Even though the chances of Basra falling appeared remote, Iran's battlefield successes suggested a looming strategic shift in the war in Iran's favor. In March, Iran resumed its attacks on Gulf shipping, scoring eight hits, all but one against non-Arab vessels.[64]

The turning point resulting in U.S. intervention was Kuwait's request in November-December 1986 for Western and Soviet protection against further Iranian attacks on its shipping.[65] Until that time, the Joint Chiefs had held stubbornly to their current force strategy under which for years they had managed to limit U.S. commitments in the region. Playing down the impact of renewed threats to shipping, the JCS cautioned against hasty action. Indeed, during interagency deliberations extending from late 1986 into early 1987, Admiral Crowe and members of his staff made the point repeatedly that while they appreciated the seriousness of the situation, they saw the Kuwaiti request as opening Pandora's Box by pressuring the United States to protect other noncombatants' shipping. As Crowe later recalled:

"I had done more agonizing over this issue than over any other since my appointment as Chairman."⁶⁶

After weighing the pros and cons, President Reagan concluded that the resumption of Iranian raids on Kuwaiti shipping, coupled with the possibility of direct Soviet intervention, left the United States no choice but to play a more active and direct role. By early March 1987, hoping to head off Soviet involvement, he and his advisors settled on a policy of escorting 11 Kuwaiti tankers reflagged as American vessels, part of a multinational effort to protect shipping in the Gulf.⁶⁷ Working out the details fell to Admiral Crowe, who arrived in Kuwait a few days later on a previously scheduled visit to the Middle East. By the time he returned to Washington, Crowe was convinced that reflagging the Kuwaiti tankers held the key not only to the maintenance of regional stability, but also to the preservation of friendly relations with the Arab world. "My conclusion," he recalled, "was that we should go into the Persian Gulf . . . because it was the best chance we had to repair our Arab policy and to make some significant headway in an area where it was absolutely crucial for us to forge the strongest ties we could manage."⁶⁸

OPERATION *EARNEST WILL*

The ensuing escort operation (*Earnest Will*) finally got underway in July 1987 and lasted until September 1988. Though undertaken on a multinational basis, it had only token contributions from other Western countries and was predominantly a U.S.-led and U.S.-directed affair. At its height, *Earnest Will* involved 27 U.S. surface vessels and 13,700 American Service personnel. It was also the first major test of the recently reconstituted joint system under the Goldwater-Nichols Act. Over the years, the Rapid Deployment Force and its successor, U.S. Central Command (USCENTCOM), had done extensive planning for ground and air operations in the Middle East. But having been unable to find a well-qualified senior naval officer for his staff, the USCINCCENT, General George B. Crist, Jr., USMC, had as yet made limited headway toward developing a maritime plan for the region. Seeking to consolidate his authority, Crist sought full control of the operation and in so doing found himself at odds with his Navy counterparts. In late August 1987, to end the squabbling, Secretary of Defense Weinberger established a new subcommand—Joint Task Force Middle East (JTFME)—headed by a naval officer, Rear Admiral Dennis M. Brooks, who exercised day-to-day responsibility for escort duties, while Crist oversaw strategic direction of the operation from his headquarters in Tampa, Florida.⁶⁹

This inauspicious introduction to the era of "jointness" under Goldwater-Nichols was soon followed by the need for a wholesale reappraisal of the U.S. role and objectives under *Earnest Will*. As originally envisioned, the operation was to have been a fairly passive enterprise focusing on escort functions. But by the time it commenced, the security situation in the Persian Gulf had deteriorated to the point that U.S. warships were becoming as much the target as commercial shipping. A case in point was the cruise missile attack against the American frigate USS *Stark* on May 17, 1987, by an Iraqi fighter that nearly sank the ship and left 37 U.S. sailors dead. Though the Iraqis promptly apologized, insisting that the attack had been a mistake, the incident underscored the dangers involved by the very presence of U.S. warships in the Persian Gulf and helped to usher in a more aggressive approach by the Joint Chiefs toward their escort responsibilities.

The Iraqi attack on the *Stark* notwithstanding, the assumption in the Pentagon and at USCENTCOM headquarters continued to be that Iran was the principal troublemaker and the most likely to come into conflict with U.S. forces. Operating on that assumption, JCS planners expected the Iranian threat to take several forms. With replacement parts and pilots in short supply, Iran had all but abandoned air attacks on shipping since the spring of 1986 and had turned to unconventional tactics carried out by Revolutionary Guards, who proved adept at hit-and-run raids using small speedboats and powerful rocket-propelled grenades. At the same time, Iran also acquired a small arsenal of short-range Chinese SILKWORM antiship missiles, which it deployed adjacent to the Strait of Hormuz and on the Faw Peninsula within range of Kuwait.[70] The most dangerous and persistent threat, however, came from Iranian antiship mines. Initially, the Iranians denied any involvement in mining operations. But on September 21, 1987, a U.S. Army helicopter-gunship, flying from a Navy frigate, strafed and disabled the *Iran Ajr*, a converted Iranian troop ship, as it was laying mines in the path of the convoying oil tankers. The next day, U.S. Navy SEALS boarded the ship and seized a sizable cache of military documents confirming Iran's involvement in mine-laying and other operations.[71]

Following the *Iran Ajr* incident, American and Iranian forces became engaged in a steadily escalating contest for control of the Persian Gulf. By the end of 1987, Iranian attacks on shipping were up 53 percent over the year before. Avoiding ships under U.S. escort, the Iranians concentrated their attacks on vessels without protection. As a result, the JCS came under mounting pressure (primarily from Saudi Arabia) to expand the scale and scope of the U.S. protection regime by providing assistance, upon request, to all nonbelligerent vessels under attack. On April 14, 1988, the on-again-off-again conflict finally boiled over when the missile frigate USS *Samuel B. Roberts* found itself in the middle of a freshly laid Iranian minefield. In attempting to escape the *Roberts* suffered heavy damage when it struck one of the mines. U.S. retaliation was inevitable.

The day after the incident, Admiral Crowe attended a breakfast meeting at the Pentagon hosted by Secretary of Defense Frank C. Carlucci to discuss retaliatory measures. Also present were Secretary of State Shultz and the President's assistant for national security affairs, Lieutenant General Colin L. Powell, USA. Feeling that the United States had exercised restraint long enough, Crowe, with Carlucci's support, urged destruction of an Iranian warship to demonstrate that "we were willing to exact a serious price." Around 11 a.m., the meeting adjourned to the White House, where President Reagan joined in. In the President's mind there was no doubt that retaliation was imperative. Moreover, he offered no objection to further military action should Iran resist or challenge U.S. forces. Around noon, Crowe placed a secure telephone call to Crist at USCENTCOM headquarters in Tampa, Florida, relaying the President's decision and setting in motion Operation *Praying Mantis*, which got underway on April 18.[72]

The immediate targets were the Sassan and Sirri gas-oil platforms in the central and southern Persian Gulf. While it was against U.S. policy to attack "economic" targets, these (like other Iranian oil platforms) were heavily fortified and served as bases for raids on shipping. Only the Sirri platform was still pumping oil.[73] In retaliation for the destruction of the oil rigs, Iranian air and naval forces counterattacked, precipitating a major naval battle. During the engagement, U.S. air and surface units, using laser-guided bombs and other advanced technologies, destroyed a missile patrol boat and several smaller craft, sank the British-built Iranian frigate *Sahand*, and severely damaged its sister ship, the *Sabalan*. By the time the engagement was over, Iran had lost half of its navy. The only U.S. loss, apparently the result of a mechanical failure, was a Cobra attack helicopter and its two-member crew.[74] Still, in assessing the overall outcome, Crowe was quite pleased. Feeling that the United States had made a much more forceful statement of its resolve this time around, he was also deeply impressed by the high degree of joint action achieved in the field.[75]

By mid-1988, with its economy in a shambles, much of its navy at the bottom of the Persian Gulf, and its air force down to a handful of flyable planes, Iran was no longer in a position to mount a serious challenge to the United States. Sensing that the worst had passed, JCS planners began to prepare for the drawdown of U.S. forces. In April 1988, for air defense purposes, the Navy added an Aegis missile cruiser, the USS *Vincennes*, to its flotilla operating in the Persian Gulf. The decision to do so was at the instigation of the NSC Staff, which wanted to avoid a repetition of the *Stark* incident, and went against the better judgment of JCS and Navy planners, who considered Aegis cruisers ill-suited to the relatively shallow "green water" environment of the Persian Gulf.[76] According to Admiral Crowe, who reluctantly supported the NSC's recommendation, the deployment of the *Vincennes* was a belated development and came about only after intelligence reports that the Iranians, having become desperate, were reconfiguring what was left of their air force to attack U.S. warships.[77]

On July 3, 1988, while on patrol duty in the Persian Gulf, the *Vincennes* shot down an Iranian civilian airliner, killing all 290 passengers and crew aboard. Unable to distinguish one type of plane from another, the *Vincennes*' radar mistook the airliner for an Iranian F–14, which had been prowling the same area the past few days. Immediately after, as if sobered by the incident, Iran and Iraq dramatically scaled back their military operations in the Persian Gulf. The last reported attack against neutral shipping by either belligerent occurred on July 20. Having fought one another almost continuously for 8 years, both sides were showing marked signs of war-weariness, especially Iran. The end of the war was anticlimactic, as Iran and Iraq both grudgingly accepted a UN-brokered cease-fire, which took effect on August 20, 1988. Escorts ended a month later, though as a precaution the Navy continued to operate a less demanding regime of protection, termed an "accompany mission," that lasted until June 1989.

Throughout *Earnest Will*, the approaching end of the Cold War undoubtedly allowed the Joint Chiefs to operate more freely and to take greater risks. A major factor in Middle East politics from the mid-1950s on, the Soviet Union was barely noticeable during the escort operation and its aftermath. Still, it was the possibility that Moscow might steal the march on the West by taking over protection of Kuwait's tankers that prodded the United State into action in the first place. Though not as strong as it was, the specter of Soviet power remained a formidable factor.

Overall, however, the demise of Soviet power was steadily reshaping JCS perceptions of American security interests and the accompanying need for military forces. For two generations, the Joint Chiefs had framed their assessments of U.S. defense requirements around the dangers posed by a nuclear-armed Soviet Union and its satellites. But by the end of the Reagan Presidency, the chiefs' image of the Communist threat had begun to change. Although they still credited the Soviet Union as having formidable military capabilities, they could not ignore the emerging changes in Soviet policy instigated by new leadership in Moscow. While it was too soon to tell with certainty how the Gorbachev reforms would play out, one clearly intended result was to loosen the Soviet military's grip on resources. Should that trend continue, it would doubtless fundamentally alter JCS perceptions of their own military requirements.

In sum, as the Reagan administration drew to a close, decades of tension and competition between East and West were starting to give way, a situation far different from only 8 years earlier. Whether the current rapprochement would last or, like "peaceful coexistence" and détente degenerate into another round of the Cold War, remained to be seen. As usual, the Joint Chiefs were cautiously optimistic, not wanting to let down their guard but aware also that change was in the air. They could sense that they were entering a period of transition but could not as yet foresee its outcome or full impact.

NOTES

1 On the background and origins of the JCS reorganization debate, see Gordon Nathaniel Lederman, *Reorganizing the Joint Chiefs of Staff: The Goldwater-Nichols Act of 1986* (Westport, CT: Greenwood Press, 1999), chaps. 1–2; Daniel Wirls, *Buildup: The Politics of Defense in the Reagan Era* (Ithaca: Cornell University Press, 1992), 79–101; and Archie D. Barrett, *Reappraising Defense Organization* (Washington, DC: National Defense University Press, 1983).

2 CM-79-78 to SECDEF, September 1, 1978, "National Military Command Structure and Departmental Headquarters Studies," U, JCS 1977/409-5.

3 Interview with General David C. Jones, USAF (Ret.), by Walter S. Poole and Steven L. Rearden, February 4, 1998, Arlington, VA.

4 James R. Locher III, *Victory on the Potomac: The Goldwater–Nichols Act Unifies the Pentagon* (College Station: Texas A&M University Press, 2002), 33–37.

5 David C. Jones, "Why the Joint Chiefs of Staff Must Change," *Armed Forces Journal International* (March 1982), 62–72.

6 Edward C. Meyer, "The JCS—How Much Reform Is Needed?" *Armed Forces Journal International* (April 1982), 82–90.

7 CM-143-82 to SECDEF, November 22, 1982, "JCS Reorganization," JCS Reorganization Notebook No. 3, JHO; see also Locher, 79–80.

8 JCS Historical Division, *Organizational Development of the Joint Chiefs of Staff, 1942–1989* (Washington, DC: Joint Secretariat, Joint Chiefs of Staff, November 1989), 60–62.

9 P.L. 98-525, October 19, 1984; H. Rpt. No. 98-1080.

10 "CJCS Guidance to the OJCS Special Study Group," January 7, 1985, JHO Collection.

11 NSDD 175, "Establishment of a Blue Ribbon Commission on Defense Management," June 17, 1985, U, available at <http://www.fas.org/irp/offdocs/nsdd/nsdd-175.htm>.

12 CM-152-86 to Service chiefs, January 29, 1986, "Nomination and Selection of Quality Officers for Assignment to OJCS," U; and CM-153-86 to Directors and Heads of Agencies, OJCS, January 29, 1986, "Selection of Officers to the OJCS," U, both in JHO 15-009.

13 See Crowe's testimony in U.S. Congress, House, Committee on Armed Services, Subcommittee on Investigations, *Hearings: Reorganization of the Department of Defense*, 99:2

(Washington, DC: GPO, 1986), 343; see also Lederman, *Reorganization of the JCS*, 72, 138 note no. 39.

14 William J. Crowe, Jr., *The Line of Fire* (New York: Simon & Schuster, 1993), 148.

15 JCSM-397-85 to SECDEF, November 13, 1985, "Joint and Service Improvement Initiatives," U, JHO 15-006; Crowe, *Line of Fire*, 155.

16 JCSM-401-85 to SECDEF, November 12, 1985, "DOD Organization," U, JHO 15-006.

17 Letter, Weinberger to Goldwater, December 2, 1985, U, JHO 15-007.

18 Crowe, *Line of Fire*, 158–159; Locher, 423–424.

19 Letter, Crowe to Aspin, August 13, 1986, U, JHO 15-0012.

20 *Congressional Quarterly Almanac, 1986*, 459. Although a major event in the history of the Department of Defense, passage of the Goldwater-Nichols Act received no mention in either Reagan's or Weinberger's memoirs.

21 See Christopher M. Bourne, "Unintended Consequences of the Goldwater-Nichols Act," *Joint Force Quarterly* (Spring 1998), 103–104.

22 Crowe, *Line of Fire*, 160.

23 CM-424-86 to DJS, November 6, 1986, "OJCS Restructuring," U; and DJSM 1915-86 to Directors and Heads of Agencies, OJCS, November 7, 1986, "OJCS Restructuring Directive," U, both in JHO 15-0012; *Organizational Development of the Joint Chiefs of Staff, 1942–1989* (Washington, DC: Historical Division, Joint Secretariat, Joint Chiefs of Staff, November 1989), 64–65.

24 Crowe, *Line of Fire*, 159.

25 CM-660-87 to SECDEF, April 6, 1987, "Duties of the VCJCS," U, JHO 15-0012; Memo, SECDEF to CJCS, April 15, 1987, "Duties of the VCJCS," U, in notebook labeled "OCJCS Restructuring Nov 1987," JHO Collection.

26 CM-465-86 to CNO, December 5, 1986, "Implementation of the Special Operations Command," U; DJSM 226-87 to SECDEF, February 6, 1987, "Timely Advice," U, both in OCJCS Restructuring November 1987 Notebook, JHO Collection.

27 See Paul Y. Hammond, "Fulfilling the Promise of the Goldwater-Nichols Act in Operational Planning and Command," in James A. Blackwell, Jr., and Barry M. Blechman, eds., *Making Defense Reform Work* (Washington, DC: Brassey's, 1990), 127–129.

28 Crowe, *Line of Fire*, 161.

29 Richard L. Kugler, *Commitment to Purpose: How Alliance Partnership Won the Cold War* (Santa Monica, CA: RAND, 1993), 428–430; John S. Duffield, *Power Rules: The Evolution of NATO's Conventional Force Posture* (Stanford, CA: Stanford University Press, 1995), 227.

30 Though Rogers was instrumental as Army Chief of Staff and SACEUR in promoting the AirLand Battle concept, much of the inspiration behind the idea came from two successive heads of the U.S. Army's Training and Doctrine Command (TRADOC)—General William E. DuPuy and General Donn A. Starry. See John L. Romjue, "The Evolution of the AirLand Battle Concept," *Air University Review* 35 (May–June 1984), 4–15.

31 Bernard W. Rogers, "Follow-on Forces Attack: Myths and Realities," *NATO Review*, No. 6 (December 1984), 1–9.

32 William E. Odom, *The Collapse of the Soviet Military* (New Haven: Yale University Press, 1998), 75–78.

33 According to General Sir Peter de la Billière, who led British troops during *Desert Shield/Desert Storm* in 1990-1991, efforts by the British Army of the Rhine to assemble one fully operational armored brigade for that operation "turned the whole system inside-out." "Some of our armoured vehicles were old and plain worn out; others were run down and not properly maintained or else not used due to the lack of spares, money and training." General Sir Peter de la Billière, *Storm Command: A Personal Account of the Gulf War* (London: HarperCollins, 1992), 26.

34 Gordon S. Barrass, *The Great Cold War: A Journey Through the Hall of Mirrors* (Stanford, CA: Stanford University Press, 2009), 341.

35 Mikhail Gorbachev, *Memoirs*, trans. by Georges Peronansky and Tatjana Varsavsky (New York: Doubleday, 1995), 401f.

36 Joint Staff, *United States Military Posture FY 1989*, 1–9 and passim, U, JHO 13-005.

37 Ronald Reagan, *An American Life* (New York: Simon and Schuster, 1990), 615; Gorbachev quoted in Jonathan Haslam, *Russia's Cold War* (New Haven: Yale University Press, 2011), 354.

38 Maynard W. Glitman, *The Last Battle of the Cold War: An Inside Account of Negotiating the Intermediate Range Nuclear Forces Treaty* (New York: Palgrave Macmillan, 2006), 117–119.

39 Reagan Diary, October 27, 1986, available at <http://www.reaganlibrary.com/white-house-diary.aspx>.

40 Joseph P. Harahan, *On–Site Inspections Under the INF Treaty* (Washington, DC: On–Site Inspection Agency, U.S. Department of Defense, 1993), 169–175, reprints the INF Treaty. The treaty did away with 2,692 missiles and banned eight types of systems. For the NATO powers, these were the Pershing II, the BGM–109 ground–launched cruise missile, and the Pershing 1A. For the Soviet Union, they were the SS–20, SS–4, SS–5, SS–12, and SS–23. Two missiles that had been tested but not deployed were also banned because of their ranges: the U.S. Pershing 1B and the Soviet SSC–X-4 cruise missile.

41 Joseph D. Douglass, Jr., *The Soviet Theater Nuclear Offensive* (Washington, DC: Office of Director of Defense Research and Engineering, Net Technical Assessment, 1976), examines the origins of this strategy and his philosophic underpinnings.

42 Haslam, 359.

43 Akhromeyev quoted in George P. Shultz, *Turmoil and Triumph: My Years as Secretary of State* (New York: Charles Scribner's Sons, 1993), 1012.

44 Barrass, 212–214, 342–343; Odom, 134–135.

45 Hugh A. Williams, "Flexible Response and the INF Treaty: What Next? (Study Project, U.S. Army War College, Carlisle Barracks, PA, March 14, 1988), 1, 13–17.

46 Crowe, *Line of Fire*, 264.

47 "Remarks on the Departure of General Secretary Mikhail Gorbachev," December 10, 1987, *Reagan Public Papers, 1987*, 1498–1499.

48 Reagan, *An American Life*, 697–705; Raymond L. Garthoff, *Détente and Confrontation: American-Soviet Relations from Nixon to Reagan* (Washington, DC: Brookings, 1994, rev. ed.).

49 Shultz, 1086–1087.

50 Tom Rogers, *The Soviet Withdrawal from Afghanistan: Analysis and Chronology* (Westport, CT: Greenwood Press, 1992), 31–36.

51 Brian L. Davis, *Qaddafi, Terrorism, and the Origins of the U.S. Attack on Libya* (New York: Praeger, 1990), 18.

52 NSDD 179, July 20, 1985, "Task Force on Combatting [sic] Terrorism," available at <http://www.fas.org/irp/offdocs/nsdd/nsdd-179.htm>.

53 Diary Entry, April 7, 1986, Douglas Brinkley, ed., *The Reagan Diaries* (New York: HarperCollins, 2007), 402–403. Also see Charles G. Cogan, "The Response of the Strong to the Weak: The American Raid on Libya, 1986," *Intelligence and National Security* 6 (July 1991), 618.

54 Stanik, *Operation Eldorado Canyon*, 176–205, summarizes the operation. For criticism of the operation as a joint endeavor, see John F. Lehman, *Command of the Seas* (New York: Charles Scribners' Sons, 1988), 300–301; and Davis, 120.

55 Quoted in Margaret Thatcher, *The Downing Street Years* (New York: HarperCollins, 1993), 444.

56 U.S. Congress, *Report of the Congressional Committees Investigating the Iran–Contra Affair*, 100:1 H. Rpt. No. 100-433 and S. Rpt. No. 100-216 (Washington, DC: GPO, 1987), 64–65.

57 PL 99-500, October 18, 1986. See also NSDD 248 (sanitized), "Central America," available at <http://www.fas.org/irp/offdocs/nsdd/nsdd-248.htm> (accessed July 19, 2011).

58 Robert Kagan, *A Twilight Struggle: American Power and Nicaragua, 1977–1990* (New York: Free Press, 1996), 588–589.

59 Brinkley, 587–588.

60 See James A. Baker, III, with Thomas M. DeFrank, *The Politics of Diplomacy: Revolution, War and Peace, 1989–1992* (New York: G.P. Putnam's Sons, 1995), 48–53.

61 Michael A. Palmer, *Guardians of the Gulf: A History of America's Expanding Role in the Persian Gulf, 1833–1992* (New York: Free Press, 1992), 116–117.

62 Entry, June 11, 1981, Brinkley 25.

63 Entry, February 11, 1987, Ibid., 474.

64 Nadia El-Sayed El-Shazly, *The Gulf Tanker War: Iran and Iraq's Maritime Swordplay* (London: Macmillan, 1998), 224–225.

65 Palmer, 122–123.

66 Crowe, *Line of Fire*, 180.

67 Caspar Weinberger, *Fighting for Peace* (New York: Warner Books, 1990), 397; Shultz, 925–926.

68 Crowe, *Line of Fire*, 181.

69 Palmer, 132.

70 U.S. intelligence identified these missiles as the HY–2, a Chinese copy of the Soviet-made SS–N-2 STYX. The Joint Staff referred to these missiles as the SILKWORM, a generic term used to encompass several slightly different models.

71 Weinberger, *Fighting for Peace*, 414–415; El–Shazly, 248.

72 Crowe, *Line of Fire*, 200–201.

73 Hans S. Pawlisch, "Operation Praying Mantis," Appendix A in Palmer, *On Course to Desert Storm*, 141–146, places output of the Sirri platform at around 180,000 barrels per day.

74 David B. Crist, *Gulf of Conflict: A History of U.S.–Iranian Confrontation at Sea*, Policy Focus No. 95 (Washington, DC: Washington Institute for Near East Policy, 2009), 8–9.

75 Crowe, *Line of Fire*, 201–202.

76 Palmer, 146.

77 CM-1485-88 to SECDEF, August 18, 1988, "Formal Investigation into the Circumstances Surrounding the Downing of Iran Air Flight 655 on 3 July 1988," (declassified), 887/546 (July 28, 1988); and <http://www.dod.mil/pubs/foi/reading_room/172.pdf>.

General Colin L. Powell, USA, Chairman of the Joint Chiefs, 1989–1993

Chapter 16

ENDING THE COLD WAR

Reagan and Gorbachev met for the last time in New York City in December 1988. By then the two leaders had developed an easy collaboration that both hoped would carry over into the presidency of Reagan's recently elected successor, George H.W. Bush. A former member of Congress, Director of Central Intelligence, ambassador to China, and Reagan's vice president for 8 years, Bush came to the White House with more practical experience in national security affairs than any President since Eisenhower. As part of his agenda while in New York, Gorbachev addressed the UN General Assembly and used the occasion to announce that the Soviet Union would unilaterally reduce its armed forces by half a million men and withdraw 50,000 troops and 5,000 tanks from Eastern Europe over the next 2 years. Moscow, Gorbachev insisted, wanted military forces only for defensive purposes and would use them for nothing else. A dramatic, headline-grabbing gesture, Gorbachev's announcement convinced Secretary of State George Shultz that the Cold War was more than drawing to a close. Indeed, Shultz insisted: "It was over."[1]

While Shultz's declaration may have been premature, it aptly captured the prevailing mood. After decades of tension and confrontation, the prospect of establishing a peaceful modus vivendi between East and West was too appealing for anyone, including the Joint Chiefs, to ignore. Practically no one expected the Soviet Union to disappear or its Warsaw Pact allies to lay down their arms. But with Gorbachev continuing to tender the olive branch, the opportunities for normalizing relations, settling differences in a peaceful atmosphere, and creating new partnerships seemed measurably improved.

POLICY IN TRANSITION

Like others in Washington, the Joint Chiefs were hard pressed to draw a fully coherent picture of the future from the rapid changes taking place in East-West relations. Typically cautious, they believed that relaxed tensions with the Soviet Union offered opportunities to improve relations but were reluctant to let down their guard. Their attitude at the outset of the Bush administration remained essentially the same as it had been during the last few years of Reagan's Presidency when the motto had been "Trust but verify." The Bush White House was of a similar persuasion, eager to explore the settlement of outstanding issues yet leery of taking too much for granted. As the new national security advisor, Lieutenant General Brent Scowcroft, USAF (Ret.), recalled: "I was suspicious of Gorbachev's motives and skeptical of his prospects."[2]

Scowcroft's concerns were not unfounded. True, there had been dramatic improvements in East-West relations since Gorbachev's advent and the signing of the 1987 INF Treaty. But since then, progress in the strategic arms reduction talks and parallel negotiations aimed at limiting conventional forces in Europe had been negligible. Gorbachev's pledge to withdraw 50,000 troops from Europe may have sounded like a major concession, but to the Joint Chiefs of Staff and other military experts it would do little to alter the overall strategic balance, which remained heavily weighted toward the Warsaw Pact. Despite denials by Gorbachev, reports reaching the West also pointed to a high priority Soviet program to develop a new range of biological weapons.[3] Meanwhile, Moscow continued to pursue policies in other areas that were inimical to U.S. interests. Even as it withdrew its troops from Afghanistan, the Soviet Union still poured heavy amounts of assistance into propping up a pro-Communist regime in Kabul. Likewise, it remained a firm ally of the Sandinistas in Nicaragua who, with the help of Cubans and East Germans, continued to export Communist revolution throughout Central America.

Thus, even though the Cold War might have appeared to be over, the Bush administration found itself up against problems that suggested an ongoing, albeit lower-keyed, competition with the Soviet Union. Neither friend nor foe, Moscow fell awkwardly in between. Pointing to the "challenges and uncertainties" that the waning Cold War presented, President Bush decided to launch a comprehensive review of basic U.S. policy (designated NSR 12) shortly after taking office.[4] Among other things, he wanted to know how he should balance policy toward Moscow with the steady decline of support for defense spending, a reflection of expectations in Congress and with the public at large that as East-West relations improved, the United States could reduce the size of its armed forces. Actually, the process of reaping a "peace dividend" was well underway. From consuming a post-Vietnam high of 6.6 percent of the country's GNP in fiscal years 1986 and 1987, national defense had declined to 5.8 percent by the time President Bush entered the Oval Office. When he left in 1993, it would be down to 4.7 percent, the lowest since the end of demobilization immediately following World War II.[5]

Within the Pentagon, a debate quickly developed between the Office of the Secretary of Defense and the Joint Staff over how and where to allocate resources to meet the "challenges and uncertainties" mentioned in the President's directive. OSD wanted to maintain the force structure more or less within its current configuration, with a continuing focus on Europe, while the Joint Staff wanted to strike a balance with other regions of the world. Assuming a low level of threat to Europe and a reduced force posture in years to come, JCS planners sought to make better use of available resources by shifting from the Cold War strategy of "forward defense," with forces deployed at static points along the Soviet Union's periphery, to a strategy of "forward presence" emphasizing flexibility to move forces around and to insert them as needed in the event of regional contingencies.[6]

As these debates were taking place, events in Eastern Europe were acquiring a dynamic of their own, bringing down one Communist regime after another over the course of 1989 and culminating in the toppling of the infamous Berlin Wall that November. Unable to keep up with the rapid changes sweeping Europe, the Bush administration suspended work on NSR 12 and several other reviews it had requested on the future of U.S.-Soviet relations until things settled down. Rather than relying on recapitulations of past policies, President Bush wanted fresh ideas and new insights.[7] As Colin Powell later remarked, NSR 12 failed to measure up and became "doomed to the dustbin."[8] All the same, not all was lost. Out of the give and take connected with the project at the Pentagon emerged a new National Military Strategy for 1992–1997 (NMS 92-97), which Admiral Crowe, the Chairman of the Joint Chiefs, submitted to Secretary of Defense Richard B. Cheney and President Bush in late August 1989. Though not as far reaching a change as the Joint Staff had originally intended, the new strategy—described by the Chairman as "forward defense through forward presence"—clearly downplayed prior commitments to Europe and stressed instead the role of force projection and flexible response to deal with regional crises and instability and to preserve worldwide U.S. influence.[9]

POWELL'S IMPACT AS CHAIRMAN

Presentation of the new National Military Strategy was one of Admiral Crowe's last formal functions as Chairman. On October 1, 1989, he relinquished his duties to General Colin L. Powell, USA, the first African-American to become Chairman, and at age 52 the youngest CJCS. A product of the Reserve Officers' Training Corps program at City College of New York, Powell had served two tours in Vietnam, earning two purple hearts, and had decided to make the Army his career. A rising star, his military duty for the next two decades alternated between field assignments and high-profile jobs in Washington either at the Pentagon or the White House. During the Reagan years, he served as military assistant to Secretary of Defense Weinberger and as the President's assistant for national security affairs from 1987 to 1989. Promoted to general in April 1989, he served briefly as head of U.S. Army Forces Command (FORSCOM), then a JCS specified command, at Fort McPherson, Georgia, before President Bush named him as Crowe's successor, passing over about a dozen more senior officers.[10]

With Powell's appointment as Chairman, the 1986 Goldwater-Nichols Act finally came of age. While Crowe had done a faithful job of implementing the law, his tenure had straddled two stools, from the corporate decisionmaking practices that had existed prior to Goldwater-Nichols, to the new era that vested primary authority and responsibility in the CJCS. Embracing an evolution-not-revolution philosophy, Crowe had made changes slowly in order to gain the Service chiefs' cooperation and confidence in the new system. Though he had restructured the Joint Staff to meet Goldwater-Nichols requirements, his alterations were relatively minor and basically involved reshuffling existing offices and personnel. In February 1989, seeing room for improvement, the Director of the Joint Staff, Lieutenant General Hansford T. Johnson, USAF, initiated an in-depth review of Joint Staff functions, looking to reduce Service influence while broadening the scope of Joint Staff participation in DOD affairs. The immediate results, however, were minimal. Overall, the Joint Staff continued to operate much as it had, as a long-range planning and strategic advisory body dominated by inter-Service committees whose officers' primary loyalty remained to their respective Services.[11]

Under Powell the emphasis within the Joint Staff shifted to addressing more current affairs and to providing up-to-date joint assessments to assist the Chairman and the Secretary of Defense in the policy process. Determined to exercise the powers given him under Goldwater-Nichols, Powell siphoned off the best officers from the Services. In so doing he vastly enhanced the stature, influence, and effectiveness of the Joint Staff over the Service staffs and within the interagency system.[12] With representation at practically every level, the Joint Staff was assured "a seat at the table" in every major policy discussion and could assert its prestige and power on a range of issues extending beyond those of the Chairman's personal interest. In sharp contrast to the ponderous methods associated with it in years past, the post-Goldwater-Nichols Joint Staff as Powell redesigned it acquired a reputation for incisive and fast responses. The upshot was a more visible, active, and aggressive Joint Staff with institutionalized influence placing it on a par with OSD, the State Department, the CIA, and other established agencies in the policy process. By the time he returned to civilian life, Powell considered it "the finest military staff anywhere in the world."[13]

Like Crowe, Powell placed high priority on developing effective working relationships with the Service chiefs and his deputy, the Vice Chairman. The serving Vice Chairman when Powell took office was General Robert T. Herres, USAF, who opted for early retirement in 1990. Both he and his successor, Admiral David E. Jeremiah, USN, were able and respected officers. A former astronaut, Herres had been first head of the United States Space Command, while Jeremiah was a former naval task force commander in the Mediterranean and fiscal advisor to the Secretary of the Navy. In theory, they functioned as the Chairman's alter ego. But like all deputies, they operated in their boss's shadow and performed whatever chores he might assign, more often than not the less glamorous administrative tasks.

The situation with respect to the Service chiefs was more delicate and complicated. With the strength of Goldwater-Nichols behind him, Powell knew that he was under no obligation to seek a corporate consensus before making recommendations. But after friction developed over his handling of the base force plan (see below), he realized that it was preferable to have the chiefs' cooperation and support than their opposition. Taking the lesson to heart, he met with them over 50 times during Operation *Desert Shield/Desert Storm* but held most of the meetings in his private office rather than in the "tank," thereby removing all doubt as to who was in charge. Attempting to establish an air of collegiality, he sought to work with the Service chiefs as a team and often referred to the JCS as the "six brothers." Yet he was also not averse to acting on his own when he deemed it necessary and thought it more important to win the approval of the Secretary of Defense and the President.[14]

According to journalist Rick Atkinson, Powell was "the most politically deft" CJCS since Maxwell Taylor.[15] Having been Weinberger's protégé and Reagan's national security advisor, Powell knew the ins and outs of power as well as anyone and moved easily in the rarified atmosphere of high-level policymaking. Under Bush, he was welcomed immediately into the President's "Core Group" of close friends and advisors.[16] One of the assets he brought with him as Chairman was a personal familiarity with many senior members of the Bush administration, including the President himself. Even though Bush wanted his administration to be distinct and separate, not merely an extension of his predecessor's, there were still many familiar faces from Reagan's presidency. Powell was on a first-name basis with practically all of them. As much as anything, Powell's influence derived from the thoroughgoing sense of professionalism he projected and what President Bush described as the Chairman's "quiet, efficient" manner.[17]

At the Pentagon, Powell's most difficult challenge was to develop a productive partnership with his immediate superior, Secretary of Defense Richard B. Cheney. A former congressman from Wyoming, Cheney impressed Powell as incisive, smart, and tough. Yet even though the two generally worked well together most of the time, there were stresses and strains in their relationship which, according to one account, left "an intellectual divide and a residue of mistrust" between them that lasted for years.[18] Cheney took a narrow view of the Chairman's advisory role and on more than one occasion rebuked Powell for offering what he regarded as unsolicited political opinions. "I was not the National Security Advisor now," Powell recalled; "I was only supposed to give *military* advice."[19]

Indeed, in dealing not only with Powell but with other senior officers, Cheney insisted on close civilian control and oversight of the military. Shortly after taking office, he publicly reprimanded Air Force Chief of Staff General Larry D. Welch for "freelancing" to gain a congressional committee's support for the Peacekeeper missile. Later, in September 1990, in part at Powell's instigation, he fired Welch's successor of less than 3 months, General Michael J. Dugan, for "poor judgment" stemming from comments Dugan made to the press about Iraq's recent invasion of Kuwait and how the United States should respond. Aware of the Secretary's sensitivities, Joint Staff action officers became increasingly cautious in their public remarks and learned to double check whatever they were working on with OSD to avoid any appearance of an "end run" around Cheney's authority.[20]

While Powell left his mark as Chairman in many ways, one of his most well-known contributions was the "doctrine" that bore his name concerning the use of military power. Modeled on six "tests" that Secretary of Defense Weinberger had enumerated in 1984, the Powell Doctrine laid out broad guidelines to help shape any decision committing U.S. forces to combat. Weinberger's purpose had been to preempt critics and allay their concerns that the Reagan administration's proactive use of military power might lead, as in Vietnam, to open-ended commitments or "unwinnable" wars.[21] For Powell, the function of the guidelines he developed was more personal. Having witnessed the debacle in Vietnam first-hand, he resolved that the lessons of that war should not be lost. Powell was no pacifist, but his caution in committing U.S. troops to combat often frustrated and irritated his superiors. Some called him the "reluctant warrior." As a professional soldier Powell believed that military force should be applied in careful and deliberate ways, with the full support of Congress and the American public, toward achieving identifiable political objectives, and that once involved in a conflict the United States should use all power at its disposal to bring the campaign to a swift and successful conclusion.[22]

Powell's thoughts on these matters had been evolving for 20 years and came to fruition with his service as Chairman of the Joint Chiefs, first in the aftermath of the Panama operation in 1989 and, later, in connection with the liberation of Kuwait. Though the JCS never formally endorsed the Powell Doctrine, parts of it found their way into an updated version of the National Military Strategy issued in 1992. Powell wanted to include a statement that the ability to use "overwhelming force," as during the operations in Panama and Kuwait, was the most effective deterrent in a regional crisis. At the White House, however, the prevailing sentiment was that Powell's prescription went too far. "I was strongly opposed to the Powell doctrine," recalled Scowcroft. "I thought it precluded using force unless we went all out. I thought it was nonsense."[23] At the suggestion of Under Secretary of Defense Paul D. Wolfowitz, Powell toned down his rhetoric and called instead for the application of "decisive force," a somewhat less explicit concept. Yet as far as Powell was concerned, the fundamental strategic purpose remained the same.[24]

THE BASE FORCE PLAN

One of Powell's most significant contributions as Chairman was his "base force" blueprint for the post-Cold War defense establishment. Although the Joint Chiefs had considerable experience in downsizing after previous wars, they had yet to find a formula that avoided fierce inter-Service rivalry and competition for dwindling resources, accompanied by a precipitous drop-off in the effectiveness of the Armed Forces. Past build-downs had invariably yielded low morale among the Services and a defense establishment of either hollow capabilities, as after World War II and Vietnam, or a seriously unbalanced force structure, as after Korea, that had severely constrained the plausible range of military options in crises. As they looked to the future, Cheney and Powell agreed that the post-Cold War demobilization should be different, and that it should retain the essential elements of a balanced, robust military.[25]

Developing the base force went hand in hand with fashioning a military strategy adapted to the emerging post-Cold War spectrum of threats. While Crowe had begun the process with the submission of NMS 92-97, his assessments still reflected a fairly rigid Cold War outlook, stressing preparations for global and regional conflicts. Powell's first task was to interject greater flexibility into strategic planning. Expecting regional contingencies in Southwest Asia, the Far East, and Latin America to predominate, he downplayed the danger of a global war and made a leap of faith that the Soviet threat would steadily diminish. At the time, there was considerable uncertainty in the Intelligence Community over whether Gorbachev would remain in power and much speculation that sooner or later a conservative reaction would bring his authority and reforms to an end. Indeed, by 1990 there were signs that in response to these pressures, Gorbachev was veering toward a more conservative stance and that the process of reform and restructuring was losing its momentum.[26] Powell assumed, however, that even though Gorbachev might waver from time to time, he would stay the course. Convinced that the Soviet Union was changing for the better, Powell believed that the Gorbachev reforms were practically irreversible and that the net effects would be a progressive weakening of centralized Communist Party authority, a decline in Soviet military power, and eventually the transformation of the Soviet Union into a federation or commonwealth-type state. One clear sign that Soviet power was on the wane was the disestablishment of the Warsaw Pact in the summer of 1991. In light of this and other evidence of diminishing Soviet authority, Powell anticipated a reduced need by the United States for either a large arsenal of expensive strategic weapons for deterrence purposes or costly ground and air forces built around fighting a war of attrition in Europe.[27]

During the early stages of planning the base force, estimated reductions for U.S. forces remained in flux. Projected manpower cutbacks ranged from a low of 10

Figure 16–1.

	FY 1986 (Reagan Buildup)	FY 1991 (Actual at End of Cold War)	Projected Base Force by FY 1999
Comparison of Projected Base Force and Actual Convential Capabilities, FYs 1986–1999			
Active Duty Personnel	2.2 million	2 million	1.6 million
Army Active Divisions	18	16	12
Air Force Active Divisions TFWs	24	22	15
Navy Carriers*	13	12	13
Other Navy Combatants	363	307	259
USMC Divisions/Wing Teams	3/3	3/3	3/3

*Total is number of carriers on active duty; does not include one ship normally in service life extension and/or nuclear refueling overhaul and one training carrier.

Source: *1991 Joint Military Net Assessment* (March 1991), chapter 3.

percent envisioned by the Office of the Secretary of Defense, to as much as 25 percent in planning papers generated by the Joint Staff. As it turned out, the JCS figure proved the more accurate.[28] Based on his estimate of future strategic requirements, Powell saw no Service emerging unscathed, though he expected the cutbacks to fall most heavily on the Army and the Air Force. Anticipating strenuous objections from the Services (not to mention the "leaks" to the press that would inevitably follow), Powell avoided discussing these matters in detail with his JCS colleagues prior to briefing the Secretary of the Defense and the President.[29]

By late November, Powell had a green light from the Secretary and the President for further planning and had completed a preliminary round of consultations with his budget and resource advisors, the Service chiefs, and the combatant commanders. By then, the Berlin Wall had fallen and Communism was in open retreat across Eastern Europe. Even the most die-hard skeptics were coming around to the view that the Cold War was over and that the time was rapidly approaching to make corresponding adjustments in the U.S. force posture. Still, there were legitimate differences of opinion among the Service chiefs and the CINCs over where to cut and how far to go.[30] Powell realized that with the power and authority he possessed under Goldwater-Nichols, he had no need to consult with anyone other than the President, the Secretary, and the NSC. But as an experienced military bureaucrat, he also recognized that without the Service chiefs and the CINCs behind him, he was unlikely to get the cooperation he needed to carry his plan forward.

One of Powell's main concerns as planning progressed was to avoid reductions imposed arbitrarily by either the OMB or Congress. The most serious challenge came from Senator Sam Nunn of Georgia, a prominent Democrat and chairman of the Senate Armed Services Committee. Focusing his public career on defense matters, Nunn had been instrumental in drafting the Goldwater-Nichols Act and had played a key role in a companion measure (the Nunn-Cohen Act) to bolster special operations forces by mandating the creation of a unified command for that purpose.[31] Rumored to have his eye on a run for the Presidency, Nunn repeatedly accused the Bush administration of being slow to recognize the benefits of the Cold War's demise. Nunn was well aware of the strong sentiment in Congress in favor of cutting defense and sought to turn it to his advantage. Urging fellow Democrats not to act rashly, he laid out an alternative strategic concept for the post-Cold War era which he termed "flexible readiness—high readiness for certain forces and adjustable readiness for others." Elaborating his views in a series of speeches between late 1989 and the spring of 1990, Nunn called for a large-scale pull-back of U.S. troops from Europe, greater reliance on tactical nuclear weapons for deterrence purposes, and increased emphasis on Reserve capabilities.[32]

Toward the end of April 1990, with Nunn nipping at his heels, Chairman Powell confirmed in a speech to the Council on Foreign Relations in Washington that, in response to the changes taking place in Eastern Europe and elsewhere, the Bush administration was reexamining its long-term military requirements. Shortly thereafter, he told the *Washington Post* that he was looking at reductions in force strength of up to 25 percent over the next 5 years.[33] Predictably, cuts of such magnitude encountered objections from the Service chiefs, who had already agreed to significant reductions as part of the normal budget process. The base force cuts would be on top of that. But through continuous reworking of the figures and augmentations to the force structure here and there, Powell was able to overcome their resistance and produce a broadly acceptable plan.[34]

Accompanied by Secretary Cheney and Under Secretary of Defense Paul Wolfowitz, the Chairman briefed President Bush on June 26, 1990, on the development thus far of the base force plan and the strategic concept behind it. After a lengthy discussion Bush approved the plan and indicated he wanted to highlight it in a public speech. Delayed because of a mix-up between the White House and the Pentagon over who was responsible for drafting the speech, Bush finally unveiled his administration's new defense strategy in an appearance at the Aspen Institute in Colorado on August 2, 1990, the same day Iraqi troops invaded and occupied

Kuwait. Though Bush offered few specifics, he confirmed that cutbacks of 25 percent in conventional forces were on the way by the end of the decade and that under the forward presence concept "regional contingencies" would replace Europe as the focus of future U.S. military planning. He also indicated that sooner or later there would be cutbacks in strategic forces as well, but implied that for the time being the requirements of preserving an "effective deterrent" while negotiating a Strategic Arms Reduction Treaty (START) with the Soviet Union would take precedence in determining the size and configuration of the strategic arsenal.[35]

While preparations to implement the base force plan were ongoing throughout the fall of 1990 and on into the winter of 1991, the emergency in Kuwait and the Bush administration's decision to mount a military challenge to the Iraqi invasion left JCS planners in the awkward position of overseeing a major buildup in the Middle East even as they were preparing for general reductions in force levels. Budget estimates forwarded to Congress in February 1991 reflected some of these downward adjustments. As more details appeared, the vision grew of a permanent post-Cold War defense establishment of 1.6 million uniformed personnel (down from 2.2 million at the height of the Reagan buildup) organized into an Active-duty Army of 12 divisions, an Air Force of 15 tactical fighter wings, a Navy of 272 combatant vessels (including 13 carriers), and a Marine Corps of 3 division-air wing teams.[36]

For some, especially those reluctant to admit that the Cold War was over, the Gulf War was a clear warning against large defense cuts. But for Chairman Powell, it was a distraction from the unavoidable process of adjusting to a new security environment in which large defense establishments would play a diminishing role. Once the Kuwait emergency was over, Powell expected calls from Congress and the public for a "peace dividend" to intensify. The base force was the most realistic way Powell saw of providing the expected cuts while avoiding the pitfalls of previous demobilizations and preserving a credible long-term defense posture. No one, least of all the Service chiefs, saw it as the ideal solution. But as regional contingencies and humanitarian assistance missions replaced the threat of a large-scale conflict in Europe as the country's top security concerns, it became harder and harder to justify the maintenance of a defense establishment comparable in size and capabilities to that of the past.

Despite the time and energy invested in developing it, the base force concept proved relatively short-lived. Under the planning done by the Chairman and his aides, force structure targets were to be reached between fiscal years 1995 and 1997, with the overall structure firmly in place by FY99 (see figure 16–1). But with the change of administrations in 1993 came pressure to take a fresh look at the country's defense posture and to achieve larger reductions. The result was the Clinton administration's bottom-up review (BUR), something the new President had promised during the campaign.

Resting on a strategic concept similar to that of the base force, the BUR continued to stress the importance of effective capabilities for regional conflicts, but envisioned force cuts of one-third or more and comparable savings in spending based on FY90 levels.[37] A more ambitious agenda than Powell's, the BUR's goals also proved more difficult to achieve without producing shortfalls in capabilities which Joint Staff planners saw as increasing the level of risk in executing the approved military strategy.[38]

OPERATIONS IN PANAMA

As Powell grappled with shaping a new force structure, the kinds of post–Cold War problems he expected Washington to face were already beginning to appear. One was the uneasy situation in Panama, where the United States had enjoyed a military presence and well-established security interests for nearly a century. At the center of the controversy was Panamanian strongman Manuel Antonio Noriega, who came to power following the 1981 death of General Omar Torrijos in a suspicious airplane crash. A career soldier, Noriega had been Torrijos' military intelligence chief and boasted that one of his jobs was to provide liaison between the CIA and Cuban president Fidel Castro.[39] In August 1983, Noriega enhanced his position by promoting himself to general and becoming the de facto head of state. Shortly thereafter, he pressured the legislature into converting the National Guard into the Panama Defense Forces (PDF), over which he alone exercised authority. As his power grew, so did graft, corruption, illegal drug trafficking, and the repression of political opponents.

Throughout Noriega's rise to power, the Joint Chiefs' primary concerns were the security of the Panama Canal and the integrity of the extensive network of U.S. military installations in the former Canal Zone (CZ), where U.S. Southern Command (USSOUTHCOM) had its headquarters. While the 1977 Panama Canal Treaty had ended U.S. ownership and control of the canal, the United States and Panama continued to share joint responsibility for its defense until the end of 1999. After that, any further presence of U.S. forces in Panama would be by the agreement of both parties. Economically, there was little to justify continuing the U.S. military presence in Panama. The canal was too narrow to accommodate modern supertankers and other large ships, and by the 1980s its revenues had fallen into a steady decline, much to the consternation of the Panamanian government. But as long as there remained a leftist insurgency in nearby El Salvador and a Soviet and Cuban presence in Nicaragua, the JCS balked at giving up their base of operations. Now was not the time, the chiefs believed, to cut and run.

Despite Noriega's unsavory reputation and brutish behavior, the Joint Chiefs were cautiously confident that they could do business with him. But as the political climate in Panama continued to deteriorate, they became less and less optimistic. Aware that many in Washington were having second thoughts about backing him, Noriega turned to Libya, Nicaragua, and Cuba for economic and military assistance.[40] In response to PDF harassment of U.S. personnel, the commander of US-SOUTHCOM, General Frederick F. Woerner, Jr., USA, became openly critical of Noriega and his regime. Though advised by both Crowe and Powell (who was still at the White House serving as National Security Advisor) to tone down his rhetoric, Woerner persisted in attacking Noriega. Persuaded that Woerner had become a political liability, President Bush named General Maxwell R. Thurman, USA, as his successor. In early July 1989, without consulting Crowe, who was out of town, Secretary of Defense Cheney arranged for Army Chief of Staff General Carl E. Vuono to go to Panama to deliver the news to Woerner that he was to be relieved.[41]

By then, President Bush knew that sooner or later he would have to seek Noriega's removal from power. Approved policy (NSD 17) sanctioned by the National Security Council in July 1989 authorized the Joint Chiefs of Staff and the Secretary of Defense to develop plans for asserting U.S. treaty rights in Panama and to keep Noriega and his supporters off balance. Authorized operations fell into four categories based on an escalating scale of risks and visibility, all aimed in one way or another at grinding down Noriega's power and authority. Only as a last resort would the United States undertake direct military action to overthrow Noriega's regime.[42] Much of the preparatory work and logistical planning for these operations fell under Powell's aegis while he headed FORSCOM at Fort McPherson, Georgia. Thus, as he made ready to take up new duties as Chairman of the Joint Chiefs, Powell was already well versed in the plans and preparations that would eventuate in Noriega's downfall.

Rather that resorting to military intervention, the Bush administration would have preferred that the Panamanians take matters into their own hands and remove Noriega themselves. However, there were few people left in Panama by then who were willing to risk defying Noriega's authority. One of the exceptions was a respected Panamanian officer, Major Moisés Giroldi Vega, a senior member of Noriega's security detail who had become disenchanted with the regime. At some point, Giroldi's wife made contact with the CIA and sought American help for her husband in staging a coup to topple Noriega.[43] Giroldi originally scheduled the coup for October 1, 1989, but because of changes in Noriega's schedule he delayed acting until 2 days later. By then, Thurman, the newly arrived SOUTHCOM commander, had become suspicious of the whole affair, as had his superiors in Washington, including General Powell. When at last Giroldi did act, elite PDF units loyal to Noriega promptly intervened to rescue their leader. By that evening they had routed the plotters and Giroldi had been tortured and executed.

In the aftermath of the failed October 3 coup, a reign of terror descended on Panama as Noriega dramatically increased repression of the civilian opposition and carried out a blood-purge of dissident elements in the PDF. Reliable reports estimated that he executed as many as 70 soldiers and arrested 600 more.[44] Heavily criticized for not giving Giroldi more credence and support, the Bush administration began active preparations for toppling Noriega's government under a joint military intervention plan called Blue Spoon. Although Powell as always was uneasy about the use of force and the casualties that were bound to result, he was increasingly convinced that a military solution might be the only viable option for ending Noriega's control. Insisting that the job be done thoroughly, Powell favored the application of overwhelming military power, not only to assure Noriega's downfall but to neutralize his primary source of support—the PDF—and "pull it up by the roots."[45]

Despite preparations to intervene, neither Bush nor Powell was eager for a showdown. Remembering earlier interventions, Bush wanted to avoid a repetition of the failed 1980 Iran hostage rescue mission or a recurrence of the debilitating inter-Service rivalry that had hampered the 1983 Grenada invasion.[46] No less concerned than Bush that the intervention should succeed, Powell paid meticulous attention to the planning process and insisted on numerous rehearsals to make sure U.S. forces were fully trained and prepared. Gaining in complexity, Blue Spoon called for a closely coordinated all-arms attack using around 25,000 troops, supported by four separate combatant commands. In contrast, Noriega had at most 4,000 effective fighters, backed by 8,000 paramilitaries. By mid-December 1989, about half of the U.S. ground troops allocated to the operation were already in-country, with the rest on 72-hour alert at bases in the United States, awaiting airlift. Thurman wanted as much firepower as possible to be in place before action commenced, and toward that end he arranged to have Sheridan light tanks and Apache attack helicopters brought in under the cover of darkness, then concealed them at secure secret locations.[47]

Even with the United States poised to strike, Powell declined to recommend a timetable for launching operations. Preferring to bide his time, he hoped that American economic and political sanctions would nudge Noriega into stepping down without recourse to military action. But as the standoff continued, Noriega's defiance only grew stronger. On December 15, 1989, he delivered a fiery speech to the Panamanian National Assembly, after which the lawmakers adopted a resolution proclaiming a state of war "while [U.S.] aggression lasts." The next evening, members of the PDF shot and killed an American Marine lieutenant riding in a car that ran a roadblock, beat up a U.S. Navy officer who witnessed the incident, and threatened to rape his wife. Convinced that Noriega had "gone over the line," Powell held an emergency meeting with Cheney and Wolfowitz on the morning of Sunday, December 17. All agreed the time had come to intervene, whereupon Cheney arranged a meeting with the President that afternoon. Remembering the mistake he made with the base plan, Powell wanted to make sure he had the support of the Service chiefs before going to the White House, and later that morning he invited them to his official quarters at Fort Myer, adjacent to the Pentagon. Following an impromptu briefing and a review of the latest intelligence, all agreed that Blue Spoon was a sound plan. The only reservations were those expressed by the Commandant of the Marine Corps, General Alfred M. Gray, Jr., who regretted that it did not give the Marines a larger role.[48]

The meeting with the President that afternoon lasted nearly 2 hours and produced no surprises. Besides Powell, Cheney, and President Bush, the only others to attend were Scowcroft, his deputy Robert M. Gates, Secretary of State James A. Baker III, and Marlin Fitzwater, the President's press secretary. Like everyone else, Bush was fed up with Noriega and wanted him removed before he killed or roughed up more Americans, seized hostages, or launched a surprise attack on U.S. installations. According to Baker's recollections, there was very little if any debate over the merits of invading Panama. Instead, discussion focused on the mechanics of the operation, clearing it with congressional leaders, and the myriad diplomatic and logistical details linked to the invasion. Earlier, echoing views they heard repeatedly from Capitol Hill, Baker and others at the State Department had been urging more forceful action against Panama. Now that Powell had come around to their point of view, they felt vindicated and somewhat smug. "After years of reluctance," Baker later wrote, "the Pentagon was ready to fight."[49]

Three days later, during the early hours of December 20, the attack commenced, with Navy special forces, Army Rangers, and Air Force "stealth" fighters spearheading the assault against key strategic installations. Now called *Just Cause*, the operation proceeded in methodical fashion to suppress PDF resistance. Fighting around the *Comandancia*, Noriega's headquarters, was the most intense of all. But by the next day, except for occasional skirmishes, the conflict was over and Guillermo Endara, whose election as president earlier in the year Noriega had nullified, was installed in office. Given the size of the overall effort, U.S. casualties were relatively light: 23 killed and 312 wounded. Panamanian losses were 297 killed, 123 wounded, and 468 detained.[50] Unable to flee the country, Noriega initially hid in a brothel, then took sanctuary in the Papal *Nunciatura* in Panama City. Quickly wearing out his welcome there, he surrendered in early January 1990 and was returned to Miami, Florida, where he was jailed under a 1988 warrant for drug trafficking.

A complex and difficult operation to mount, *Just Cause* was the Joint Chiefs' most all-encompassing joint venture under the new Goldwater-Nichols law to that point. To be sure, there were some complaints that it had been "an Army-run show from start to finish."[51] Others, however, praised it as a model of inter-Service collaboration. "*Just Cause*," said one senior commander afterwards, "was a joint operation in every sense of the word."⁵² Its success stemmed not only from the availability and use of overwhelming force to subdue Noriega and his followers, but also from the meticulous advance planning, streamlined command and control, and improved coordination at all levels—all products to one degree or another of the Goldwater-Nichols legislation. Unlike the haphazard Grenada operation, where the Marines invaded one half of the island and the Army the other with limited coordination between attacking units, the United States went into Panama in a unified effort, using inter-Service task forces to achieve designated objectives. While similar results might been have been achieved under the Joint Chiefs' old corporate decisionmaking system, there doubtless would have been longer debates, less assurance of effective inter-Service cooperation, and in the end higher casualties. As the first real test under Goldwater-Nichols, the new JCS system rose to the challenge.

THE CFE AGREEMENT

Part of the success behind the Panama operation was that the United States was able to carry it out with virtually no worry of interference from the Soviet Union, even while Moscow continued to have strong ties to nearby Cuba and Nicaragua. But as the Cold War drew to a close, the Soviets, heeding Gorbachev's lead, seemed to offer fewer challenges, as if they were no longer in a position to resist. Most striking of all was a more relaxed and flexible Soviet approach toward negotiations. To be sure, the Soviets did not give way easily, nor did their interpretations of accords always match those of the West. But for the first time, they began to show an uncommon interest in harmonizing differences sooner rather than later, a sharp departure from past negotiating practices. For the Joint Chiefs as for others in Washington, it was a novel experience that was in many ways hard to comprehend.

Among the notable accomplishments were those in the field of arms control, which for decades had been the Cold War's most contentious diplomatic battlefield. Even with the Cold War winding down, the JCS remained as uneasy and suspicious of arms control as ever. But over the years they had learned to accommodate themselves and to fit strategy and programs within arms control confines. Building on the momentum of the 1987 INF Treaty, President Reagan hoped to conclude reduction agreements for conventional and strategic forces before leaving office but did not have time to complete his mission. What he bequeathed to his successor was a half-finished agenda: a "mandate," approved jointly by NATO and Warsaw Pact leaders in January 1989, laying out a work plan for achieving limitations on conventional forces in Europe (CFE); and a draft Strategic Arms Reductions Treaty that aimed at a 50 percent cut in offensive strategic arms.

For the incoming Bush administration and for the Joint Chiefs as well, President Reagan had been moving too fast. In surveying the scene, Scowcroft thought Reagan had "rushed to judgment about the direction the Soviet Union was heading" under Gorbachev and had lost his sense of priorities. Instead of paying attention to the "strategic aspects of arms control," Scowcroft believed, Reagan and his advisors became absorbed in trying to promote Gorbachev's success at home and ended up "placing emphasis on reductions as a goal in itself."[53] By and large, the Joint Chiefs agreed. The first order of business was to determine whether progress was feasible in the CFE arena, which was the subject of resumed negotiations in Vienna in March 1989. Previously known as the Mutual and Balanced Force Reduction (MBFR) talks, these negotiations had dragged on inconclusively since 1973, a tribute to both sides' perseverance and latent optimism if nothing else. Energized by Gorbachev's pledge to withdraw 50,000 Soviet troops from Eastern Europe, the CFE talks received a further boost in May 1989, when the Warsaw Pact agreed in principle to accept a NATO proposal calling for equal levels of heavy weapons, a long-standing Western goal. A year and a half later emerged the CFE Treaty, signed in Paris in November 1990 amid growing euphoria over improved East-West relations. By then, popular discontent had swept Communist governments from power throughout Eastern Europe, the Warsaw Pact seemed to be on its last legs, and Gorbachev had endorsed the need for political pluralism in the Soviet Union.

In light of the sweeping changes taking place in Eastern Europe at the time, the impact of the CFE Treaty was largely symbolic. With or without an agreement, NATO and the Warsaw Pact were disarming posthaste anyway. What the treaty provided were guideposts, coupled with provisions for on-site inspections to make sure that both sides duly complied. Dealing only with military hardware from the Atlantic to the Urals (ATTU), the treaty capped total deployment in the 2 alliances at 40,000 battle tanks, 40,000 artillery pieces, 60,000 armored combat vehicles, 13,600 combat aircraft, and 4,000 attack helicopters.[54] But since NATO's combat holdings were already at or below the treaty's levels in several categories, the JCS expected its restraints to have a limited effect on curbing Western capabilities.[55] To accompany the treaty, there was a joint declaration proclaiming "the end of the era of division and confrontation" which the two sides promised to replace with "new partnerships and . . . the hand of friendship."[56]

While many commentators heaped praise on the CFE Treaty, the Joint Chiefs of Staff reserved judgment. Shortly before the treaty was signed, the Soviets withdrew large amounts of military equipment behind the Urals rather than proceeding with destruction as called for in the agreement. On the day before the signing ceremony, they tabled new data indicating the sudden discovery of three "coastal defense divisions" subordinate to the Soviet Navy. Since the CFE agreement did not cover naval forces, the Soviets argued that none of the arms assigned to these divisions (5,400 pieces) should count against the allowed Eastern Bloc total.[57] As British historian Jonathan Haslam observed, "The [Soviet] General Staff were digging in their heels."[58] Suspicious of Moscow's intentions, the Joint Chiefs of Staff gave the CFE Treaty a tepid recommendation during testimony before Congress in the summer of 1991. Attempting to put the best face possible on the deal, Chairman Powell called it "a major success story for the Atlantic Alliance" that would "strengthen stability and security in Europe" and help establish "a stable and secure balance of conventional armed forces . . . at much, much lower levels." His JCS colleagues, however, offered notably more restrained endorsements. All the same, the treaty represented greater progress toward limiting conventional forces than anything else to that point, and on that basis alone it stood out as a major contribution toward ending the Cold War.[59]

START I AND ITS CONSEQUENCES

With the CFE talks finally bearing fruit, the Bush administration turned its attention to the unfinished Strategic Arms Reduction Talks Treaty. Deeming Reagan's goal of a 50 percent cutback in offensive arms excessive and probably unattainable, the Bush White House, with JCS concurrence, set its sights on lesser objectives.[60] But with the Cold War abating, there was far less political pressure either at home or abroad than in years past to demonstrate progress on controlling strategic arms. Thus, in addressing the problem the Bush administration avoided seeking wholesale changes to what had already been agreed upon and decided to wait until follow-on talks (START II) to launch any major initiatives. At the same time, however, senior Bush administration figures saw a clear link between effectively addressing arms control issues and preserving the U.S. leadership role with its friends and NATO allies. "If we performed competently in arms control," Scowcroft believed, "alliance confidence in our ability to manage the broader relationship would soar."[61]

While working on the base force plan, Powell skirted the issue of reductions in strategic forces on the assumption—confirmed by President Bush in his Aspen Institute speech—that the principal sizing mechanism for the strategic arsenal would be a finished START agreement. Thus, Powell had no choice other than to treat estimates of strategic capabilities as highly tentative. Since reaching a post-Vietnam peak in FY 1985, U.S. spending on strategic forces had fallen steadily, so it stood to reason that the trend would continue for the foreseeable future. Like the cutbacks in conventional forces, Powell expected reductions in strategic forces to level off around the middle of the decade and stabilize by the end. Even before factoring in arms control, he estimated that to stay within projected spending limits, it might be necessary to eliminate the entire air-breathing leg of the strategic triad including the B–2 stealth bomber, a proposal that drew sharp objections from the Air Force.[62] Bowing to political realities, Powell revised his estimates and came up with projections of a strategic force by the end of the decade comprising 18 Trident missile submarines, 550 ICBMs, and about 250 manned bombers, including 50 B–2s.[63]

The trouble in reaching a START agreement had less to do with overall numbers of delivery vehicles than with the characteristics and performance of weapons, the continuing proliferation of MIRVed systems, and sublimits on air- and sea-launched cruise missiles. These issues had vexed arms controllers and military planners for years and came no closer to permanent resolution in START I than they had during earlier negotiations. To help facilitate progress, President Bush authorized what amounted to two sets of negotiations: the formal talks held in Geneva, and parallel discussions between Secretary of State Baker and Soviet Foreign Minister Eduard Shevardnadze. It was largely through the latter that the START I agreement emerged. In the past, the use of back-channel negotiations to broker deals had been a major source of irritation to the Joint Chiefs. But owing to the changes in lines of authority brought about by Goldwater-Nichols, coupled with the regular and direct access that Powell enjoyed to the Oval Office, there were rarely any serious problems of this sort during the Bush years.

A major difference between the Reagan and Bush administrations was the waning enthusiasm of the latter for the Strategic Defense Initiative and its corresponding effect on gaining Soviet cooperation on reaching an offensive strategic arms agreement. By the time the Bush administration took office, it was increasingly clear that support for SDI in Congress was declining and that, on technical grounds alone, an effective system of strategic defense was still decades away. Under consideration for possible validation were no fewer than six competing technologies.[64] In assessing SDI's long-term prospects, neither Crowe nor Powell saw it playing a significant role in foreseeable American defense plans. Both endorsed continuing research and development but reserved judgment on full-scale production and deployment.[65] Weighing one thing against another, Bush concluded that "a shield so impenetrable" that it would obviate the "need for any kind of other defense" was too expensive and impractical.[66] By deciding to downgrade SDI and turn it back into an R&D program, Bush removed a source of intense friction in Soviet-American relations and made it easier to negotiate a START agreement.[67]

The first big breakthrough in the START negotiations came in February 1990 when, in a sharp turnaround, the Soviets indicated their readiness to accept U.S. loading rules and verification procedures dealing with air- and sea-launched cruise missiles. What prompted the Soviets to drop their previous objections is unclear, though it probably had something to do with Gorbachev's desire for a further improvement in U.S.-Soviet relations in order to increase Moscow's chances of obtaining economic aid from the West. Whatever the reason, it seemed at the time that a START agreement was near at hand. But by April, when Shevardnadze visited Washington for further discussions, the Soviets had retreated from their earlier position and now demanded new conditions and more restrictions. From the increased presence of senior military officers on the Soviet delegation and their apparent influence, the signs were unmistakable that Gorbachev's strategy of accommodation with the West was under attack at home and that the conservatives were striving to regain a larger voice in Soviet policy. As one observer described it, Secretary of State Baker "swallowed hard" and went back to the bargaining table.[68] By then, keeping Gorbachev in power had become as important to Bush and his advisors as it had been to Reagan, and in some ways it overshadowed the particulars of any agreement. "We in the Bush Administration," Baker recalled, "knew we could not reform the Soviet Union. But we realized nonetheless that we could assist the process."[69]

Still, it took more than a year of further negotiations before a START agreement reached final form. Signed on July 31, 1991, the START I Treaty required the United States and the Soviet Union to cap their strategic warheads at 6,000, with sublimits on various missile types, and to reduce the number of strategic launch vehicles on each side by about one-third, to 1,600 from 2,250 (the limit allowed under SALT II). For the United States, which had fewer delivery vehicles to begin with, the reductions were more like 25 percent, while for the Soviets they were closer to 35 percent overall and more than 50 percent in heavy ICBMs, the mainstay of the Soviet strategic arsenal. Under a separate "political agreement" dealing with long-range nuclear sea-launched cruise missiles, the two sides embraced controls that generally accorded with American preferences. For verification purposes, the treaty relied on on-site inspections, regular exchanges of test data, and national technical means. According to Powell and Cheney, the thrust of the agreement was to move both parties away from land-based ICBMs, which might be used precipitously in a crisis, and to encourage greater reliance on less destabilizing systems such as ballistic missile submarines and "slow flyers" like cruise missiles.[70]

Nearly 10 years in the making, the START I Treaty was a historic achievement—the first offensive strategic arms accord that actually mandated force reductions. But while it was generally applauded in the West, it met with stiffening resistance in Moscow, where the consensus among conservatives was that Gorbachev had gone too far in making concessions. On top of the CFE treaty, the recent collapse of the Warsaw Pact, and Gorbachev's penchant for political and economic reform, the START I agreement was the last straw. In August 1991, while Gorbachev was vacationing in the Crimea, hard-line Communists attempted a coup. Observing events from Washington, Powell was initially alarmed that the plotters might succeed in installing a reactionary regime. But by the second day, his worries began to subside as evidence appeared that the coup had little or no support from either the KGB or the military rank and file.[71] Rallying behind Boris Yeltsin, head of the Supreme Soviet of the Russian Federation (i.e., Russia's state president), supporters of the regime formed a phalanx to protect the Russian parliament building where Yeltsin had his headquarters. The coup leaders, unable to generate significant popular backing for their cause, soon lost heart and the revolt was over within 4 days. Gorbachev immediately returned to Moscow to claim victory, but from that point on it was Yeltsin's power and authority that were on the rise. By the end of the year, Gorbachev was out of a job, the Soviet Union had dissolved, and a federation of former Soviet states had taken its place.

As the Soviet Union was breaking up, a debate was taking place in Washington between the Pentagon and the White House over how the United States should respond. To show his solidarity with the reformers and to keep the Soviet Union's large arsenal of nuclear weapons from falling into the wrong hands, Bush proposed seeking immediate additional cuts in strategic nuclear arms, a so-called START-plus agreement. Skeptical whether the time was right in view of the unsettled political situation in Eastern Europe, Secretary of Defense Cheney declared such measures to be "premature" and perhaps "imprudent." Meanwhile, Powell and the Joint Chiefs submitted a list of less ambitious suggestions, including a lowering of the alert status of U.S. strategic bombers and the removal of short-range nuclear missiles from surface ships and attack submarines. More discussions followed, culminating in late September 1991 in a televised address by the President outlining his START-plus plan to remove all remaining U.S. short-range nuclear missiles from Europe (those under 500 kilometers which the INF Treaty did not cover), cancel further work on a rail-garrison version of the Peacekeeper missile program, and seek a complete ban on all remaining U.S. and Soviet MIRVed ICBMs.[72]

As part of his initiative, President Bush also announced that the Strategic Air Command (SAC), long the symbol and repository of American nuclear power, would stand down and that a new U.S. Strategic Command (USSTRATCOM) would replace it. This change had been in the making for some time and grew out of the recognition among the Joint Chiefs and the combatant commanders that as Cold War tensions relaxed and the defense budget shrank, there was less justification for a single command devoted exclusively to strategic operations. A key figure in creating the new organization was SAC's last commander in chief, General George Lee Butler, USAF, who as director of strategy and plans (J-5) on the Joint Staff had been instrumental in helping Powell develop the base force plan. Butler believed that SAC suffered from an outdated mission focus that equated "strategic" with "nuclear" operations and that the new command should have a broader vision of its responsibilities combining functions previously assigned to SAC with similar conventional and nuclear tasks performed by other commands. As usual, there were lengthy debates and considerable competition among the Services for authority and influence within the new organization. After sorting out the various proposals, Powell recommended and President Bush approved a revision to the Unified Command Plan that took effect on June 1, 1992. Now a unified rather than a single-Service "specified" command, as SAC was, USSTRATCOM consolidated elements of the old Strategic Air Command with components drawn from the former Atlantic command, Pacific Command, and U.S. Space Command.[73]

Many people believed that the Cold War began with the advent of the atomic bomb in 1945 and gathered momentum as both sides sought to outdo each other in nuclear weapons. If so, the 1991 START I agreement, more than anything else, marked the end of the Cold War and the onset of a new era in which the United States and the remnants of the Soviet Union began the laborious process of turning back the clock and doing away with their nuclear arsenals. Having been key participants in the buildup, the Joint Chiefs of Staff were now in the forefront of the process of disarming. Testifying in the summer of 1992 in support of the START I agreement, General Powell lauded it as "a critical foundation" for further reductions in strategic arms and, as such, a major step from "a confrontational to a cooperative relationship" between East and West. This time, in sharp contrast to the lukewarm endorsement they had given the CFE Treaty the year before, the Service chiefs enthusiastically praised the START I agreement as being in the country's best interests.[74]

The chiefs' change of attitude doubtless had a lot to do with the collapse of the Soviet Union and, with it, the dissolution of the Soviet armed forces, once one of the most formidable military organizations in history. As it became apparent that the Soviet state would not survive the abortive coup of August 1991, the military also knew its days were numbered. Under a deal reached that December, the leaders of the former Soviet republics—soon to be the Confederation of Independent States (CIS)—agreed to preserve unified command and control of the armed forces insofar as feasible, including the strategic rocket forces. But it was too little too late to keep the old organization intact, and as the year ended, the Soviet armed forces along with the Soviet Union itself formally ceased to exist. A rump establishment, the CIS armed forces continued to function, but with no practical way of exercising authority, it was out of business in a year and a half as Russia, the Ukraine, and the other former Soviet states set up their own ministries of defense.[75]

The downfall of the Soviet Union sealed the end of Cold War. By then, as an ongoing institution, the Joint Chiefs of Staff had seen it all, from the uneasy collaboration between Washington and Moscow in World War II, down through the collapse of cooperation after the war, the dark days of the Korean conflict, the tense moments of the

Cuban Missile Crisis, the agony of Vietnam, and the decades of costly competition in strategic nuclear arms. With these experiences before them, Powell and the Joint Staff had done their best to prepare the U.S. military for the expected transition into the post-Cold War world. But they scarcely imagined the scale and scope of the changes that would actually take place. As the Cold War ended, it ushered in a new era that was in some ways more dangerous and certainly less predictable than the one it replaced.

NOTES

1. Comments by Shultz at a conference in Princeton, NJ, February 1993, in William C. Wohlforth, ed., *Witness to the End of the Cold War* (Baltimore: Johns Hopkins University Press, 1996), 91.

2. George Bush and Brent Scowcroft, *A World Transformed* (New York: Alfred A. Knopf, 1998), 13.

3. David E. Hoffman, *The Dead Hand: The Untold Story of the Cold War Arms Race and Its Dangerous Legacy* (New York: Doubleday, 2009), 349–351.

4. NSR 12, March 3, 1989, "Review of National Defense Strategy," (declassified), available at <http://www.fas.org/irp/offdocs/nsr12.pdf>.

5. U.S. Department of Defense, *National Defense Budget Estimates for FY 1992* (Washington, DC: Office of the Comptroller, March 1991), 147.

6. Lorna S. Jaffe, *The Development of the Base Force, 1989–1992* (Washington, DC: Joint History Office, Office of the Chairman of the Joint Chiefs of Staff, July 1993), 3–4.

7. Robert Gates, *From the Shadows: The Ultimate Insider's Story of Five Presidents and How They Won the Cold War* (New York: Touchstone, 1996), 440. See also David Rothkopf, *Running the World: The Inside Story of the National Security Council and the Architects of American Power* (New York: Public Affairs, 2005), 273–275; and Michael R. Beschloss and Strobe Talbott, *At the Highest Levels: The Inside Story of the End of the Cold War* (Boston: Little, Brown, 1993), 24–25.

8. Colin Powell with Joseph E. Persico, *My American Journey* (New York: Random House, 1995), 437.

9. Jaffe, 4.

10. *The Chairmanship of the Joint Chiefs of Staff, 1949–1999* (Washington, DC: Joint History Office, Office of the Chairman of the Joint Chiefs, 2000), 159–165, gives a synopsis of Powell's career.

11. DJSM-147-89 to VDJS, February 11, 1989, "Joint Staff"; SM-508-89 to DJS, June 13, 1989, "Examination of Joint Staff Responsibilities and Structure"; and MFR by Colonel James A. Moss, Jr., USAF, October 15, 1989, "1989 Review of Joint Staff Functions, Structure and Administrative Support," U, all in Notebook "1989 Review of Joint Staff Functions," JHO Notebook Collection, No. 172.

12. Charles A. Stevenson, "The Joint Staff and the Policy Process" (Paper for 1997 Annual Meeting of the American Political Science Association, Washington, DC, August 28–31, 1997), 7–9, 14–15.

13. Powell, *My American Journey*, 445.

14 Jaffe, 49–50; Powell, *My American Journey*, 438–439; and Peter J. Roman and David W. Tarr, "The Joint Chiefs of Staff: From Service Parochialism to Jointness," *Political Science Quarterly* 113 (1998), 104–106.

15 Rick Atkinson, *Crusade: The Untold Story of the Persian Gulf War* (Boston: Houghton Mifflin, 1993), 123.

16 Peter W. Rodman, *Presidential Command* (New York: Knopf, 2009), 181.

17 Bush and Scowcroft, 23.

18 James Mann, *Rise of the Vulcans: The History of Bush's War Cabinet* (New York: Viking, 2004), 184.

19 Powell, *My American Journey*, 464 (emphasis in original).

20 Bob Woodward, *The Commanders* (New York: Simon & Schuster, 1991), 76–80; Powell, *My American Journey*, 404–406, 476–478; Joint Staff Memo for Distribution, April 28, 1989, "Yr Paper—Strategy Review," U, JHO 16-0076.2; R.W. Apple, Jr., "Confrontation in the Gulf: The General's Error," *The New York Times*, September 19, 1990.

21 Caspar Weinberger, *Fighting for Peace* (New York: Warner Books, 1990), 397, 401–402, 433–445.

22 Powell, *My American Journey*, 434; Powell, "US Forces: The Challenges Ahead," *Foreign Affairs* (Winter 1992), 32–45. See also Walter LaFeber, "Colin Powell: The Rise and Fall of the Powell Doctrine," in Anna Kasten Nelson, ed., *The Policy Makers: Shaping American Foreign Policy from 1947 to the Present* (Lanham, MD: Rowman & Littlefield, 2009), 153–177.

23 Scowcroft quoted from an interview in Jon Western, "Warring Ideas: Explaining U.S. Military Intervention in Regional and Civil Conflicts" (Ph.D. diss., Columbia University, May 2000), 324.

24 U.S. Department of Defense, *National Military Strategy of the United States, 1992* (Washington, DC: GPO, January 1992), 10 (unclassified); Jaffe, 48.

25 A collaborative effort, the development of the base force plan reflected inputs from a variety of sources. While Powell remained in overall charge of the effort, giving it a high degree of personal attention, he relied heavily on advice and analytical inputs provided by Lieutenant General George Lee Butler, USAF, and Major General John D. Robinson, directors of the Joint Staff responsible for strategy and resource allocation, respectively. At the same time, to avoid any misunderstandings later, Powell maintained an ongoing dialogue with Under Secretary of Defense Paul Wolfowitz, who advised Cheney on strategy and policy matters.

26 Raymond L. Garthoff, *The Great Transition: American-Soviet Relations and the End of the Cold War* (Washington, DC: Brookings, 1994), 437–444, looks at the difficulties Gorbachev faced and the shifts in policy that took place at that time.

27 Don M. Snider, *Strategy, Forces and Budgets: Dominant Influence in Executive Decision Making, Post-Cold War, 1989–91* (Carlisle Barracks, PA: Strategic Studies Institute, 1993), 7–13.

28 Jaffe, 9–10; Eric V. Larson, David T. Orletsky, and Kristin Leuschner, *Defense Planning in a Decade of Change* (Santa Monica, CA: RAND, 2001), 9.

29 Powell later characterized his failure to brief the Service chiefs prior to briefing the Secretary and the President as a "mistake." See Powell, *My American Journey*, 438–440.

30 Snider, 13–18.

31 John Partin, *United States Special Operations Command History* (MacDill AFB, FL: HQ USSOCOM/ SOCS-HO, 2002), 4–5; SM-54-87 to DJS, January 16, 1987, "Implementation Planning for Activation of USSOCOM," U, Enclosure B to JCS 2542/175.

32. Quote from "Sen. Nunn on Vision of Military," *The New York Times*, April 20, 1990. See also "Nunn Says U.S. Should Negotiate Deeper Cuts on Troops in Europe," *The New York Times*, January 1, 1990; and "Nunn Opens a Double Attack In Military Spending Debate," ibid., March 23, 1990.

33. *The Washington Post*, May 7, 1990.

34. Powell, *My American Journey*, 451–455; Jaffe, 30–35.

35. "Remarks at the Aspen Institute Symposium in Aspen, Colorado," August 2, 1990, *Bush Public Papers, 1990*, 1089–1094.

36. Force levels from *1991 Joint Military Net Assessment* (March 1991), chap. 3, 5–8, JHO 14-010. In its general outline, the force structure proposed in Powell's base force bore exceedingly close similarities to the force structure incorporated into the FY50 military budget, the first unified defense budget prepared by the JCS after World War II.

37. Les Aspin, *Report on the Bottom-Up Review* (Washington, DC: Department of Defense, October 1993), 107–109.

38. See Larson et al., 41–81.

39. Deborah Hart Strober and Gerald S. Strober, eds., *Reagan: The Man and His Presidency* (Boston: Houghton Mifflin, 1998), 164–165.

40. Ronald H. Cole, *Operation Just Cause: The Planning and Execution of Joint Operations in Panama: February 1988–January 1990* (Washington, DC: Joint History Office, 1995), 6.

41. Lawrence A. Yates, *The U.S. Military Intervention in Panama: Origins, Planning, and Crisis Management, June 1987–December 1989* (Washington, DC: Center of Military History, 2008), 223–224; Woodward, 96–97.

42. NSD 17, July 22, 1989, "US Actions in Panama," summarized in Yates, 232; and Cole, 11–12.

43. Yates, U.S. *Military Intervention in Panama*, 249; James A. Baker, III, *The Politics of Diplomacy: Revolution, War and Peace, 1989–1992* (New York: G. Putnam's Sons, 1995), 185.

44. Frederick Kempe, *Divorcing the Dictator: America's Bungled Affair with Noriega* (New York: G. Putnam's Sons, 1990), 396.

45. Powell quoted in Cole, *Operation Just Cause*, 14.

46. Cole, *Operation Just Cause*, 29.

47. Yates, 270; Cole, *Operation Just Cause*, 19, 37–38.

48. Yates, 274–76; Herspring, *The Pentagon and the Presidency*, 307; Woodward, 162–167; Powell, *My American Journey*, 422–423.

49. Baker, *Politics of Diplomacy*, 189.

50. U.S. losses from JS Form 136, J-1 to DJS, July 18, 1990, "Congressional Request . . . Regarding American Casualties during the Invasion of Panama," SJS 1778/473-00; Panamanian losses from Cole, *Operation Just Cause*, 65. Cole gives U.S. wounded as 322, an apparent misprint.

51. Bernard E. Trainor, "Jointness, Service Culture, and the Gulf War," *Joint Force Quarterly* (Winter 1993–1994), 71.

52. LTG Carl W. Stiner, USA, quoted in Cole, *Operation Just Cause*, 71.

53. Bush and Scowcroft, 12.

54. 'Treaty on Conventional Armed Forces in Europe," November 19, 1990, and "White House Fact Sheet," ca. November 19, 1990, *Public Papers of George Bush, 1990*, 1640–1641.

55 See the statement by USAF/CoS Gen Merrill A. McPeak, July 16, 1991, in U.S. Congress, Committee on Foreign Relations, Subcommittee on European Affairs, *Hearings: The CFE Treaty*, 100:1 (Washington, DC: GPO, 1991), 130 (hereafter cited as *CFE Treaty Hearings*).

56 "Joint Declaration of Twenty-Two States," November 19, 1990, *Public Papers of George Bush, 1990*, 1644.

57 Lambert W. Veenendaal, "Conventional Stability in Europe in 1991: Problems and Solutions," *NATO Review* 47, no. 4 (August 1991), 3–8. See also Bush and Scowcroft, 500.

58 Jonathan Haslam, *Russia's Cold War: From the October Revolution to the Fall of the Wall* (New Haven: Yale University Press, 2011), 379.

59 Quote from Powell testimony, July 16, 1991, *CFE Treaty Hearings*, 86; Service chiefs' statements, ibid., 129–131.

60 See NSD 40, May 14, 1990, "Decisions on START Issues," U, available at <www.fas.org/irp/offdocs/nsd/nsd40.pdf> (accessed July 19, 2011).

61 Bush and Scowcroft, 40.

62 Jaffe, 23, 39,

63 *1991 Joint Military Net Assessment*, chap. 3, 5–6.

64 Steven L. Rearden, "Congress and the Strategic Defense Initiative, 1983–1989" (Study Prepared for the Strategic Defense Initiative Organization, January 31, 1992), chap. 7, U; and Sanford Lakoff and Herbert York, *A Shield in Space? Technology, Politics, and the Strategic Defense Initiative* (Berkeley: University of California Press, 1989), 116.

65 In his National Military Strategy for 1992–1997, Crowe recommended that the United States "actively pursue a strategic defense program" to determine its "feasibility." Powell concurred in the need for further R&D but endorsed a "smaller and less expensive" program than SDI oriented toward providing protection against limited attacks. See Colin L. Powell, *The National Military Strategy of the United States* (Washington, DC: GPO, January 1992), 6–7.

66 "President's News Conference," January 27, 1989, *Bush Public Papers, 1989*, 26.

67 Beschloss and Talbott, 117–118.

68 Don Oberdorfer, *The Turn: From the Cold War to a New Era—The United States and the Soviet Union, 1983–1990* (New York: Poseidon Press, 1991), 407. See also Garthoff, 422–423.

69 Quoted in Gordon S. Barrass, *The Great Cold War: A Journey Through the Hall of Mirrors* (Stanford, CA: Stanford University Press, 2009), 349–350.

70 U.S. Congress, Senate, Committee on Foreign Relations, *Hearings: The START Treaty*, 100:2 (Washington, DC: GPO, 1992), 122 (hereafter cited as *START Treaty Hearings*).

71 Powell, *My American Journey*, 538–539.

72 Beschloss and Talbott, 445–446; Bush and Scowcroft, 541–542; "Address to the Nation on United States Nuclear Weapons," September 27, 1991, *Bush Public Papers, 1991*, 1220–1224.

73 Ronald H. Cole et al., *The History of the Unified Command Plan, 1946–1999* (Washington, DC: Joint History Office, 2003), 92–95.

74 *START Treaty Hearings*, 114, 144–147.

75 William E. Odom, *The Collapse of the Soviet Military* (New Haven: Yale University Press, 1998), 370–374.

Night Attack, by Mario Acevedo (Courtesy of the Center of Military History)

Chapter 17

STORM IN THE DESERT

As the Cold War drew to a close, other problems took its place. None was more threatening to American interests than Iraqi dictator Saddam Hussein's invasion of Kuwait in early August 1990. The Joint Chiefs of Staff had long viewed the Middle East and Southwest Asia as potential trouble spots, and over the years they steadily became more mindful of the region's difficulties. Indicative of the growing importance they attached to the Middle East was their decision in 1983 to create a regional planning organization, the U.S. Central Command (USCENTCOM). While maintaining a limited U.S. presence in the area, USCENTCOM conducted combined training exercises with friendly countries, bolstered diplomatic support for U.S. interests, and coordinated multilateral protection of international shipping. Assuring unfettered access to the Persian Gulf oil fields was normally USCENTCOM's top concern. But with the Soviet threat to Europe and an unstable situation on the Korean peninsula still claiming priority, the JCS had refused to allocate significant resources to the region on a permanent basis and had dealt with it in ad hoc fashion as the need arose.

The demise of the Cold War combined with Saddam Hussein's covetous designs on his oil-rich neighbor, Kuwait, changed JCS perceptions of U.S. security requirements in Southwest Asia. As the Soviet threat to Europe receded, the JCS also adopted a more relaxed outlook toward the Far East where improved relations with China pointed to a more stable geopolitical environment. As a result, the Joint Chiefs felt more comfortable earmarking assets for regional contingencies elsewhere in line with the emerging "forward presence" doctrine. Though Southwest Asia was not the only place that caught their eye, it loomed larger than the others because of its strategic location, economic importance to the West, and growing potential for trouble.

ORIGINS OF THE KUWAIT CRISIS

Following the UN-brokered armistice ending the Iran-Iraq War in the summer of 1988, the United States intensified its efforts to broaden relations with Baghdad, always the U.S.-favored party in the conflict. Shortly after taking office, the Bush administration launched a comprehensive review of U.S. policy toward the Persian Gulf (NSR 10), focusing on U.S. interests there, the role of the Soviet Union, relations with Iran, Iraq, Saudi Arabia, and the other Gulf states, and the level of U.S. military involvement. The key issue raised in NSR 10 was whether U.S. interests in the region—economic, political, and military—remained vital in view of the changed strategic environment there and, if so, whether the existing investment of U.S. power and resources reflected that importance.[1]

The review confirmed that major changes in the strategic environment of the Persian Gulf over the past decade mandated greater American interest and involvement, and recommended that the United States bolster regional peace and stability through closer cooperation and collaboration with friendly governments. Step-by-step improvements in U.S.-Iraqi relations were crucial to the success of this policy. While aware that problems with Saddam were bound to arise, the Bush administration was cautiously optimistic that it could moderate his behavior and increase U.S. influence in Iraq through carefully targeted economic, political, and military assistance. In exchange for U.S. help, Saddam should be prepared to give up his chemical and biological weapons, curb his nuclear ambitions, break his ties with terrorist organizations, and stop meddling in the internal affairs of Lebanon and other Mideast countries.[2]

Saddam, however, had his own agenda, which involved nothing less than establishing an Iraqi hegemony across the region. Bloodied but undefeated in the war with Iran, the Iraqi dictator was at the pinnacle of his power and prestige, a formidable, dangerous, and unpredictable figure who had the largest and most powerful military force in the region at his disposal. Aiming to regain some of the oil export market he lost to other Gulf producers during the conflict, Saddam accused neighboring Kuwait, Saudi Arabia, and the other Gulf states of undercutting Iraq's recovery by surreptitiously increasing oil production and driving down prices, even though these countries had been among his staunchest allies in the recent conflict. According to former Russian Foreign Minister Yevgeny Primakov, who knew Saddam personally, the Iraqi leader assumed that he had a more or less free hand, based on U.S. help against Iran, and could do virtually as he pleased without risking American retaliation as long as Iran remained under the control of a radical anti-Western regime.[3]

Meanwhile, the United States launched a progressive military draw-down in the Persian Gulf. With Operation *Earnest Will* coming to a close, the Joint Chiefs saw no justification for the sizable air and naval forces they had assembled to escort neutral shipping at the height of the Iran-Iraq war in 1987–1988. By the summer of 1989 USCENTCOM's presence in the Gulf was essentially back to its pre-escort level—a handful of naval vessels backed by the intermittent presence of a carrier battle group in the Indian Ocean and North Arabian Sea. Whether the retention of a larger U.S. naval presence in Southwest Asia would have assured greater stability, deterring Iraq from aggression against Kuwait, remains an open question. Saddam's ruthless drive to dominate Middle East politics and his insatiable ambitions would have been hard to check in any case. Nonetheless, as U.S. forces withdrew, the odds increased that they would be back again sooner or later. The retreat may have been unavoidable, but it left the Joint Chiefs, among others, decidedly uneasy and created a political and military vacuum in the region that Saddam was only too happy to fill.[4]

During the summer of 1990 Saddam steadily increased the pressure on Kuwait. While complaining that his neighbor was pumping excessive oil and driving down prices, Saddam precipitated a border dispute with Kuwait, the same pretext he used for going to war with Iran in 1980. He also became highly critical of the United States and stepped up menacing rhetoric and gestures toward Israel by deploying Scud ballistic missiles aimed at Tel Aviv. Still committed to the constructive engagement policy, the Bush administration hoped to diffuse the situation and elicit cooperative behavior from Saddam with pledges of nonlethal military assistance, loans, and credit guarantees to help finance grain imports and to rebuild Iraq's battered economy. Much to Saddam's irritation, however, the proffered assistance was slow to materialize.[5]

Increasingly belligerent, Saddam began massing forces along Iraq's common frontier with Kuwait in a show of gunboat diplomacy. While the Intelligence Community declined to rule out the possibility of an invasion, it could find no hard evidence that Saddam was preparing an attack. Indeed, the absence of Iraqi logistical support led General Powell and analysts on the Joint Staff to suspect that Saddam was bluffing and was more interested in eliciting concessions from Kuwait and its neighbors than in starting another war.[6] Following the Chairman's lead, JCS action officers dealing with the Middle East shied away from recommending anything remotely resembling a military response without first exploring other options and ascertaining clear-cut political objectives. But with tensions building, a military confrontation seemed increasingly unavoidable. On July 25, 1990, Saddam summoned April Glaspie, the U.S. Ambassador to Baghdad, to an impromptu interview. Professing friendship for the United States, Saddam expounded at length on his desire for a peaceful resolution of the dispute with Kuwait but did not rule out military action. In return, Glaspie assured him that President Bush was also interested in a peaceful outcome but also wanted close U.S. relations with Iraq. Subsequently, critics of the Bush administration pounced on Ambassador Glaspie's remarks as a virtual invitation for Saddam to invade Kuwait. Whether Saddam viewed them in that light is unclear. More than likely, he had already made up his mind to attack Kuwait and in summoning Glaspie, was trying to gauge how the United States would respond.[7]

While continuing to give lip service to a diplomatic solution, Saddam moved more units into position and by the end of July had approximately 140,000 troops and 2,000 Soviet-made T-72 tanks and other armored vehicles along the border with Kuwait. On August 2, 1990, he launched his attack. The invaders met light resistance and within a few days were in full control of the country, which Saddam proceeded to annex. Demanding that Saddam withdraw his forces immediately, President Bush declared that Iraqi aggression "will not stand."[8] But despite a tough declaratory policy, the administration had no firm plan of action. For the time being, containing Saddam's aggression and deterring him from attacking neighboring Saudi Arabia were the administration's only firm objectives. Only time would tell whether the United States would be willing to go further and take steps to evict Iraqi forces from Kuwait.

FRAMING THE U.S. RESPONSE

Even though General Powell and the Joint Staff had been closely monitoring the situation in the Middle East for some time, looking at alternative contingency plans as they went along, Saddam's invasion still caught them by surprise and unprepared. Like almost everyone else in Washington at the time, they expected the confrontation between Iraq and Kuwait to end peacefully. As Lieutenant General George Lee Butler, USAF, director of J-5, described the state of mind in the Joint Staff, "We had the warning from the intelligence community—we refused to acknowledge it."[9] When the Iraqis attacked Kuwait, the Joint Chiefs of Staff had few forces in or near the vicinity of the Persian Gulf and were only beginning to take steps to get more there. Most of the planning done prior to the Iraqi invasion centered on OPLAN 1002-90, an updated version of a Cold War-era USCENTCOM plan to defend Iran against a Soviet invasion. Arguing that the threat of Iraqi aggression now outweighed the danger of a Soviet attack, General H. Norman Schwarzkopf, USA, Commander in Chief of Central Command (USCINCCENT), had requested JCS permission to shift the geographic focus of OPLAN 1002-90 to reflect a possible Iraqi invasion of either Kuwait or Saudi Arabia. In December 1989 the Joint Chiefs gave Schwarzkopf permission to proceed.[10]

While the detailed work of revising OPLAN 1002-90 had just begun by the time Iraq invaded Kuwait, its broad outlines were fairly clear and well known. Basically, OPLAN 1002-90 envisioned war on a grand scale, with the mobilization and deployment of 200,000 U.S. ground troops and supporting air and naval units taking on an Iraqi force of comparable if not larger size and capabilities. With a strength of over one million men, the Iraqi Army was one of the largest in the world. But it relied heavily on conscripts armed with older models of Eastern Bloc and Chinese weapons. The core of Iraq's defense establishment consisted of eight elite Republican Guard divisions (expanded to 12 divisions following the invasion of Kuwait) commanded by officers who had sworn personal allegiance to Saddam. Made up of volunteers, the Republican Guard carried more up-to-date weapons than the regular army and constituted Saddam's most effective and reliable force. Military and political analysts in the West generally considered it a key prop of Saddam's regime. Iraq's air component, though strong on paper with over 800 planes, had few experienced pilots and operated under a defensive doctrine that limited its range and effectiveness. On the other hand, Iraq's air defenses, though somewhat outdated, were rated among the best in the world, built around sophisticated low-level anti-aircraft artillery and portable surface-to-air missiles.[11]

The greatest dangers Iraq posed sprang from the uncertainties surrounding its capabilities for chemical, biological, and nuclear warfare, known collectively as "weapons of mass destruction," or WMD. Available delivery means included short-range Scud missiles, aerial bombs, artillery shells, rockets, and spray tanks mounted on aircraft. Saddam's desire to make Iraq a nuclear power was well known. Even though the Israelis dealt his program a major setback by destroying the Tuwaitha atomic reactor in 1981, rumors persisted that he was continuing to explore ways of acquiring atomic bombs and might have stockpiled enough fissionable material for a small arsenal. Biological weapons were also of interest to Saddam but seemed to hold less promise and appeal than chemical weapons. During the 1980s, Saddam mounted poison gas attacks against local insurgencies and Iranian troop formations. Since then, he had continued to replenish his chemical weapons stockpile, threatening to use it against anyone who got in his way.

In surveying what they were up against, senior members of the Bush administration were understandably wary. By far the most cautious was the Chairman of the Joint Chiefs, General Powell. Convinced that Saddam should be contained, Powell readily agreed to rush reinforcements to the Middle East to block the Iraqis from moving against Saudi Arabia (Operation *Desert Shield*). But he initially opposed offensive operations aimed at liberating Kuwait, a much larger and more complicated task which, based on preliminary estimates, would require substantially more troops and eight months to a year of preparation. In view of the risks involved, he was prepared to treat Kuwait as expendable and concentrate on protecting Saudi Arabia. "I think we'd go to war over Saudi Arabia," he told Schwarzkopf, "but I doubt we'd go to war over Kuwait."[12] Recalling the popular backlash against Vietnam, Powell believed that any attempt to liberate Kuwait by force would need full congressional and public support. Without that, he saw little hope of success. As an alternative to military action, Powell endorsed a regime of economic, political, and diplomatic sanctions against Iraq and was prepared to wait up to 2 years for them to have an effect.[13]

Powell's strategy of restraint contrasted sharply with the emerging determination in the White House to restore the status quo ante one way or another as quickly as possible. Like the Chairman, President Bush hoped to avoid going to war. But he had less confidence than Powell in the efficacy of sanctions and felt that the longer the West delayed in acting, the more entrenched Saddam would become. Applying a historical perspective, Bush saw a "direct analogy" between the invasion of Kuwait and Nazi Germany's aggression against Poland in World War II. Prodded by Scowcroft, who considered Powell overly cautious, the President moved steadily toward a policy of liberation through military action and looked to Cheney to manage the details and bring the Joint Chiefs of Staff into line. "Cheney recognized early that sooner or later it would come to force," Bush recalled. "Dick was probably ahead of his military on this."[14]

During the early days of the crisis, as the administration sought to define its position, Powell and Cheney seemed to go separate ways. Resisting hasty decisions and commitments, Powell played for time and tried to focus the debate on political objectives and whether military action was in the best interest of the United States. Cheney became frustrated and insisted that Powell concentrate more on developing and refining military options.[15] "Colin," he said, "you're Chairman of the Joint Chiefs. You're not Secretary of State. You're not the National Security Advisor anymore. And you're not Secretary of Defense. So stick to military matters." Looking back, Powell agreed that Cheney was right, but he gave way grudgingly and offered military advice that was almost always framed, as only Powell could do, around its potential political impact during the ensuing planning process and buildup of forces.[16]

OPERATIONAL PLANNING BEGINS

Despite the Goldwater-Nichols reforms, operational planning for *Desert Shield–Desert Storm* encountered many of the problems the Joint Chiefs had experienced during crises in the past. This included initial confusion and uncertainty, followed by largely improvised responses, with inputs from several sources at the same time. While Powell was gradually turning the Joint Staff into an unrivaled planning and staff-action organization, he had yet to complete the process. Thus, the door remained open for the Services' planning staffs to make inputs, often on their own initiative. With limited staff available and his own plans in flux, Schwarzkopf desperately needed help from wherever he could get it. The result was a rather chaotic period at the outset of the crisis that saw planning diverge along two separate lines, one running through the Joint Staff where Powell's influence predominated, the other through a wholly separate Air Staff planning cell known as Checkmate. Eventually, these lines converged at Schwarzkopf's USCENTCOM headquarters, where they became integrated into an overall strategic concept. But in their origins and purpose, they reflected two sharply different military philosophies for coping with the crisis.

Powell and the Joint Staff initially occupied the stronger and more influential position owing to their statutory role and increasing preeminence within military planning circles. After the extraordinary success of the Panama invasion, few dared to gainsay the Joint Staff's growing skill for organizing and coordinating joint operations. While the President had not yet fully made up his mind about Kuwait, those close to him could sense the drift in his thinking. As a precaution, in addition to the defensive actions taken under Operation *Desert Shield* at the outset of the crisis, Secretary Cheney ordered the CJCS and USCENTCOM to develop an offensive option that would be available to the President in case Saddam Hussein chose to engage in further aggression or other unacceptable behavior, such as killing Kuwaiti citizens or foreign nationals in Kuwait or Iraq.[17] As characterized by one account, the Joint Staff's earliest response resembled "a typical cold-war, limited-option sort of thing."[18] Using OPLAN 1002-90 as their guide, Joint Staff planners initially estimated that evicting the Iraqis could be done with a force not much larger than that being organized at the time for Operation *Desert Shield*—about 200,000 troops plus supporting air and naval units. Powell, however, found these estimates insufficient. With his eye on avoiding a military confrontation, the Chairman hoped to intimidate Saddam and convince him through a combination of sanctions and a highly visible military buildup to back down without a fight. Should that approach fail, he wanted to be prepared to conduct "a full-scale air, land, and sea campaign" that would quickly overwhelm Saddam, just as he had overwhelmed Noriega. "We had learned a lesson in Panama," Powell contended. "Go in big and end it quickly." With these as Powell's planning guidelines, Joint Staff estimates of the required force varied almost daily and became practically open-ended.[19]

Initially, Powell operated under very few constraints. Looking at the military possibilities and various options, a consensus developed early on in Washington that the United States would need sizable forces to counter Saddam and that the build-down under the base-force plan, only recently announced by the President, should be put on hold. Yet as projected force requirements for the Middle East began to mount, they pointed to increased expenditures that left senior administration officials decidedly uneasy. Hoping to defray some of the "staggering" expense, as

Secretary of State Baker described it, the Bush administration actively solicited contributions of money and/or troops from around the world to create a multinational coalition to liberate Kuwait. Eventually, nearly fifty countries agreed to provide assistance in one form or another. But even with those inputs, there was still a high likelihood that the United States would bear the brunt of the costs.[20]

Cheney never presumed to challenge Powell's professional expertise, but as Secretary of Defense, his first concern was to weigh the financial impact of the operation. It was on that basis that he began to take a closer look at the proposals coming out of the Joint Staff. The Goldwater-Nichols Act may have streamlined the advisory process, making it more timely and responsive, but it also inadvertently created barriers to the flow of military ideas and information reaching the Secretary, the President, and the NSC. Though he continued to rely heavily on Powell and the Joint Staff, Cheney decided to shop for other views as well. As one military analyst described it, "Cheney adroitly and informally bypassed Powell for additional military opinions to assure himself of differing views.... This technique did not sit well with Powell and, although he never challenged Cheney's right to solicit advice from others, it angered him."[21]

The most attractive alternative to a large-scale buildup on the ground was increased reliance on airpower. Actually, Powell and Cheney were both skeptical of strategies built around airpower and could not find much evidence that the air campaigns of previous wars had been either very successful or decisive. In years past, even some airpower enthusiasts would have agreed. But since Vietnam, as the Air Force shed its dependence on nuclear weapons and turned to reviving its conventional capabilities, its confidence in the efficacy of airpower rose steadily. By the end of the Cold War, with the advent of improved planes employing stealth technology, increasingly reliable precision-guided munitions, and more effective command and control using high-speed computers and space-based satellites, the chances of a conventional bombing campaign having a decisive impact on future wars seemed more assured than ever. Little by little, as interest at the White House in developing an airpower-oriented strategy began to grow, views on airpower around the Pentagon likewise began to change.[22]

Powell concurred that airpower had a major role to play, and in the immediate aftermath of the Iraqi invasion both he and Schwarzkopf turned to airpower as their most readily available and effective means of deterring Saddam from further aggression or punishing him if he should make a move against Saudi Arabia.[23] Of the forces rushed to the Middle East under Operation *Desert Shield*, Joint Staff planners put major emphasis on large Air Force deployments of combat aircraft and aerial reconnaissance planes as the bulk of the initial "package." All the same, Powell resisted the notion, popular in some quarters of the Air Force, that a carefully orchestrated air campaign could practically win a war alone.[24] To Powell's consternation, the Air Force Chief of Staff, General Michael J. Dugan, openly suggested such a possibility shortly after Iraq invaded Kuwait. During the return flight from a fact-finding trip to the Middle East in August 1990, Dugan regaled reporters with his views, which subsequently appeared in the *Washington Post*. While making the Iraqis "look like a pushover" with airpower, Powell recalled, Dugan further suggested that American military planners were "taking their cue from Israel" on how to deal with Saddam, a remark that was sure to antagonize many Arabs. Cheney agreed that Dugan's behavior was "dumb, dumb, dumb" and promptly fired him for "poor judgment." The ignominious departure meant that Dugan's tenure as Chief of Staff lasted only 3 months.[25]

Even though airpower advocates had lost one of their strongest and most influential spokesmen, their cause remained very much alive. Hints of growing interest in airpower at the White House doubtless fueled the process. Soon to emerge as the initial architect of the air campaign against Iraq was Air Force Colonel John A. Warden III, who headed a planning cell in the Air Staff known as Checkmate. Trained as a fighter pilot, Warden served in Vietnam and during the 1970s and 1980s steadily refined his views on the role and application of airpower. Some regarded him as the most innovative thinker the Air Force had produced since Billy Mitchell after World War I. Basically, Warden took issue with the AirLand Battle doctrine, the dominant military concept since Vietnam, which urged closer coordination between ground and air forces, with the aim of using airpower to achieve decisive maneuver on the ground. In Warden's scheme of things, air superiority should take precedence; once achieved, "in many circumstances it alone can win a war."[26]

Amid rising tensions in the Middle East, Warden emerged as the leading spokesman for increased reliance on airpower in the expected showdown with Saddam. One of Warden's admirers was Secretary of the Air Force Donald B. Rice, a former president of the RAND Corporation (originally an Air Force think tank) and an ardent proponent of airpower. If previous U.S. involvement in the Persian Gulf had been primarily a Navy show and toppling Noriega predominantly an Army affair, Rice and like-minded others wanted the looming conflict with Iraq to be first and foremost an air war. Warden and his staff (a group comprised initially of about two dozen young Air Force officers) were eager to oblige. Within days of Iraq's invasion of Kuwait, they received an urgent request from the Vice Chief of Staff of the Air Force to provide General Schwarzkopf with advisory assistance. Expecting to be called upon sooner or later, Warden had initiated work the day before on an outline plan called "Instant Thunder" for strategic air operations against Iraq. As described by Air Force historian Richard G. Davis, "Instant Thunder" was "a stand-alone war-stopper" that called for a concerted 6-day effort designed to incapacitate the Iraqi leadership and destroy its key military capabilities.[27]

While Powell duly acknowledged Checkmate's contributions, terming them "the heart of the *Desert Storm* air war," he took issue with the single-Service approach and around mid-August directed that Army, Navy, and Marine Corps officers be included in Warden's organization. Thenceforth, Checkmate's papers and reports bore the logo of the Joint Staff, and its activities acquired the appearance, if not always the reality, of jointness under the Directorate of Operations (J-3).[28] The spirit of Goldwater-Nichols notwithstanding, inter-Service coordination, especially with the Navy, remained tenuous throughout the crisis. As eager as the Air Force was to leave its mark, the Navy disliked having its carrier-based aircraft placed under a joint tasking system and would have preferred to operate on its own.[29] During the conflict, applying its own priorities as the opportunity arose, the Navy withheld as many as a third of its aircraft to protect its carriers. Of the Navy planes that did participate in offensive operations, only a limited number were equipped to deliver the precision-guided munitions that were crucial to the execution of Warden's strategic bombing concept. The Navy's most significant contribution to the air campaign was its Tomahawk land-attack cruise missiles (TLAMs). Launched from surface ships and attack submarines, the low-flying TLAMs were ideal for daytime attacks against highly defended targets and could also be used when adverse weather grounded fighter-bombers.[30]

Checkmate's direct involvement in shaping the air war was relatively short-lived. At Schwarzkopf's request, Warden flew to Riyadh and on August 20 briefed Lieutenant General Charles A. Horner, USAF, Schwarzkopf's air deputy and US-CENTCOM's acting forward commander. Horner accepted Checkmate's target scheme but rejected Warden's "airpower alone" strategy because it ignored the large number of Iraqi troops and tanks poised on the border with Saudi Arabia.[31] Asserting control from there on out, Horner created his own Special Planning Group for air operations, a multi-Service unit (later expanded to include NATO and Saudi representatives), and placed Brigadier General Buster C. Glosson, USAF, in charge. Dubbed the "Black Hole," it operated in utmost secrecy out of the basement of the Royal Saudi Air Force headquarters in downtown Riyadh. Throughout the crisis, Glosson was in constant contact with Warden and drew heavily on Checkmate for advice, ideas, and personnel. But from that point on, primary responsibility for air war planning became an inter-Service operation, with Checkmate, the Joint Staff, and Glosson's Black Hole organization in Saudi Arabia working in unison.[32]

Checkmate's eclipse brought a fundamental change of philosophy that steered planning for the air campaign back into line with Powell's view of airpower as a supporting element of the ground war. On that point, Powell and Schwarzkopf—both Army officers—thought exactly alike. "Instant Thunder" disappeared and in its place emerged a more conventional plan for an integrated air-ground campaign. Though still built around Warden's phased sequence of attacks and basic target scheme, Schwarzkopf's integrated approach took a larger range of military and related targets into account. As the target list grew, so did the need for aircraft, intelligence, and logistical support. What Warden and his colleagues in Checkmate had originally envisioned as an intensive 6-day bombing and interdiction campaign turned into plans for a month or more of round-the-clock air operations aimed not just at driving the Iraqis out of Kuwait but at eliminating Saddam and his armed forces as a future threat to the region.

THE ROAD TO WAR

By late September 1990, working closely with Schwarzkopf, Powell had assembled a plan to defend Saudi Arabia and was gradually developing a military strategy to expel the Iraqis from Kuwait, starting with an intense air campaign, should sanctions and diplomacy fail. Major elements of the *Desert Shield* force were now in place, while the remainder were either en route to Saudi Arabia or being fitted out for deployment. Whether more would follow remained to be seen. Although Powell had repeatedly discussed the various options in general terms with Cheney and the President, he had yet to receive a clear signal of the President's intentions. As a result, final preparations remained in limbo. Privately, the President was increasingly reconciled to a military showdown. Frustrated by Saddam's intransigence in the face of efforts by Gorbachev and others to broker a settlement, Bush saw the chances of a peaceful resolution steadily slipping away and now looked on the looming confrontation as "a moral crusade." Rumors had already begun to spread that should armed intervention become necessary, the JCS expected a minimum of 10,000 casualties and up to 50,000 if Saddam used chemical and biological weapons. Even though public and congressional opinion generally endorsed the administration's "get tough" approach toward Saddam, the prevailing sentiment leaned more toward sanctions than the exercise of military power. Among leaders on Capitol Hill, reliance on air and sea capabilities received preference over a potentially bloody ground campaign.

Realizing that the country was in no mood for a war if one could be avoided, President Bush continued to defer a final decision on military action. Before making further commitments, he wanted a clearer picture of what it would take to defeat Saddam and arranged with Powell for a formal briefing at the White House on October 11, 1990.[33] Schwarzkopf had recently moved his headquarters from Tampa to Riyadh and, pleading that his plans were still gestating, wanted to come to Washington to explain the situation and lead the briefing himself. At Powell's insistence, however, he stayed behind and designated his chief of staff, Major General Robert B. Johnston, USMC, to lead the USCENTCOM delegation. The day before going to the White House, Powell held a dry-run presentation at the Pentagon for Cheney, the Service chiefs, and senior members of the Joint Staff. Glosson summarized the progress on the air war while an Army lieutenant colonel gave the briefing on the ground campaign. Afterwards, Powell drew Glosson aside and admonished him for making the air war look too easy. For the presentation the next day, Powell wanted Glosson to "tone it down" and curb his estimates of the outcomes. "Be careful over at the White House tomorrow," Powell said. "I don't want the President to grab onto that air campaign as a solution to everything."[34]

The White House briefing on October 11 revealed a military planning process at midstream. Glosson's toned-down presentation notwithstanding, it was clear that planning for the air campaign was well ahead of preparations for the ground war, which was now designated Phase IV in the planned sequence of operations. Utilizing forces and equipment currently deployed, Phase IV was basically a single-corps thrust into the middle of the Iraqi defenses, a strategy that one senior OSD official mocked as the "charge of the light brigade into the wadi of death."[35] While bypassing Iraqi strong points, the proposed attack would still encounter key Iraqi ground units. Heavy casualties were almost certain.[36] As Scowcroft remembered the briefing, it "sounded unenthusiastic, delivered by people who didn't want to do the job. . . . I was appalled with the presentation and afterwards I called Cheney to say I thought we had to do better."[37]

Like many of Roosevelt's meetings with the Joint Chiefs of Staff during the early days of World War II, the White House briefing on October 11, 1990, was a largely exploratory affair. If Powell's underlying purpose was to dissuade Bush from hasty action, he was eminently successfully. "The briefing made me realize," Bush recalled, "we had a long way to go before . . . we had the means to accomplish our mission expeditiously, without impossible loss of life."[38] But the episode also deepened the rift between Powell and Cheney and made the Secretary of Defense more aware than ever that he needed an alternative to the CJCS as a source of advice. Disappointed in what Powell and Schwarzkopf came up with, Cheney established a special advisory unit in OSD headed by retired Army Lieutenant General Dale A. Vesser. A former Director of J-5 and currently Assistant Deputy Under Secretary of Defense for Resources and Plans, Vesser had been involved in deployment planning for *Desert Shield* almost from the outset. His new tasking from the Secretary was to double check the planning coming out of the Joint Staff and USCENTCOM and to look into alternative strategic concepts.[39]

Shortly after the ill-starred White House briefing, at the urging of the President and the Secretary of Defense, Powell flew to Saudi Arabia in hopes of finding a "more imaginative" Phase IV strategy. He carried assurances from the President that Schwarzkopf could have "whatever forces he needed to do the job."[40] Earlier, to augment his planning staff, USCINCCENT requested help from the Jedi Knights, an elite Army planning team from the Command and General Staff College at Fort Leavenworth, Kansas. To overcome the defects in the earlier concept, they proposed a strategy that promised a higher degree of success with fewer casualties through a flanking maneuver west of the Iraqi defenses in Kuwait. Though bolder and more innovative, the new plan would also require more troops, more heavy armor, and additional air and sea support. By the time Powell arrived, Schwarzkopf had already given the plan his enthusiastic blessing and had a request in hand for at least another mechanized corps. Powell cautioned that it might be necessary to secure "a clear mandate from Congress and the American people" before bringing more forces into the Gulf or committing them to combat. But his immediate concern was to reassure Schwarzkopf that, as the President had indicated, he could have whatever he needed to complete his mission. "If we go to war," the Chairman said, "we will not do it halfway."[41]

Returning to Washington, Powell held a series of briefings starting with Cheney and the Service chiefs to present the new strategy and its force requirements, now approaching half a million troops. While acknowledging that the new plan needed work, he still saw it as a significant improvement. By and large, the Service chiefs agreed. The sole exception was General Merrill A. McPeak, who succeeded Dugan as Air Force Chief of Staff. Suspecting that the available intelligence had inflated Iraqi capabilities, McPeak doubted the need for the massive ground build-up that Powell and Schwarzkopf were planning and saw it mainly as an attempt by the Army to embellish its role at Air Force expense. But his efforts to dissuade Powell were apparently half-hearted and he soon gave up, realizing that the momentum was against him.[42]

On October 30, Powell personally presented the new strategy to the President and his core group of advisors. Powell recalled that as he ran down the list of force requirements, there were gasps and gulps from practically everyone in the room except the President. Scowcroft thought the proposed augmentations were "so large that one could speculate they were set forth by a command hoping their size would change [the President's] mind about pursuing a military option."[43] Bush, however, was unfazed. Remembering Glosson's briefing of a few weeks before, he inquired about the increased use of airpower in lieu of ground forces but found the Chairman more adamantly opposed than ever. "Mr. President," he said, "I wish to God that I could assure you that airpower alone could do it, but you can't take that chance."[44]

To speed up deployment of the heavy armor Schwarzkopf requested, Powell proposed withdrawing VII Corps from Germany (comprising half of the Army's strength in Europe) and moving it en masse to Saudi Arabia. Assuming all went well, U.S. forces would be in a position to commence offensive air operations around the middle of January 1991 and launch a ground attack a month later. Only a few years earlier, with the Soviet threat hanging over Western Europe, the unilateral withdrawal of U.S. forces from Germany on this scale was utterly unthinkable. But in the light of recent events—the pending CFE Treaty and the collapse of Communist power in Eastern Europe—the situation changed.[45]

On November 8, President Bush announced a significant augmentation in the number of troops being sent to the Persian Gulf, setting off a political battle in Washington that lasted into the new year.[46] At issue was the 1973 War Powers Act, a legacy of Vietnam, which curbed the President's authority to commit to combat without explicit approval from Congress. Bush and Scowcroft both scoffed at the law, arguing that it infringed on the President's duties as Commander in Chief and was therefore unconstitutional. Powell, however, took the matter more seriously and welcomed an open airing of the issues. During the preparations for the Panama operation, he had not paid much attention to gathering congressional support, mainly because he found sentiment on Capitol Hill to be ahead of the administration on the need for intervention.[47] A large-scale war in the Middle East involving the call-up of Reserves, with possibly thousands of U.S. casualties, was another matter. Echoing positions taken by the Joint Chiefs from the early days of the Reagan administration on, Powell wanted congressional preferences clearly on record before taking military action against Saddam. The upshot was a vigorous debate in Congress culminating on January 12, 1991, in the adoption of resolutions by both houses authorizing the President to use force to expel the Iraqis from Kuwait in accordance with UN directives. At long last, Powell had the mandate he wanted.

FINAL PLANS AND PREPARATIONS

President Bush's decision to augment the U.S. buildup in the Persian Gulf set the stage for the largest U.S. military campaign since Vietnam—the liberation of Kuwait, also known as Operation *Desert Storm*. Like the 1944 D-Day invasion of Europe, *Desert Storm* was both a joint and combined operation. As such, it tested not only the Bush administration's diplomatic skills in coalition-building, but also its progress toward fulfilling the goals of the Goldwater-Nichols Act. While not the resounding display of "jointness" that some hoped it would be, the overall operation still reflected an increased level of inter-Service cooperation and collaboration, a positive sign that the Goldwater-Nichols reforms were slowly but surely taking hold.

At the heart of the American-led effort to liberate Kuwait was an unusual set of command and control arrangements. From his temporary headquarters in Riyadh, General Schwarzkopf exercised broad strategic direction over an international coalition that grew to 700,000 troops representing 28 countries by the time military action commenced early in 1991. His direct operational control (OPCON) extended to about two-thirds of the total, mostly U.S. and British forces. French forces operated independently but coordinated closely with USCENTCOM. Egyptian, Syrian, and other Islamic forces invited to participate in military operations did so with the understanding that they would be subject to Saudi OPCON. A tricky arrangement in theory, it worked remarkably well in practice. By the time the ground offensive began in February 1991, the coalition had effectively evolved into two combined commands—the Western allies under Schwarzkopf, and the Islamic members under the senior Saudi commander, Prince Khalid bin Sultan.[48]

Final planning and preparations for *Desert Storm* took place through Schwarzkopf's USCENTCOM organization. Like other combatant commands under the Joint Chiefs, USCENTCOM operated at the top with an integrated military staff but functioned through Service-oriented subcommands for ground, sea, air, and amphibious operations.[49] The only one of those that approached truly joint-combined status during *Desert Shield–Desert Storm* was Horner's air component, U.S. Air Forces Central Command (CENTAF), which from September 1990 on included Navy, Marine, and British representatives.[50] Among his duties, Horner functioned as Joint Forces Air Component Commander (JFACC), in which capacity he had authority to plan the air war, but not Service-specific command for anything other than Air Force assets.[51] Still, his control of coalition air assets exceeded that of any U.S. commander in either the Korean or Vietnam Wars.[52] Despite its joint appearance, CENTAF retained a distinctly Air Force perspective that heavily influenced the use of intelligence, targeting priorities, and the allocation of resources for the air campaign—all sources of friction to some degree with the other Services, which had their own views on how airpower should be applied. The Navy, which operated under less rigid planning procedures than the Air Force, found CENTAF's methods especially onerous.[53] As a rule, CENTAF either worked around those problems or relied on informal agreements to paper over them. Though not always the ideal solution, these ad hoc agreements seemed to avoid any serious misunderstandings. One of the earliest and most successful compromises, dating from September 1990, was the agreement reached between CENTAF and the

Marine Corps, under which the Marines allocated roughly half their combat planes in-theater to CENTAF-directed strategic operations in exchange for assurances of B–52 and Air Force tactical support of their ground operations.[54]

While providing overall strategic direction, Schwarzkopf was determined to avoid micromanaging field operations as he and Powell often complained McNamara and President Johnson did in Vietnam, to the detriment of the war effort. Preferring a system of decentralized command, he allowed his subordinates maximum freedom of action as long as they adhered to USCINCCENT's overall strategy. That applied to planning for the air war as well as for the ground campaign and resulted in less than ideal coordination between USCENTCOM's component commands. The upshot was that Schwarzkopf personally assumed operational control of all ground forces in the Kuwait Theater of Operations (KTO) but was still unable, once the fighting began, to achieve much more than nominal synchronization between USCENTCOM's advancing Army (ARCENT) and Marine Corps (MARCENT) components.[55]

Despite his reputation for fastidious planning and attention to detail, Powell left Schwarzkopf more or less alone once they had an agreed plan of action. Describing him as "testy by nature" and "short-tempered," Powell acknowledged that Schwarzkopf could be difficult to work with. But he had the utmost confidence in the USCINCCENT's leadership and wanted to protect the longstanding American tradition that accorded commanders independence and initiative in the field, a concept he thought the Vietnam experience had assailed. In effect, Powell extended this doctrine a step further by applying it to the planning process. Using his CJCS position as a buffer, he allowed Schwarzkopf to move ahead with final preparations for *Desert Storm* with minimal interference from the "armchair strategists" in Washington.[56]

On December 19, 1990, Powell and Cheney arrived in Riyadh for 2 days of briefings, the final review before the President approved launching *Desert Storm*. Back in Washington, there was growing pressure from Secretary of the Air Force Rice and officers on the Air Staff to suspend preparations for a ground assault and to rely exclusively on airpower to defeat the Iraqis. At issue was Europe's overburdened transportation network, which was causing intermittent disruptions in redeploying VII Corps' heavy equipment from Germany to the Middle East.[57] Seizing the opportunity, Rice launched an eleventh-hour effort to derail the ground offensive and arranged for Warden to conduct a special briefing for the Secretary of Defense on December 11 to persuade Cheney that an airpower-alone strategy could crush Iraqi resistance and win the war. Giving a heavy-handed performance, Warden insisted that a concerted air campaign could cut the strength of the Republican Guard in half and with enough time and bombs reduce Iraqi armor and artillery in the KTO by 90 percent. Cheney was noncommittal, but as he and Powell arrived in Riyadh they knew they faced some hard decisions.[58]

Much of what they heard covered familiar ground. While Horner defended the particulars of the air campaign as currently planned, Schwarzkopf did the same for the ground war. Wanting to leave no stone unturned, Cheney peppered both commanders with tough questions and eventually asked them point blank whether Warden and other airpower enthusiasts were right in claiming that air strikes could take the Republican Guard down by 50 percent. Horner and Schwarzkopf acknowledged that computer analysis deemed it feasible and that Glosson and his staff were operating with that goal in mind. But with the moment of truth fast approaching, they conceded that it was a tall order and that nothing like it had ever been tried. While offering a generally positive assessment, Horner made no secret of his doubts.[59]

As for the ground offensive, Schwarzkopf offered assurances that despite delays, the buildup was moving ahead and would continue under cover of the air strikes. He estimated that he would be ready to launch his land attack (G-Day) sometime between mid-February and March 1. Ground combat would entail several interrelated operations. XVIII Airborne Corps and a French division would attack to the west and cut off Iraqi forces in the KTO. VII Corps and British units would conduct the main Coalition effort and attack to the east of XVIII Corps, engaging and destroying the Republican Guard. Finally, along the coast, U.S. Marines and Arab units would launch a combined offensive to hold enemy forces and eventually open the way for retaking Kuwait City. Schwarzkopf expected to have Kuwait back in safe hands in 2 weeks and spend another 4 weeks consolidating his victory. What would happen after that was apparently not discussed.[60]

Seeing no better alternative, the Secretary of Defense approved USCINC-CENT's plans and returned to Washington where he and Powell discussed them further with the President. While lauding the professionalism of the air campaign planners, Cheney admitted to being less impressed with preparations for the ground war. Though there was still the debate in Congress to contend with, Bush agreed to go ahead with scheduling the air offensive but determined that the actual start of the land campaign would require a subsequent Presidential decision in February. Only a few weeks earlier, Bush had listened to what he characterized as an "upbeat briefing" by McPeak on the air campaign and may have hoped it would rule the day and avoid the need for a bloody confrontation on the ground. Powell, as always, remained skeptical, but everyone involved realized that the time for planning and for theoretical discussions was fast drawing to a close.[61]

On January 15, 1991, President Bush approved a general statement of war aims (NSD 54) authorizing U.S. military action in accordance with various UN resolutions. Despite the enormous force the United States and its coalition partners were assembling, the stated objectives in the President's directive were limited to bringing about "Iraq's withdrawal from Kuwait" and restoring the region to the status quo prior to the invasion. Only if Saddam resorted to the use of chemical, biological, or nuclear weapons, carried through on threats to mount a terrorist campaign against the United States and its allies, or adopted a scorched earth policy by destroying Kuwait's oil fields, should steps be taken to replace his regime.[62] In contrast, US-CENTCOM's preparations for military action both on the ground and in the air—plans approved at the highest levels—envisioned a much more ambitious agenda that included not only the restoration of Kuwait's sovereignty but also the de facto disarmament of Iraq and the annihilation of Saddam's most formidable military forces, the Republican Guard. Under the air campaign, U.S. forces planned to "fragment and disrupt Iraqi political and military leadership," a goal sometimes described as "decapitating" the Iraqi government. In short, there would be no holding back. If the opportunity presented itself, Schwarzkopf and his field commanders had tacit authority to go all the way and eradicate Saddam's regime.[63]

LIBERATING KUWAIT: THE AIR WAR

Operation *Desert Storm* commenced during the early hours of January 17, 1991, with an attack by Army Apache helicopters against enemy radar installations in western Iraq. As the Iraqi installations burned, more than one hundred coalition fighter-bombers swept through the "hole" in the enemy radar fence bound for various targets across the country. Almost simultaneously, a squadron of Air Force Stealth F–117s using precision-guided bombs struck key command, control, and communications nodes in Baghdad, while British Tornados bombed key airfields with special munitions designed to incapacitate the runways. There soon followed additional attacks from conventional air-launched cruise missiles (CALCMs) delivered by B–52s based in the United States and Tomahawk missiles fired from Navy vessels in the Persian Gulf and Red Sea. All in all, it was a dazzling display of joint and combined airpower and the most closely coordinated operation of its kind in history. Five hours into the air campaign, a voice identified as Saddam Hussein's declared over state radio: "The great duel, the mother of all battles, has begun."

Coalition air and missile strikes continued with only occasional let-up until the cessation of hostilities on February 28, 1991. Though a few Iraqi jets made it into the air to offer a challenge, most stayed on the ground. Some pilots flew their planes to sanctuary in neighboring Iran. Initially, the bombing campaign adhered closely to the targeting and phased sequence of attacks as recommended by Warden's Checkmate organization and as subsequently modified by Glosson's Special Planning Group. Directed against 12 separate target sets, the intended goals of the air campaign were to assure coalition forces' air superiority, cripple Saddam's political and military command and control, disrupt essential industries and public services, isolate Iraqi forces in Kuwait and eventually defeat them, and deny Iraq the wherewithal to carry out future aggression or to pose a threat with nuclear, biological, or chemical weapons. In pursuit of those objectives, coalition forces flew nearly 65,000 combat sorties during the war, with 75 percent of them directed against Iraqi forces in the KTO.[64]

Shortly after the air war began, planners came under unexpected political pressure to amend their objectives. The day after the air campaign commenced, Saddam made good on a threat to launch Scud missiles armed with high-explosive warheads against Israel. Six hit Tel Aviv and two landed on Haifa, doing little physical damage but having immense psychological impact.[65] Since the onset of the crisis, the Bush administration did everything it could to dissuade the Israelis from becoming involved and now faced the prospect of Israeli retaliation unless U.S. forces took out the Scuds. With an effective range of only 500 miles, a relatively small warhead (between 200 and 500 pounds), and limited accuracy, the Scud missile, in Horner's opinion, was "militarily insignificant." Only if the Iraqis armed their Scuds with chemical or biological agents did Horner or other military planners see a serious danger. Weighing one thing against another, CENTAF planners downplayed the Scud threat. After destroying the fixed sites targeted at the outset of the bombing campaign, they looked to the Army's Patriot missile defense system to cope with the problem.[66]

Following the attacks on Israel, however, Schwarzkopf and Horner came under mounting pressure from Washington to divert more air assets than they had intended to neutralize the Scuds. Intelligence was sketchy and proved to be on the low end, but as a working estimate planners assumed an Iraqi arsenal of 600 Scud missiles (and variants), 36 mobile launchers, and 28 fixed launchers in 5 complexes in western Iraq.[67] The mobile systems proved the most vexing. Out of roughly 2,000 sorties per day during the early stages of the air campaign, Schwarzkopf estimated that US-CENTCOM and its allies diverted approximately a third of their assets to the mobile "Scud hunt," largely to no avail other than to placate the Israelis.[68] Meanwhile, NATO reassigned four Patriot antimissile batteries to Israel, while the Joint Chiefs established a special planning cell within the U.S. Embassy in Tel Aviv, headed by a senior Joint Staff intelligence officer, to coordinate with the Israelis.[69] As a rule, Schwarzkopf had a low professional opinion of special operations forces and used them sparingly. But to help get the air campaign back on track, he called in the Joint Special Operations Command (JSOC), which deployed a 400-man unit to western Iraq in late January 1991. Joined by British commandos, the JSOC teams scoured the Iraqi desert for mobile Scuds and claimed a dozen "kills," though none were confirmed.[70]

According to after-action reports, the hunt for the elusive Scuds caused preplanned attacks against some targets to be postponed but did not significantly degrade the effectiveness of the air campaign. Equally if not more detrimental to the air war was a weather front that stalled over Iraq on the third day of the conflict, disrupting operations for the next 3 days and resulting in the cancelation of some attacks. But by the tenth day of the offensive (D+10), the coalition had achieved undisputed air superiority over Iraq, permitting operations at high and medium altitudes with "virtual impunity." From that point on, coalition aircraft went about their tasks with systematic thoroughness.[71]

After the war, the air campaign's role in Iraq's defeat became a hotly debated issue. For those who had been around long enough, it conjured up memories of the contentious strategic bombing controversy after World War II (see chapter 3). Most assessments gave the air campaign mixed marks. On the plus side, it was without doubt a striking success in demonstrating the capabilities of new technologies (especially Stealth fighter-bombers and precision-guided munitions) in crippling Iraq's communications and war-supporting infrastructure. But it was less effective in undermining Saddam's leadership and eliminating the residual capabilities of his armed forces. Intelligence on Saddam's chemical, biological, and nuclear programs proved so poor that many key installations that were carefully hidden remained untouched. While air bombardment destroyed thousands of Iraqi tanks and other vehicles, about half of the losses occurred during the Iraqi Army's headlong retreat in the face of advancing coalition ground forces. The goal of a 50 percent reduction in effective Iraqi military strength through airpower prior to launching the ground war was never achieved.[72] A large part of the explanation for the air campaign's shortcomings was the brief duration of the war. Hence, even in areas where airpower achieved all of its objectives, it still fell below expectations. "It was prudent to have done so," observed the authors of the Gulf War Airpower Survey, "but attacking oil refineries and storage in Iraq bore no significant military results due to the swift collapse of the Iraqi Army." The same was essentially true of strategic attacks against Iraq's electrical power grid and other public services.[73]

Yet without the air war, the liberation of Kuwait doubtless would have taken far longer at far greater cost. Assured by their superiors that the air campaign would last no more than a week, many Iraqi units found the month-long bombing intolerable and surrendered at the first opportunity when the ground campaign began.[74] As an exercise in jointness, the air war was probably the most successful and effective single part of the campaign. Air Force planners played the leading role in orchestrating the air war and in overseeing its execution. The Air Force also provided more planes than any other Service and flew the largest number of sorties—three and half times more than the Navy and over 60 percent of the total for the conflict.[75] As the dominant Service in the air war, the Air Force tended to impose its judgments and values on the other Services and coalition partners. Friction, especially with the Navy, became virtually inevitable. But by the same token, there was a predisposition on the part of all involved to compromise and cooperate as the need arose. In a very real sense, there was no other choice. Mounting the air campaign was the most complex and technically demanding aspect of the Gulf War. It created an operational environment in which success was directly dependent on effective joint collaboration.

PHASE IV: THE GROUND CAMPAIGN

While the United States and its allies achieved air superiority against Iraq with relatively little difficulty, indications were that they would have a much tougher time overcoming resistance on the ground in Phase IV. Evicting an estimated half million Iraqi troops from Kuwait, many of them heavily dug in and experienced in trench warfare from years of conflict with Iran, was a daunting prospect. More ominous was the possibility that Saddam might employ chemical or biological weapons against advancing coalition forces. Assessments, both official and unofficial, ranged from a few hundred to tens of thousands of American casualties. Preparing for the worst, USCENTCOM's medical staff expected as many as 20,000 U.S. killed and wounded.[76] Though Scowcroft, McPeak, and a few others considered these estimates of Iraqi capabilities exaggerated, most policymakers and planners were too cautious not to take them seriously; hence the willingness of Bush and Cheney to follow Powell's advice and expedite a massive buildup of land armies.

In pushing for the buildup, Powell's purpose had been twofold: to intimidate Saddam into capitulating without a fight or, failing that, to apply overwhelming force that would crush Iraqi resistance with as few losses as possible to the United States and its allies. The air war was the critical first step, but under the strategy embraced by Powell and Schwarzkopf it was never an end in itself. Though both lauded the role of airpower, neither saw it as decisive. As in Panama, they expected the fate of Kuwait to be decided on the ground.

Thinking along similar lines, Saddam was confident that his forces could ride out an air bombardment and effectively resist a ground assault.[77] Drawing on his experience in positional warfare against Iran, Saddam created a layered defense with elaborate trenches, sand berms, and mine fields to slow the attackers' advance and inflict heavy casualties. Bolstering his strategic reserve, he quietly began removing his Republican Guard divisions from Kuwait in September 1990 and redeployed them to rear echelon positions. Regular army infantry replaced them. Time and again during the war with Iran, Saddam used similar battlefield tactics. Once the thrust of the attacker's offensive was apparent and had been reduced by the forward units, the reserve force made up of Republican Guard divisions would move in for the kill and destroy the enemy. A successful strategy against the limited capabilities of the Iranians, it proved considerably less effective against the coalition's relentless air bombardment, heavy armor, mechanized artillery, and other sophisticated weapons.[78]

Coalition ground forces had limited contact with the opposing Iraqis prior to launching their main offensive in late February 1991. Up to then, the largest and most intense engagement was the battle of Khafji, a coastal Saudi town just south of the Kuwaiti border. Believing that the coalition was massing its forces there for a thrust up the coast, Saddam ordered a division-sized preemptive attack against Khafji on January 29, 1991. Heavy fighting raged for two days. In the end both sides claimed victory—the Iraqis for having requited themselves reasonably well in the face of overwhelmingly stronger opposition and the coalition for inflicting heavy losses on the invaders and driving them back to their lines using intense air, artillery, and naval bombardment. Militarily, the battle had little impact on the course of the war. But it did much to bolster the morale of Saudi forces who had taken part in the fighting and convinced Schwarzkopf that Iraqi combat skills were overrated.[79]

By the time the main attack to liberate Kuwait commenced on February 23–24, Schwarzkopf had at his disposal one of the most impressive arrays of conventional firepower ever assembled including all the best of the Reagan buildup, from the planes, helicopters, and missiles flying overhead, to the tanks and armored personnel carriers on the ground, to the ships offshore. Since the Iraqis were armed largely with Soviet tanks and other Eastern bloc weapons, some in the press likened *Desert Storm* to a Cold War proxy conflict. In line with the Bush administration's pending base force reorganization plan, many U.S. units and their equipment were slated for immediate demobilization once Kuwait was liberated. *Desert Storm* was to be their last hurrah.

Under the weight of this awesome force, Iraqi resistance crumbled faster than anyone expected and the fighting was over in 100 hours. Some Iraqi units held their ground and offered credible resistance, but many gave up quickly and surrendered or deserted the battlefield. It turned out that allied intelligence had consistently overestimated the size and capabilities of the Iraqi Army, so when the showdown came it was almost anticlimactic. Instead of the half million or more Iraqi troops in Kuwait as originally believed, there were probably between 200,000 and 220,000. Prewar intelligence also credited the Iraqis with 800 more tanks and 600 more artillery pieces than they had.[80] Enemy casualties were likewise far fewer than the 10,000 that were widely reported. A revisionist account, intentionally aimed at deflating such claims, asserted that there were as few as 4,500 Iraqi military losses during both the air and ground wars. This conjecture, based on selective anecdotal evidence, is probably too low. But remembering the unfavorable publicity and sordid controversy arising from McNamara's enemy "body count" in Vietnam, Powell suppressed the issuance of official figures on Iraqi losses.[81]

Like the air war, the ground campaign fell short of achieving some of its key objectives due in large part to its relatively brief duration. The greatest disappointment was the coalition's failure to destroy the Republican Guard, one of the cornerstones of Saddam's political and military power. Eliminating the Guard as an effective fighting force was a declared objective in NSD 54 and was the responsibility of the all-mechanized VII Corps commanded by Lieutenant General Frederick M. Franks, Jr., USA, which spearheaded the main assault. Brought in on short notice from Germany, VII Corps was organized, trained, and equipped to operate against the Warsaw Pact along a fairly static front in Central Europe and did not have much time to acclimate itself to the faster pace of desert warfare. "I do not want a slow, ponderous pachyderm mentality," Schwarzkopf declared. "I want VII Corps to *slam* into the Republic Guard."[82] Though Franks did what he could to pick up the tempo, it was still not fast enough to suit the USCINCCENT. Ultimately, in combination with ongoing air attacks, VII Corps inflicted heavy equipment losses on some of the Republican Guard's best units, including the elite Medina, Hammurabi, and Tawakalna divisions. Franks declared it "a victory of staggering battlefield dimensions."[83] Confirming Franks' assessment, Powell told President Bush that, based on initial reports, U.S. forces were "crucifying" the enemy.[84] Later, however, Powell learned that much of the Republican Guard never committed to battle and that three divisions escaped essentially intact to the safety of fallback positions near the Iraqi city of Basra.[85]

Failure to destroy the Republican Guard meant that Saddam remained a credible and dangerous military power. As a result, instead of a prompt withdrawal from the Persian Gulf, the United States became entangled for more than a decade in a low-intensity conflict using air and naval power to contain Saddam's rogue regime and police the region. While toppling Saddam was never an overt objective of *Desert Storm* (indeed, some Islamic governments would never have joined the coalition if it was), it was always one of the Bush administration's preferred outcomes. An elusive goal, it would continue to haunt American foreign policy until the combined U.S.-British invasion of Iraq in 2003 finally brought down Saddam's government.

THE POST-HOSTILITIES PHASE

On March 3, 1991, Schwarzkopf and senior officers of the U.S.-led coalition met with Iraqi generals at Safwan airfield just inside Iraq to conclude a ceasefire. Looking back, Bush and Scowcroft acknowledged that they agreed to halt the war based on mistaken information that the Republican Guard had been largely destroyed and that air strikes had rendered Saddam's WMD research and production facilities inoperable. By the time they learned otherwise, it was too late to reconsider. Saddam's politico-military base of power remained secure. Still, they insisted that they had done the right thing by bringing the war to a prompt conclusion. The Bush administration had achieved its declared aim of evicting the Iraqis from Kuwait, but as the fighting subsided it faced an unexpected backlash of "bad press" arising from reports of civilian casualties, televised bomb damage in Baghdad, and pictures of destroyed enemy tanks and assorted vehicles along the "highway of death" out of Kuwait City. President Bush wanted the United States to emerge from the war with improved relations and a favorable image in the Arab world, and it served his purposes better to limit further carnage.[86]

After the war, there was much second-guessing that by ending the conflict too soon the United States and its partners had passed up the opportunity to topple Saddam. Army planners attached to USCENTCOM had in fact sketched out a plan for a march on Baghdad if the opportunity arose. But the concept they proposed lacked defined objectives and assumed that the mere presence of U.S. forces nearing the city would be enough to compel Saddam to capitulate and step down. How U.S. forces would respond if Saddam refused was unclear. Not surprisingly, the plan received a cool reception followed by a curt rejection at Schwarzkopf's headquarters.[87] Weighed against *Desert Storm*'s initial accomplishments, moreover, U.S. and coalition casualties were incredibly light, and no one was eager to incur more. While some in the Air Force would have preferred additional time to test their theories about the role and impact of airpower, most were satisfied that they made large strides toward proving their case. With enemy resistance collapsing on all fronts, Powell and Schwarzkopf concurred that the Iraqi Army was a spent force and that a ceasefire would be in the interest of all concerned.[88]

Compared to the meticulous planning that went into the military preparations for the Gulf War, planning for the postwar period was sketchy and haphazard. According to

Charles W. Freeman, Jr., the U.S. Ambassador to Saudi Arabia, the Bush administration downplayed long-term political planning lest leaks "unhinge the huge and unwieldy coalition" the United States had so painstakingly put together to fight the war.[89] As a result, preparing for the postwar period was not a high priority on anyone's agenda. Still, to some extent it was unavoidable. Undertaken on a close-hold basis, postwar planning became largely an interagency distillation of views by the NSC Deputies Committee, where the Vice Chairman, Admiral David E. Jeremiah, represented the JCS.

In early February 1991, while testifying on Capitol Hill, Secretary of State Baker presented the gist of the deputies' deliberations to that point. One proposal under active consideration was to create a permanent Arab peacekeeping force backed by an increased U.S. naval presence in the Persian Gulf. During preliminary discussions of this and other issues affecting postwar security arrangements, Joint Staff (J-5) planners opposed an increased U.S. military presence in Southwest Asia on the grounds that it would divert resources from other missions and go against promises the United States made to the Saudis and other Arab governments that Western forces would promptly withdraw from the region once Kuwait was liberated. As the deputies' deliberations progressed, however, a consensus emerged that there was no alternative other than for the United States to assume a larger, more active postwar role in Gulf affairs. While the UN was likely to have overall responsibility, the United States, operating through USCENTCOM, had the only reliable organization in place with the necessary resources to police the region, assure the delivery of humanitarian aid to refugees displaced by the war, and assist Kuwait with its reconstruction. The deputies agreed that to the extent feasible the U.S. presence should be discrete and inconspicuous. For planning purposes, they were looking at the prepositioning of supplies and equipment for several Army brigades that could be quickly airlifted to the Middle East in case of renewed trouble, the permanent stationing of an Air Force tactical fighter wing somewhere in the Persian Gulf, additional units of Marines afloat offshore at all times, and an unspecified increase in naval forces with more frequent carrier visits to the region.[90]

The rest of Baker's plan traversed familiar ground and envisioned regional arms control agreements to curb the proliferation of conventional arms and prevent Iraq from reviving its WMD capability, a program of regional economic development, renewed energy conservation to lessen U.S. dependence on Middle East oil, and last but not least a revived peace process between Israel and the Palestinians. The formidable agenda looked good on paper. But as the Secretary of State acknowledged, the plan was still very tentative. To succeed it would need the full cooperation of all involved, including the Iraqis. Hardest of all would be a revived Arab-Israeli peace process that did not include substantial prior concessions from Israel.[91]

Efforts by the Deputies Committee to clarify a postwar strategy for the Middle East were still underway when the coalition's senior military officers met with their Iraqi counterparts in early March to sign the Safwan ceasefire accords. Some of Cheney's aides wanted the ceasefire to include tough restrictions on Iraq's military capabilities and full Iraqi disclosure of all WMD research sites. But the Joint Staff saw no need for such detail and argued successfully that specific guidance would only complicate Schwarzkopf's mission of negotiating an effective truce.[92] Modeled on a recently adopted UN Security Council resolution (S/RES 686), the ceasefire imposed limited constraints on enemy forces and left Iraq's military establishment essentially intact. Toward the end of the Safwan meeting, the Iraqis requested permission to use helicopters, which they insisted were essential for communication purposes owing to the damage coalition bombing had caused to ground transportation systems. Schwarzkopf was without instructions on the matter and, treating it as a reasonable request, agreed. He soon regretted his decision.[93] Shortly after the truce, Iraqi armed forces began using helicopter gunships to help suppress rebellions that had broken out among dissident Shiites in southern Iraq and Kurds in the north. Press accounts exaggerated the role the helicopters actually played, but the impression in the West was that the coalition had seriously blundered by not banning them.

Thus, despite the setback in Kuwait, Saddam remained as defiant and dangerous as ever and a source of continuing tensions in the Persian Gulf. All the same, the most lasting impression from the Gulf War was that it was a stupendous military triumph for the United States. Shaking off the stigma of Vietnam, U.S. forces had put on an awesome display of military power that achieved stated objectives with stunning efficiency and effectiveness. The Powell Doctrine of applying overwhelming force against the enemy had again prevailed, probably saving countless American lives. Not since World War II had the American public's confidence in the military and its leadership been so high. Much of the adulation fell on Schwarzkopf, whose gruff, no-nonsense manner, and commanding bearing made him an instant celebrity. Yet others basked in approbation as well. Indeed, for the vast majority of the Persian Gulf veterans it was an exhilarating experience as they returned home to tickertape parades and open-arm welcomes, honors that had eluded Vietnam veterans.

An unintended side effect of the Gulf War was the impetus it gave to reassessing the nature of future conflicts. In orchestrating such a lopsided victory, American planners exploited the latest military technologies to the fullest and in so doing made the defeat of Saddam's forces look easy—maybe too easy. Underlying the American success was a phenomenon known as the revolution in military affairs (RMA), which helped give the United States swift military dominance over Iraq. Dating from theoretical studies initiated in the 1970s, RMA stressed the interaction of new forms of communications, improved battlefield management techniques, and the application of "smart" weaponry to gain superiority over the enemy. As the "lessons" of RMA's application in the Gulf War emerged, the notion took hold in some circles that future wars could be short, precise, and relatively painless. No member of the JCS, least of all General Powell who had done as much as anyone to engineer the victory, seriously believed that. But as the conflict ended, it seemed that a new era in warfare might be near at hand.

A further result of the war was the growing recognition that "jointness" had been an integral part of the victory. Iraq's defeat had come about not merely by superior force of arms but through carefully coordinated planning and the joint application of military power. While Service planning staffs played key roles at the outset of the crisis in shaping both the air and ground campaigns, and while the conflict had not always gone as scripted (especially during the ground war phase), it was clear by the war's end that joint direction and control had a major impact in shaping the outcome. Indeed, in Powell's estimation, the Gulf War saw the "triumph of joint operational art."[94] That jointness would be a prerequisite to the success of future military operations as resources continued to contract was almost certain. Implementation of the Goldwater-Nichols Act may not yet have been in full stride as its authors intended, but things were moving inexorably in that direction.

NOTES

1. NSR 10, "US Policy Toward the Persian Gulf," February 22, 1989 (declassified), Bush Presidential Records, available at <http://bushlibrary.tamu.edu/research/pdfs/nsr/nsr10.pdf>.

2. NSD 26, October 2, 1989, "US Policy Toward the Persian Gulf," U, Bush Presidential Records, available at <http://bushlibrary.tamu.edu/research/pdfs/nsd/nsd26.pdf>.

3. Yevgeny Primakov, *Russia and the Arabs*, trans. Paul Gould (New York: Basic Books, 2009), 314–315.

4. Powell mentions the military's apprehension in his Oral History, PBS "Frontline," aired January 9, 1996, transcript, 3–4, available at <www.pbs.org/wgbh/pages/frontline/gulf/oral/decision.html>.

5. NSD 26, October 2, 1989, loc. cit.; James A. Baker III, *The Politics of Diplomacy: Revolution, War and Peace, 1989–1992* (New York: G.P. Putnam's Sons, 1995), 266–267.

6. Powell Oral History, PBS "Frontline," transcript, 3.

7. Baker, *Politics of Diplomacy*, 272.

8. "Remarks and an Exchange with Reporters on the Iraqi Invasion of Kuwait," August 5, 1990, Bush Public Papers, 1990, 1102.

9. Quoted in Matthew M. Aid, *The Secret Sentry: The Untold History of the National Security Agency* (New York: Bloomsbury Press, 2009), 192.

10. Robert H. Scales, Jr., *Certain Victory: United States Army in the Gulf War* (Washington, DC: Office of the Chief of Staff, United States Army, 1993), 43.

11. Alexander S. Cochran et al., *Gulf War Airpower Survey*, Vol. I., Part I, *Planning* (Washington, DC: GPO, 1993), 203–207 (series hereafter cited as *GWAPS*); and U.S. Department of Defense (DOD), *Conduct of the Persian Gulf War: Final Report to Congress* (Washington, DC: GPO, 1992), 9–16.

12. Quoted in H. Norman Schwarzkopf, with Peter Petre, *It Doesn't Take a Hero* (New York: Bantam Books, 1993), 344.

13. Michael R. Gordon and Bernard E. Trainor, *The Generals' War: The Inside Story of the Conflict in the Gulf* (Boston: Little, Brown, 1995), 130–131.

14. George Bush and Brent Scowcroft, *A World Transformed* (New York: Knopf, 1998), 354, 375.

15. Dick Cheney, with Liz Cheney, *In My Time: A Personal and Political Memoir* (New York: Threshold Editions, 2011), 185.

16. Colin L. Powell, *American Journey* (New York: Random House, 1995), 465–466.

17. DOD, 65.

18. Richard T. Reynolds, *Heart of the Storm: The Genesis of the Air Campaign Against Iraq* (Maxwell AFB, AL: Air University Press, 1995), 16.

19. Powell, *American Journey*, 479, 487.

20. Baker, *Politics of Diplomacy*, 288.

21. Bernard E. Trainor, "Jointness, Service Culture, and the Gulf War," *Joint Force Quarterly* (Winter 1993–94): 72.

22. Richard G. Davis, *On Target: Organizing and Executing the Strategic Air Campaign Against Iraq* (Washington, DC: Air Force History and Museums Program, 2002), 3–9.

23. Diane T. Putney, *Airpower Advantage: Planning the Gulf War Air Campaign, 1989–1991* (Washington, DC: Air Force History and Museums Program, 2004), 24.

24. See Powell's testimony of December 3, 1990, in U.S. Congress, Senate, Committee on Armed Services, *Hearings: Crisis in the Persian Gulf Region—U.S. Policy Options and Implications*, 101:2 (Washington, DC: GPO, 1990), 662–663; and Putney, 263–264.

25. Powell, *American Journey*, 476–477.

26. John A. Warden III, *The Air Campaign: Planning for Combat* (Washington, DC: National Defense University Press, 1988), 169. Also see Richard P. Hallion, *Storm over Iraq: Airpower and the Gulf War* (Washington, DC: Smithsonian Institution Press, 1992), 116–117.

27. Thomas A. Keaney and Eliot A. Cohen, *Gulf War Airpower Survey: Summary Report* (Washington, DC: GPO, 1993), 36–37; Richard G. Davis, "Strategic Bombardment in the Gulf War," in R. Cargill Hall, ed., *Case Studies in Strategic Bombardment* (Washington, DC: Air Force History and Museums Program, 1998), 546–547.

28. *GWAPS*, I, Pt. I, 114; DOD, 65.

29. Edward J. Marolda and Robert J. Schneller, Jr., *Shield and Sword: The United States Navy and the Persian Gulf War* (Washington, DC: Naval Historical Center, 1998), 184.

30. Putney, 344.

31. *GWAPS*, I, Pt. I, 125–126, and Pt. II, 158.

32 Davis, "Strategic Bombardment in the Gulf War," 545.

33 Bob Woodward, *The Commanders* (New York: Simon & Schuster, 1991), 303.

34 Powell quotations from Putney, 221.

35 Henry Rowen quoted in Gordon and Trainor, 144.

36 Frank N. Schubert and Theresa L. Kraus, eds., *The Whirlwind War: The United States Army in Operations DESERT SHIELD and DESERT STORM* (Washington, DC: Center of Military History, 1995), 107.

37 Bush and Scowcroft, 381; Cheney Oral History Interview, no date, PBS "Frontline," transcript, 4, available at <www.pbs.org/wgbh/pages/frontline/gulf/oral/cheney>.

38 Bush and Scowcroft, 381.

39 Cheney, "Frontline" Oral History, 5; Putney, 228; Gordon and Trainor, 144–145. One of Vesser's tasks was to evaluate the so-called "Western Excursion," a proposal developed in OSD to occupy western Iraq and from there to launch or threaten an attack on Baghdad. Arguing that it was logistically unsupportable, USCENTCOM strenuously opposed the plan and it eventually died. Still, it was a lingering source of friction between OSD and USCENTCOM. See Schwarzkopf, 428–429.

40 DOD, 66; also see Schwarzkopf, 419.

41 Scales, *Certain Victory*, 131–133; Powell quotes from Schwarzkopf, 426–427.

42 Rick Atkinson, *Crusade: The Untold Story of the Persian Gulf War* (Boston: Houghton, Mifflin, 1993), 123–124; Woodward, 313 314.

43 Bush and Scowcroft, 431.

44 Powell Oral History, PBS "Frontline," transcript, 2; see also Powell, *American Journey*, 488–489.

45 Schubert and Kraus, 107–110; Bush and Scowcroft, 393–395.

46 "President's News Conference on the Persian Gulf Crisis," November 8, 1990, Bush Public Papers, 1990, 1580–1581.

47 See Powell, *American Journey*, 419–20.

48 DOD, 42–45; Schubert and Kraus, 130.

49 USCENTCOM also included a fifth combat component command for special operations, but lacking a full-blown organization it functioned in a limited capacity and had no operational role. A separate organization, the U.S. Joint Special Operations Command (JSOC), carried out special operations during *Desert Storm*.

50 Davis, "Strategic Bombardment in the Gulf War," 545.

51 DOD, 101–102.

52 James A. Winnefeld and Dana J. Johnson, *Joint Air Operations: Pursuit of Unity in Command and Control, 1942–1991* (Annapolis: Naval Institute Press and RAND Corp., 1993), 126.

53 See Marolda and Schneller, 183–190.

54 Putney, 175, 274–294 and passim; Gordon and Trainor, 311–312.

55 Atkinson, 67–68; John R. Ballard, *From Storm to Freedom: America's Long War with Iraq* (Annapolis: Naval Institute Press, 2010), 52–53; and Trainor, "Jointness, Service Culture, and the Gulf War," 73.

56 Powell, *American Journey*, 503.

57 See James K. Matthews and Cora J. Holt, *So Many, So Much, So Far, So Fast: United States Transportation Command and Strategic Deployment for Operation Desert Shield/ Desert Storm* (Washington, DC: Joint History Office and Research Center, U.S. Transportation Command, 1995), 167–169, 174–175.

58 Gordon and Trainor, 186–190; Putney, 262–263.

59 Putney, 305–309.

60 DOD, 230–231, 243.

61 Davis, *On Target*, 161; Bush and Scowcroft, 431–432; DOD, 70.

62 NSD 54, "Responding to Iraqi Aggression in the Gulf," January 15, 1991, Bush Presidential Records, NSC Collection; and <http://bushlibrary.tamu.edu/research/pdfs/nsd/nsd54.pdf> (accessed July 19, 1911).

63 DOD, 95–96.

64 Davis, "Strategic Bombardment in the Gulf War," 528; DOD, 253.

65 During the war, Saddam also launched Scud attacks against coalition positions in Saudi Arabia; 39 Scuds struck Israel and 44 hit Saudi Arabia. The most deadly attack came on February 25, 1991, when a Scud landed on barracks in Dhahran killing 25 U.S. military personnel and injuring another 100.

66 Atkinson, 85–90; Putney, 267–270. Developed originally as an antiaircraft missile, the Patriot was upgraded in 1988 to provide a limited capability against tactical ballistic missiles. The Gulf War was its first practical test.

67 DOD, 97.

68 Schwarzkopf, 486. Coalition air crews reported destroying around 80 mobile Scud launchers, nearly all of which were later found to have been decoys. A few actual launchers may have been destroyed but probably not more than a dozen. See *GWAPS Summary Report*, 83–90.

69 DOD, 168.

70 Gordon and Trainor, 244–246.

71 DOD, 124–129.

72 *GWAPS*, II, Pt. 2, 202–220.

73 *GWAPS Summary Report*, 77 (quote) and passim.

74 See Perry D. Jamieson, *Lucrative Targets: The U.S. Air Force in the Kuwaiti Theater of Operations* (Washington, DC: Air Force History and Museums Program, 2001), 144–145 and passim.

75 Table 64, "Total Sorties by U.S. Service/Allied Country by Mission Type," *GWAPS*, V, 232.

76 Woodward, 349.

77 Aid, 193.

78 DOD, 83–84.

79 Schwarzkopf, 496.

80 *GWAPS*, II, Pt. 2, 218–220.

81 See John G. Heidenrich, "The Gulf War: How Many Iraqis Died?" *Foreign Policy* 90 (Spring 1993): 108–125.

82 Schwarzkopf, 502. Emphasis in original.

83   Tom Clancy with Fred Franks, Jr., *Into the Storm: A Study in Command* (New York: G.P. Putnam's Sons, 1997), 447.

84   Powell quoted in Atkinson, 469.

85   Scales, *Certain Victory*, 300–301, 314–315; DOD, 281; Ballard, 74.

86   Gordon and Trainor, 412; Bush and Scowcroft, 488–489.

87   Gordon and Trainor, 452–454.

88   Schwarzkopf, 542–543; Powell, 519–525. Shortly after the war, in a televised interview, Schwarzkopf changed his mind and indicated that he would have preferred to continue fighting a few more days but neglected to mention what specific objectives he had in mind.

89   Freeman interviewed March 31, 1998 and quoted in Andrew Cockburn and Patrick Cockburn, *Out of the Ashes: The Resurrection of Saddam Hussein* (New York: HarperCollins, 1999), 33.

90   Baker testimony, February 6, 1991, U.S. Congress, House, Committee on Foreign Affairs, *Hearings: Foreign Assistance Legislation for Fiscal Years 1992–93*, 102:1 (Washington, DC: GPO, 1991), Pt. 1, 6–7 and passim; Baker, *Politics of Diplomacy*, 412–413.

91   Baker, *Politics of Diplomacy*, 412–414.

92   Gordon and Trainor, 444.

93   Schwarzkopf, 566–567.

94   Gordon and Trainor, 464.

Chapter 18

CONCLUSION

Like the defeat of Germany and Japan in World War II, the victory over Iraq in 1991 proved to be a watershed in the history of American military policy and strategy. The biggest military operation mounted by the United States since the Vietnam War, *Desert Shield/Desert Storm* was also exceedingly complex and difficult to execute. One of the keys to its success was the coordinated planning and direction provided by the Chairman of the Joint Chiefs, General Colin Powell, and the officers of the Joint Staff, working in collaboration with the military Services; the theater commander, General Norman Schwarzkopf; and the allies who made up the anti-Iraq coalition. The result was not only an awesome display of American-led military power, but also a reaffirmation that joint planning and joint direction of components in the field were increasingly essential to success in modern warfare.

What may seem to have been a relatively easy victory was far from preordained. Rather, it was the product of a long and complicated process, with antecedents reaching back to the creation of the Joint Chiefs of Staff in World War II. Established in January 1942 to expedite wartime planning and strategic coordination with the British, the Joint Chiefs operated initially under the direct authority and supervision of the President, performing whatever duties he assigned in his capacity as Commander in Chief. After the war, as part of the 1947 reorganization of the Armed Services under the National Security Act, the JCS acquired statutory standing with a list of assigned duties and became a corporate advisory body to the President, the Secretary of Defense, and the National Security Council. The corporate nature of the Joint Chiefs' advisory role ended upon passage of the 1986 Goldwater-Nichols Act, which transferred the tasks and duties previously performed collectively by the JCS to the Chairman. But in retaining the Joint Chiefs of Staff as an organized entity, the new law affirmed that they should continue to hold "regular" meetings and act as "military advisors" to the Chairman.

Prior to Goldwater-Nichols, the role, influence, and reputation of the Joint Chiefs of Staff waxed and waned. World War II undoubtedly marked the high-water mark of JCS authority and influence. Operating without a formal charter, they exercised a virtual monopoly on national security, oversaw the formulation of strategy, maintained essential military liaison with America's allies, and provided general direction for a broad array of key war-related activities. Despite their wide-ranging mandate, however, the JCS never became a fully functioning general staff. The greatest weakness of the JCS system, then as later, was its composition as a committee of coequal Service chiefs. Expected in theory to rise above their individual concerns, they were all too susceptible to inter-Service pressures and rivalries, a legacy of the separateness between the Services in years past and a harbinger of things to come. With the Army focused on the war in Europe and the Navy concentrating on the Pacific, two sets of interests invariably competed for manpower and industrial production, resulting in disagreements over strategy and the allocation of resources. With the emergence of the Army Air Forces as a de facto separate Service, reaching consensus decisions became even more difficult. Fortunately, the level-headed influence of Admiral William D. Leahy, Chief of Staff to the Commander in Chief, and President Franklin D. Roosevelt's imposing presence prevented these quarrels and rivalries from getting out of hand. Yet given the personalities involved and the entrenched institutional interests each JCS member represented, it was remarkable that they accomplished what they did.

As World War II drew to a close, the role and functions of the Joint Chiefs began to change. In addition to their planning and advisory duties, the JCS acquired oversight responsibilities for the various unified and specified commands that sprang from the 1946 Unified Command Plan (UCP). An extension of the World War II practice of creating "supreme commands," the UCP affirmed that joint planning and joint operational control should go hand in hand. However, the most far-reaching changes affecting the chiefs' functions were those arising from the postwar debate over unification of the Armed Services. Embracing a War Department proposal, President Harry S. Truman sought to abolish the JCS and replace them with a single chief of staff and a closely unified structure overseen by a civilian secretary of defense. Opponents of unification, led by the Navy, championed a less centralized system. Arguing the need for improved coordination in lieu of outright unification, they opposed the single chief of staff concept and urged a loosely knit committee-style system that included preserving the JCS more or less unchanged. The ensuing compromise under the National Security Act of 1947 leaned toward the Navy's model and kept the JCS intact, subject to the direction, authority, and control of the Secretary of Defense. The Joint Staff, which had been an integral part of JCS operations during the war, also acquired statutory standing, but with a ceiling of only one hundred officers it was a mere shadow of its former self and was soon swamped with more work than it could handle.

The next few years were a period of painful adjustment for the Joint Chiefs of Staff. Promising "evolution, not revolution" to ease the transition, the first Secretary

CONCLUSION

of Defense, James Forrestal, took a go-slow approach to unification. Seeing himself as a coordinator, he looked to the Joint Chiefs for much-needed assistance and leadership in keeping the Services in line and in recommending the most effective and efficient allocation of resources. A daunting task, it tested the chiefs' patience with one another practically to the breaking point. Despite the menacing behavior of Moscow and several "war scares," the chiefs were often at odds over the assignment of Service functions and the choice of weapons and strategy for coping with the Soviet threat. As more and more of the disputes became public, they left the JCS with a tarnished image and a growing reputation as a committee of quarrelsome military bureaucrats intent on protecting vested interests at the expense of the common good.

Whether a more centralized system with stronger authority at the top could have avoided these early difficulties is open to question. While it might have helped, it would not have solved the underlying problem—a fundamental difference of opinion within the defense establishment on how to arm, train, and prepare for future wars. New technologies—the atomic bomb premier among them—and rapid advances in aviation, missiles, electronics, and other fields created fresh opportunities for the Services and new ways of looking at military strategy. But with money in short supply, inter-Service competition and friction displaced rational discussion. Treating one another as rivals rather than partners, the Services scrambled to lay claims to military functions that would guarantee them continuous future funding.

Early efforts to improve JCS performance met with mixed success. While Congress welcomed greater military efficiency and effectiveness, it refused to tamper with the basic JCS corporate structure lest it acquire the traits of a "Prussian-style general staff." Moving cautiously, Congress agreed in 1949 to add a full-time JCS Chairman who was without Service responsibilities and to double the size of the Joint Staff. While the Chairman's powers were initially narrowly defined, his designation as the Nation's senior military officer heightened his stature and prestige well beyond his legal authority. The first JCS Chairman, General Omar Bradley, USA, was initially guarded in exercising his authority and in offering advice. But as he became more familiar with what was expected of him, Bradley realized that he had no choice and had to become more actively involved. Adopting a procedure that other Presidents would copy, Truman directed that only the CJCS attend NSC meetings on a regular basis. As a result, it became almost routine for the Secretary of Defense, the President, and the National Security Council to work directly with or through the Chairman, a practice that further enhanced his de facto role as spokesman for the military. In dealing with the Service chiefs, however, the Chairman's powers to resolve disputes remained limited. He could coax and cajole and sometimes engineer compromises, but he could not compel cooperation. To preserve JCS credibility, Chairmen often resorted to advancing their own interpretation of JCS advice rather than trying to compose differences and achieve a common view.

Meanwhile, the intensification of the Cold War, new U.S. commitments under the North Atlantic Treaty, and the emergence in the summer of 1949 of the Soviet Union as a nuclear power increased pressure on the United States to strengthen its defense posture. Driven by domestic budgetary considerations and recent breakthroughs in nuclear weapons design, the evolving U.S. strategy downplayed the role of conventional forces and stressed air-atomic retaliation by the Air Force's long-range bombers in case of Soviet aggression. Not everyone agreed that this was a sound course to follow, certainly not the Navy, which had its own competing view of national security built around a proposed fleet of flush-deck "super carriers." But as an all-around solution to the country's defense needs, the air-atomic strategy was irresistible. An intimidating threat, it was technologically feasible, commanded strong bipartisan support in Congress, and could be priced to fit practically any reasonable spending limit the White House might impose.

Following the outbreak of the Korean War, the brakes on military spending came off as the Truman administration launched a "peacetime" military buildup of unprecedented proportions. Under the guidelines laid down in NSC 68, defense planning pointed to a "year of maximum danger" in anticipation of which each Service received roughly an equal allocation of resources, an expensive but expeditious approach that allowed the JCS to go about their business amid reduced competition and rivalry. But as costs climbed and the expected showdown with the Soviets failed to materialize, attention shifted to developing a more stable defense posture for the "long haul." The process accelerated with the change of administrations in 1953. Finding the Joint Chiefs unable to agree on what should be done, President Dwight D. Eisenhower took matters into his own hands and gave defense policy a "new look." Convinced that atomic energy held the key, he developed a long-term deterrence posture resting on the threat of "massive retaliation" by the Air Force, backed by general purpose forces armed increasingly with tactical and battlefield nuclear weapons.

Eisenhower assumed that sooner or later the JCS would come around to his view that low-yield nuclear weapons represented the new "conventional" weapons and were suitable for limited warfare. Toward that end, both Admiral Arthur Radford and General Nathan Twining, the first two Chairmen he appointed, did what they could to elicit cooperation from the skeptical Service chiefs. Presented with repeated opportunities to test the President's theory during the Indochina and Formosa crises, they declined. For them as for others, crossing the nuclear threshold was becoming almost synonymous with all-out war. Since the objectives were invariably in Asia, there were awkward racial implications as well. Nonetheless, the

CONCLUSION

JCS accepted tactical nuclear weapons as an integral part of the American arsenal and encouraged NATO to follow suit as a means of offsetting the numerical Soviet advantage in conventional forces. NATO's "new approach" mirrored the new look on a lesser scale and relaxed pressures on the European allies to maintain sizable and expensive general purpose forces. But it also left NATO more dependent than ever on a nuclear response as its first line of defense, a problem that would dog the Alliance down to the dying days of the Cold War.

Despite strenuous efforts to hold down military spending, the Eisenhower administration achieved limited savings. Faced with unexpected increases in Soviet capabilities, it became involved in a steadily escalating strategic arms competition with the Soviet Union, first in long-range bombers and later in intercontinental ballistic missiles. Though the Air Force's monopoly on strategic bombers was well established, the missile field was wide open and soon produced a free-for-all competition among the Services that required direct intervention by the Secretary of Defense. Meanwhile, the Joint Chiefs continued to endorse across-the-board force proposals that exceeded available funding. Unable to overcome their "splits" and recommend an integrated statement of requirements, they eventually adopted a catch-all approach that lumped Service requirements together in no particular order of priority under the Joint Strategic Objectives Plan (renamed the Joint Strategic Planning Document in the late 1970s), which critics likened to a Christmas "wish list."

Frustrated by the disarray among his military advisors, Eisenhower sought further changes to the National Security Act aimed at improving JCS performance. Under revised legislation passed in the summer of 1958, the Chairman acquired about as much power and authority as he could reasonably exercise while still operating within the traditional JCS corporate structure and the consensus-based advisory system. At the same time, however, the new law bestowed additional responsibilities and authority on the Secretary of Defense that diminished the JCS role. From that point on, the Secretary of Defense and those around him—not the Chairman, the Service chiefs, or the Joint Staff—would be the center of military policy and decisionmaking, the galvanizing force, as it were, within the Department of Defense (DOD).

The nadir of JCS influence came during the 1960s as Secretary Robert S. McNamara took charge of the Pentagon and the Vietnam War. Given a free hand by Presidents Kennedy and Johnson to reform DOD, McNamara imposed a tight and highly sophisticated system of planning, programming, and budget management that gave the Office of the Secretary of Defense more control of the military than ever before. By the time he finished, the JCS had become a marginalized institution. Though McNamara insisted that he wanted the closest possible cooperation and collaboration with the Joint Chiefs, he did not hesitate to act unilaterally if it suited his needs or he perceived the chiefs to be dragging their heels. Pushing the doctrine of "flexible response," he made reducing military dependence on nuclear weapons his first order of business, a goal popular with some in the military and with a growing number of civilian military theorists. But it was less appealing to planners on the Joint Staff and their counterparts in Europe who had to cope with limited resources to offset overwhelming Soviet superiority in conventional forces. Extending his writ into areas previously the exclusive domain of the JCS, he challenged prevailing assumptions about strategic requirements and established new targeting criteria, limiting them mainly to the needs of a retaliatory (second-strike) "assured destruction" capability. To curb future costs and growth in nuclear forces, McNamara capped the size of the U.S. strategic offensive arsenal (a ceiling which, in terms of launchers, remained more or less intact until the end of the Cold War) and practiced unilateral restraint in the acquisition and deployment of both anti-missile defense systems and of new weapons, especially those he deemed to have "first-strike" potential.

To the Joint Chiefs, the constraints McNamara imposed seemed almost certain to bring about parity if not inferiority in strategic forces vis-à-vis the Soviet Union as well as weakening deterrence and inviting Soviet aggression. But from Kennedy's Presidency on, JCS access to and influence within the Oval Office fell off sharply, limiting the chiefs' influence over defense policy and the weapons acquisition process. As a result of the Bay of Pigs debacle, Kennedy lost practically all trust in JCS advice and appointed his own in-house consultant on military affairs, retired Army Chief of Staff General Maxwell D. Taylor. A personal friend of Kennedy's, Taylor returned to active duty to become Chairman of the Joint Chiefs on the eve of the Cuban Missile Crisis and was the only JCS member who participated regularly in high-level meetings during that episode.

Taylor was the first Chairman to see himself almost exclusively as a "trusted agent" for the President and the Secretary of Defense. With the possible exception of Admiral Radford, previous Chairmen had adopted a middle-of-the-road approach, acting both as spokesmen for the "military viewpoint" (i.e., their Service colleagues) and as the administration's representative to the military. Once the Cuban Missile Crisis was behind him, however, Taylor devoted his time as Chairman to bringing the chiefs into line with White House and OSD preferences. A thankless task, it produced mixed results and diminished his stature and respect in the eyes of his JCS colleagues. The CJCS during Johnson's Presidency, General Earle G. Wheeler, USA, served both as a go-between for the JCS and the White House, and as an Oval Office advisor who eventually gained access to the President's inner circle. Subsequent Chairmen generally followed Wheeler's lead, though they

CONCLUSION

sometimes found it hard to tell where their responsibilities as JCS spokesmen ended and those of trusted agents of the President or Secretary began. Until Goldwater-Nichols redefined the CJCS's role and responsibilities, Chairmen customarily functioned as a little of both. None, however, came even close to matching the level of influence exercised collectively by the JCS in World War II.

The most trying times for the Joint Chiefs were during the Vietnam War. Finding their views and recommendations consistently rejected as too extreme, they gave in to a military strategy of graduated responses that they regarded as ineffectual and doomed to fail. That they dutifully went along with the Johnson administration's conduct of the war reflected not only their professionalism and dedication, but also their underlying belief that sooner or later the President, the Secretary of Defense, and the other civilians running the war would see the light, accept the JCS view, and initiate the necessary changes. But by the time that opportunity arose, public and congressional opinion had turned so strongly against the war that ramping up the conflict, as the JCS favored, was utterly unthinkable. In the wake of the Viet Cong's Tet offensive in early 1968, the JCS were about the only ones left in Washington who still rated the war as winnable.

As the Vietnam War wound down, the JCS struggled to adjust to the realities of a country that had lost confidence in its military and was increasingly skeptical of the anti-Communist containment policies of the past. Among the various consequences of the conflict, none was more profound than the breakdown of the bipartisan Cold War consensus that had governed and sustained American foreign policy since World War II. Opposition to Vietnam by a large and vocal sector of the American public had realigned the political landscape, while the emergence of the Great Society gave domestic programs a growing claim on resources in direct competition with the military's. The result was a greater-than-expected retrenchment in post-Vietnam military spending. During the 1950s and 1960s, the Defense Department had routinely consumed around 10 percent of the country's gross national product; from the early 1970s on, it was lucky to get half of that. Yet despite the severe cutbacks, competition among the Services for funds was less intense than after previous wars, thanks in large part to McNamara's programmatic and procedural changes, which now pre-allocated the bulk of the military budget around functional categories that changed little from year to year.

The most serious military problem facing the Joint Chiefs in the aftermath of Vietnam was the surge in Soviet offensive strategic power. Given the limited support in Congress for new defense programs, the Nixon administration turned to adroit diplomacy—détente with the Soviet Union and the quasi-alliance with China—to shore up the precarious American position. Forced to adjust, the JCS became reluctant converts to the virtues of arms control, a key pillar of détente, as a means of curbing the threat posed by Soviet strategic forces. While they had shown a fleeting interest in the Baruch Plan for international control of atomic energy immediately after World War II, the JCS had since been among the most consistent skeptics and critics of arms control and disarmament. But with the Soviets steadily gaining in strategic nuclear power, and with little prospect that the United States would take up the challenge, the chiefs were compelled to reassess their position.

Indeed, no issue caused the Joint Chiefs more headaches during the later decades of the Cold War than the strategic arms control negotiations with the Soviet Union. While the Joint Chiefs saw no choice but to go along, they were uneasy with the whole arms control process and found the initial results—a 1972 treaty severely restricting antimissile deployments and a temporary "freeze" on offensive strategic launchers—deeply disturbing and generally at odds with U.S. interests. Missile defense was an area where the United States had been technologically ahead of the Soviets all along, and with the cap on land-based missile deployments, the Soviets now enjoyed a 60-percent advantage over the United States in ICBM launchers. The United States remained ahead in the number of targetable strategic warheads, though even that advantage was slipping away as the Soviets (copying the United States) turned increasingly to arming their long-range missiles with multiple independently targetable reentry vehicles. In debating the SALT I agreements before Congress in 1972, the Joint Chiefs made their endorsement of the accords conditional upon significant improvements in the U.S. strategic posture, including a new manned strategic bomber (the B–1), a more powerful ICBM (the MX), and a fleet of Trident submarines carrying more missiles with bigger payloads. Yet even with those enhancements, the JCS knew that the strategic balance was likely to remain about the same. The days of American strategic superiority were past, and whatever advantages that position may have conferred were long gone.

A curious anomaly of the post-Vietnam period was the extent to which the country's political leaders played down the role of military power in American foreign policy while trying to find new ways of making the Department of Defense and the JCS appear more efficient and effective. The explanation for this apparent paradox lies in the obvious desire of senior policymakers to avoid complications abroad like those that led to involvement in Vietnam, while shoring up the public's weakened confidence in its Armed Forces. One means of doing so was to revive JCS participation in the policy process on something other than the ad hoc basis of the Kennedy-Johnson years when military advice was practically ignored. Starting with the revival of the NSC system under President Richard M. Nixon and his assistant for national security affairs, Henry A. Kissinger, the JCS steadily regained

CONCLUSION

a regular voice in interagency deliberations that allowed them to make inputs to decisions and to have their ideas at least heard at practically every level.

Larger, more fundamental changes in the JCS system seemed inevitable but were slow in coming due to a lack of agreement on what they should entail. President Jimmy Carter leaned toward a more streamlined system that would do away with consensus-based advice. But he gave JCS reform low priority and became preoccupied with other matters—making peace between Israel and Egypt, transferring U.S. control of the Panama Canal, and, above all, negotiating a SALT II Treaty with the Soviet Union—that required JCS acquiescence if not outright support to get through Congress. In those circumstances, Carter could ill afford to engage in a reorganization battle with the chiefs and still expect them to endorse his policies enthusiastically. Letting the reorganization issue drift, he expected to return to it in his second term but never had the opportunity.

With the advent of the Reagan administration in 1981, attention turned to rebuilding the country's military power, a task begun cautiously in the dying days of the Carter administration as relations with the Soviet Union again deteriorated. Under Reagan, bolstering the Armed Forces mushroomed into the longest and largest peacetime military expansion in American history. Still, in terms of GNP, annual military spending during the Reagan years never came close to what it was between the Korean and Vietnam wars. By now, Soviet troops were heavily engaged in Afghanistan, Communist-backed insurgencies were gaining ground from southern Africa to Central America, and the Soviets were threatening NATO with the deployment of a new generation of highly accurate and more usable intermediate-range missiles known as the SS–20. With détente dead, the Cold War was again front and center.

Despite his high regard and lavish praise for the military, President Ronald Reagan used the Joint Chiefs sparingly to help orchestrate his administration's rearmament program. The chiefs' desires for improvements in the force posture were well known and were not much different from the agenda the President and his advisors brought with them into office. Like the expansion under NSC 68, the Reagan buildup was an all-Service affair, with a slight tilt toward the Navy for power-projection purposes. Once underway, it acquired a momentum of its own under spending guidelines negotiated between OSD and the Office of Management and Budget, a practice dating from McNamara's time. The chiefs' most lasting and innovative contribution came in February 1983 when, during a routine meeting with the President, they proposed a stepped-up research and development program for ballistic missile defense to explore new space-based technologies, thus planting the seeds of the Strategic Defense Initiative. The chiefs assumed that as the progenitors of the project they would play a major role in its development and act as coordinators with the Services. But after giving SDI an enthusiastic endorsement, the President looked to Secretary of Defense Caspar W. Weinberger rather than the JCS to carry it forward.

Being well aware of the flaws and limitations of the JCS system, Reagan and Weinberger were content to work around the Joint Chiefs. Indeed, they saw nothing fundamentally wrong with the existing setup despite the ingrained culture of inter-Service rivalry and competition. By the early 1980s, power and control within the Defense Department were concentrated more than ever in the hands of the Secretary of Defense and his immediate staff. The Joint Chiefs, with their influence dimmed by Vietnam, were a relatively weak and pliable organization. Weinberger liked it that way and saw no need for changes that might dilute his authority. His critics in Congress, however, had other ideas, and with defense expenditures soaring they wanted more checks and balances within DOD. Pointing to a lengthy list of lapses in joint operations (the *Mayaguez* incident, the abortive Iran hostage rescue, the Grenada intervention, and the terrorist bombing of the Marine barracks in Beirut), they seized on proposals for improvements from a former CJCS, General David C. Jones, USAF, and revived the dormant campaign to reform the JCS. Out of the legislative action that followed emerged the Goldwater-Nichols Act in 1986.

A sharp departure from the pattern of previous defense reform measures, Goldwater-Nichols marked the triumph of congressional preferences over those of the Executive. During the debate leading to passage of the legislation, consultation between the administration and the reformers on Capitol Hill was perfunctory, strained, and limited. Many of the objections the administration raised had to do with the enormous amount of prescriptive detail that Congress wanted included to institutionalize "jointness" and root out alleged Service parochialism, much of it dealing with officer promotion and other personnel matters. Once the law was passed, there was little enthusiasm for it at OSD or the White House and even less among serving senior military officers. Realizing that it would take time to bring the Services around, the Chairman of the Joint Chiefs, Admiral William J. Crowe, Jr., adopted "evolution, not revolution" as his motto, an echo of Forrestal's sentiments toward unification four decades earlier.

Like the 1947 National Security Act, Goldwater-Nichols was a product of its times. While the earlier law drew its inspiration from the experiences of World War II, Goldwater-Nichols reflected a distinctly Cold War outlook. Addressing threats associated with the missile age, when rapid decisions based on prior planning could make all the difference, it stressed more streamlined command and control and crisp, clear-cut military planning and advice in lieu of the ponderous deliberations and sometimes ambiguous recommendations inherent in the traditional JCS corporate system. By the time Goldwater-Nichols became law, however, the Cold War was already in the initial stages of winding down, rendering the need for such reforms less acute. With the advent of new, more moderate leadership in Moscow, the conclusion of the INF Treaty, the Soviet withdrawal from Afghanistan, and the disintegration of Communist power in Eastern Europe, Washington and Moscow were on track toward a more durable modus vivendi. Increasingly, as the Cold War receded into the history books, the threats facing American military planners became less obvious and the requirements of national security more complex and subtle than coping with a heavily armed adversary like the Soviet Union.

Early tests of the Goldwater-Nichols reforms seemed to pass with flying colors, helped along by the pursuit of narrowly defined objectives—assuring the safe passage of oil tankers through the Persian Gulf for one, and overthrowing the brutish Panamanian dictator Manuel Noriega for another. Neither of those operations required more than a fraction of the enormous military power the United States amassed during the Cold War and both probably could have been carried out with equally effective results under the old JCS system. But with the benefits of the Goldwater-Nichols reforms gradually coming into play, their execution appeared to go more smoothly and efficiently.

Iraq's invasion and annexation of Kuwait in the summer of 1990 posed a bigger challenge. Yet from all outward appearances, the JCS seemed to take the matter in stride. Citing an uncommonly high level of cross-Service collaboration and integrated effort, the Chairman of the Joint Chiefs, General Colin L. Powell, USA, decreed *Desert Shield/Desert Storm* to be a model of joint operational art. Even so, there was a heavy dependence on the Services' planning staffs in shaping the air and ground campaigns and numerous instances of inter-Service friction stemming from continuing differences over doctrine and operating procedures. At the same time, Secretary of Defense Richard B. Cheney sometimes bypassed Powell and the Joint Staff and sought alternative recommendations outside the normal chain of command. Yet even if the first Gulf War was not the unqualified endorsement of Goldwater-Nichols principles that the law's proponents hoped, it amply demonstrated that the system was sound and likely to stay.

The rapid eviction of Iraqi forces from Kuwait also erased the remaining stains of Vietnam and restored the American public's confidence in its Armed Forces. One untoward consequence of the campaign, however, was that it fostered the erroneous and rather naïve belief that modern military technology could achieve wonders and that future wars could be fought quickly and successfully at limited cost and sacrifice. Underlying the American success against Iraq was the availability of overwhelming military power augmented by the Reagan buildup. Yet even before *Desert*

Shield/Desert Storm began, plans were well advanced to dismantle the Nation's huge Cold War defense establishment and replace it with a smaller, more efficient "base force." Recalling the debilitating effects of previous build-downs, the architect of the base force plan, General Powell, sought to preserve residual capabilities that would avoid the harsh and disruptive cutbacks of the past. But after the collapse of the Soviet Union in the summer of 1991, the lure of further "peace dividends" became irresistible. While the United States emerged from the Cold War as the only remaining "superpower," it was a title won by default that was soon to be accompanied by a significantly less robust military establishment.

The demise of the Cold War did not, of course, bring a cessation to threats from abroad. Likened sometimes to a marathon rather than a sprint, the challenge of preserving national security remained an ongoing problem. As the focal point of the Nation's military planning, the Joint Chiefs of Staff organization continues to play an active and prominent role in national policy. Because of the changes mandated under the Goldwater-Nichols Act, JCS participation increasingly reflects the judgments, preferences, and recommendations of the Chairman and the Joint Staff, rather than the corporate assessments of the past. All the same, the JCS remains a unique organization whose individual members can still approach the Secretary of Defense directly to discuss contentious issues. Over time JCS contributions have profoundly helped to shape the role and impact of the United States in world affairs. To be sure, the JCS system as it emerged and evolved from World War II on was hardly perfect. Yet without it, military planning would have been far different and more haphazard, and the outcomes would have been both less certain and less favorable to the protection of U.S. interests.

Glossary

AAF	Army Air Forces
ABDACOM	Australian-British-Dutch-American Command
ABM	antiballistic missile
AEC	Atomic Energy Commission
ARVN	Army of the Republic of Vietnam
BMD	ballistic missile defense
BOB	Bureau of the Budget
BUR	bottom-up review
C³I	command, control, communications, and intelligence
CBI	China-Burma-India Theater
CCS	Combined Chiefs of Staff
CDI	Conventional Defense Initiative
CFE	Conventional Forces in Europe
CIA	Central Intelligence Agency
CENTAF	U.S. Air Forces Central Command
CINCNELM	Commander in Chief, U.S. Naval Forces, Eastern Atlantic and Mediterranean
CINCPAC	Commander in Chief, Pacific
CINEUR	Commander in Chief, Europe
CINCFE	Commander in Chief, Far East
CINCUNC	Commander in Chief, United Nations Command (Korea)
CIP	counterinsurgency plan
CIS	Confederation of Independent States
CJCS	Chairman of the Joint Chiefs of Staff
CNO	Chief of Naval Operations

Glossary

COMUSMACV Commander, U.S. Military Assistance Command, Vietnam
DARPA Defense Advanced Research Projects Agency
DASA Defense Atomic Support Agency
DCI Director of Central Intelligence
DEW distant early warning
DIA Defense Intelligence Agency
DMZ demilitarized zone
DSTP Director of Strategic Target Planning
EAC European Advisory Commission
EDIP European Defense Improvement Program
EDP European Defence Community
ERP European Recovery Program
EWP emergency war plan
FAL Forces Armées de Laos
FCDA Federal Civil Defense Administration
FOFA Follow-on Forces Attack
FRG Federal Republic of Germany
FSLN Sandinista National Liberation Front
FY fiscal year
FYDP Five-Year Defense Program
GLCM ground-launched cruise missile
GNP gross national product
GPO U.S. Government Printing Office
IAEA International Atomic Energy Agency
ICBM intercontinental ballistic missile

Glossary

IOC	initial operational capability
INF	intermediate-range nuclear forces
IPCOG	Informal Policy Committee on Germany
IRBM	intermediate range ballistic missile
JAAN	Joint Action of the Army and Navy
JCS	Joint Chiefs of Staff
JFACC	Joint Forces Air Component Commander
JHO	Joint History Office
JIC	Joint Intelligence Committee
JLRSE	Joint Long-Range Strategic Estimate
JOEWP	joint outline emergency war plan
JPWC	Joint Post-War Committee
JSCP	Joint Strategic Capabilities Plan
JSM	[British] Joint Staff Mission
JSOP	Joint Strategic Objectives Plan
JSPC	Joint Strategic Plans Committee
JSPD	Joint Strategic Planning Document
JSSC	Joint Strategic Survey Committee
JSTARS	Joint Surveillance Target Attack Radar System
JSTPS	Joint Strategic Target Planning Staff
JWPC	Joint War Plans Committee
KMAG	Korean Military Advisory Group
KTO	Kuwait Theater of Operations
LTDP	Long-Term Defense Program
MB	Munitions Board
MBFR	mutual and balanced force reductions
MDAP	Mutual Defense Assistance Program

Glossary

MED	Manhattan Engineer District
MILREP	Military Representative to the President
MIRV	multiple independently targetable reentry vehicle
MLC	Military Liaison Committee
MLF	multilateral nuclear force
MRBM	medium-range ballistic missile
MRV	multiple reentry vehicle
MTDP	Medium Term Defense Plan
NAC	North Atlantic Council
NATO	North Atlantic Treaty Organization
NCA	National Command Authority
NCO	noncommissioned officer
NESC	Net Evaluation Subcommittee
NIE	national intelligence estimate
NME	National Military Establishment
NORAD	North American Air Defense Command
NPG	Nuclear Planning Group
NSA	National Security Agency
NSC	National Security Council
NSDM	National Security Decision Memorandum
NSRB	National Security Resources Board
NSTL	National Strategic Targeting List
NVA	North Vietnamese Army
OCB	Operations Coordinating Board
ODM	Office of Defense Mobilization
OMB	Office of Management and Budget

Glossary

ONI	Office of Naval Intelligence
OPEC	Organization of the Petroleum Exporting Countries
OSS	Office of Strategic Services
PDF	Panama Defense Forces
PPBS	planning, programming, and budgeting system
R&D	research and development
RDB	Research and Development Board
RMA	revolution in military affairs
RV	reentry vehicle
POL	petroleum, oils, and lubricants
RDF	Rapid Deployment Force
RDJTF	Rapid Deployment Joint Task Force
ROC	Republic of China
ROK	Republic of Korea
SAC	Strategic Air Command
SACEUR	Supreme Allied Commander Europe
SALT	Strategic Arms Limitation Talks
SAM	surface-to-air missile
SDI	Strategic Defense Initiative
SEAC	Southeast Asia Command
SEATO	Southeast Asia Treaty Organization
SHAPE	Supreme Headquarters, Allied Powers Europe
SIGINT	signals intelligence
SIOP	Single Integrated Operational Plan
SLBM	submarine-launched ballistic missile

Glossary

START	Strategic Arms Reduction Talks
SVN	South Vietnam
SWNCC	State-War-Navy Coordinating Committee
TFW	tactical fighter wing
UAR	United Arab Republic
UCP	Unified Command Plan
ULMS	undersea long-range missile system
UMT	universal military training
UN	United Nations
UNC	United Nations Command
USCENTCOM	U.S. Central Command
USCINCCENT	Commander in Chief of Central Command
USFORSCOM	U.S. Army Forces Command
USMACV	U.S. Military Assistance Command, Vietnam
USREDCOM	U.S. Readiness Command
USSOUTHCOM	U.S. Southern Command
USSTRATCOM	U.S. Strategic Command
VCJCS	Vice Chairman of the Joint Chiefs of Staff
WSAG	Washington Special Action Group
WMD	weapons of mass destruction
WSEG	Weapons Systems Evaluation Group

INDEX

Page numbers with n indicate notes.
Italicized page numbers indicate illustrations.

A–6s, 464
A–12 spy plane, 182, 183, 205n47
Able Archer 83 exercises, NATO's, 440–441
Abrams, Creighton W., *304*, 313, 319, 323, 377
Acheson, Dean
  on Berlin access rights, 216
  on Bradley as CJCS, 113
  on European versus Asian issues, 117
  Korean War onset and, 103, 104
  MacArthur and, 107, 114–115
  military spending projections and, 101
  on MTDP goals and German rearmament, 120
  on U.S. and China, 97–98
  on U.S. role in NATO, 96
  as Z Committee advisor, 99
Ad Hoc Requirements Committee, CIA and, 162
AD–70, on NATO improvements, 345, 346
Adams, Sherman, 200
Advanced ICBM (AICBM), 370. *See also* MX
advanced manned strategic aircraft (AMSA), 262, 368
Advanced Research Projects Agency (ARPA), 265
Aegis missile cruisers, 471–472
Afghanistan
  CIA and *mujahideen* of, 431
  Soviet invasion of, 335, 395, 400, 409–410
  Soviet Union and pro-Communist regime in, 480
  Soviet withdrawal from, 459, 461–462
Africa. *See* North Africa
African-Americans, Ledo Road building from Burma to China by, 35
Agnew, Spiro, 355, 378
Air Force
  airpower for *Desert Shield/Desert Storm* and, 512–513, 520–521, 524–525
  Ballistic Systems Division, 262
  on counterforce of bombers and missiles, 174–175
  intelligence estimates on Soviet missile program by, 180, 182
  on land-based aviation for antisubmarine warfare, 72
  McNamara on massive retaliation doctrine of, 248
  on McNamara's assured destruction strategy, 250
  missile development and, 175, 176, 177, 183
  NATO retardation bombing plan and, 118–119
  as New Look beneficiary, 143–144
  nuclear weapons for, 79
  on U.S. Strategic Command to replace SAC, 187
  X–16 photoreconnaissance plane and, 161
aircraft. *See also specific types of*
  for the Soviets during World War II, 19
aircraft carriers, swing strategy for, 145
AirLand Battle, 458, 474n30, 513
air-launched cruise missiles (ALCMs), 380–381
airpower. *See also* Vietnam War
  for *Desert Shield/Desert Storm*, 512–513, 520–521, 524–525
Akhromeyev, Sergei F., 460
Algeria, 156
Allen, Lew, Jr., 393, 423
Allison, Royal B., 337–338, 341–342, 376
Ambrose, Stephen E., 195
amphibious operations, 67, 69, 72
ANADYR (Soviet missiles to Cuba), 226–227, 241n74
Anderson, George W., Jr., 219, 224, 231, 233
Andrew, Christopher, 434–435
Angola, Soviet support for Marxist regimes in, 383
anti-ballistic missile (ABM) system
  as essential, JCS on, 263–264
  McElroy on Army and, 181
  McNamara on, 268
  testing of, 146
  U.S.-Soviet negotiations on, 266, 339–340
  Vietnam War and, 308
  Watkins on, 436
antisatellite (ASAT) weapons, 434
antiwar movement
  Fulbright and, 285
  Johnson and, 293
  Nixon and, 317, 325
  Vietnam War and, 297, 306, 308, 319, 543
Arab-Israeli conflict. *See also* Middle East
  Gulf War (1990) and, 529
  October War (1973) and, 346, 352, 354–357
  partitioning of Palestine and, 76–77
  Six Day War (1967) and, 308, 352–353

555

State Dept. on Baghdad Pact and, 190–191
Árbenz Guzmán, Jacobo, 196
ARCADIA meetings (1941), 1–2, 9, 34
Arias, Oscar, 466
Armed Forces
    Bush review of size of, 480
    demobilization after World War II of, 59
    JCS on Reagan buildup and capabilities of, 426–427
    nuclear weapons for, 79
    planning for post–World War II organization and composition of, 38–40
    post-*Desert Storm* public opinion on, 547–548
    post-Vietnam war modernizing of, 367–370
    post–World War II defense policy for, 61–64
    Powell's base force plan for, 485–489
    Reagan on image of, 421–425
    reorganization and reform of, 64–69
    Roosevelt's personal control of, 2–3
arms control. *See also* détente; Strategic Arms Limitation Talks
    Chinese Communists' testing H-bomb and, 267–268
    at end of Cold War, 493–495
    INF Treaty (1987), 459–461, 475n40
    JCS concerns on détente and, 336–337, 543–544
    Reagan buildup and, 438–443
    U.S. and Soviets negotiations on, 159
    U.S. scientists on ABM development and, 266
    use of term, 160
arms sales, foreign, Carter-era versus Reagan-era, 429–430
Army, U.S.
    on amphibious operations, 67, 72
    inter-Service rivalry between Navy and, 3
    MacArthur on proposed postwar Navy–Marine Corps merger, 65–66
    missile development and, 177, 183
    New Look defense budget and, 145–146
    peacetime total force (1970s) and, 367
    Plans Division, Navy's joint general staff proposal and, 4–5
    post–World War II projections for size and capabilities of, 39
    Special Forces, JFK's expansion of, for Vietnam War, 279
    Vietnam War effect on, 329
Army Air Forces (AAF)
    on air transport control, 67
    ARCADIA meetings (1941) and, 1–2
    China as base for targeting Japan by, 35, 37
    Combined Chiefs of Staff and, 2
    509th Composite Group of, 48
    Flying Tigers absorbed into, 36
    JCS consensus decisions and, 538
    MacArthur on proposed postwar Navy–Marine Corps merger and, 65–66
    Pacific Ocean Area command support by, 32
Army-Navy Petroleum Board, 7
Arnold, Henry H.
    atomic bomb and, 47, 48
    CCS at the Second Quebec Conference and, *xiv*
    Chinese affairs and, 35
    Combined Chiefs of Staff and, 2
    on post–World War II organizational reform, 39, 40
    steps down from JCS, 60
    strategic bombing of Germany and, 13
    Twentieth Air Force in the Pacific under, 30–31, 46
Asia-Pacific War (1942–1945)
    China-Burma-India Theater, 33–37
    dawn of the atomic age and, 46–52
    ending war with Japan, 43–46
    post–World War II planning beginning during, 38–43
    strategy and command in the Pacific, 29–33
    two-front war and, 29
assured destruction strategy
    controlled escalation and, 373
    McNamara on, 248–250, 265, 268
    Schlesinger on selective changes to, 374–375
    SIOP on, 371
Aswan Dam, Egypt, 191
Atkinson, Rick, 483
Atlantic Alliance. *See* North Atlantic Treaty Organization
Atlas (ICBM), 176, 202n1
atomic bomb. *See also* nuclear weapons
    development of, 46–47
    interdepartmental Interim Committee on, 48–49
    post–World War II war debates on, 62–63, 544
    Roosevelt on Soviets and, 19
    testing of, 47–48, 63–64, 78–79
    UMT in light of, 61–62
Atomic Energy Act (1954), 155
Atomic Energy Commission (AEC), 62, 69–70, 78–79, 99, 115, 234
"Atoms for Peace" speech, Eisenhower's, 160
Attlee, Clement R., 114, 252
attrition, war of, in Vietnam, 294, 296, 297, 306
Australia, 147, 306
Australian-British-Dutch-American Command (ABDACOM), 30
Autumn Forge exercises, NATO's, 440
Awarding the Channel Command (ACCHAN), NATO's, 121

B–1 bomber

INDEX

Brown as CJCS and, 377
Carter's cancellation of, 392–393, 395
fly-before-you-buy policy and, 368–369
JCS requests for, 262, 337, 341, 544
Reagan buildup and, 427–428
Rumsfeld on, 385
Schlesinger on, 374, 375
B-2 stealth bomber, 428, 496, 524
B-29s
to Britain in 1950, 118
ending war with Japan and, 45
modified for atomic bomb, 48
Pacific Ocean Area command support by, 32
as reinforcement of American airpower in Europe (1948), 78
retirement of, 143
SILVERPLATE, nuclear-delivery system of, 72
Soviet targets and, 82, 91–92n89
U.S. bombing missions from China and, 37, 53–54n34
B-36s, 82, 143
B-47 medium range aircraft, 143, 248
B-50s, 82, 143
B-52s
bombing in Vietnam by, 295, 325–326
intercontinental, New Look defense budget and, 143
JCS requests for, 262
LeMay's objection to cancellation of, 248
Soviets on ALCMs for, 380–381
transfer to Western Pacific (1965), 289–290
B-70 supersonic bomber, 248
Backfire bomber, Soviet, 380, 381
Baghdad Pact, 190–194, 352
Baker, James A., III, 492, 496, 497, 511, 529
Ball, George W., 226, 232, 289, 294
ballistic missile defense (BMD)
McNamara on, 265
Reagan's revival of, 433–438, 545–546
resource allocation controversy for, 181–182
Soviet Union and, 264, 338
Barrass, Gordon S., 357, 458
Baruch Plan, 62, 69, 160, 544
base force plan
collaborative development of, 501n25
force levels for FY 1950 compared to, 502n36
Powell and, 485–489, 501n29
Soviet Union collapse and, 548
START agreement and, 495–496
Batista, Fulgencio, 197
Baxter, James Phinney, 178
Bay of Pigs operation, 198, 213–216, 225, 542
Belgium, NATO and, 157, 259

Beneš, Eduard, 73
Berlin. *See also* West Germany
access rights to, 216–217
contingency plans for, 199–200
Live Oak plans for, 200–201
Nixon and Kissinger on solution for, 343
Soviet blockade of, 73, 77–78, 90n53
Berlin Task Force, interdepartmental, 217
Berlin Wall, *210*, 218, 219–220, 481, 486
Bermuda conference (1953), 154
Betts Report (DOD, 1964), 265
Bevin, Ernest, 78
Bikini Atoll, atomic bomb testing at, 64
biological weapons, 480, 506, 509, 522–525
Bissell, Richard M., Jr., 170n123, 183
Black Hole organization, Saudi Arabia, 514
Blandy, William H. P., 63
Blue Ribbon Commission on Defense Management, 452
Blue Spoon (against Noriega government), 491–492
Blue Steel Mk1 missile, 254
Blue Streak air-to-surface missile, Britain's, 253–254
body counts, for Vietnam War, 297
BOLERO-ROUNDUP plans for Continental invasion, 10–11
bomb shelter programs, 1950s, 178
Bomber Command, Britain's, 78, 117
Boone, Walter F., 194, 195
bottom-up review (BUR), Clinton's, 488–489
Bradley, Omar N.
defense reorganization (1958) and, 184–185
as Eisenhower political supporter, 134–135
on ending the Korean War, 139–140
Far East tour by Johnson and, 102
increased role of CJCS and, 124, 539
JCS meeting at Naval War College (1948) and, *58*
joins the JCS, 74
Korean War and, 103, 107, 109, 110, 128n56
MacArthur's dismissal and, 115–116
on nuclear weapons for Europe, 119
Truman on NSC meetings and, 113
on U.S. involvement in Asian mainland wars, 290
Brezhnev, Leonid I., 340, 355, 357, 380–381, 382, 398–399
Britain. *See also* Bomber Command, Britain's; Chiefs of Staff Committee; Churchill, Winston S.; United Kingdom
AAF on air-atomic missions from, 64
Arab-Israeli conflict and, 191–192
Baghdad Pact and, 192–193, 194
Casablanca Conference and, 13–14
Communist attack against Korea and, 119
Desert Shield/Desert Storm and, 519, 522

557

Grand Alliance during World War II and, 18–19
Live Oak (planning body) and, 200–201
Middle East presence of, 190
MLF demise and, 257
NATO and, 121, 157, 259
on NATO improvements, 345
on NATO's Conventional Defense Initiative, 475n33
North African decision and, 10
on nuclear weapons, 114, 147
on Palestine withdrawal by, 77
Skybolt program and, 253–255
Soviet blockade of Berlin and, 78
U.S. bombing Libya and, 463–464
Vietnam War and, 307
war on the periphery and, 11
Brooke, Alan F., *xiv*, 1, 16, 17
Brooks, Dennis M., 469
Brown, George S., 377, 379–380, 382, 392
Brown, Harold, 254, 392, 393–394, 395, 396, 409
Brzezinski, Zbigniew, 396, 404–405, 406, 408, 411, 412
Budget Advisory Committee, 80
Bundy, McGeorge
    Cuban missile crisis and, 227, 232
    on ending the Korean War, 140
    on JCS estrangement with JFK, 233
    JCS unhappiness with graduated pressure and, 287
    joint congressional resolution on Southeast Asia and, 285
    NSC under Kennedy and, 212
    Vietnam War strategy and, 288, 291
Bundy, William P., 288
Bureau of the Budget, 101, 137
BURIA (Warsaw Pact exercise in Berlin), 220
Burke, Arleigh A.
    on Bay of Pigs operation, 214, 215–216, 225
    on defense reorganization, 184, 185–186, 188
    on McNamara and his staff, 212
    on military operation against Cuba, 197–198
    on Nasser's strength, 191
    on swing strategy for carriers, 145
Burma, 33, 36–37
Burma Road, *28*, 35
Bush, George H.W. *See also* Operation *Desert Shield/Desert Storm*; Powell, Colin L.
    on air campaign for *Desert Storm*, 521
    base force plan under, 485–489
    briefing *Desert Shield* for, 515–516
    CFE Treaty under, 493–495
    as CIA director, 384–385
    on halting Gulf War, 528
    on Iraq's invasion of Kuwait, 510
    military policy in transition under, 479–481
    Panama operations under, 489–493
    Powell on forces for *Desert Storm* and, 517
    Powell's impact as CJCS under, 481–484
    START I and its consequences under, 495–500
Bush, Vannevar, 47, 56n79, 85–86
Butler, George Lee, 498, 501n25, 508

C4 missiles, 369, 374
Cairo Conference (SEXTANT, Nov. 1943), 37
Cambodia
    bombing halted in, 326
    invasion of (1970), as Vietnamization test, 318–319
    Mayaguez affair (1975) and, 378, 379
    secret bombing in, 317–318
    Vietnam War and, 294, 316
Canada, need for U.S. bases in, HALFMOON plan and, 75
Carlucci, Frank C., 471
Carney, Robert B., 135, 145, 151, 154, 155
Carter, Jimmy
    Iran hostage rescue mission under, 411–414
    JCS relationship with, 391–394, 545
    NATO and INF controversy under, 400–403
    Rapid Deployment Force creation under, 408–411
    SALT II and, 397–400
    strategic forces and PD-59 of, 394–397
    Third World crises under, 403–407
Carter, Rosalynn, 413
Carter Doctrine, 410
Casablanca Conference (Jan. 1943), 12–13
Casey, William J., 422
Castro, Fidel, 196, 197, 198–199, 227, 228. *See also* Cuba
Central America. *See also* Latin America; Nicaragua
    Carter administration goals for, 408
    Cold War in, 464–465, 480
    economic and military assistance under Reagan for, 429
    Reagan on Communist takeovers in, 430–431
Central Intelligence Agency (CIA). *See also* Intelligence Community
    Afghanistan's *mujahideen* and, 431
    Bay of Pigs operation and, 214
    British deception on Suez crisis for, 192
    on Chinese forces in Korea (1950), 110
    covert action against Castro and, 198–199
    creation of, 68
    JCS under Powell and, 482
    MacArthur's conflicts with, 102
    OPLAN 34A for Vietnam and, 280–281
    on Soviet bloc forces' preparedness (1956–1957), 158

INDEX

on Soviet missile capabilities in 1950s, 182
on Soviet missiles in Cuba, 227–228
on Soviet Order of Battle, 253
on Soviet Union's nuclear weapons program, 98
U–2 photoreconnaissance plane and, 161, 162–163, 170n123
U.S. overthrow of Árbenz regime and, 196
on Warsaw Pact buildup, 345
Central Treaty Organization (CENTO), 196
chain of command
    defense reorganization (1958) and, 186
    Goldwater-Nichols legislation on, 454–455
    Johnson and McNamara in Vietnam War and, 297–298
    Key West Agreement (1948) removing JCS from, 134
    Kissinger and, 316, 322, 389n67
    Nixon in Vietnam War and, 322
    in North Africa and Europe under CCS, 29–30
    in Pacific under JCS, 30–31
Chairman, Joint Chiefs of Staff (CJCS). *See also* individual CJCSs
    Defense authorization broadening powers of (1984), 451
    Eisenhower on strengthening powers of, 134, 184, 185
    evolution of role of, 542–543
    Goldwater-Nichols legislation on, 454–455, 548
    Korean War and emergence in importance of, 124
    legislation establishing role for, 82, 539
    nuclear test ban and, 236
    President's Special Committee on Indochina and, 148
    SALT II talks and, 376–377
    target selection for air war in Vietnam and, 295
    Truman on NSC meetings and, 113, 539
    U–2 photoreconnaissance plane and, 161–162
Chamberlin, Stephen J., 74
Chamoun, Camille, 194, 195
Chapman, Leonard F., Jr., 314
Checkmate (Air Staff planning cell), 511, 513, 514–515, 523
Checkpoint Charlie. *See also* Berlin Wall
    confrontation at, 219–220
chemical weapons
    Korean War and, 139
    Saddam Hussein and, 467, 506, 509, 515, 522–525
    U.S. work on, 434, 440–441
    Vietnam War and, 311
    Warsaw Pact and, 344
Cheney, Richard B.
    Panama operations (1989) and, 491–492
    planning *Desert Shield/Desert Storm*, 511, 512, 513, 520–521

on post-Cold War demobilization, 485, 487
Powell and, 483–484, 510, 516
Powell's briefing on *Desert Storm* and, 516, 517
as Sec. of Defense, 481, 547
on START I Treaty, 497
Chennault, Claire L., 36
Chiang Kai-shek. *See also* China; Nationalist Chinese; Republic of China; Taiwan
    Cairo meeting with Roosevelt, Churchill, and the CCS, 17
    CCR and JCS difficulties of working with, 37
    JCS coordination during World War II with, 9
    Nationalist regime collapse and, 95, 96–97
    retreat to interior China by, 34
    Stilwell as chief of staff for, 34–35
    Stilwell's contempt for, 35–36
    Taiwan Strait crisis (1955) and, 152
    on U.S. bombing missions from China, 53–54n34
    Wedemeyer as chief of staff for, 38
Chiefs of Staff Committee, Britain's, 1, 9, 121, 153, 190
Childs, Marquis, 341–342
China. *See also* Chiang Kai-shek; Mao Zedong; People's Republic of China
    becomes Communist People's Republic, 95–97
    Soviet engagement during World War II in, 21
    U.S. military in World War II and, 33–34
China Lobby, in Washington, D.C., 98, 349
China Theater, 37–38
China-Burma-India Theater (CBI), 33
Chinese Communists, 150, 267. *See also* Mao Zedong; People's Republic of China
Chinese Nationalists. *See* Chiang Kai-shek; Taiwan
Chou En-lai, 152, 350, 351
Christopher, Warren M., 413
Chromite attack on North Korea, MacArthur's, 107–108
Churchill, Winston S.
    ARCADIA meetings (1941) and, 1–2
    on Axis surrender, 13
    on combined unified command, 29–30
    on Continental invasion (May 1943), 15–16
    Grand Alliance during World War II and, 18–19
    on North African invasion, 11
    post–World War II plan for Germany and, 42
    on propping up Chiang, 37
    QUADRANT talks and, 16
    at Second Quebec Conference, *xiv*, 18
Clark, Mark W., 138–139
Clark, William P., 424
Claude V. Ricketts, USS, 258
Clay, Lucius D., 74, 77–78
Cleveland, Harlan, 153

Clifford, Clark M., 268, 311, 312
Cline, Ray S., 162
Clinton, Bill, 488–489
Closely Spaced Basing (CSB) plan, for MX missiles, 435
Codevilla, Angelo, 445–446n41
Colby, William E., 385
Cold Dawn (Newhouse), 338
Cold War. *See also* détente; *specific administrations*; *specific wars or conflicts*
   Lippmann's naming of, 69
Cold War, ending of
   CFE Treaty (1990) and, 493–495
   East-West relations and, 479–481
   operations in Panama and, 489–493
   Powell as CJCS and, 481–484
   Powell's base force plan and, 485–489
   START I and its consequences for, 495–500
collateral damage estimates, for Harmon Committee, 85
Collins, J. Lawton, 107, 113, 115, 134–135
color plans, Joint Army and Navy Board on, 3–4
Combined Chiefs of Staff (CCS). *See also* Quebec Conference
   Asia-Pacific War plans and, 32
   Atlantic-European strategy during World War II under, 9
   command chain in North Africa and Europe under, 29–30
   confirmation of *Overlord* and diminished involvement of, 17–18
   JCS formation and, 2, 5, 23n6
   JSSC on tripartite United Chiefs of Staff replacing, 26n76
   at the Second Quebec Conference, *xiv*
Command, Control, and Communications Systems Directorate (J-6), 455
command, control, communications, and intelligence (C_3I), 249
Commander in Chief, Far East (CINCFE), 105, 115, 116
Commander in Chief, Pacific (CINCPAC)
   McNamara, Tonkin Gulf incident and, 284–285
   OPLAN 34A for Vietnam and, 280–281
   Vietnam conflict assessment (1960) and, 278
Commander in Chief, U.S. Naval Forces, Eastern Atlantic and Mediterranean (CINCNELM), 190, 194, 208n109
Committee on the Present Danger, 422, 443n3
Communications and Electronics Directorate (J-6), 185
Communist China. *See* People's Republic of China
Communists. *See also* Cuba; Soviet Union; Warsaw Pact
   Greek insurgency and, 69
Conant, James B., 56n79
Confederation of Independent States (CIS), 499
Conference on Security and Cooperation in Europe (CSCE), 343
Congress. *See also* Goldwater-Nichols Defense Reorganization Act
   on defense spending, 426, 428, 435, 480, 487
   on *Desert Shield/Desert Storm*, 515, 518
   JCS reform during second Reagan term and, 450–453, 546
   on Kissinger's power under Ford, 385
   McNamara on defense spending and, 246
   McNamara on MLF for NATO and, 258
   mobilization of Reserves for Vietnam and, 297
   nuclear test ban and, 234
   post–World War II nuclear program oversight by, 61
   on SALT I agreements, 341
   SALT II and JCS backchannel talks with, 379
   on Strategic Defense Initiative, 496
   on Vietnam War, 285–286, 325
   Vinson and, 185
contras, Nicaraguan, 424, 431, 465–466
controlled response, assured destruction strategy and, 248–249
Conventional Defense Initiative (CDI), NATO's, 457–458
Conventional Forces in Europe (CFE) Treaty, 494–495, 518. *See also* Mutual and Balanced Force Reduction Talks
Coolidge, Charles A., 184–185
Cooper-Church amendment, 320
counterforce/no-cities doctrine, 249, 374–375
counterinsurgency plan (CIP), for South Vietnam, 278
Counterterrorist Joint Task Force (CTJTF), 414
countervailing strategy (1978), 396–397
covert operations
   Berlin crisis and, 201
   in Cuba, RFK and, 225
   in Egypt to undermine Soviet influence, 192
   under Eisenhower, 198–199, 202
   in Indochina, direct military involvement versus, 148
   to prevent Soviet presence in Western Hemisphere, 196
   Taylor as MILREP and, 213
   in Third World conflicts, 383, 403
   in Vietnam, 279, 281, 286, 321
Crist, George B., Jr., 469
cross-Channel operation, World War II, 15, 21, 46
Crowe, William J., Jr.
   as CJCS, *448*, 482
   Congress on JCS reorganization and, 452–453

INDEX

implementation of Goldwater-Nichols by, 455–457, 531
on INF Treaty, 460–461
on Iranian aggression in Persian Gulf, 468–469, 471
Libyan terrorism and, 463
NMS 92–97 and, 481
on Strategic Defense Initiative, 496
cruise missiles, 380–381, 403, 514
Cuba
    Burke on military action against, 197–198
    Ethiopian–Somalia conflict and, 404
    Grenada and, 429
    missile crisis, 228–233, 243n105, 542
    missile crisis origins, 224–228, 240–241n71
    revolution (1959) in, 196
    Sandinistas in Nicaragua and, 408
    Soviet combat brigade in (1979), 400
    State Department halts shipments (1958) to, 197
Cunningham, Andrew B., xiv
current force strategy, Middle East policy during Reagan era and, 432
Cushman, Robert E., Jr., 377
Cutler, Robert, 178
Cypress, Greece and Turkey tensions over, 308
Czechoslovakia, 73, 90n53, 269, 336

D5 missiles, 369
damage limitation debate, Soviet nuclear stockpile and, 261–262
Davis, Richard G., 513–514
Davy Crockett (spigot mortar), 146
DC 6/1 (NATO's organizing defense plan), 117
DC 13 (Medium Term Defense Plan), 119
de Gaulle, Charles, 257, 258–259
Deane, John R., 20–21
Decker, George H., 219, 222
DEFCON 3, 357
Defense, Secretary of. *See also* Defense Department; Office of the Secretary of Defense; *specific individuals*
    Eisenhower on ballistic missiles as responsibility of, 176
    Eisenhower's relationships with, 136
    Forrestal on JCS chair as, 82
    on functional responsibility of Services for missile development, 177
    on FY 1952 military budget, 110
    Key West Agreement on advise on Service functions for, 203n19
    National Military Establishment under, 68
    NSA's 1958 amendments on authority of, 246, 541
    review of NESC targeting priorities and, 188
    SACEUR communication with, 120

on U–2 program oversight committee, 170n123
Defense Advanced Research Projects Agency (DARPA), 265, 433
Defense Atomic Support Agency (DASA), 188, 234
defense budget. *See* military budget
Defense Department. *See also* War Department
    budget authority FYs 1981–1989 of, 427
    Eisenhower on aerial reconnaissance over Soviet Union and, 161
    on fly-before-you-buy acquisitions, 368
    McNamara system for, 245–247
    reorganization (1958) of, 184–186, 206n63
Defense Intelligence Agency (DIA), 338, 345, 355
Denfeld, Louis E., *58*, 74, 81, 85, 86, 99
Dennison, Robert L., 85, 230
dense pack, for MX missiles, 435, 436, 438, 447n64
Dereliction of Duty (McMaster), 327
Desert One, 412, 413–414
Desert Storm. *See* Operation *Desert Shield/Desert Storm*
Desoto Patrol Program, Gulf on Tonkin incident and, 284, 286
détente (1972–1979). *See also* strategic stability
    arms control and, 543–544
    China quasi-alliance and, 347–351
    Middle East involvement and, 351–358
    Moorer's summary of, 358
    overview of, 335–336
    SALT I and, 336–342
    shoring up NATO and, 342–346
Diem, Ngo Dinh, 278, 279, 281
Dien Bien Phu, 1954 siege at, 148
Dill, John, xiv, 5, 22–23n3
Director of Strategic Target Planning (DSTP), 189
Directorate for Operational Plans and Joint Force Development (J-7), 455
disarmament negotiations. *See* arms control
Discoverer spy satellite, 205n53, 218
Dobrynin, Anatoly, 232, 338–339, 378
Dominican Republic crisis (1965), 292
domino theory, 293
Donovan, William J., 7
DOWNFALL plan for Japan, 45
draft Presidential memorandums (DPMs), 246, 247
DROPSHOT (long-range plan 1950–51), 111
dual-basing, McNamara on, 259
Dubcek, Alexander, 336
Dugan, Michael J., 484, 513
Dulles, John Foster
    Bay of Pigs operation and, 214, 215–216
    as Latin America specialist, 196
    on liberation doctrine, 421
    on massive retaliation policy, 142

on military budget (1956), 157
on missile gap and national intelligence estimates, 182–183
NATO's New Approach and, 154
on nuclear weapons against Chinese Communists, 151
President's Special Committee on Indochina and, 148
DuPuy, William E., 474n30

East Germany, 158, 201, 217, 343. *See also* Warsaw Pact
Easter Offensive, by North Vietnamese Army, 323, 340, 351
Eastern Europe, 42, 157–158, 220, 481, 486. *See also* Warsaw Pact
Eden, Anthony, 192, 193
Egypt. *See also* Arab-Israeli conflict; Suez Canal
  Anglo-French-Israeli coalition invasion into, 193
  covert operations in, 192
  Desert Shield/Desert Storm and, 519
  HALFMOON plan and, 75
  Soviet support for, 352, 353, 354
Eighth Air Force, 66–67
Eighth Army, 107–108
Einstein, Albert, 47
Eisenhower, Dwight D.
  on Bay of Pigs operation, 214
  defense spending under, 245–246
  on Indochina conflict, 278
  JCS service by, 60, 65, 66, 74, 82
  Laos crisis and, 221, 239n45
  as military officer in World War II, 17–18, 30
  on MLF in Grand Design for European union, 256, 272n50
  as NATO Supreme Allied Commander, Europe, 118–123
Eisenhower, Dwight D., first term
  arms race curbing under, 158–163
  elected president (1952), 133–134
  on ending the Korean War, 137–140
  Indochina as test for New Look, 146–149
  on JCS reorganization, 134–137
  NATO's conventional posture and, 156–158
  New Approach in Europe and, 152–156
  New Look security strategy under, 140–146, 540
  Taiwan Strait confrontation, 149–152
Eisenhower, Dwight D., second term
  Berlin pressured by Soviets, 199–202
  Cuba, Castro, and Communism, 196–199
  Gaither Committee report and, 177–179
  on JCS reorganization, 184–185
  Middle East defense under, 190–196
  missile gap and BMD controversies during, 179–183
  missile program evolution, 174–177
  reorganization and reform (1958–1960), 183–190
  Sputnik I, Third World issues and, 173–174
Eisenhower, Milton, 196
Eisenhower Doctrine (1957), 194, 195
Ely, Paul, 148–149
Embick, Stanley D., 11
Endara, Guillermo, 492
Eniwetok, nuclear weapons testing at, 78
Enterprise (nuclear-powered carrier), 144
ERASER (nonnuclear alternative to HALFMOON), 75–76
Ermarth, Fritz W., 441
Essex-class attack carriers, 31–32
Ethiopia, war with Somalia and, 404
Europe. *See* Eastern Europe; North Atlantic Treaty Organization; Western Europe; *specific countries*
Europe, war in (1942–1945)
  North African decision and, 10–12
  origins of joint planning and, 2–9
  preparing for *Overlord*, 15–18
  second front debate and JCS reorganization during, 12–15
  two-front war and, 29
  unified command after, 65
  U.S. collaboration with Soviet Union during, 18–22
European Advisory Committee (EAC), 42
European Defence Community (EDC), 120, 123, 155
European Defense Improvement Program (EDIP), 345, 346
European Recovery Program (ERP), 69, 75
executive agent system, in Asia-Pacific during World War II, 30–31
Executive Committee (ExCom), Kennedy's, 229, 230
Executive Order 9877, delineating Service roles and missions, 67, 69

F–18s (carrier-based bombers), 464
F–111 (medium-range fighter-bomber), 368, 464
Fairchild, Muir S., 11
Faisal II, King of Iraq, 194–195
FAL (Forces Armées de Laos), 221–222, 223–224
Farouk, King of Egypt, 191
Fat Man bomb, 48, 50
Fechteler, William M., 134–135
Federal Civil Defense Administration (FCDA), 178
Federal Republic of Germany (FRG). *See* West Germany
Felt, Harry D., 221–223, 224
Fermi, Enrico, 47
Fifteenth Air Force, SAC and, 66–67
5412 Committee, 198

INDEX

Finletter Commission, 72
Fitzwater, Marlin, 492
509th Composite Group, 48, 66–67
Five-Year Defense Program (FYDP), 247
FLAMING DART reprisal raids against Viet Cong, 289
FLEETWOOD plan, for unified defense budget, 75, 80
flexible-response force posture. *See also* Follow-On Forces Attack concept; graduated pressure or response
    Berlin access rights and, 217
    JFK on, 211, 247–248
    McNamara on, 542
    NATO and, 251–253, 342–343
    NATO's MC 14/3 strategy on, 260–261
    Nitze on, 179
    U.S. reserves for NATO versus Vietnam War and, 259–260
fly-before-you-buy acquisitions (1970s), 368–369
Flying Tigers, 36
Follow-On Forces Attack (FOFA) concept, 458
force de frappe, France's, 256–257, 258
Force Structure, Resource, and Assessment Directorate (J-8), 455
Ford, Gerald R.
    becomes President, 378
    Mayaguez affair (1975) and, 383–384
    personnel changes under, 384–385
    SALT II and, 379, 380–382
Foreign Affairs, "The Sources of Soviet Conduct" (Kennan, 1947), 70–71
foreign policy, Reagan buildup and, 429–432
Formosa. *See* Taiwan
Formosa Strait. *See* Taiwan Strait
Forrestal, James V.
    on absence of presiding JCS chair, 82
    on coordinating role of JCS after World War II, 60
    on custody and control of nuclear weapons, 79
    death of, 82–83
    on ERASER as alternative to HALFMOON, 75–76
    inter-Service bickering and, 69
    on JCS as key to unification law, 71–73, 538–539
    on McCloy on atomic bomb to Truman, 49–50
    military budget for FY 1950 and, 76, 79–81, 89n46
    push to increase military budget by, 73, 74
    reliance on U.S. nuclear monopoly by, 98
    as Sec. of Defense meeting with JCS, 58
    Truman on holding military budget limits by, 76
    unification bill and, 67
    unified command in Pacific and, 65–66
    Weapons Systems Evaluation Group and, 85, 86
Forrestal, Michael, 288

Forrestal-class super carriers, 144–145
Foster, John S., Jr., 372–373, 374
Foster, William C., 178
Fracture Jaw plan, 311
France
    Communist attack against Korea and, 119
    deferring action on the EDC, 155
    Desert Shield/Desert Storm and, 519
    Live Oak (planning body) and, 200–201
    Nasser's nationalization of Suez Canal and, 192–193
    NATO and military commitments at end of 1950s by, 156
    NATO buildup in Europe and, 122–123
    on nuclear weapons against expansion of Chinese Communist power, 147
    opposition to MLF by, 256–257
    Pleven Plan, European Defence Community and, 120
    Soviet blockade of Berlin and, 78
    Treaty of Paris (1952) and, 123
    war-weariness for Indochina conflict and, 146–147
Franklin D. Roosevelt (carrier), 356
Franks, Frederick M., Jr., 527
Freedman, Lawrence, 263
Freeman, Charles W., Jr., 529
FRG (Federal Republic of Germany). *See* West Germany
FSLN (Sandinista National Liberation Front), 407–408
Fulbright, J. William, 285–286
"Functions of the Department of Defense and Its Major Components" (DOD Directive 5100.1, 1958), 186

Gabriel, Charles A., 423, 435
Gaddis, John Lewis, 142
Gaither, H. Rowan, 178
Gaither Committee, 177–179
Galahad commando unit (Merrill's Marauders), 33
Galosh (Soviet air defense system), 264, 265
Gardner, Trevor, 175
Gast, Philip C., 412
Gates, Robert M., 492
Gates, Thomas S., Jr., 136, 182–183, 188
Gayler, Noel, 325, 378–379
General Advisory Committee, AEC's, 62
Geneva conference(s), 162, 171n134, 277, 358
Geneva Convention, on repatriation of POWs (1949), 138
Germany. *See also* West Germany
    ARCADIA meetings (1941) on defeat of, 9
    Casablanca Conference (Jan. 1943) on combined bombing of, 13
    Informal Policy Committee on, 43

post-World War II treatment of, 42–43
Soviet troops in (1948), 73
Gilpatric, Roswell L., 218, 219, 232
Giroldi Vega, Moisés, 490–491
Glaspie, April, 507–508
Glass, Henry E., 213
Glosson, Buster C., 514, 516, 517, 521, 523
Goldwater, Barry, 286
Goldwater-Nichols Defense Reorganization Act (1986)
  backchannel talks and, 496
  CJCS authority under, 537
  controversial features of, 454–455
  Desert Shield/Desert Storm and, 518–519
  diminished JCS role in passage of, 453
  flow of military ideas and information and, 512
  implementation of, 455–457
  Joint Task Force Middle East under, 469–470
  Operation *Just Cause* under, 492–493
  Packard Commission and, 452
  planning procedures under, 112
  post-Vietnam War JCS role and, 327, 546–547
  Powell as CJCS and, 482
  special operations and, 384
Goodpaster, Andrew J., 345, 360n42
Gorbachev, Mikhail S.
  on armed forces reductions, 479, 480
  coup against, 497–498
  Powell on reforms by, 485
  as reform-minded Soviet leader, 449
  Saddam Hussein and, 515
  Soviet Union restructuring by, 459–462
  START negotiations and, 496–497
graduated pressure or response
  air war in Vietnam and, 309, 310
  JCS and, 543
  Kennedy on, 211
  McMaster on JCS acceptance of, 327
  McNamara on, SIGMA I–64 testing of, 283
  SIGMA II–64 testing of, 287
  Taylor on, 282–283
  as Vietnam War strategy, 287–289
Graham, Daniel O., 434
Grand Alliance, 18–19, 61. *See also* Britain; Soviet Union; United States
Gray, Alfred M., Jr., 492
Gray, Gordon, 178
Great Debate (1951), on U.S. commitment to NATO, 121
Great Society, Johnson's, 250, 293, 543
Grechko, Andrei, 382
Greece, 69, 121–122, 123, 308
Green Berets, 279

Greene, Wallace M., Jr., 282, 301n61
Greenfield, Kent Roberts, 7
Grenada invasion (1983), 429, 430
Gribkov, Anatoli I., 240–241n71
Gromyko, Andrei, 380
ground-launched cruise missiles (GLCMs), 403
Groves, Leslie R., 47, 48, 51
Gruenther, Alfred M., *58*, 154, 162, 184–185
Guam Doctrine, Nixon's, 343, 366
Guantanamo Bay, U.S. naval base at, 198
Guatemala, 196
Gulf of Tonkin incident, 284–292
Gulf War (1990). *See* Iraq; Kuwait; Operation *Desert Shield/Desert Storm*
Gulf War Airpower Survey, 524

Haig, Alexander M., Jr., 346, 357, 422
Haile Selassi, 404
HALFMOON plan, for unified defense budget, 75–76, 80
Halloween Massacre (1975), 384–385
Halperin, Morton H., 263, 337
Halsey, William F., 31
Handy, Thomas T., 1
Hanyok, Robert J., 285
hard-knock option
  for Dominican Republic crisis (1965), 292
  for Vietnam War, 288, 289, 290, 291, 323
Harmel, Pierre, 261
Harmel Report, on NATO's MC 14/3, 261, 342
Harmon, Hubert R., 84–85
Harriman, W. Averell
  on Laos crisis, 223
  MacArthur's dismissal and, 115
  named Ambassador to the Soviet Union, 20
  named special assistant for national security affairs, 113
  nuclear test ban and, 235
  on waning Soviet interest in military collaboration, 21
Harrison, William K., Jr., 138
Haslam, Jonathan, 495
Hawaiian Conference (1966), 296
Hayes, Grace Person, 32
Hayward, Thomas D., 393, 423
Herres, Robert T., 455, 482
High Frontier, 434
High Level Group (HLG), 402
Himalayas ("the Hump"), supplies and equipment for China and, 35
Hiroshima, bombing of, 50, 51
Hitch, Charles J., 254
Hizballah terrorists, airliner hijacking by (1985), 463

INDEX

Ho Chi Minh Trail, 296, 309, 311, 319–320, 321
Hoffmann, Stanley, 335
Hollis, Leslie, C., *xiv*
Holloway, James L., Jr., 195
Holloway, James L., III, 208n115, 377, 393, 413
Honolulu conference (1965), 292
Hopkins, Harry, 8, 10
Horner, Charles A., 514, 519, 521
Hound Dog missile, 254
House Armed Services Committee, 450, 453
Hull, Cordell, 8
Hull, John E., 86–87
human rights, Carter on, 395, 416n18
Hungarian rebellion (1956), 158, 193
Hussein, King of Jordan, 194, 195
Huyser, Robert E., 406
hydrogen bomb (H-bombs), 98–99, 141, 144, 159, 174

I Corps region, South Vietnam's, 307, 308, 310–311
Ia Drang Valley conflict, Vietnam (1965), 296
ICBMs. *See* intercontinental ballistic missiles
Iklé, Fred C., 422
IL–28 fighter-bombers, Soviet, 243n105
Inch'on Operation, MacArthur and, 107–108
Independence (carrier), 355–356
India, 33, 159
India-Burma Theater, 37–38
Indochina. *See also* Vietnam War
   crises of 1954–1955 in, 135
   French and Communist Viet Minh struggle in, 146–147
   JCS debate on involvement in war in, 147–148
Informal Policy Committee on Germany (IPCOG), 43
initial operating capability (IOC), for Soviet ICBMs, 175
Instant Thunder plan against Iraq, 513–514, 515
Institute for Defense Analyses (IDA), 369
Intelligence Community. *See also specific intelligence units*
   accuracy of intelligence on Soviet capabilities and, 142
   British deception on Suez crisis for, 192
   on China's political dominance in Asia, 348
   concerns on 1948 Soviet buildup in Germany, 73–74
   on Cuba and the Soviet Union in late 1950s, 197
   on Gorbachev, 485
   JCS interpretations of same data differently from, 180–181
   overlooking Far East before North Korea attacked South Korea, 102
   on Saddam's forces against Kuwait, 507
   on South Vietnam collapse, 382
   on Soviet bloc forces' preparedness (1956–1957), 157–158, 160–161
   on Soviet ICBM arsenal (late 1960s), 338
   on Soviet missile in 1950s, 175, 182
   on Soviet missile in 1960s, 264
   on Soviet missiles in Cuba, 227
   on Soviet Union's nuclear weapons program, 98
   on Warsaw Pact buildup, 345
Intelligence Directorate (J-2), 185
intercontinental ballistic missiles (ICBMs)
   Air Force development of, 183, 337
   Carter on reductions in, 398
   Gaither Committee report on, 177–179
   intelligence on Soviet development of, 180–181
   interim SALT agreement on, 359n21
   Jackson on SALT agreements and, 341
   JCS estimates on quantity needed for assured destruction strategy, 249–250, 270n23
   JCS on frozen numbers for, 340–341, 544
   land-based, assured destruction strategy and, 248
   post-SALT I Soviet testing of, 342, 375
   Powell's estimates for, 496
   on SALT I agenda, 338
   small, single-warhead, Scowcroft Commission on, 438
   Soviet, surprise attacks during 1950s by, 144
   Soviet launch of (1957), 173
   Soviet R&D on, 264
   START I Treaty on, 497
   U.S. concerns on 1950s Soviet development of, 174–177
Interim Committee, interdepartmental, on atomic bomb policies and use, 48–49, 50
intermediate-range ballistic missiles (IRBMs). *See also* SS–4; SS–5; Thor
   Air Force development of, 183
   French development of, 257
   land-based, Army testing of, 146
   Soviet testing of, 175
intermediate-range nuclear forces (INF)
   NATO on modernization of, 402, 440–441
   Treaty mandating elimination of (1987), 459–461, 475n40
International Atomic Energy Agency (IAEA), 160
inter-Service rivalry
   during and after World War II, 3
   conferences on (1948 and 1949), 69
   Crowe on overdramatization of, 452–453
   Desert Shield/Desert Storm and, 514, 517, 525
   fundamental differences of opinion and, 539
   JCS in Vietnam era and, 328
   in JCS under Radford and Eisenhower, 135–137

Jones on JCS organizational structure and, 450
of mid- to late-1950s, 173–174
missile development and, 176–177, 183–184
MX missile and, 370
over joint strategic targeting, 187–190
peacetime total force (1970s) and, 366–367
Rapid Deployment Force and, 409
Rapid Deployment Joint Task Force and, 431
Reagan buildup and, 426, 546
rearmament for Korean War and, 112
Richardson Committee study on post-World War II organizational reform and, 40
Vietnam War strategy and, 308
during World War II, 3, 538
Zumwalt on inter-Service cooperation and, 315
Iran. *See also* Iran-Iraq war
antiship mines in Persian Gulf of, 470–471
as anti-Soviet U.S. ally, 405–406
Baghdad Pact and, 190
hostage rescue mission (1979–1980), 411–414
Khomeini's rise to power in, 406–407
U.S. embassy seizure in, 409
U.S. on Baghdad Pact and, 191
U.S. relations with (1970s), 352
Iran Ajr (minelayer), 470
Iran-contra affair (1986), 424, 465–466
Iran-Iraq war, 432, 467–468, 472, 505
Iraq. *See also* Iran-Iraq war; Operation *Desert Shield/Desert Storm*
Baghdad Pact and, 190
defense establishment of, 509
Kuwait invasion by, 487–488, 505, 508
post-hostilities phase of Gulf War and, 528–531
rebellion (1958) against monarchy of, 194–195, 352
Soviet support for, 352, 383
U.S. on Baghdad Pact and, 191
USS *Stark* attack by, 470
Ismay, Hastings, *xiv*, 124
Israel. *See also* Arab-Israeli conflict
Lebanon invasion by, 429
Nasser and, 191, 192–193
Saddam Hussein's belligerence against, 507, 523–524, 534n65
U.S. arms sales to, 352, 353
USCENTCOM excluded from involvement with, 431
USS *Liberty* attacked by, 353
Italy, 15–16, 257

Jackson, C.D., 148
Jackson, Henry M., 265–266, 341, 375, 376, 379, 381
Japan
ARCADIA meetings (1941) on defeat of, 9
atomic bomb development and use against, 46–52
ending war with, 43–46
War Plan Orange against, 3, 31–32
Jedi Knights (Army planning staff), 517
Jeremiah, David E., 482–483, 529
Jews, partitioning of Palestine for homeland for, 76–77
Joe 1 (Soviet's first nuclear device), 98
Johnson, Hansford T., 482
Johnson, Harold K., 286, 290–291, 301n61, 328
Johnson, Louis
becomes Sec. of Defense, 83
briefed on ROK Army strength, 103
on military spending projections under NSC 68, 101–102
Revolt of the Admirals (1949) and, 99–100
steps down as Sec. of Defense, 107, 128n56
on U.S. troops sent to Korea, 104
on weapons effects and WSEG studies, 85
as Z Committee advisor on U.S. atomic energy program, 99
Johnson, Lyndon B. *See also* McNamara, Robert S.; Vietnam War
air war in Vietnam and, 295
arms sales to Israel under, 353
on call-up of Reservists for NATO, 260
Dominican Republic crisis and, 292
on ending Vietnam War, 311
Glassboro summit with Kosygin and McNamara and, 267
on JCS strategy for Vietnam, 329
McNamara as Sec. of Defense under, 245
MLF demise and, 258
on not seeking reelection, 305
SALT and, 336
strategic review of Vietnam War (1966) and, 296
Vietnam bombing suspended by, 312–313
Vietnam War chain of command and, 297–298
Johnston, Robert B., 516
Joint Action of the Army and Navy (JAAN), 4, 23n16
Joint Administrative Committee, 14
Joint Analysis Directorate (JAD), 206n73
Joint Army and Navy Board, U.S., 3, 5
Joint Chiefs of Staff (JCS). *See also* Chairman, Joint Chiefs of Staff; détente; inter-Service rivalry; National Security Council
Bay of Pigs operation and, 214–216
on catalog of commitments military budget would support, 80
chain of command in Vietnam War and, 297–298
on China as threat, 348
Cold War and niche for, 124–125
contingency plans for U.S. troops supplied to NATO by, 122
Cuban missile crisis and, 229–231, 232–233

INDEX

divergences from CCS organization, 6–7
early meetings and work of, 5–6
formation of, 1–2
Goldwater-Nichols legislation on, 454–455, 482
Intelligence Community and, 180–181
Johnson's deescalation of Vietnam War and, 312–313
Kennedy administration and, 211–213
Korean War onset and, 103–104
Laos crisis and, 221–224, 240n55
leadership at Tehran Conference (Nov. 1943) by, 17–18
legal status of, 8–9, 67–68, 538
MacArthur and, 106–107
McNamara on increased defense spending and, 246
NCA amendments (1948) on diminished role for, 82, 541–542
North African decision and, 10–11
nuclear test ban and, 234
as obstacle to Eisenhower's plans, 141–142, 150
organization chart (1942), *6*
organization chart (1947), *68*
organization chart (1959), *187*
organization chart (1987), *456*
Pacific strategy during World War II under, 9
on post-World War II atomic bomb use, 63, 64
post-World War II defense establishment and, 51–52
on post-World War II military readiness, 61–62
Reagan's meetings with, 424
Reagan's Strategic Defense Initiative and, 433, 436–438
rearmament for Korean War and, 112–113
redefined mission (1947) of, 68
reorganization after Casablanca Conference of, 13–14
SACEUR communication with, 120
showdown with McNamara over air war in Vietnam and, 309–310
strategic bombing of Germany and, 13
Tonkin Gulf incident and, 284–285
TRIDENT Conference and, 14–15
Truman on unification debate and, 60
Vietnam War and, 286–292, 326–329
World War II influence of, 7–8, 537–538
Joint Forces Air Component Commander (JFACC), 519
Joint Intelligence Committee (JIC), 4, 5, 70, 72
Joint Intelligence Group, 72
Joint Logistics Committee, 6, 14, 72
Joint Logistics Group, 72
Joint Long-Range Strategic Estimate (JLRSE), 112
Joint New Weapons Committee, 6
Joint Outline Emergency War Plan (JOEWP), 111, 114
Joint Planning Committee, U.S., 4
Joint Post-War Committee (JPWC), 42
Joint Program for Planning, 246
Joint Psychological Warfare Committee, 6
Joint Special Operations Command (JSOC), 524, 533n49
Joint Staff Mission (JSM), Britain's, 2
Joint Staff Planners (JPS). *See also* Joint Strategic Plans Committee
 on ending war with Japan, 45
 McNarney's JCS reorganization and, 14
 Operation *Crossroads* under, 63
 staffing for, 5
 Unified Command Plan and, 66
 war in the Pacific and, 44
 Wedemeyer on State Dept. liaison with, 41
Joint Strategic Capabilities Plan (JSCP), 112, 136, 194
Joint Strategic Committee, U.S., 4
Joint Strategic Objectives Plan (JSOP)
 under Carter, 393
 McNamara's use of systems analysis and, 246–247
 Nixon and Kissinger on reordering priorities in, 347–348
 rearmament for Korean War and, 112
 as Service requirements without priorities, 136, 541
 timeframe for, 269n5
Joint Strategic Planning Document (JSPD), 393, 541
Joint Strategic Plans Committee (JSPC), 72, 83, 84–85, 138–139. *See also* Joint War Plans Committee (JWPC)
Joint Strategic Survey Committee (JSSC)
 on atomic bomb's military and strategic impact, 51
 Casablanca Conference on combined bombing of Germany and, 13
 creation of, 11
 first objectives of, North African decision and, 12
 FY 1952 military budget review and, 100
 JPWC under, to work with State and EAC, 42
 on naval blockade of Cuba, 228
 part-time inter-Service offices on, 72
 on replacing CCS with tripartite United Chiefs of Staff, 26n76
 Service functions report (1946) by, 67
 as State Dept. and JCS liaison, 41
 to study Marshall's proposal for post-World War II organizational reform, 39–40
 on Weapons Systems Evaluation Group, 85–86
 World War II membership on, 6
Joint Strategic Target Planning Staff (JSTPS), 186–187, 188, 189, 206n78, 252
joint task force (JTF 116), 221–222
Joint Task Force Middle East (JTFME), 469
Joint War Games Agency, 206n73, 283
Joint War Plans Committee (JWPC), 14–15, 41, 45, 46, 71. *See also* Joint Strategic Plans Committee

jointness. *See also* Goldwater-Nichols Defense Reorganization Act
   Checkmate and, 514
   Desert Storm as, 518–519, 525, 531, 547–548
   Goldwater-Nichols legislation on, 454, 546
   Gulf War (1990) and, 531
   Kuwait shipping escort operation and, 470
   Libyan bombing raid as display of, 464
Jones, David C.
   as CJCS, *390*, 392–393
   on countervailing strategy (1978), 396
   on increasing the military budget, 421
   Iran hostage rescue mission and, 411, 413
   on JCS organizational structure, 450, 546
   joins the JCS, 377
   on SALT II, 399–400
   on Steadman group's reorganization plan for JCS, 394
   on stealth technology, 395
   on three near-simultaneous conflicts' planning, 427
   Weinberger's relationship with, 422–423
Jordan, 195, 352, 357
Jordan, Hamilton, 413
JSTARS (surveillance and tracking system), 458
Jupiter (medium-range ballistic missiles, MRBMs)
   Army and Navy development of, 176
   Khrushchev on Cuban missile crisis and, 232–233
   with NATO host countries (1950s), 252
   retirement of, 272n46
   Soviet IRBMs in Cuba comparable to, 228–229
   testing of, 146
   in Turkey, Soviets' complaints about, 226
   Wilson assigns to Air Force, 177

Karch, Frederick J., 277
Kassim, Abdul-Karim, 195
KC–135 jet tankers, 143
Kelley, Paul X., 410, 423
Kennan, George F., 70, 80, 100
Kennedy, John F. *See also* McNamara, Robert S.
   assassination of, 245
   Bay of Pigs operation under, 213–216
   Berlin under siege and, 216–220
   Cuban missile crisis origins under, 224–228, 241n84
   Cuban missile crisis under, 228–233
   defense spending increases under, 246
   on flexible response doctrine, 211
   JCS estrangement with, 233–234
   Laos crisis under, 221–224
   Macmillan meeting on Skybolt or Polaris with, 254–255
   on McNamara's FYDP, 247
   nuclear test ban and, 234–236
   Vietnam commitment under, 278–279, 280, 281
Kennedy, Robert F., 213, 215–216, 222, 225, 232, 305
Kenney, George C., 31
Key West Agreement (1948), 113, 134, 203n19
KGB, 434–435, 440–441. *See also* Soviet Union
Khalid bin Sultan, Prince, 519
Khe Sanh battle, Vietnam War, 305, 307–308, 310–311
Khomeini, Ruhollah (Ayatollah), 407
Khrushchev, Nikita S.
   on Berlin and the West, 199
   deploying missiles to Cuba, 226–227
   Eisenhower meetings with, 201
   on liberation wars in Latin America, 225–226
   on McNamara's counterforce/no-cities doctrine, 249
   on MRBMs in Turkey versus Cuba, 232, 243n105
   nuclear test ban and U.S. relations with, 234, 235
   on Open Skies proposal, 162
   Powers' U–2 mission and summit with Eisenhower and, 183
   propaganda and deception campaign on missile development by, 180
   Vienna meeting with JFK and, 217
   Vietnam War and, 296
   on wars of national liberation, 278
Killian, James R., Jr., 161, 162, 173
Kim Il-song, 103
King, Ernest J.
   atomic bomb development and, 47
   Casablanca Conference (Jan. 1943) and, 13
   on casualty estimates for ending war with Japan, 46
   CCS at the Second Quebec Conference and, *xiv*
   on charter for JCS during World War II, 8
   Combined Chiefs of Staff and, 2
   ending war with Japan and, 44–45, 49
   on JSSC's objective for landing in Europe, 12
   Leahy appointment as Chief of Staff to Commander in Chief and, 7
   MacArthur friction with Nimitz and, 32
   on major offensive in Central Pacific, TRIDENT Conference and, 14, 15
   on numerical superiority for Pacific Ocean Area command, 31–32
   on post–World War II organizational reform, 39, 40
   QUADRANT (Aug. 1943) talks on *Overlord* and, 16
   on Russia's role in Europe during World War II, 9
   steps down from JCS, 60
   on Truman's decision to use atomic bomb, 50
   on unified combined commands, 30
   working in harmony with Marshall during World War II by, 3
Kinkaid, Thomas C., 31

INDEX

Kissinger, Henry A.
- on air campaign against North Vietnam, 322–323
- on ALCMs discussed at Vladivostok, 381
- backchannel contacts by, 316, 322
- on chain of command, 389n67
- on Christmas bombing in Vietnam, 325
- on concessions to PRC involving Taiwan, 349–350, 351
- on détente, 335
- dual roles in Nixon administration of, 363n87
- ending Vietnam War and, 313
- Ford and, after Nixon's resignation, 378, 379
- Ford on role of, 385
- Laird's plan for Vietnamization and, 318
- on LAM SON 719 operation, 319, 320
- on MIRV limitation for SALT II, 379
- on new strategic weapons after Vietnam, 375
- NSC system and, 544–545
- October War (1973) and, 355
- policy process under, 315
- on reorienting U.S. strategy on PRC, 348
- Rowny and, 376
- SALT I backchannel talks with Dobrynin and, 339, 378
- Schlesinger rivalry with, 378, 383
- Schlesinger standoff on Israeli assistance with, 356
- secret talks with China and, 348
- on Shah of Iran, 405–406
- on Sino-Soviet relations (1970s), 361n46
- on Soviet ICBM arsenal, 338
- on targeting practices in SIOP, 371
- on Vietnamization, 319

Kistiakowsky, George B., 182, 189
Knox, Frank, 8
Korean Military Advisory Group (KMAG), U.S., 103
Korean War (1950–1953)
- Chinese intervention in, 110
- Cold War and outbreak of, 96
- ending, Eisenhower and, 137–140
- impact of Chinese intervention in, 111–113
- as JCS turning point, 124
- MacArthur, Inch'on Operation, and, 105–108
- onset of, 102–105
- Truman meeting with MacArthur and, 109
- Truman on NSC 68 and, 108–109

Kosygin, Alexei N., 267
Kuwait. *See also* Operation *Desert Shield/Desert Storm*
- appeal for shipping protection by, 468–469
- Iraq's invasion of, 487–488, 505, 508
- origins of 1990 crisis in, 506–508
- shipping escorts for, 469–472, 547

Kvitsinskiy, Yuli, 440, 447n58

Kwantung army, Japanese, 21
Kyes, Roger M., 148

Laird, Melvin R.
- on air campaign in Vietnam War, 321, 322, 323
- on Kissinger and targeting doctrine, 372
- on Kissinger's circumvention of authority, 316
- on Soviet ICBM arsenal (late 1960s), 338
- on targeting doctrine revisions to Nixon, 373
- on Trident missile submarines, 370
- on U.S. support for NATO, 345
- in Vietnam, ca. 1969, *304*
- Vietnamization strategy and, 318

LAM SON 719 operation and, 319–321
Land, Edwin H., 162
land-based aviation for antisubmarine warfare, 67, 69, 72
Landon, Truman H., 100
Lansdale, Edward G., 278–279
Laos, 221–224, 286, 294, 326
Latin America. *See also* Central America
- Eisenhower's anti-Communist policies in, 196–197
- Soviet focus on, 225–226

Lavelle, John D., 322
Lawrence Livermore research laboratory, 433
Lawton, Frederick J., 101
Leahy, William D.
- atomic bomb development and, 47
- aversion to political-military affairs by, 40–41, 55n49
- CCS at the Second Quebec Conference and, *xiv*
- on charter for JCS during World War II, 8
- as Chief of Staff to Commander in Chief, 7, 538
- on ending war with Japan, 45
- interdepartmental Interim Committee on atomic bomb and, 49
- as JCS chair during World War II, 82
- opposition to unification by, 64–65
- post–World War II JCS service by, 60
- on post–World War II organizational reform, 40
- Roosevelt's meeting with MacArthur and Nimitz and, 44
- on Truman's decision to use atomic bomb, 50

Lebanon, 194, 195, 429, 430
Ledo Road, from Burma to China, 35
Lehman, John F., Jr., 453
LeMay, Curtis E.
- on ABM, 263–264
- becomes Air Force Chief of Staff, 224–225
- Berlin crisis and, 219
- Cuban missile crisis and, 230, 232, 233–234
- McNamara's flexible response force posture and, 248

NATO retardation bombing plan and, 119
Operation *Crossroads* under, 63
as SAC commander, 79
on Taylor and McNamara's plan for Vietnam, 283
on troops and bombing after Tonkin Gulf incident, 286
Lemnitzer, Lyman L.
  Berlin crisis and, 217, 219
  on Cuban surveillance, 228
  on 8000-man force versus Win Plan, 279–280
  on Laos crisis, 221
  Laos crisis and, 223–224
  on McNamara and his staff, 212
  McNamara's FYDP and posture statement from, 247
  on nuclear weapons for NATO, 252
  steps down as CJCS, 224
  Taylor as MILREP and, 213
lend-lease program, 21, 34–35, 37, 54n35
liberation doctrine, Reagan on, 421
Liberty, USS, 353
Libya, 383, 429, 430, 463–464
Lilienthal, David E., 99
limited employment options, in revised targeting doctrine (1973), 373
Limited Test Ban Treaty (1963), 235, 266
Lincoln, George A., 178–179
Linebacker, 323–324, 329, 351
Linebacker II, 325–326, 329, 351
Lippmann, Walter, 69
Lisbon Conference (1952), on NATO force goals, 123
Little Boy bomb, 48, 50
Live Oak (planning body), 200–201, 219
Logistics Directorate (J-4), 185
log-rolling process, 136
Lon Nol, 318–319
long-range bombers, 248, 341
Long-Term Defense Program (LTDP), NATO's, 401, 457
Los Alamos, New Mexico, 48, 433
Lovett, Robert A., 96, 134

MacArthur, Douglas
  on Chinese forces inside Korea's borders, 110
  Collins's reports on Korean War and, 113–114
  command of war under, 31
  as Commander in Chief, Far East, 105–106
  ending war with Japan and, 44–45
  friction between Nimitz and, 32
  Intelligence Community conflicts with, 102
  on Republic of China, 349
  Southwest Pacific Area command under, 30–31
  Truman meeting on Wake Island 1950 with, *94*, 109–110
  Truman's dismissal of, 113–116
  unified command in Pacific and, 65–66
Macmillan, Harold, 222, 254–255
Maddox, USS, 284
Mahan, Erin R., 257
The Malaise of Soviet Society (British intelligence report), 428–429
Malinovskiy, Rodion, 226
Manhattan Engineer District (MED) (Manhattan Project), 47–50, 51, 62, 86
Mansfield, Mike, 344
Mao Zedong, 34, 36, 37, 95–96, 351. *See also* People's Republic of China
March Crisis of 1948, 73–76, 90n53
Marine Corps
  on amphibious operations, 67, 69, 72
  arrival in Vietnam, 277, 290
  CENTAF compromise with, 519–520
  coequal status with Service chiefs for, 394
  Johnson on Vietnam mission for, 291–292
  post-World War II projections for size and capabilities of, 39
  proposed post-World War II merger with Navy, 65–66
  Truman on, 59–60
Mark 12A warhead, 385, 396
Marshall, George C.
  on alternative to atomic bomb, 50
  atomic bomb development and, 47
  becomes Sec. of Defense, 107
  cautions Wedemeyer on role in China, 38
  CCS at the Second Quebec Conference and, *xiv*
  on charter for JCS during World War II, 8
  Chiang Kai-shek and, 97
  Chinese affairs and, 35
  Combined Chiefs of Staff and, 2
  on Continental invasion in 1942 or 1943, 10
  Dill's collaboration with, JCS development and, 5
  on direct link between President and JCS, 7
  on European Recovery Program, 69
  interdepartmental Interim Committee on atomic bomb and, 49
  MacArthur friction with Nimitz and, 32
  MacArthur's dismissal and, 115–116
  on MacArthur's operations in North Korea, 109
  military budget for FY 1950 and, 81
  on post-World War II organizational reform, 39, 40
  relaying Deane's sobering assessment of the Soviets, 20
  Roosevelt and, 24n28
  Roosevelt's North African decision and, 10–11
  on Soviet collaboration during World War II, 22

INDEX

Truman's reliance on, 60
on universal military training, 61–62
on U.S. support for NATO, 121
on waning Soviet interest in military collaboration, 21
working in harmony with King during World War II by, 3
Marshall Plan, 100
Martin, Joseph W., Jr., 115
Masaryk, Jan, 73
massive retaliation doctrine, 248–249, 251, 258, 261
Matsu Islands, Taiwan Strait, 135, 150, 151–152
Mayaguez affair (1975), 383–384
MC 14/1 (NATO's forward strategy, 1952), 121–122, 251
MC 14/2 (NATO's strategy on tactical weapons, 1957), 157, 251
MC 14/3 (NATO's flexible-response strategy, 1966–1967), 258–261, 342
MC 48 (NATO's New Approach strategy), 152–156, 541
MC 48/3 (NATO's implementation strategy, 1966–1967), 260–261
McCain, John S., Jr., 319
McCarthy, Eugene, 305
McCloy, John J., 49–50
McConnell, John P., 264, 290, 309, 368
McDonald, David L., 233
McElroy, Neil H., 136, 181, 182, 184, 185, 186
McFarlane, Robert C., 463, 468
McMaster, H.R., 283, 327
McNamara, Robert S.
  AMSA opposition by, 368
  Bay of Pigs operation and, 215, 225
  Berlin Task Force and, 217, 218
  chain of command in Vietnam War and, 297–298
  Cuban missile crisis and, 228, 229, 230, 231, 232
  damage limitation debate and, 261–267
  disdain for JCS's position on air campaign by, 309
  Honolulu conference (1965) on Vietnam War and, 292
  JCS and, 214, 249–250, 328, 541–543
  Laos crisis and, 222, 223–224
  on LeMay on the JCS, 233–234
  MLF demise and, 255–258
  NATO's MC 14/3 strategy and, 258–261
  reconfiguring strategic force posture under, 247–251
  refused expanding bombing in Vietnam, 309–310
  as Sec. of Defense, 211–212, 213, *244*
  Sentinel and seeds of SALT, 267–269
  on SIGMA II–64 testing graduated pressure hypothesis, 287
  Skybolt affair and, 253–255
  on stalemate in Vietnam, 305–306
  steps down as Sec. of Defense, 311
  systems analysis used by, 211, 245–247
  target selection for air war in Vietnam and, 295, 296
  Trident missile development and, 369
  U.S. involvement in Vietnam and, 282
  Vietnam assistance policy and, 280
  Vietnam War strategy and, 288, 291, 294
McNarney, Joseph T., 14, 80
McNaughton, John T., 288
McPeak, Merrill A., 517, 521, 525
Mediterranean
  combined unified command for campaigns in, 29–30
  U.S. Navy during Reagan years in, 428
  Voroshiloff questions Brooke on campaigns in, 17
Mediterranean Command (CINCAFMED), NATO's, 121
Medium Term Defense Plan (MTDP), 119, 120, 123
medium-range ballistic missiles (MRBMs), 256, 272n50
Merchant, Livingston, 257
Merrill's Marauders (Galahad commando unit), 33
Meyer, Edward C., 423, 451
Meyer, John C., 325, 326
Middle East. *See also* Persian Gulf; *specific countries*
  Carter Doctrine on, 410–411
  Eisenhower Doctrine on, 194, 195
  Operation *Blue Bat* in, 194–196
  Palestine partitioning for Jewish state in, 76–77
  politico-military vacuum in, 193–194
  Reagan Doctrine and, 429
  Soviet influence in, 190, 383
  Soviet-American relations and (1970s), 351–352, 353–354, 355–356, 357, 358
  Suez crisis in, 191–193
  U.S. Commander in Chief for, 208n109
Midgetmen ICBMs, 438
military budget
  Carter's proposed reductions in, 392, 395
  for *Desert Shield/Desert Storm*, 511–512
  Eisenhower on stable level for, 141, 541
  Eisenhower's reduction of rate of growth in, 143
  for FY 1950, gap between requirements and Truman's limits on, 79–80
  for FY 1950, integrated defense plan and, 76
  for FY 1950, international situation and, 76–77
  for FY 1950, submitted by Forrestal, 81
  for FY 1952, Johnson's hold-the-line spending policy for, 99–100
  for FY 1952, Korean War onset and, 108–110
  for FY 1952, Reagan's compared with, 444n19
  for FY 1981 and FY 1982, Reagan buildup and, 425–429

JCS concerns on NATO budget and, 96
Johnson as JCS Chairman and, 83
Johnson's Great Society and, 250, 293, 543
March Crisis of 1948 and, 73–74
McNamara's assured destruction strategy and, 250
McNamara's use of systems analysis and, 246–247
Nike-X (1966) and, 266
NSC 68 projecting military threat in 1950s and, 101, 540
Powell's base force plan and, 495–496
under Reagan, 545
rearmament for Korean War and, 111
Rumsfeld on growth in, 385
supplemental, Kennedy's call for, 217–218
Truman's limitation on, integrated statement of service and, 71–73
unified, HALFMOON plan for, 75
unified, omitted from NSA (1947), 89n46
Vietnam War and allocation of, 306–307, 365–366
Vietnam War as portion of, 308
Military Committee, NATO's, 120
Military Liaison Committee, AEC's, 62, 70
military personnel policy. *See also* military budget
Goldwater-Nichols legislation on, 455, 546
military technologies. *See also specific weapons*
Gulf War (1990) and, 530–531
reorganization and challenges of, 65
stealth aircraft, 395
Unified Command Plan and, 66–67
Vessey on strategic defense and, 436–437
MILL POND (interagency plan for Laos), 221–222
mines, Iranian antiship, 470–471
mining North Vietnamese ports, 289, 323, 329, 351
Minuteman (intercontinental ballistic missile), 176, 248, 374
Minuteman III (MIRVed ICBM), 263
missile gap, 180–183, 218
missile program. *See also specific types of missiles*
JSTPS and growth of, 186–187
Momyer, William W., 310
Mondale, Walter, 413
MONGOOSE (covert operations in Cuba), 225
Montgomery, Bernard Law, 18, 120, 124
Moorer, Thomas H.
on air attacks against North Vietnam, 321, 322–323
on Christmas bombing in Vietnam, 325, 326
as CJCS, 314, *334*
final posture statement to Congress (1974) by, 358
Foster's study panel on targeting and, 372
Kissinger's secret rendezvous with Zhou Enlai and, 350
LAM SON 719 operation and, 319–320
on modernizing U.S. strategic arsenal, 341

October War (1973) and, 355, 356, 357
on peacetime total force (1970s), 367
SALT I agreement and, 339
Morgenthau, Henry, Jr., 42
Mountbatten, Louis, 36–37
Mozambique, 383
mujahideen in Afghanistan, U.S.-backed, 431, 461
multilateral nuclear force (MLF), 252, 255–258
multiple independently targetable reentry vehicles (MIRVs), 262–263, 379
multiple reentry vehicles (MRVs), 262
Munitions Board (MB), of National Military Establishment, 68
Murphy, Daniel J., Sr., 356
MUSKETEER (British-French-Israeli military operation against Egypt), 192–193
mustard gas, Truman on, 139
Mutual and Balanced Force Reduction (MBFR) Talks, 343, 376, 439–440, 494
mutual assured destruction (MAD), 341, 433
mutual assured survival, 434
Mutual Defense Assistance Program (MDAP), 96, 117, 125n3
MX (intercontinental ballistic missile)
Carter on deployment of, 396
development problems for, 370
as freeze movement target, 435
JCS requests for, 262, 368, 544
Reagan buildup and, 427–428, 438
Rumsfeld on development of, 385
Schlesinger on NSDM 242 and, 374, 375

Nagasaki, bombing of, 50, 51
Nassau agreement (1962), 254–255, 257
Nasser, Gamal Abdul, 191–192, 193
National Command Authority (NCA), 186, 454
national intelligence estimate (NIE 11-5-57), 175
national intelligence estimate (NIE 11-8-59), 175
National Military Establishment (NME), 67–68, 82. *See also* Defense Department
National Military Strategy for 1992–1997 (NMS 92-97), 481, 484, 485
National Security Act (1947)
on accountable civilians, 134
integration of military budget omitted from, 89n46
on JCS duties as corporate advisory body to President, 537
1948 amendments to, increased JCS size under, 72
1948 amendments to, on Chairman, JCS, 82
1958 amendments to, Sec. of Defense authority and, 246, 541
statutory standing for JCS under, 67–68, 538
National Security Agency (NSA), 175, 227, 355, 408
National Security Council (NSC)

INDEX

Bundy on JCS participation in meetings of, 287
creation of, 68
on Cuban invasion after Bay of Pigs, 225
on curbing expansion of Chinese Communist power, 147
to develop national objectives for military requirements, 76
disarmament debates within, 160
as Eisenhower's high-level policy forum, 136–137
on Foster's targeting panel's findings, 372–373
Iran hostage rescue mission and, 413
Kennedy's changes to, 212–213
Kissinger's circumvention of, 378
Korean War and changes in, 124
national security policy analysis, Service requirements and, 80
Nixon's use of, 315, 331n43
on nuclear weapons for Armed Forces, 79
planning FY 1952 military budget and, 111
Review Group, Kissinger as head of, 316
State-Defense review of FY 1952 military budget for, 100–101
Truman and enhanced role during Korean War for, 112–113
national security decision memoranda (NSDM), Nixon's, 316
National Security Decision Memorandum (NSDM 95), 345
National Security Decision Memorandum (NSDM 109), 219
National Security Decision Memorandum (NSDM 133), 345
National Security Decision Memorandum (NSDM 242), 371–373
National Security Decision Memorandum (NSDM 245), 379
National Security Resources Board (NSRB), 68
National Security Study Memoranda (NSSM), Nixon's, 316, 347
National Strategic Targeting List (NSTL), 189
national technical means, arms control verification using, 337
Nationalist Chinese. *See* Chiang Kai-shek; Taiwan
Navy, U.S.
air offensive against Soviet war-making capacity (1948) and, 75
CENTAF in *Desert Storm* and, 519
Combined Chiefs of Staff and, 2
inter-Service rivalry between Army and, 3
on Johnson as JCS Chairman, 83
on land-based aviation for antisubmarine warfare, 67, 69, 72
missile development and, 176, 177, 183
New Look defense budget and, 144

nuclear weapons for, 79
post-World War II projections for size and capabilities of, 39
proposed post-World War II merger with Marines, 65–66
Reagan buildup and expansion of, 428
Special Projects Office, 262
on Trident missile submarines, 369–370
Navy Department, 3, 6, 40
Navy General Board, 4–5
Navy Plans Division, 4–5
Net Evaluation Subcommittee (NESC), 188
Netherlands, 157, 259
Neutrality Act (1937), U.S., 34
neutron bomb, 402, 457
New Frontier, JFK's, 211, 236n1
New Look (national security policy). *See* Eisenhower, Dwight D., first term
New Zealand, 147, 306
Newhouse, John, 337–338
Nicaragua
Iran-contra affair (1986) in, 424, 465–466
logistical but not military support for contras in, 431
Sandinistas in, 407–408, 467
Nichols, Bill, 453
Nike-X ballistic missile defense system, 264–268
Nike-Zeus ballistic missile defense system, 181, 183
NIKE-ZEUS interceptor missile, 146
Nimitz, Chester W.
command of war under, 31
divided command at Pearl Harbor and, 66
ending war with Japan and, 44–45
friction between MacArthur and, 32
joins the JCS, 60
leaves the JCS, 74
Pacific Ocean Area command under, 30–32
on single Pacific command, 65–66
Nitze, Paul H.
ex-official SALT talks and, 339–340
Gaither Committee report and, 178–179
JCS on ABM and, 268
on military power in foreign policy, 100
military spending projections under NSC 68 by, 101
NSDM 109 and, 219
Reagan's buildup and, 422
on walk in the woods formula for arms control, 440, 447n58
Nixon, Richard M. *See also* détente; Vietnam War
air campaign against North Vietnam and, 322–323, 324
China and, 348, 350–351

on détente, 335–336
on hawks in U.S. after end of détente, 385
on Jackson pressure on SALT negotiating team, 342
JCS and, 313–314, 544–545
Laird's plan for Vietnamization and, 318
on LAM SON 719 operation, 319, 320–321
Middle East involvement and, 354, 355, 356
modernizing the strategic deterrent under, 367–370
on peace with honor in Vietnam, 305
peacetime total force under, 365–367
policy process under, 315
SALT I under, 336–342
SALT II under, 375–379
on SIOP limitations, 371
on Soviet ICBM arsenal, 338
targeting doctrine revised under, 371–375
on Vietnamization, 319
Watergate scandal and, 346, 355, 356, 377–378
winding down Vietnam War under, 316–321
on Year of Europe (1973), 345–346
Noriega, Manuel Antonio, 489–492, 547
Normandy invasion, planned invasion of Japan versus, 46
Norstad, Lauris
  B-29s in Europe and, 78
  on Berlin access rights, 216, 218
  contingency plans for Berlin and, 200–201
  Executive Order on Service functions and, 69
  JCS meeting at Naval War College (1948) and, 58
  multilateral nuclear force concept and, 255
  NATO retardation bombing plan and, 118–119
  NATO's New Approach and, 154
  on nuclear weapons for NATO, 169n98, 252
  unification bill and, 67
North, Oliver L., 424, 468
North Africa, World War II and, 10–12, 29–30, 76. *See also* Mediterranean
North American Air Defense Command (NORAD), 144
North Atlantic Council (NAC), 117–118, 120, 123
North Atlantic Treaty (1949), 95, 96
North Atlantic Treaty Organization (NATO)
  Carter on conventional forces for, 401
  conventional posture in late 1950s of, 156–158
  creation of, 75
  at a crossroads (1952), 124
  defense planning for, 117
  Desert Storm and, 523
  flexible-response force posture and, 251–253
  France on MLF and, 256–257
  Gorbachev's restructuring of Soviet Union and, 460
  on limiting conventional forces in Europe, 493–495

modernization by (late 1970s), 402–403
morale and discipline (mid-1970s) in, 346
as multilateral nuclear force, 255–256, 272n50
Mutual and Balanced Force Reduction Talks and, 439–440
Mutual Defense Assistance Program of, 96
New Approach and nuclear weapons for, 152–156, 541
Nixon and Kissinger on revitalization of, 335
Nixon's Israeli assistance during Watergate and, 356
one-and-a-half war strategy and support for, 366
Reagan buildup and, 457–458
recognizing upgrades needed by, 345–346, 360n42
Soviet influence in Middle East and, 190
U.S. commitment at end of 1960s to, 342–343
U.S. disagreements on need for support of, 344–345
U.S. Great Debate on commitment to, 121
Vietnam War and, 306
Vietnam War and U.S. power and influence within, 258–259
Warsaw Pact as threat to, 343–344
Warsaw Pact buildup and, 400–401
North Korea. *See also* Korean War
  invasion of South Korea by, 102–105
  repatriation concerns of POWs from, 138
North Vietnam. *See also* Vietnam War
  Laos crisis and, 223
  mining ports of, 289, 323, 329, 351
  Tonkin Gulf incident and, 284–285
North Vietnamese Army (NVA). *See also* Ho Chi Minh Trail
  air defenses of, 295
  air strikes on, 323–325
  base camps in Cambodia and Laos of, 294, 317
  Cambodian invasion (1970) and, 318–319
  Khe Sanh siege by, 310
  LAM SON 719 operation and, 320
  Nixon on withdrawal of, 316
  war of attrition and, 297
North Vietnamese Communist Party, 296
NSC 68 (State-Defense review of FY 1952 military budget), 100–101, 108–109, 540
NSC 162/2 (nuclear weapons for use as other munitions), 142
NSD 54 (*Desert Storm*), 522, 527
NSDD 32 (Reagan's national security policy), 428
NSDD 99, 431
NSR 10 (Bush's review of Persian Gulf policy), 506
NSR 12 (Bush's review of national security), 480, 481
Nuclear Non-Proliferation Treaty (NPT), 266–269
Nuclear Planning Group (NPG), NATO's, 258, 402
nuclear weapons. *See also* arms control; atomic bomb; Soviet Union; *specific types of*

INDEX

air war in Vietnam and, 295
allies' objections to use in Western Pacific of, 147
assured destruction strategy under McNamara for, 248
atmospheric testing of, 159, 162–163
B-29s in Europe implying threat of, 78
Brezhnev's ultimatum to Nixon on Israel's October War (1973) and, 357
Carter's control of, 396
China's development of, 348
freeze movement during Reagan era against, 434–435
French development of, 256–257
JCS reluctance to use in Indochina, 149
JSPC's plan to end Korean War using, 138–139, 164n12
as key to future security, Eisenhower on, 141, 142, 540–541
Korean War and potential use of, 104
Laos crisis and consideration of, 222–223, 240n55
MacArthur on tactical use in Korea for, 114
McNamara's assured destruction strategy and, 250–251, 542
for NATO, flexible-response force posture and, 251–253
for NATO, JCS in 1960s and, 259–260, 272n61
for NATO, Norstad on, 169n98
for NATO retardation purposes (1951), 118
NATO's forward strategy on conventional weapons versus, 122
NATO's intermediate-range, 402, 440–441
NATO's New Approach and, 152–156, 541
production increases under Truman and Eisenhower of, 158–159
Reagan buildup and strategic balance with Soviet Union on, 427–428
Saddam's, intelligence on, 524
Soviet, deployed to Cuba, 226–228, 240–241n71
stockpile of U.S. versus Soviet, 159, 169–170n113
strategic plans for, as classified, 371–372
Taiwan Strait confrontation and, 150–151
test ban, Kennedy and, 234–236, 243n110
Truman on restoring production of, 69–70
Vietnam War contingency planning and, 288
Warsaw Pact's buildup of, 344
Nuclear Weapons Employment Policy (NUWEP-80), 397
Nunn, Sam, 487
Nunn-Cohen amendment, 384
Nuri al-Said, 195

October War (1973), 346, 353–358
Office of Defense Mobilization (ODM), 178
Office of Management and Budget (OMD), 367. See also Stockman, David A.
Office of Naval Intelligence (ONI), 231
Office of Strategic Services (OSS), 6–7, 21, 102
Office of the Secretary of Defense (OSD). See also Defense, Secretary of
on ballistic missile defense, 182
defense reorganization (1958) and, 184–185, 186
JCS authority in Vietnam era and, 328
JCS under Powell and, 482
JCS under Reagan and, 424–425
Kissinger bypassing of, 316
McNamara's FYDP and posture statement from, 247
peacetime total force (1970s) and, 367
planning FY 1952 military budget and, 111
reorganization of 1953 and, 183–184
SALT II talks and, 378–379
on Soviet ICBM arsenal, 338
on three near-simultaneous conflicts' planning, 426–427
OFFTACKLE plan for European defense, 117–118
Ogarkov, Nikolai V., 441, 460
oil embargo (1973), 356
Okinawa, 44–45
OMEGA (Anglo-American plan for Middle East), 192
one-and-a-half war strategy, 348, 366, 426–427
Open Skies proposal, Eisenhower's, 160, 162
Operation *Badr* (Egyptian), 355
Operation *Blue Bat* (Lebanese intervention), 195–196
Operation *Broadaxe* (deception plan against Japan), 46
Operation *Coronet* (invasion near Tokyo), 45, 46
Operation *Crossroads*, 63–64
Operation *Desert Shield/Desert Storm*
air war phase of, 522–525
coordinated planning and direction of, 537
final plans and preparations for, 518–522
framing U.S. response to Iraq's invasion of Kuwait, 508–510
ground campaign in, 525–528
operational planning for, 510–515
origins of Kuwait crisis and, 505–508
post-hostilities phase of, 528–531
road to war and, 515–518
Operation *Dominic* (atmospheric nuclear weapons testing), 234–235, 243n110
Operation Duck Hook, 317
Operation *Eagle Claw*, 412, 414
Operation *Earnest Will*, 469–473
Operation *Eldorado Canyon*, 464
Operation *Husky* (against Sicily), 12
Operation *Just Cause* (against Noriega government), 492–493

575

Operation *Matterhorn* (B-29s to China), 37, 53–54n34
Operation *Nougat* (underground nuclear tests), 234
Operation *Olympic* (invasion of southern Japan), 45, 46
Operation *Overlord*, 16, 17
Operation *Praying Mantis*, 471
Operation *Rolling Thunder*
   air campaign against North Vietnam, 290, 292, 294–295, 307
   graduated rules and, 309
   Johnson's modification to, 310
   Linebacker compared to, 323–324
   Moorer on, 314
Operation *Staunch*, 467–468
Operation *Torch* (against North Africa), 10–11, 12
Operational Plans and Interoperability Directorate (J-7), 455
Operations Coordinating Board (OCB), 137, 212
Operations Directorate (J-3), 185–186
OPLAN 34A for Vietnam, 280–281
OPLAN 1002-90 plan for Middle East, 508–509
Organization of American States (OAS), 197
Organization of the Petroleum Exporting Countries (OPEC), 406

Pacific. *See also* Asia-Pacific War
   JCS strategy during World War II in, 9
   strategy and command in, 29–33
   unification challenges in, 65–66
Pacific Ocean Area command, 30–32
Packard, David, 368, 452
Pahlavi, Mohammed Reza (Shah of Iran), 405–406, 411
Pakistan, 190, 191
Palestine Liberation Organization, 429
Palestinian terrorist groups (copy), 383
Palmer, Bruce, Jr., 293, 298
Panama Canal, 196, 404, 407, 489
Paris, Treaty of (1952), 123
Pate, Randolph McC., 186
Pathet Lao (Laotian Communists), 221, 223–224
Patterson, Robert P., 67, 73, 89n46
Peacekeeper program demise, 438
peacetime total force. *See also* Truman, Harry S., peacetime challenges for
   Nixon administration on, 366
   Reagan buildup and, 426
Pearl Harbor, 66, 160
Penkovskiy, Oleg, 218
pentomic divisions, 145–146, 155, 166n51
People's Liberation Army, 110
People's Republic of China (PRC). *See also* Chinese Communists; Mao Zedong; Zhou Enlai

   JSTPS comprehensive target list against countries of, 189
   Korean War onset and, 109
   Laos crisis and, 223
   MacArthur's overconfidence against, 113–114
   military divisions operating in Korea (1950) of, 110
   Nixon and Kissinger's quasi-alliance with, 347–351, 543–544
   Nixon on rapprochement with, 313, 316
   North Vietnamese Army support by, 362n70
   repatriation concerns of POWs in Korean conflict from, 138
   Taiwan Strait confrontation and, 152
Perle, Richard N., 422
Pershing II (P–II) ballistic missiles, 403
Persian Gulf. *See also* Operation *Desert Shield/Desert Storm*; *specific countries*
   Bush administration review of U.S. policy on, 506
   Carter Doctrine on cooperative security framework for, 410
   Kuwait shipping escort operation in, 469–473, 547
   Reagan and tensions in, 467–469
   Soviet threats during Reagan era in, 431–432
   troops for, 517–518
   U.S. military draw-down by 1989 in, 506–507
   U.S. Navy during Reagan years in, 428
Personnel Directorate (J-1), 185, 377
Philippines, 44, 45
Phoumi Nosavan, 221
Picher, Oliver S., 185
PINCHER studies on potential U.S. war with Soviet Union, 71
Pipes, Richard E., 422
planning, programming, budgeting system (PPBS), McNamara's, 246
Planning Board, interagency, 137, 139, 212
Plans and Policy Directorate (J-5), 185, 206n73, 214
plausible deterrence, NATO's New Approach and, 155
Pleiku, South Vietnam, Viet Cong strike against, 287, 289
Pleven Plan, French-sponsored, 120
Pogue, Forrest C., 50
Poland, 429. *See also* Eastern Europe
Polaris A–3 missile, multiple warheads on, 262
Polaris fleet ballistic missile system
   Cuban missile crisis and, 232–233
   Eisenhower's New Look and development of, 144, 176, 183
   integration of, with other strategic forces, 187, 188–189
   McNamara's acceleration of, 248
   missile testing, 234–235

INDEX

for NATO, JFK on, 256
  as Skybolt substitute, British agree to, 254–255
political-military affairs, JCS and, 40–43, 55n49
poodle blanket paper, 219
Portal, Charles, *xiv*, 1
Poseidon (submarine-launched MIRV), 262–263
posture statements, 247, 270n8, 358
Postwar Foreign Policy Advisory Committee, U.S., 41
Potsdam Conference (TERMINAL, July-Aug. 1945), 43, 50
Pound, Dudley, 1
Powell, Colin L.
  base force plan and, 485–489, 495–496, 501n25, 501n29
  briefing Bush on *Desert Storm*, 515–516
  as CJCS, *478*, 481–484
  Desert Shield/Desert Storm coordination and, 537, 547
  Desert Shield/Desert Storm planning and, 510–513, 514–515, 520–521
  on *Desert Storm* ground campaign, 525
  on Iranian aggression in Persian Gulf, 471
  on Iraqi army, 527, 528
  on Iraq's invasion of Kuwait, 509–510
  on NSR 12, 481
  Panama operations (1989) and, 490, 491–492
  Schwarzkopf and, 517, 520
  on START I Treaty, 497, 499
  on Strategic Defense Initiative, 496
  on War Powers Act (1973), 518
Powell, Jody, 413
Powell Doctrine, 484
Power, Thomas S., 249, 270n23
Powers, Francis Gary, 183, 205n52
precision-guided munitions (PGMs), 323–324
Presidential directive (PD–59), 397
President's Military Representative (MILREP), 213. *See also* Taylor, Maxwell D.
President's Special Committee on Indochina, 148
Primakov, Yevgeny, 506
prisoners of war (POWs), 138, 325
program packages, McNamara's use of systems analysis and, 246
Project Defender, 265, 437
Project SILVERPLATE, 48
Project SOLARIUM, 141
public opinion. *See also* antiwar movement
  on defense spending, 480
Pueblo incident (1968), 383
Pustay, John S., 412

Qaddafi, Muammar, 429, 430, 462–464
Qingdao, China, U.S. military base at, 97–98
Quebec Conference (OCTAGON, Sept. 1944), *xiv*, 42
Quebec Conference (QUADRANT, Aug. 1943), 16, 32–33, 36–37, 38
Quemoy Islands, Taiwan Strait, 135, 150, 151–152
Quick Reaction Alert Force, SACEUR's, 233
Quinlan, Michael, 260

Radford, Arthur W.
  as CJCS, *132*, 135–136, 144, 540, 542
  cooling off Taiwan Strait tensions and, 152
  defense reorganization (1958) and, 184–185
  Dien Bien Phu siege (1954) and, 148–149
  French requests for assistance for Dien Bien Phu siege and, 148
  JCS meeting and, *58*
  Korea fact-finding tour with Eisenhower and, 139
  on military budget reductions (1956), 157, 158
  on Nasser, 191
  on New Look defense policy, 142
  OMEGA plan for Middle East and, 192
  on Soviet development of ICBMs, 174
  on Taiwan Strait intervention, 150–152
  U–2 reconnaissance over Soviet Union and, 162
  unification bill and, 67
Radford, Charles E., 316
Rainbow plans, 4
Ramgarh Training Center, India, 36
RAND Corporation, 188, 248
Rapid Deployment Force (RDF), 408–411, 414, 421, 431
Rapid Deployment Joint Task Force (RDJTF), 405, 410, 431
Reagan, Ronald. *See also* Reagan buildup
  arms control plans of, 493–494
  Central American showdown under, 464–466
  changes in Soviet Union and, 449, 472–473
  debating JCS reorganization under, 449–453
  Goldwater-Nichols legislation under, 454–457, 474n20
  Gorbachev and, 459–462, 479
  JCS role in SDI concept development by, 436–437, 545–546
  Kuwait shipping escort operation and, 469–472
  NATO resurgent under, 457–458
  Persian Gulf tensions and, 467–469
  terrorism and confrontation with Libya, 462–464
Reagan buildup
  Armed Forces' image and, 421–425
  arms control and, 438–443
  forces and budgets and, 425–429, 545
  military power and foreign policy and, 429–432
  Strategic Defense Initiative and, 432–438, 545–546

Reagan Doctrine, 429
REAPER (mid-range plan), 111–112
Reed, Thomas C., 422
Reorganization Plan Number 6, 134
Republic of China (ROC). *See* Taiwan
Republic of Korea (ROK), 102–105, 306, 395. *See also* Korean War
Republican Guard, Iraq's, 509, 526, 527–528
Research and Development Board (RDB), 68, 86
Reserves, U.S., 259–260, 297, 307
retardation bombing, NATO defense plan on, 118, 130n96
Revolt of the Admirals (1949), 99, 135
revolution in military affairs (RMA), 530–531
Rhee, Syngman, 103, 108, 140
Rice, Donald B., 513, 520
Richardson, James O., 40
Rickover, Hyman G., 369
Ridgway, Matthew B.
   on Dien Bien Phu air support, 149
   joins the JCS, 135
   Korean War and, 114, 116–117
   as NATO and U.S. Supreme Allied Commander, Europe, 123–124
   NATO's New Approach and, 154, 155
   on New Look defense budget, 145
   on Taiwan Strait intervention, 151–152
   on U.S. in Indochina war, 147–148
Roberts, Samuel B., USS, 470–471
Roberts, William L., 103
Robertson, Walter S., 152
Robinson, John D., 501n25
Rockefeller, Nelson A., 134, 160, 161–162
Rockwell International, 368–369
Rodman, Peter W., 315
Rogers, Bernard W., 458, 460
Rogers, William P., 344
Roosevelt, Franklin D. *See also* Asia-Pacific War; Europe, war in
   ARCADIA meetings (1941) and, 1–2
   on Axis surrender, 13
   on combined unified command, 29–30
   Grand Alliance during World War II and, 18–19
   JCS and, 2–3, 7, 8–9
   loans to China under, 34
   MacArthur and Nimitz in Pearl Harbor with, 44
   Manhattan Project and, 47
   Marshall and, 24n28
   North African decision and, 10–12
   political-military affairs during World War II and, 41
   post-World War II plan for Germany and, 42
   at Second Quebec Conference, *xiv*

   Top Advisory Group of, 47, 56n79
Rosenberg, David Alan, 159
Rostow, Walt W., 279, 280, 285, 288
Rowny, Edward L., 376, 377, 399, 422, 441
Royal Lao Government (RLG), 221
Rumsfeld, Donald H., 384–385
Rusk, Dean, 217, 218, 222, 227, 230, 232, 289, 295
Russell, Richard, Jr., 265–266
Russia. *See* Soviet Union
Ryan, John D., 314, 322

SA–2 surface-to-air missile (SAM), 183, 227, 228
SAC ZEBRA, 119
Sadat, Anwar, 354, 355, 357–358
Saddam Hussein, 467, 505, 506, 507–508, 522, 525–528, 530
Safeguard (missile defense system), 339
Safwan ceasefire accords, for Gulf War, 528, 530
Sakharov, Andrei, 398
SAMOS (Satellite and Missile Observation System), 205n53
Samuel B. Roberts, USS, 470–471
Sandinista National Liberation Front (FSLN), 407–408, 430, 465, 466, 467, 480
SANDSTONE nuclear weapons testing, 78–79
Saudi Arabia
   Saddam Hussein's threats against, 506, 514, 534n65
   U.S. on protection of, 508, 509–510, 512, 515
   U.S. relations with, 352, 357
Schlesinger, Arthur M., Jr., 213
Schlesinger, James R.
   Ford and, 379, 384
   Kissinger and, 356, 378, 383
   on NATO preparedness, 346
   NSDM 242 interpretation by, 373–375, 396
   SALT II talks and, 378, 381
   as Sec. of Defense, *364*
   on South Vietnam collapse, 382
Schlesinger doctrine, 373–375, 396
Schmidt, Helmut, 402
Schroeder, Gerhard, 306
Schwarzkopf, H. Norman
   conventional firepower for Gulf War under, 526–527
   Desert Shield/Desert Storm coordination and, 519, 537
   Desert Shield/Desert Storm planning, 510–511, 515–516
   on *Desert Storm* ground campaign, 525
   on forces for *Desert Storm*, 517, 521
   on Iraqi army as spent force, 528, 535n88
   micromanaging concerns of, 520
   on OPLAN 1002-90 plan for Middle East, 508

INDEX

Saddam's attacks on Israel and, 523–524
Scowcroft, Brent
    on arms control, 494, 495
    on Gorbachev, 480
    on Gulf War, 510, 525, 528
    Panama operations (1989) and, 492
    Peacekeeper program demise and, 438
    on Powell Doctrine, 484
    on Powell's briefing to Bush on *Desert Storm*, 516
Scud missiles, 523–524, 534n65
Security Resources Panel (SRP), 178. *See also* Gaither Committee
Sentinel (ballistic missile defense system), 268, 339. *See also* Nike-X ballistic missile defense system
Service chiefs. *See also* inter-Service rivalry; unification debate
    CJCS on SALT II and, 376–377
    Commandant of the Marine Corps and, 394
    Eisenhower and, 148, 151–152, 157, 540
    on flexible response doctrine, 282–283
    Goldwater-Nichols legislation and, 453, 454, 455–456, 482–483
    JCS executive agent system and, 30–31
    Johnson on ending Vietnam War and, 311, 312
    Kennedy's ExCom and, 229
    on nuclear weapons, 149, 200, 201
    Powell on forces for *Desert Storm* and, 517
    Powell on Goldwater-Nichols legislation and, 483, 486–487
    Powell on Panama operations (1989) and, 492
    Reagan buildup and, 425
    Reagan's Strategic Defense Initiative and, 436
    SALT I and, 235, 236
    on Skybolt program, 254
    on START I Treaty, 499
    on Taylor as CJCS, 225, 283
    Truman on NSC meetings and, 113, 539–540
    Weinberger and, 423
Service Secretaries, 8
Seventh Fleet, 105, 349
SEXTANT. *See* Cairo Conference
Shah of Iran, 405–406, 411
Sharp, Grant, 288–289, 290, 292, 297, 307, 310
Shchukin, Aleksandr, 339, 340
Shepherd, Lemuel C., Jr., 135, 151
Sherman, Forrest P., 67, 69, 107, 144–145
Shevardnadze, Eduard, 496, 497
Shultz, George P., 422, 463, 471, 479
Sicily, Operation *Husky* against, 12
signals intelligence (SIGINT) intercept program, 106–107, 109, 221, 285
Sihanouk, Norodom, 317, 318
SILKWORM antiship missiles, Chinese, 470, 475n70

Single Integrated Operational Plan (SIOP), 189, 218, 248–249, 250, 371
Six Day War of June 1967, 308, 352–353
Sixth Fleet, 190, 195, 356
Skybolt air-to-surface missile, 253–255
SLEDGEHAMMER plans for Continental invasion, 10
Slessor, John, 3
Sloan, Stanley R., 153
Sloss, Leon, 396
smart bombs, for Vietnam War, 323–324
Smith, Walter Bedell, 148
Somalia, Ethiopian war with, 404
Somoza, Anastasio, 407–408, 465
Sonnenfeldt, Helmut, 315
Sorensen, Theodore C., 213, 232
"The Sources of Soviet Conduct" (Kennan, 1947), 70–71
South Korea, 102–105, 306, 395. *See also* Korean War
South Vietnam. *See also* Vietnam War
    collapse of (1975), 382–383
South Vietnamese Army (ARVN)
    LAM SON 719 operation and, 319–321
    setbacks (1963) for, 280, 281
    Vietnamization and, 318–319
Southeast Asia. *See also specific countries*
    end of Cold War and JCS outlook on, 505
    joint congressional resolution on U.S. in, 285–286
    Laos crisis and widening conflict in, 222–223
Southeast Asia Command (SEAC), 36–37
Southeast Asia Treaty Organization (SEATO), 190
    Field Force, 222, 223
Southwest Pacific Area command, 30–31
Soviet Union. *See also* Cuba; Eastern Europe; *specific leaders and ministers of*
    Afghanistan invasion by, 335, 395, 400, 409–410
    Arab-Israeli conflict and, 354, 355, 356, 357
    armed forces reductions in, 479, 480
    arms sales to Egypt by, 191
    atomic bomb and, JSSC's concerns on, 51
    atomic bomb testing (1949) in, 96, 98, 540
    ballistic missile development in 1960s by, 264
    Berlin pressured by, 199
    Berlin Wall confrontation by, 219–220
    British and U.S. planners on World War II role of, 9
    British intelligence on malaise of, 428–429
    Carter's view versus JCS's view of, 394–395
    CFE Treaty (1990) and, 494–495
    containment strategy for, 70–71
    coup against Gorbachev in, 497–498, 499
    coup in Czechoslovakia and, 73–74
    Cuba's alignment with, 196

Ethiopian war with Somalia and, 404
on Follow-On Forces Attack concept, 458
freeze movement in the UK and, 434–435
Gorbachev's restructuring of, 459–462
ICBMs of, 144
JSTPS comprehensive target list against, 189
Kuwait shipping escort operation and, 472
Live Oak plans for Berlin and, 200–201
MC 14/3 as message to, 260
Middle East influence by, 190, 352
on MIRV limitation for SALT II, 379
missile buildup by, détente and, 337
Nixon on détente with, 313, 316, 317, 543–544
Nixon's trip to China and, 350
NSC 68 projecting military threat in 1950s by, 100–101, 540
nuclear test ban and, 234–236
Order of Battle, McNamara's analysis of, 252–253
Panama operations (1989) and, 493
post-SALT I testing by, 342, 375
post-World War II unsettled relations with, 61, 75, 89n39
post-World War II Western powers' disputes and, 42
SALT I backchannel talks and, 338–339
SALT II talks and, 398
on Schlesinger doctrine, 375–376
South Vietnam's collapse and, 383
SS-20 missile of (1976), 401
START negotiations and, 496–497
strategic nuclear power of U.S. in early 1970s and, 251
Suez crisis and, 193–194
TCP on preemptive strike capability by, 161–162
thermonuclear weapons testing (1953) in, 141
U.S. on surprise attack threats by, 161–163
Vietnam War and, 296
on Vladivostok concessions, 382
World War II collaboration with, 18–22, 50–51
Spaatz, Carl, 60, 66–67, 74
Spaatz-Tedder Agreement (1946), 64
special operations, 384, 533n49
Special Operations Review Group (SORG), 413–414
Sprague, Robert C., 178
Sputnik satellites, 173, 180
SR–71 spy plane, 205n47
SS–4 (medium-range ballistic missiles, MRBMs), Soviet, 227, 228, 232
SS–5 (intermediate-range ballistic missiles, IRBMs), Soviet, 228
SS–20, Soviet land–based triple–warhead mobile missile, 401, 402, 440, 460, 545
Stalin, Josef, 17, 18–19, 22, 98, 103, 138
Standing Group, NATO's, 120, 154

Standing Liaison Committee, U.S., 41
Standley, William H., 20
Star Wars, 433
Stark, Harold R., 2, 5
Stark, USS, 470
Starry, Donn A., 474n30
START-plus agreement, 498. *See also* Strategic Arms Reduction Talks
State, Secretary of, 170n123
State Department
   on Baghdad Pact and Arab-Israeli tensions, 190–191
   Batista regime in Cuba and, 197
   British on negotiating end to Vietnam War and, 307
   under Eisenhower, JCS and, 137
   FY 1952 military budget review and, 100–101
   JCS under Powell and, 482
   JSSC as liaison between JCS and, 41
   MacArthur and, 105
   Policy Planning Committee's national security policy analysis of, 80
State-War-Navy Coordinating Committee (SWNCC), 43
Steadman, Richard C., 393–394
Stennis, John C., 265–266, 309
Stethem, Robert, 463
Stevenson, Adlai E., 230
Stilwell, Joseph W., 34–36, 37, 97
Stimson, Henry L., 8, 30, 42, 48, 49, 51, 56n79
Stinger antiaircraft missiles, for Afghanistan, 461
Stockman, David A., 425, 426
Stoler, Mark A., 15
Strategic Air Command (SAC)
   air offensive against Soviet war-making capacity (1948) and, 75
   atomic attack plan against Chinese Communist forces and, 147
   atomic-capable aircraft of, 82, 91–92n89
   Britain' Bomber Command and, 78
   Bush on USSTRATCOM as replacement for, 498–499
   Cuban missile crisis and, 230
   inter-Service rivalry over 1958 reforms and, 187–188
   LeMay's transformation of, 82
   McNamara's assured destruction strategy and, 250
   NATO defense plan and, 117, 118, 119
   New Look defense budget and, 143–144
   nuclear weapons for, 79
   U–2 flights over Cuba and, 228
   Unified Command Plan and, 66–67
Strategic Arms Limitation Talks (SALT I)
   criticism in U.S. on, 340–341
   damage limitation debate and, 262

INDEX

détente starting with, 335, 336–342
    interim agreement, 359n21
    JCS on B–1 bomber and, 544
    Limited Test Ban Treaty (1963) and, 235
    Nixon administration and, 269
Strategic Arms Limitation Talks (SALT II)
    Carter and, 397–400
    JCS's work with OSD on, 378–379
    modernizing the strategic deterrent and, 375–378
    Reagan on, 441, 447n64
    Vladivostok mini-summit and, 380–382
Strategic Arms Reduction Talks (START), 441–442, 447n64, 460, 461, 488, 495–500
strategic bombing. *See also* strategic forces; strategic stability
    appeal of, 81–82
    of Germany during World War II, 13
    as Soviet deterrent in Europe (1948), 78
Strategic Defense Initiative Organization (SDIO), 437
Strategic Defense Initiative (SDI), 432–438, 456, 496
strategic forces. *See also* strategic stability
    Carter's PD-59 on, 394–397
    McNamara on restructuring of, 247–251
    Polaris fleet ballistic missile system and, 187, 188–189
    Reagan buildup and, 427–428
Strategic Rocket Forces (SRF), 181
strategic stability (1970s). *See also* détente
    Ford administration after Vladivostok and, 381–386
    modernizing the strategic deterrent, 367–370
    peacetime "total force" and, 365–367
    SALT II start and, 375–378
    targeting doctrine revisions and, 371–375
    Vladivostok mini-summit and, 378–381
Strat-X study, 369
Studies, Analysis, and Gaming Agency (SAGA), 206n73, 377
submarine-launched ballistic missiles (SLBMs), 248, 262–263, 338, 340–341, 359n21
submarines. *See* Poseidon; Trident missile submarines
Suez Canal, 191–192, 352
Sultan, David I., 38
Supreme Allied Commander, Europe (SACEUR), NATO, 118–119, 120, 123–124, 218
Supreme Allied Commander, North Atlantic (SACLANT), 120–121
Supreme Headquarters, Allied Powers Europe (SHAPE), 120. *See also* North Atlantic Treaty Organization
surface-to-surface missile complex, in Banes, Cuba, 227
Symington, W. Stuart, 67, 309
Syria, 352, 383, 519

systems analysis, 211, 246, 252–253, 309
Szilard, Leo, 47

Tachen Islands, Taiwan Strait, 150, 151
Taiwan, 44, 97, 139, 150, 349, 350–351. *See also* Chiang Kai-shek
Taiwan Strait, 105, 135, 149, 150, 151–152
Taiwan Strait Patrol, 349
Tallinn Line (Soviet air defense system), 264
target selection, 295, 296, 371–375, 396–397. *See also* Joint Strategic Target Planning Staff; Vietnam War
Taylor, Maxwell D.
    as Ambassador to Vietnam, 286
    Berlin crisis and, 200, 217, 219
    as CJCS, 224–225, 264
    Cuban issues and, 215–216, 229, 230–231
    Honolulu conference (1965) and, 292
    McNamara's FYDP and posture statement from, 247
    on New Look defense budget, 145, 195
    nuclear test ban and, 235, 236
    as President's Military Representative, 213, 225, 542
    on Skybolt, 254
    Vietnam War strategy and, 279, 280, 282–283, 291
Teapot Committee (1954), 175
Technological Capabilities Panel (TCP), 161
technology. *See* military technologies
Tehran Conference (Nov. 1943), 17, 18, 21, 41
terrorists
    Libyan, 463–464
    Palestinian, 383
Test Ban Treaty, Kennedy administration and, 337
Tet offensive, 305, 310, 543
Thailand, 224, 306
thermonuclear weapons, 99, 141, 144, 267, 348, 374. *See also* hydrogen bomb
Thieu, Nguyen Van, 324
Third World
    Carter and Soviet encroachment in, 403–408
    nationalism and discontent of 1950s in, 173
    Soviet focus on, 225–226
    U.S. power erosion in Vietnam and, 383
Thompson, Llewellyn E., 232
Thompson, Wayne, 295
Thor (intermediate-range ballistic missile, IRBMs), 146, 176, 228–229, 252, 272n46
Thorneycroft, Peter, 254
three near-simultaneous conflicts' planning, 426–427
313th Bombardment Wing of 509th Composite Group, 48
Thurman, Maxwell R., 490
Titan (intercontinental ballistic missile), 176

Tomahawk land-attack cruise missiles (TLAMs), 514
Tonkin Gulf incident, 284–292
Top Advisory Group, Roosevelt's, 47, 56n79
Torrijos, Omar, 489
Towers, John H., 23n6
Treasury Department, 137
Treaty of Friendship and Cooperation, Soviet-Egyptian, 354
Treaty of Paris (1952), 123
triangular infantry divisions, U.S. Army, 146
TRIDENT Conference (May 1943), 14–15
Trident missile submarines
    Carter's support for, 396
    development of, 369–370, 385
    JCS requests for, 337, 341, 368, 544
    Powell's estimates for, 496
    Schlesinger on, 374
trip wire theory, 251, 252
TROJAN (emergency war plan), 84
Truman, Harry S., military challenges for
    atomic bomb and, 48–49, 50, 62, 63–64
    Chinese intervention in Korean War and, 111–113
    on ending the Korean War, 138
    European defense and security issues and, 116–125
    on H-bomb and NSC 68, 98–102
    on JCS organizational structure, 538
    Korean War onset and, 102–105
    Korea's Inch'on Operation under, 105–108
    MacArthur and, 94
    MacArthur and military buildup for Korean War under, 108–110
    MacArthur's dismissal and, 113–116
    nuclear weapons' production under, 158–159
    pressures to expand military power under, 95–98
Truman, Harry S., peacetime challenges for
    as Commander in Chief, 59–60
    defense budget for FY 1950, 76–81
    defense policy in transition, 61–64
    reorganization and reform and, 64–69
    Soviet Union and, 22
    strategic bombing controversy, 81–87
    war plans, budgets, and March crisis of 1948, 69–76, 89n45
Tsar Bomba (King of Bombs), 234
Tsingtao, China, U.S. military base at, 97–98
Tuesday lunch, Vietnam War and, 295, 312
Turkey, 69, 121–122, 123, 190, 191, 308
Turner Joy, USS, 284, 285
Twentieth Air Force, 30–31, 45
Twining, Nathan F.
    on Berlin contingency plans, 200
    as CJCS, *172*, 540

defense reorganization (1958) and, 184–185, 186, 187–188
    on Dien Bien Phu air support, 149
    joins the JCS, 135
    on Lebanese intervention, 195
    on missile gap, 181, 182–183
    NATO's New Approach and, 154
    on Taiwan Strait islands, 151
    on U–2 flights, 182
    unified command decision (1960) and, 188
two-and-a-half war strategy, 347–348
Tyuratam ICBM test facility, Soviet Union, 182

U–2 photoreconnaissance plane
    development of, 161, 170n123
    Eisenhower on, 180
    Gaither Committee report and, 179
    intelligence gathering by, 162–163, 192, 227
    Powers shot down in, 183, 201, 205n52
U–boats, German, 11, 13
ULTRA radio intercepts by U.S., 49
undersea long-range missile system (ULMS). *See* Trident missile submarines
unification debate, 39, 64–65, 72–73, 135, 538–539. *See also* Combined Chiefs of Staff; inter-Service rivalry; universal military training
Unified Command Plan (UCP), 66–67, 498, 538
United Arab Republic (UAR), 194
United Chiefs of Staff, 26n76
United Kingdom. *See also* Britain
    HALFMOON plan and, 75
United Nations Command (UNC), in Korea, 107–108, 110, 114–115, 140
United Nations commander (CINCUNC), 105, 138, 139
United Nations (UN)
    Baruch Plan to ban atom bomb debate at, 62–63
    disarmament negotiations before, 159–160
    Middle East peacekeeping force and, 77
    postwar Persian Gulf affairs and, 529
    Republic of China expulsion by General Assembly, 349
    Roosevelt on postwar planning and creation of, 41
    Security Council resolution on Gulf War ceasefire, 530
United States. *See also specific administrations*
    Grand Alliance during World War II and, 18–19
    Middle East presence of, 190
    NATO commitments, European concerns on reductions in, 157
    Operation *Overlord* confirmation and leadership by, 17–18
    preparation for Casablanca Conference by Britain versus, 13–14

INDEX

Soviet blockade of Berlin and, 77–78
United States, USS (super carrier), 83
unity of command. *See* unification debate
universal military training (UMT), 39, 54n41, 61–62
U.S. Air Forces Central Command (CENTAF), 519–520
U.S. Central Command (USCENTCOM), 405, 431–432. *See also* Schwarzkopf, H. Norman
U.S. European Command (USEUCOM), 199–200
U.S. Military Assistance Command, Vietnam (COMUSMACV), 280, 284, 288, 294–295, 297–298, 309
U.S. Readiness Command (USREDCOM), 405, 410
U.S. Strategic Command (USSTRATCOM), 498–499
U.S. Strike Command (USSTRICOM), 405
U.S.-Soviet Standing Consultative Commission, 340

Valenti, Jack J., 287
Vance, Cyrus R., 397, 398, 413
Vandenberg, Hoyt S., *58*, 74, 85, 104, 115, 134–135
Vandenberg Resolution (1948), 75
Vaught, James E., 412
Vesser, Dale A., 516, 533n39
Vessey, John W., Jr., 321, *420*, 423–424, 432, 435–437, 451, 452
Vice Chairman of the Joint Chiefs of Staff (VCJCS), 454, 455
Viet Minh, later Viet Cong. *See also* Vietnam War
  Dien Bien Phu siege (1954) by, 148
  France ending war with, 146–147
  French conflict with, NATO buildup and, 122–123
  history of U.S. combat with, 277–278
  objectives and commitment to war by, 296
  Operation *Rolling Thunder* against, 290, 292, 294–295, 307
  Pleiku, South Vietnam, strike by, 287, 289
  resilience of, 306
Vietnam War. *See also* McNamara, Robert S.; Operation *Rolling Thunder*; target selection
  American power and influence within NATO and, 258, 259
  as American war, 281–284
  balance sheet for, 326–329
  bombing after LAM SON 719 operation, 321–324
  Chinese restraint with (1970s), 351, 362n70
  Christmas bombing campaign (1972), 324–326
  JCS reassessment after, 335, 336
  Johnson and advisors consider bombing in, 287–288, 289
  Johnson's curtailing of bombing in, 311, 312–313
  military budget and, 250
  Nixon, the JCS, and the policy process, 313–316

as quagmire, 292–298
  roots of U.S. involvement in, 277–281
  Senate hearings on bombing in, 309, 310
  as stalemate, 305–310
  support for bombing in, 288–289
  Tet and its aftermath, 310–313
  Tonkin Gulf incident and aftermath and, 284–292
  winding down, under Nixon, 316–321
Vietnamization
  Cambodian invasion (1970) and, 318–319
  ceasefire and regrouping by NVA and, 327
  Johnson on, 313
  LAM SON 719 operation and, 319–321
  Nixon's plan for, 316
  U.S. withdrawal and, 317–318
VII Corps, 518, 520, 521, 527
Vincennes, USS (Aegis missile cruiser), 471–472
Vinson, Carl, 185, 205n62
Vladivostok mini-summit (1974), 380–382, 388n47
Vogt, John W., 323
Voroshiloff, Klementy, 17
Vulcan bombers, Britain's, 253
vulnerabilities panel, Rockefeller and, 162
Vulture operation, 148–149
Vuono, Carl E., 490

walk in the woods formula, for arms control, 440, 447n58
Walker, Walton H., 107–108
Wallace, Henry A., 56n79, 59
Wallop, Malcolm, 434, 445–446n41
War Department, 2, 3, 6, 10–11, 40, 42, 47, 48–49. *See also* Defense Department
war gaming, 188, 206n73, 283, 287
War Plan Orange, 3, 31
War Powers Act (1973), 518
Warden, John A., III, 513, 514, 520–521, 523
Warsaw Pact. *See also specific countries of*
  Berlin Wall confrontation and, 220, 239n43
  on conventional forces in Europe, 493, 494
  creation of, 158
  Czechoslovakia invasion by (1968), 269, 336
  disestablishment of, 485
  on Follow-On Forces Attack concept, 458
  JSTPS comprehensive target list against, 189
  Mutual and Balanced Force Reduction Talks and, 439–440
  NATO on, 343–344, 400–401
Washington Special Action Group (WSAG), 355–356, 357, 363n87
Watergate scandal, 346, 355, 356, 377–378, 379
Watkins, James D., 423, 435, 436, 453

583

weapons effects study (1949), 83, 84–85
weapons of mass destruction (WMD), 147, 509, 528, 529, 530
Weapons Systems Evaluation Group (WSEG), 83, 85–86, 181–182, 188, 248, 377
Wedemeyer, Albert C., 13, 37–38, 41, 58
Weinberger, Caspar W.
   on arms control, 438–439
   on Goldwater-Nichols legislation, 474n20
   on JCS organizational structure, 451, 453
   JCS's relationship with, 424–425, 546
   Joint Task Force Middle East and, 469
   Jones and, 422–423
   military budget and, 425–427, 428
   Reagan's Strategic Defense Initiative and, 437
Weinraub, Bernard, 393
Welch, Larry D., 484
Wells, H.G., 313
West Germany. *See also* Berlin; Berlin Wall
   France on rearmament of, 123
   French nuclear weapons and, 257
   Live Oak (planning body) and, 201
   MLF demise and, 257
   NATO and, 345, 346
   Nixon and Kissinger on East Germany and, 343
   rearmament of, 119–120, 156–157
   Treaty of Paris (1952) and, 123
   U.S. on strengthening ties with, 258
   on Washington moving NATO away from nuclear deterrence, 401
Western Europe, in World War II, 9, 19–20
   cross-Channel operation and, 10–11, 15–16, 21, 46
Western Excursion proposal, on Iraq, 533n39
Westmoreland, William C.
   air campaigns under, 294–295, 311
   chain of command in Vietnam War and, 297–298
   on graduated response, 288
   ground war in South Vietnam and, 292, 306, 307–308, 312
   on Ia Drang Valley conflict (1965), 296
   joins the JCS, 314
   on LAM SON 719 operation, 320
   on peacetime total force (1970s), 367
   on U.S. bases in Vietnam, 290
Weyand, Fred C., 377, 382
Weyland, Otto P., 67
Wheeler, Earle G.
   arms control negotiations and, 337–338
   becomes Army Chief of Staff, 224
   becomes CJCS, 264, 283–284
   JCS reorganization and, 184
   on Middle East peace settlement, 353
   military budget under McNamara and, 247
   Nixon and, 314
   as senior Johnson advisor, 311–312, 542
   Vietnam War strategy and, 288, 290, 292, 293, *304*, 308, 311
White, Thomas D., 184, 212, 224
whiz kids (McNamara's advisors), 245, 309
"Why the Joint Chiefs of Staff Must Change" (Jones, 1982), 450
Wickham, John A., Jr., 423
Willson, Russell, 11
Wilson, Charles E., 136, 139, 157, 162, 176, 177
Wilson, Henry Maitland, 23n3
Win Plan, for Vietnam War, 280
Wizard ballistic missile defense system, 181
Woerner, Frederick F., Jr., 490
Wolfowitz, Paul D., 484, 487, 491–492, 501n25
Wooldridge, E.T., 129n77
World War I, 11
World War II. *See also* Asia-Pacific War; Europe, war in
   JCS authority and influence during, 537–538
   planning, JCS on politico-military affairs and, 40–41
   planning Armed Forces organization and composition after, 38–39
   planning for Germany after, 42–43
Wright, Jim, 466

X–16 photoreconnaissance plane, 161
XVIII Airborne Corps, 521
XX Bomber Command, 33, 53–54n34

Yalta Summit Conference (Feb. 1945), 21, 43
Yeltsin, Boris, 498
Yemen, Soviet ties with, 383
Yom Kippur War (1973). *See* October War
York, Herbert F., 189

Z Committee, 99
zero-zero option, in arms control talks, 440
Zhou Enlai, 152, 350, 351
Zumwalt, Elmo R., Jr., 314–315, 356, 377, 378–379

ABOUT THE AUTHOR

Tara Parekh (NDU Press)

Dr. Steven L. Rearden holds a Bachelor of Arts degree from the University of Nebraska and a Ph.D. in history from Harvard University. His association with the Joint History Office dates from 1996. He has written and published widely on the history of the Joint Chiefs of Staff and the Office of the Secretary of Defense, and was co-collaborator on Ambassador Paul H. Nitze's *From Hiroshima to Glasnost: At the Center of Decision—A Memoir* (Grove Weidenfeld, 1989).